Bioadhesive Drug Delivery Systems

DRUGS AND THE PHARMACEUTICAL SCIENCES

A Series of Textbooks and Monographs

98. Bioadhesive Drug Delivery Systems: Fundamentals, Novel Approaches, and Development, *edited by Edith Mathiowitz, Donald E. Chickering III, and Claus-Michael Lehr*

ADDITIONAL VOLUMES IN PREPARATION

Transport Processes in Pharmaceutical Systems, *edited by Gordon Amidon, Ping I. Lee, and Elizabeth M. Topp*

Peptide and Protein Drug Analysis, *edited by Ronald E. Reid*

Protein Formulation and Delivery, *edited by Eugene McNally*

New Drug Approval Process, Third Edition: The Global Challenge, *edited by Richard A. Guarino*

Excipient Toxicity and Safety, *edited by Myra Weiner and Lois Kot-koskie*

The Clinical Audit in Pharmaceutical Development, *edited by Michael R. Hamrell*

Pharmaceutical Emulsions and Suspensions, *edited by Francoise Nielloud and Gilberte Marti-Mestres*

Bioadhesive Drug Delivery Systems

Fundamentals, Novel Approaches, and Development

edited by

Edith Mathiowitz
Brown University
Providence, Rhode Island

Donald E. Chickering III
Acusphere, Inc.
Cambridge, Massachusetts

Claus-Michael Lehr
Saarland University
Saarbrücken, Germany

MARCEL DEKKER, INC.   NEW YORK · BASEL

ISBN: 0-8247-1995-6

This book is printed on acid-free paper.

Headquarters
Marcel Dekker, Inc.
270 Madison Avenue, New York, NY 10016
tel: 212-696-9000; fax: 212-685-4540

Eastern Hemisphere Distribution
Marcel Dekker AG
Hutgasse 4, Postfach 812, CH-4001 Basel, Switzerland
tel: 41-61-261-8482; fax: 41-61-261-8896

World Wide Web
http://www.dekker.com

The publisher offers discounts on this book when ordering in bulk quantities. For
more information, write to Special Sales/Professional Marketing at the headquarters
address above.

Current printing (last digit):
10 9 8 7 6 5 4 3 2 1

PRINTED IN THE UNITED STATES OF AMERICA

Preface

Our aim in writing *Bioadhesive Drug Delivery Systems* was to provide a comprehensive reference on bioadhesion covering basic concepts and methods for characterizing bioadhesive materials, novel biological approaches designed to improve vehicle targeting or enhance uptake, and practical topics addressing product development for clinical applications. Although the goal of bioadhesion research may be a marketable device, what is ultimately important is an understanding of the fundamental concepts involved in the mechanisms of adhesion, the biological interactions, and the practical application of bioadhesive carriers. In compiling the various chapters, our target was not to resolve the issues involved in bringing a product from bench to market, or to provide a "magic formula" to reduce costs during the process, but to expose the researcher, student, or industrial scientist to the wonderful possibilities of engineering, evaluating, and manufacturing bioadhesive materials. Some researchers may consider this book an excellent starting point to familiarize themselves with the field, while others may find it a source of new ideas for developing or characterizing innovative bioadhesive systems.

The book is divided into four parts: (I) Fundamentals of Bioadhesion, (II) Methods of Evaluating Bioadhesive Interactions, (III) Novel Concepts and Strategies for Bioadhesive Delivery Systems, and (IV) Development Issues of Bioadhesive Drug Delivery Systems: Products and Clinical Trials. Each chapter is devoted to a specific topic or concept and contains relevant literature reviews of the subject, supported by novel discoveries or ideas of the contributing authors.

Part I, Fundamentals of Bioadhesion, reviews both physicochemical fundamentals and biological aspects of bioadhesion. Here we discuss the theories and concepts developed to describe adhesive interactions and explain the relevant forces associated with bioadhesive bonding. Topics covered include mechanical and chemical bonding, polymer–mucus interactions, the effect of surface energy in bioadhesion, the role of polymer hydration or water movement, and mucus rheology. In addition, the anatomy and physiology of target tissues as well as the molecular and intracellular mechanisms that may contribute to bioadhesion are discussed. Specific chapters are devoted to biochemical properties of mucus and glycoproteins, cell adhesion molecules, and cellular interactions with two- and three-dimensional surfaces.

Part II, Methods of Evaluating Bioadhesive Interactions, explains the more common techniques for bioadhesive analysis that have been adapted from traditional materials testing. Chapters are dedicated to unique and innovative systems specifically designed for characterizing adhesive interactions in biological settings. These topics include the use of microbalances and magnetic force transducers, the use of atomic force microscopy, direct measurements of molecular level adhesions, and methods to measure cell–cell interactions.

Part III, Novel Concepts and Strategies for Bioadhesive Delivery Systems, highlights the possibilities and goals of employing bioadhesive materials as drug carriers. The effects of prolonged residence time, minimized interfacial boundaries, and cellular interactions are discussed. Particular attention is devoted to receptor mediated bioadhesion, pharmaceutical transport from bioadhesive carriers, diffusion or penetration enhancers, and lectin-targeted vehicles.

Part IV, Development Issues of BDDS: Products and Clinical Trials, discusses a unique area that has not been covered in any previous bioadhesion text. The purpose of this section is to provide an illustrative overview of clinical bioadhesive applications. Chapters offer examples of vaginal, nasal, buccal, ocular, and transdermal drug delivery using bioadhesive carrier materials. Issues involved in product development, clinical testing, and production are described.

By dividing the text into four parts, we hope to introduce the reader to the various aspects and considerations involved in developing bioadhesive controlled-release systems, starting from pure scientific concepts based on theoretical ideas and ending with significant examples of specific applications. Bioadhesive polymers offer unique carrier characteristics for many pharmaceutics. They can be tailored to adhere to either the dermis or any mucosal tissue including those found in the eye and mouth, and throughout the respiratory, urinary, and gastrointestinal tracts. These materials can improve localization of delivered agents, enhance local bioavailability, decrease adverse systemic effects, and improve drug absorption and transport. Using bioadhesive materials, it may be possible to reformulate existing compounds to produce new and useful products while decreasing the overall cost in development. The focus of this text is on understanding the basic mechanisms of bioadhesion and on how this knowledge can be applied toward engineering efficient bioadhesive carrier systems for delivering therapeutic agents.

Edith Mathiowitz
Donald E. Chickering III
Claus-Michael Lehr

Contents

Contributors

Yohko Akiyama   DDS Research Laboratories, Pharmaceutical Research Division, Takeda Chemical Industries, Ltd., Yodogawa-ku, Osaka, Japan

Trevor I. Armstrong   Department of Pharmaceutical Sciences, Wyeth-Ayerst Research, Gosport, England

Bodo Asmussen   LTS Lohmann-Therapie Systeme GmbH, Andernach, Germany

Susan Bardocz   The Rowett Research Institute, Bucksburn, Aberdeen, Scotland

Maria Cristina Bonferoni   Department of Pharmaceutical Chemistry, University of Pavia, Pavia, Italy

Gerrit Borchard*   Department of Biopharmaceutics and Pharmaceutical Technology, Saarland University, Saarbrücken, Germany

Susi Burgalassi   Department of Bioorganic Chemistry and Biopharmaceutics, University of Pisa, Pisa, Italy

Barry James Campbell   Department of Medicine, University of Liverpool, Liverpool, Merseyside, England

Carla Marcella Caramella   Department of Pharmaceutical Chemistry, University of Pavia, Pavia, Italy

Gerardo P. Carino   Department of Molecular Pharmacology, Physiology, and Biotechnology, Brown University, Providence, Rhode Island

---

*Current affiliation: Leiden/Amsterdam Center for Drug Research, Leiden University, Leiden, The Netherlands.

C. James Chen   Department of Molecular Pharmacology, Physiology, and Biotechnology, Brown University, Providence, Rhode Island

Patrizia Chetoni   Department of Bioorganic Chemistry and Biopharmaceutics, University of Pisa, Pisa, Italy

Donald E. Chickering III   Department of Research and Development, Acusphere, Inc., Cambridge, Massachusetts

Karsten Cremer   Department of Research and Development, LTS Lohmann Therapie-Systeme GmbH, Andernach, Germany

A. (Bert) G. de Boer   Division of Pharmacology, Leiden/Amsterdam Center for Drug Research, Leiden University, Leiden, The Netherlands

Bas J. de Leeuw   Department of Strategy and Business Environment, Rotterdam School of Management, Rotterdam, The Netherlands

James H. Easson   Department of Biopharmaceutics and Pharmaceutical Technology, Saarland University, Saarbrücken, Germany

Stanley W. B. Ewen   Department of Pathology, University Medical School, Aberdeen, Scotland

S. Gehring   TopoMetrix GmbH, Darmstadt, Germany

Jian-Hwa Guo   Aqualon Division, Hercules Incorporated, Wilmington, Delaware

Eleonore Haltner   Department of Biopharmaceutics and Pharmaceutical Technology, Saarland University, Saarbrücken, Germany

Uwe Hartmann   Institute of Experimental Physics, Saarland University, Saarbrücken, Germany

Benjamin A. Hertzog   Department of Molecular Pharmacology, Physiology, and Biotechnology, Brown University, Providence, Rhode Island

Michael Horstmann   Department of Research and Development, LTS Lohmann-Therapie Systeme GmbH, Andernach, Germany

Lisbeth Illum   DanBioSyst UK Ltd, Nottingham, England

Jules S. Jacob Department of Molecular Pharmacology, Physiology, and Biotechnology, Brown University, Providence, Rhode Island

Dieter Jahn Institute for Organic Chemistry and Biochemistry, Universität Freiburg, Freiburg, Germany

H. E. Junginger Department of Pharmaceutical Technology, Leiden/ Amsterdam Center for Drug Research, Leiden University, Leiden, The Netherlands

P. Koschinski TopoMetrix GmbH, Darmstadt, Germany

A. F. Kotzé Department of Pharmaceutics, Potchefstroom University for Christian Higher Education, Potchefstroom, Republic of South Africa

Claus-Michael Lehr Department of Biopharmaceutics and Pharmaceutical Technology, Saarland University, Saarbrücken, Germany

Xiaoling Li School of Pharmacy and Health Sciences, University of the Pacific, Stockton, California

Henrik L. Luessen* Corporate Development Department, LTS Lohmann Therapie-Systeme GmbH, Andernach, Germany

Yoshiharu Machida Department of Clinical Pharmacy, Hoshi University, Ebara, Shinagawa-ku, Tokyo, Japan

Edith Mathiowitz Department of Molecular Pharmacology, Physiology, and Biotechnology, Brown University, Providence, Rhode Island

Walter Müller Department of Research and Development, LTS Lohmann-Therapie Systeme GmbH, Andernach, Germany

Naoki Nagahara DDS Research Laboratories, Pharmaceutical Research Division, Takeda Chemical Industries, Ltd., Yodogawa-ku, Osaka, Japan

Tsuneji Nagai Department of Pharmaceutics, Hoshi University, Ebara, Shinagawa-ku, Tokyo, Japan

---

*Current affiliation: OctoPlus, B.V., Leiden, The Netherlands.

James W. Piper   George W. Woodruff School of Mechanical Engineering, Georgia Institute of Technology, Atlanta, Georgia

A. Plückthun   Institute of Biochemistry, University of Zurich, Zurich, Switzerland

Arpad Pusztai   Department of Nutrition, The Rowett Research Institute, Bucksburn, Aberdeen, Scotland

Julie L. Richardson   Department of Pharmaceutical Research and Development, Pfizer Ltd., Sandwich, Kent, England

Robert Ros   Laboratory for Micro- and Nanotechnology, Paul Scherrer Institute, Villigen, Switzerland

Silvia Rossi   Department of Pharmaceutical Chemistry, University of Pavia, Pavia, Italy

Marco Fabrizio Saettone   Department of Bioorganic Chemistry and Biopharmaceutics, University of Pisa, Pisa, Italy

Camilla A. Santos   Department of Molecular Pharmacology, Physiology, and Biotechnology, Brown University, Providence, Rhode Island

James Schneider*   Department of Chemical Engineering and Materials Science, University of Minnesota, Minneapolis, Minnesota

Dietlind Schumacher   Department of Neuroanatomy, University of Hamburg, Hamburg, Germany

Udo Schumacher   Department of Neuroanatomy, University of Hamburg, Hamburg, Germany

Falk Schwesinger   Institute of Biochemistry, University of Zurich, Zurich, Switzerland

Amir H. Shojaei**   School of Pharmacy and Health Sciences, University of the Pacific, Stockton, California

---

Current affiliation: Carnegie Mellon University, Pittsburgh, Pennsylvania.
**Current affiliation*: Department of Pharmaceutical Sciences, School of Pharmacy, Texas Tech University Health Sciences Center, Amarillo, Texas.

John D. Smart School of Pharmacy and Biomedical Sciences, University of Portsmouth, Portsmouth, England

Robert A. Swerlick Department of Dermatology, Emory University School of Medicine, Atlanta, Georgia

M. Thanou Department of Pharmaceutical Technology, Leiden/Amsterdam Center for Drug Research, Leiden University, Leiden, The Netherlands

Louis Tiefenhauer Laboratory of Micro- and Nanotechnology, Paul Scherrer Institute, Villigen, Switzerland

Matthew Tirrell Department of Chemical Engineering and Materials Science, University of Minnesota, Minneapolis, Minnesota

J. Coos Verhoef Division of Pharmaceutical Technology, Leiden/Amsterdam Center for Drug Research, Leiden University, Leiden, The Netherlands

Cheng Zhu George W. Woodruff School of Mechanical Engineering, Georgia Institute of Technology, Atlanta, Georgia

E. zur Mühlen TopoMetrix GmbH, Darmstadt, Germany

Bioadhesive Drug Delivery Systems

1
Definitions, Mechanisms, and Theories of Bioadhesion

Donald E. Chickering III
Acusphere, Inc, Cambridge, Massachusetts

Edith Mathiowitz
Brown University, Providence, Rhode Island

I. BIOADHESION AND MUCOADHESION

The term *bioadhesion* refers to any bond formed between two biological surfaces or a bond between a biological and a synthetic surface. In the case of bioadhesive drug delivery systems, the term bioadhesion is typically used to describe the adhesion between polymers, either synthetic or natural, and soft tissues (i.e., gastrointestinal mucosa). Although the target of many bioadhesive delivery systems may be a soft tissue cell layer (i.e., epithelial cells), the actual adhesive bond may form with either the cell layer, a mucous layer, or a combination of the two. In instances in which bonds form between mucus and polymer, the term *mucoadhesion* is used synonymously with bioadhesion. In general, bioadhesion is an all-inclusive term used to describe adhesive interactions with any biological or biologically derived substance, and mucoadhesion is used only when describing a bond involving mucus or a mucosal surface.

II. MECHANISMS OF BIOADHESION

The mechanisms responsible for the formation of bioadhesive bonds are not completely clear. In order to develop ideal bioadhesive drug delivery systems, it is important to describe and understand the forces that are responsible for adhesive bond formation. Most research has focused on analyzing bioadhesive interactions between polymer hydrogels and soft tissue. The process involved in the formation of such bioadhesive bonds has been described in three steps: (a) wetting and swelling of polymer to permit intimate contact with biological tissue, (b) interpenetration of bioadhesive polymer chains and entanglement of polymer and mucin chains, and (c) formation of weak chemical bonds between entangled chains (1,2). It has been stated that at least one of the following polymer characteristics are required to obtain adhesion: (a) sufficient quantities of hydrogen-bonding chemical groups (—OH and —COOH), (b) anionic surface charges, (c) high molecular weight, (d) high chain flexibility, and (e) surface tensions that will induce spreading into the mucous layer (3). Each of these characteristics favors the formation of bonds that are either chemical or mechanical in origin.

A. Chemical Bonds

Types of chemical bonds include strong primary bonds (i.e., covalent bonds), as well as weaker secondary forces such as ionic bonds, van der Waals interactions, and hydrogen bonds. As described in this text, both types of interactions have been exploited in developing bioadhesive drug delivery systems. Although systems designed to form covalent bonds with proteins on the surface of epithelial cells may offer strength advantages, three factors may limit the usefulness of such permanent bonding. First, mucous barriers may inhibit direct contact of polymer and tissue. Second, permanent chemical bonds with the epithelium may not produce permanently retained delivery devices because most epithelial cells are exfoliated every 3 to 4 days. Third, biocompatibility of such binding has not been thoroughly investigated and could pose significant problems (4). For these reasons, many have focused on developing hydrogel, mucoadhesive systems that bond through either van der Waals interactions or hydrogen bonds. Although individually these forces are very weak, strong adhesions can be produced through numerous interaction sites. Therefore, polymers with high molecular weight and high concentrations of reactive, polar groups (such as —COOH and —OH) tend to develop intense mucoadhesive bonds (5–7).

B. Mechanical or Physical Bonds

Mechanical bonds can be thought of as physical connections between surfaces—similar to interlocking puzzle pieces. Macroscopically, they involve the inclusion of one substance in the cracks or crevices of another (8). On a microscopic scale, they can involve physical entanglement of mucin strands with flexible polymer chains and/or interpenetration of mucin strands into a porous polymer substrate. The rate of penetration of polymer strands into mucin layers is dependent on chain flexibility and diffusion coefficients of each. The strength of the adhesive bond is directly proportional to the depth of penetration of the polymer chains. Other factors that influence bond strength include the presence of water, the time of contact between the materials, and the length and flexibility of the polymer chains (4).

III. THEORIES ON BIOADHESION

The study of adhesion is not a new science. The same theories of adhesion that were developed to explain and predict the performance of glues, adhesives, and paint can be and have been applied to bioadhesive systems. In general, five theories have been adapted to the study of bioadhesion: the electronic, absorption, wetting, diffusion, and fracture theories. Some are based on the formation of mechanical bonds, whereas others focus on chemical interactions.

A. The Electronic Theory

The hypothesis of the electronic theory relies on the assumption that the bioadhesive material and the target biological material have different electronic structures. On this assumption, when the two materials come in contact with each other, electron transfer occurs in an attempt to balance Fermi levels, causing the formation of a double layer of electrical charge at the bioadhesive–biologic material interface. The bioadhesive force is believed to be due to attractive forces across this electrical double layer. This system is analogous to a capacitor: the system is charged when the adhesive and substrate are in contact and discharged when they are separated (9). The electronic theory has produced some controversy regarding whether the electrostatic forces are an important *cause* or the *result* of the contact between the bioadhesive and the biological component (10).

B. The Adsorption Theory

The adsorption theory states that the bioadhesive bond formed between an adhesive substrate and tissue or mucosa is due to van der Waals interactions, hydrogen bonds, and related forces (11,12). Although these forces are individually weak, the sheer number of interactions can as a whole produce intense adhesive strength. The adsorption theory is the most widely accepted theory of adhesion and has been studied in depth by both Kinloch (8) and Huntsberger (13,14).

C. The Wetting Theory

The ability of bioadhesive or mucus to spread and develop intimate contact with its corresponding substrate is one important factor in bond formation. The wetting theory, which was developed predominantly in regard to liquid adhesives, uses interfacial tensions to predict spreading and in turn adhesion (3,15,16). The study of surface energy of both polymers and tissues to predict mucoadhesive performance has been given considerable attention (17–20).

As an example, Fig. 1 schematically represents a bioadhesive gel (BG) spreading over soft tissue in the gut. The contact angle (Q), which should be zero or near zero for proper spreading, is related to interfacial tensions (g) through Young's equation:

$$g_{tg} = g_{bt} + g_{bg} \cos Q$$

where the subscripts t, g, and b stand for tissue, gastrointestinal (GI) contents, and bioadhesive polymer, respectively. For spontaneous wetting to occur, Q must equal zero and, therefore, the following must apply (3):

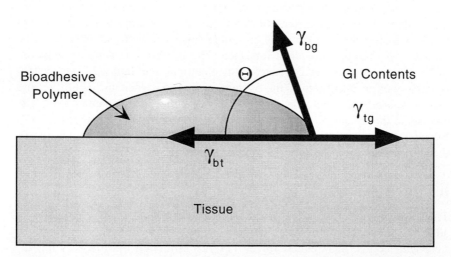

Figure 1   Schematic diagram showing the interfacial tensions involved in spreading a bioadhesive polymer over GI mucosa.

$g_{tb} \geq g_{bt} + g_{bg}$

The spreading coefficient, $S_{b/t}$, of a bioadhesive over biological tissue in vivo can be used to predict bioadhesion and can be determined as follows:

$S_{b/t} = g_{tg} - g_{bt} - g_{bg}$

For the bioadhesive to displace GI luminal contents and make intimate contact with the biological tissue (i.e. spreading), the spreading coefficient must be positive. Therefore, it is advantageous to maximize the interfacial tension at the tissue–GI contents interface (g_{tg}) while minimizing the surface tensions at the other two interfaces (g_{bt} and g_{bg}) (3).

It is theoretically possible to determine each of the parameters that make up the spreading coefficient. The interfacial tension of the tissue–GI contents interface (g_{tg}) can be determined in vitro using classical Zisman analysis (21,22). The interfacial tension at the BG–GI contents interface (g_{bg}) can be determined experimentally using traditional, surface tension–measuring techniques, such as the Wilhelmy plate method. Lastly, it has been shown that the BG-tissue interfacial tension (g_{bt}) can be calculated as follows (23,24):

$g_{bt} = g_b + g_t - 2F(g_b g_t)^{1/2}$

where values of the interaction parameter (F) can be found in published papers (25,26). Extensive studies have been conducted to determine the surface tension parameters for several biological tissues (g_t) and many commonly used biomaterials (g_b) (20).

The BG-tissue interfacial tension (g_{bt}) has been shown to be proportional to the square route of the polymer-polymer Flory interaction parameter (c):

$g_{bt} \sim c^{1/2}$

When c is small, the bioadhesive and biological components are similar structurally. This results in increased spreading and, therefore, greater adhesive bond strength (27,28).

Besides the spreading coefficient, another important parameter that may indicate the strength of an adhesive bond is the specific work of adhesion (W_{bt}). According to the Dupré equation, this is equal to the sum of the surface tensions of the tissue and bioadhesive, minus the interfacial tension (29):

$W_{bt} = g_b + g_t - g_{bt}$

Thus, using the wetting theory, it is possible to calculate spreading coefficients for various bioadhesives over biological tissues and predict the intensity of the bioadhesive bond. By measuring surface and interfacial tensions, it is possible to calculate work done in forming an adhesive bond. Both spreading coefficients and bioadhesive work directly influence the nature of the bioadhesive bond and therefore provide essential information for the development of bioadhesive drug delivery systems.

D.   The Diffusion Theory

The concept that interpenetration and entanglement of bioadhesive polymer chains and mucous polymer chains produce semipermanent adhesive bonds (Fig. 2) is supported by the diffusion theory. It is believed that bond strength increases with the degree of penetration of the polymer chains into the mucous layer (30).

Penetration of polymer chains into the mucus network, and vice versa, is dependent on concentration gradients and diffusion coefficients. Obviously, any cross-linking of either component tends to hinder interpenetration, but small chains and chain ends can still become entangled. It has not been determined exactly how much interpenetration is required to produce an effective bioadhesive bond, but it is believed to be in the range of 0.2–0.5 μm. It is possible to estimate penetration depth (l) with the following:

$$l = (tD_b)^{1/2}$$

where t is time of contact and D_b is the diffusion coefficient of the bioadhesive material in mucus. The bond strength for a given polymer is believed to be attained when the depth of penetration is approximately equal to the end-to-end distance of the polymer chains (31).

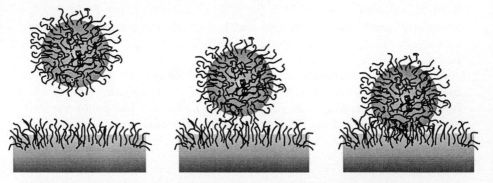

Figure 2   Mechanical bonding through interpenetration of bioadhesive and mucus polymer chains.

For diffusion to occur, it is important to have good solubility of one component in the other; the bioadhesive and mucus should be of similar chemical structure. Therefore, the strongest bioadhesive bonds should form between biomaterials whose solubility parameters (d_b) are similar to those of the target mucus glycoproteins (d_g) (8).

Thus, the diffusion theory states that interpenetration and entanglement of polymer chains are responsible for bioadhesion. The more structurally similar a bioadhesive is to its target, the greater the mucoadhesive bond will be.

E. The Fracture Theory

Perhaps the most applicable theory for studying bioadhesion through mechanical measurements has been the fracture theory. This theory analyzes the forces required to separate two surfaces after adhesion. The maximum tensile stress (s_m) produced during detachment can be determined by dividing the maximum force of detachment, F_m, by the total surface area (A_0) involved in the adhesive interaction:

$$s_m = F_m/A_0$$

In a uniform single-component system, fracture strength (s_f), which is equal to the maximum stress of detachment (s_m), is proportional to fracture energy (g_c), Young's modulus of elasticity (E), and the critical crack length (c) of the fracture site, as described in the following relationship (32):

$$s_f \sim (g_c E/c)^{1/2}$$

Fracture energy (g_c) can be obtained from the sum of the reversible work of adhesion, W_r (i.e., the energy required to produce new fracture surfaces), and the irreversible work of adhesion, W_i (i.e., the work of plastic deformation at the tip of the growing crack), where both values are expressed per unit area of the fracture surface (A_f):

$$g_c = W_r + W_i$$

The elastic modulus of the system (E) is related to stress (s) and strain (e) through Hooke's law:

$$E = \left[\frac{\sigma}{\varepsilon}\right]_{\varepsilon \to 0} = \left[\frac{F/A_0}{\Delta l/l_0}\right]_{\Delta l \to 0}$$

In this equation, stress is equal to the changing force (F) divided by the area (A_0), and strain is equal to the change in thickness (Δl) of the system divided by the original thickness (l_0) (33).

One critical assumption in the preceding analysis is that the system being investigated is of known physical dimensions and composed of a single uniform bulk material. Considering this, the simple relationship obtained cannot be applied to analyze the fracture site of a multicomponent bioadhesive bond between a polymer microsphere and either mucus or mucosal tissue. For such analysis, the equations must be expanded to accommodate dimensions and elastic moduli of each component (34). Furthermore, to determine fracture properties of an adhesive union from separation experiments, failure of the adhesive bond must be assumed to occur at the bioadhesive interface (31). However, it has been demonstrated that fracture rarely, if ever, occurs at the interface but instead occurs close to the interface (33,35).

Although these limitations exist, because the fracture theory deals only with analyzing the adhesive force required for separation, it does not assume or require entanglement, diffusion, or interpenetration of polymer chains. Therefore, it is appropriate for calculating fracture strengths of adhesive bonds involving rigid or semirigid bioadhesive materials, in which the polymer chains may not penetrate the mucous layer.

REFERENCES

1. D Duchêne, F Touchard, NA Peppas. Pharmaceutical and medical aspects of bioadhesive systems for drug administration. Drug Dev Ind Pharm. 14(2 & 3): 283–318, 1988.
2. G Ponchel, D Duchêne. Development of a bioadhesive tablet. In: M Szycher, ed. High Performance Biomaterials. Lancaster, PA: Technomic Publishing Company, 1991, pp 231–242.
3. NA Peppas, PA Buri. Surface, interfacial and molecular aspects of polymer bioadhesion on soft tissue. J Controlled Release 2:257–275, 1985.
4. KV Ranga Rao, P Buri. Bioadhesion and factors affecting the bioadhesion of microparticles. In: M Szycher, ed. High Performance Biomaterials. 1988, Lancaster, PA: Technomic Publishing Company, 1988, pp 259–268.
5. JL Chen, GN Cyr. Compositions producing adhesion through hydration. In: RS Manly, ed. Adhesion in Biological Systems. New York: Academic Press, 1970, pp 163–181.
6. H Park, JR Robinson. Physico-chemical properties of water insoluble polymers important to mucin/epithelial adhesion. J Controlled Release, 2:47–57, 1985.
7. D Duchêne, G Ponchel. Principle and investigation of the bioadhesion mechanism of solid dosage forms. Biomaterials 13:709–715, 1992.
8. AJ Kinloch. Review: The science of adhesion. Part I. Surface and interfacial aspects. J Mater Sci 15:2141–2166, 1980.

9. BV Derjaguin, VP Smilga. The electronic theory of adhesion. In: Adhesion: Fundamentals and Practice. University of Nottingham, England: Gordon & Breach, 1966.

10. BV Derjaguin, YP Toporov, VM Muller, IN Aleinikova. On the relationship between the molecular component of the adhesion of elastic particles to a solid surface. J Colloid Interface Sci 58:528–533, 1977.

11. RJ Good. Surface energy of solids and liquids: Thermodynamics, molecular forces, and structure. J Colloid Interface Sci 59:398–419, 1977.

12. D Tabor. Surface forces and surface interactions. J Colloid Interface Sci 58: 2–13, 1977.

13. JR Huntsberger. Mechanisms of adhesion. In: RL Patrick, ed. Treatise on Adhesion and Adhesives. New York: Marcel Dekker, 1966, pp 119–149.

14. JR Huntsberger. Mechanisms of adhesion. J Paint Technol 39:199–211, 1967.

15. AG Mikos, NA Peppas. Measurement of the surface tension of mucin solutions. Int J Pharm 53:1–5, 1989.

16. A Baszkin, JE Proust, P Monsenego, MM Boissonnade. Wettability of polymers by mucin aqueous solutions. Biorheology 27:503–514, 1990.

17. CM Lehr, HE Boddé, JA Bowstra, HE Junginger. A surface energy analysis of mucoadhesion. II: Prediction of mucoadhesive performance by spreading coefficients. Eur J Pharm Sci 1:19–30, 1993.

18. CM Lehr, JA Bowstra, HE Boddé, HE Junginger. A surface energy analysis of mucoadhesion: Contact angle measurements on polycarbophil and pig intestinal mucosa in physiologically relevant fluids. Pharm Res 9:70–75, 1992.

19. RT Spychal, JM Marrero, SH Saverymuttu, TC Northfield. Measurement of the surface hydrophobicity of human gastrointestinal mucosa. Gastroenterology 97:104–111, 1989.

20. DH Kaelble, J Moacanin. A surface energy analysis of bioadhesion. Polymer 18:475–482, 1977.

21. RP Campion. The influence of structure on autoadhesion (self-tack) and other forms of diffusion into polymers. J Adhes 7:1–23, 1974.

22. CT Reinhart, NA Peppas. Solute diffusion in swollen membranes. Part II. Influence of crosslinking on diffusive properties. J Membr Sci 18:227–239, 1984.

23. LA Girifalco, RJ Good. A theory for the estimation of surface and interfacial energies. I. Derivation and application to interfacial tension. J Phys Chem 61: 904–909, 1957.

24. BO Bateup. Surface chemistry and adhesion. Int J Adhes Adhesives July: 233–239, 1981.

25. S Wu. Surface and interfacial tensions of polymer melts. II. Poly(methyl methacrylate), poly(n-butyl methacrylate), and polystyrene. J Phys Chem 74:632–638, 1970.

26. RJ Good. Spreading pressure and contact angle. J Colloid Interface Sci 52: 308–313, 1975.

27. E Helfand, Y Tagami. Theory of interface between immiscible polymers. J Chem Phys 57:1812–1813, 1972.

28. E Helfand, Y Tagami. Theory of interface between immiscible polymers. II. J Chem Phys 56:3592–3601, 1972.

29. PC Hiemenz. Principles of Colloid and Surface Chemistry. 2nd ed. JJ Loagowski, ed. Undergraduate Chemistry. Vol 4. 1986. New York: Marcel Dekker, 1986, p 815.

30. SS Voyutskii. Autohesion and Adhesion of High Polymers. 1st ed. HF Mark, EH Immergut, eds. Polymer Reviews. Vol 4. New York: John Wiley & Sons, 1963, p 272.

31. AG Mikos, NA Peppas. Systems for controlled release of drugs. V. Bioadhesive systems. S T P Pharmacol 2:705–716, 1986.

32. HW Kammer. Adhesion between polymers. Acta Polym 34:112, 1983.

33. G Ponchel, F Touchard, D Duchêne, N Peppas. Bioadhesive analysis of controlled-release systems. I. Fracture and interpenetration analysis in poly(acrylic acid)–containing systems. J Controlled Release 5:129–141, 1987.

34. DH Kaelble. A relationship between the fracture mechanics and surface energetics failure criteria. J Appl Polym Sci 18:1869–1889, 1974.

35. JJ Bikerman. The Science of Adhesive Joints. 2nd ed. New York: Academic Press, 1968, p 349.

2

The Role of Water Movement and Polymer Hydration in Mucoadhesion

John D. Smart
University of Portsmouth, Portsmouth, England

I. INTRODUCTION

Bioadhesion is defined as the attachment of synthetic or biological macro-molecules to a biological tissue (Peppas and Buri, 1985). When applied to a mucosal epithelium bioadhesive interactions occur primarily with the mucus layer, and this phenomenon is referred to as *mucoadhesion* (Gu et al., 1988). Mucoadhesive dosage forms have the potential to optimize local drug delivery and systemic drug absorption. In order to maximize the performance of these dosage forms it is important to understand the nature of the adhesive interactions, and one proposed mechanism is discussed in this chapter.

In mucoadhesion, one of the adhering surfaces is a mucous membrane. Mucous membranes line the walls of various body cavities such as the gastrointestinal and respiratory tracts. They are either single-layered epithelium (e.g., the stomach, small and large intestine, and bronchi) or multilayered stratified epithelium (e.g., in the esophagus, vagina, and cornea). The former contain goblet cells, which secrete mucus directly onto the epithelial surfaces; the latter contain, or are adjacent to tissues containing, specialized glands such as salivary glands that secrete mucus onto the epithelial surface. Mucus is present as either a gel layer adherent to the mucosal surface or a luminal soluble or suspended form. The major components of all mucus gels are mucin glycoproteins, lipids, inorganic salts, and water, the latter accounting for more than 95% of the gel's weight, making it a highly hydrated

11

system (Marriott and Gregory, 1990). The mucin glycoproteins are the most important component of the mucus gel, resulting in its characteristic gel-like, cohesive and adhesive properties (Duchene et al., 1988; Gu et al., 1988). The thickness of this mucus layer varies on different mucosal surfaces, from 50 to 450 μm in the stomach (Kerss et al., 1982; Allen et al., 1990) to 0.7 μm in the oral cavity (Sonju et al., 1974). The major functions of mucus are those of protection and lubrication (they act as antiadherents). A more detailed description of mucus and mucous membranes can be found in Sec. 1b of this volume.

The largest group of mucosal-adhesive materials are hydrophilic macromolecules containing numerous hydrogen bond–forming groups (Smart et al., 1984). The presence of carboxyl or amine groups in particular appears to favor adhesion. These are called "wet" adhesives in that they are activated by moistening. However, unless water uptake is restricted, they may overhydrate to form a slippery mucilage (Chen and Cyr, 1970). These hydrogel-forming materials are nonspecific in action and show stronger adhesion to dry inert surfaces than to those covered with mucus (Mortazavi and Smart, 1995).

In considering the mechanism of mucoadhesion, several "scenarios" for in vivo mucoadhesive bond formation are possible:

> Case 1. Dry or partially hydrated dosage forms in contact with surfaces with substantial mucus layers (typically particulates administered into the nasal cavity).
> Case 2. Fully hydrated dosage forms in contact with surfaces with substantial mucus layers (typically particulates administered into the gastrointestinal tract).
> Case 3. Dry or partially hydrated dosage forms in contact with surfaces with thin or discontinuous mucus layers (typically tablets or patches placed into the oral cavity or vagina).
> Case 4. Fully hydrated dosage forms in contact with surfaces with thin or discontinuous mucus layers (typically aqueous semisolids or liquids administered into the oral cavity or vagina).

It is unlikely that one mechanism of mucoadhesion could be used to describe the adhesion processes in all of these cases. However, in many descriptions of the interactions between mucoadhesive materials and a mucous membrane, two basic steps have been identified (e.g., Duchene et al., 1988, Gu et al., 1988).

> Step 1. Contact stage: An intimate contact is formed between the mucoadhesive and mucous membrane.

Step 2. Consolidation stage: Various physicochemical interactions occur to consolidate and strengthen the adhesive joint, leading to prolonged adhesion.

A. Step 1: The Contact Stage

The mucoadhesive and the mucous membrane initially have to form a close contact with each other. In some cases these two surfaces can be readily brought together, e.g., placing and holding a delivery system within the oral cavity or vagina, depositing a particle within the respiratory tract. In other cases, such as oral drug delivery, adsorption of a particulate system from suspension onto the gastrointestinal mucosa would be a prerequisite for the adhesion process. The forces promoting adsorption of small formulations such as microparticles may be sufficient to hold them on the mucosal surface until displaced by mucus, or cell, turnover (Helliwell, 1993; Lehr, 1994). Adsorption will occur as a result of a reduction in surface free energy as two surfaces are lost and a new interface is formed. The effect of surface energies generally on the mucoadhesive process has been investigated by several workers (Esposito et al., 1994; Lehr et al., 1992a; Rillosi and Buck ton, 1995), although the adsorption of biological molecules in vivo would be expected to alter the surface properties of a test material rapidly.

B. Step 2: The Consolidation Stage

It has been proposed that if strong or prolonged adhesion is required, for example, with larger formulations exposed to stresses such as blinking or mouth movements, then a second, "consolidation" stage is required. The mucoadhesive point can be considered to contain three regions (Fig. 1): the mucoadhesive, the mucosa, and the interfacial region, consisting at least initially of mucus. Adhesive joint failure will normally occur at the weakest component of this joint (Fig. 2). For weak adhesives this would be the

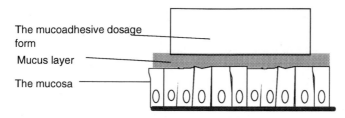

The mucoadhesive dosage form

Mucus layer

The mucosa

Figure 1 The three regions within a mucoadhesive joint.

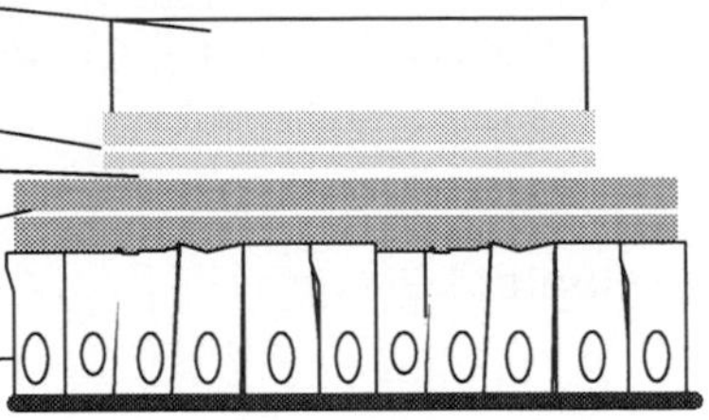

The mucoadhesive dosage form

Fracture through the hydrated gel layer

Fracture through the interface

Fracture through the mucus gel layer

The epithelium

Figure 2 Possible regions of mucoadhesive joint failure.

mucoadhesive-mucus interface; for stronger adhesives this would initially be the mucus layer but later may be the hydrating mucoadhesive material [on application of a constant tensile stress to compacts of mucoadhesive polymers, joint failure was found by Mortazavi and Smart (1994) to be a cohesive failure of the swelling polymer for all but the weakest adhesives]. The strength of the adhesive joint will therefore depend on the cohesive nature of the weakest region, and for strong mucoadhesion when a substantial mucus layer is present, an increase in the cohesive nature of this gel is necessary (however, the undesirability of obtaining such strong adhesion that the weakest component of the adhesive joint becomes the epithelial cells is obvious).

There are essentially two theories of how this gel strengthening occurs. One is based on a macromolecular interpenetration effect, which has been dealt with on a theoretical basis by Mikos and Peppas (1990). In this theory, based largely on that described by Voyutskii (1963) for compatible polymeric systems, the mucoadhesive molecules interpenetrate and bond by secondary interactions with mucus glycoproteins (Fig. 3). Evidence for this is provided by an attenuated total reflection Fourier transform infrared (ATR-FTIR) study by Jabbari et al. (1993). In their study a thin cross-linked film of poly(acrylic acid) was formed on an ATR crystal. A mucin solution was placed into contact with this film and ATR-FTIR spectra collected over a period of time. Deconvolution of these spectra revealed a peak after 6 minutes at 1550 cm^{-1} (which manifested itself as a small shoulder in the original spectrum), which was attributed to mucin dimeric carboxylic C$=$O stretching (Fig. 4), and it was proposed that this indicated the presence of interpenetrating mucin molecules within the poly(acrylic acid) film. Further indirect evidence for interpenetration is based on the rheological effects of mixing mucus with mucoadhesive gels (Hassan and Gallo, 1990; Mortazavi et al., 1992; and discussed later in this volume). Rheological synergism, an increase in the resistance to elastic deformation (i.e., mucous gel

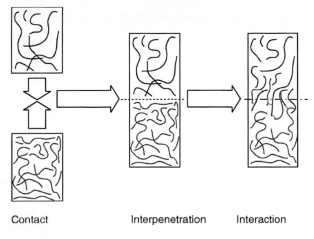

Contact   Interpenetration   Interaction

Figure 3 The interpenetration theory; three stages in the interaction between a mucoadhesive polymer and mucin glycoprotein.

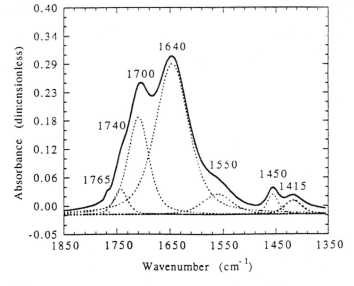

Figure 4 The deconvolution of an ATR-FTIR spectrum of a cross-linked poly(acrylic acid) in contact with pH 7 buffered mucin solution after 6 minutes. The solid lines represent the original spectrum and the dashed curves the deconvoluted peaks. (From Jabbari et al., 1993.)

strengthening), is evident, and this would undoubtedly help consolidate the adhesive joint.

Several objections to the interpenetration theory exist. Mucoadhesion arises very quickly, within a matter of seconds. For two large macromolecules to interpenetrate several micrometers in order to consolidate the adhesive joint within this time is unlikely. The rheological synergy study suggests that as soon as mucus and mucoadhesive interpenetrate, they are likely to interact and form a surface gel layer that will substantially inhibit any further interpenetration. In an electron microscopy study by Lehr et al (1992b), no evidence of interpenetration could be seen in the micrometer range. The evidence of interpenetration given by Jabbari and coworkers is less convincing if it is considered that the C$=$O dimer stretching, proposed to be indicative of mucin ingress, could have arisen from low-molecular-weight contaminants or degradation products within the commercial mucin sample. Mucin, like most biological macromolecules, is prone to degradation, and the commercial sources in particular show little of the rheological properties attributed to native mucus (Madsen et al., 1996). No attempt to characterize the continuity and integrity of the poly(acrylic acid) film on the ATR crystal was completed, and it is possible that this technique was measuring the appearance of mucin in discontinuities within the film. Mucoadhesive materials also adhere much better to solid surfaces such as Perspex (where the opportunity for macromolecular interpenetration and secondary interactions is clearly minimal), and the presence of mucus appears to inhibit, rather than promote, adhesion (Mortazavi and Smart, 1995). Other workers have also identified infrared peaks around the 1550 cm^{-1} region in carboxyl group–containing polymers and attributed these to (COO^{-}) groups (Tsibouklis et al., 1991), which would start to appear in great abundance in a hydrating poly(acrylic acid) film.

The theory proposed in this chapter is that consolidation arises from the ability of dry or partially hydrated mucoadhesive materials (case 1 and 3) to swell and dehydrate mucus gels, and it is water movement, rather than macromolecular interpenetration, that drives the consolidation of the adhesive joint.

II. THE MECHANISM OF HYDROGEL HYDRATION

Swelling is a consequence of the affinity of polymeric components for water. Polymers swell because of an imbalance between the chemical potential of solvent within the polymer and that in the surrounding medium (Kalal, 1983). Thus solvent moves as a result of polymer "osmotic pressure" until an equilibrium is achieved and the internal and external chemical potentials are equivalent. A theoretical treatment of the swelling of polyelectrolyte networks is given by Khokhlov et al. (1993). For low-molecular-weight hydrophilic polymers the equilibrium state is a solution; for high-molecular-weight or cross-linked polymers it can be a water-swollen gel; thus a dry hydrophilic polymer can physically change from a glassy solid, to a rubbery (gel-like) state and then perhaps on to a solution during the swelling process (Peppas and Korsmeyer, 1987). The extent and rate of swelling are affected by the degree of cross-linking and chain length. If the surrounding medium contains solute, the rate of swelling decreases, particularly if the solute is large and cannot enter the hydrogel network. Refugo (1976) demonstrated that similar hydrogels for which equilibrium swelling had been achieved in water are dehydrated in dextran solutions. The dextran molecule used had an average size of about 27 nm but the average hydrogel pore diameter was only 1.2 nm, so interpenetration was considered highly unlikely. Partial dehydration of the hydrogel resulted from a balancing of the chemical potentials throughout the system.

Similar events would be predicted when a gel is brought into contact with a second gel. Water movement occurs between gels until equilibrium is achieved. A polyelectrolyte gel, such as a poly(acrylic acid), will have a strong affinity for water, therefore a high "osmotic pressure" and a large swelling force (Khare et al., 1992; Silberberg-Bouhnik et al., 1995).

III. EVIDENCE OF GEL HYDRATION AND MUCUS DEHYDRATION IN MUCOADHESION

We have investigated the water transfer from a mucus gel to various dry putative mucoadhesive materials (Mortazavi and Smart, 1993). Dry 250-mg compacts of test materials were wrapped in dialysis tubing and placed into contact with 1.5 g of crude mucus (Fig. 5). These were stored in sealed cells

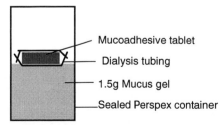

      Mucoadhesive tablet

      Dialysis tubing

      1.5g Mucus gel

      Sealed Perspex container

Figure 5   Apparatus used to assess water movement between a mucoadhesive compact and mucus gel. (From Mortazavi and Smart, 1993a.)

at 37°C and at set time intervals the weight gain of each compact was found and compared with a non–mucus-containing control (Fig. 6). A substantial weight gain after only 1 minute was recorded for the well-known mucoadhesive materials Carbopol 934P and polycarbophil (2.28 and 1.53 g cm^{-2}, respectively). It was apparent that these materials were rapidly hydrating and swelling and extracting the necessary water from the mucus gel. When Carbopol 934P gels were used in these systems, the dry compacts and 25% gels both gained water from mucus, whereas the 4% gel lost water and the 12% gel neither gained nor lost water (Fig. 7). Thus the water movement expected from the theory of gel hydration is clearly evident. This can be seen using a light microscope as described by Mortazavi and Smart (1993b). A dry 50-mg Carbopol 934P compact was placed onto a little white soft paraffin on a microscope slide and 0.2–0.3 g (degassed) mucus gel was placed in contact with the edge of the compact. After 1 minute a gel layer

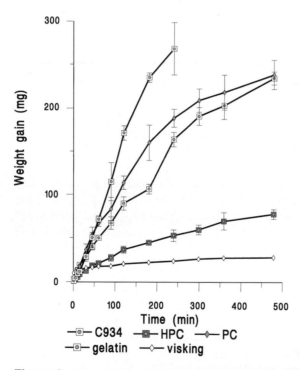

Figure 6 The weight gain of dry compacts of Carbopol 934 (C934), hydroxypropylcellulose (HPC), polycarbophil (PC), and gelatin when placed in contact with a homogenized porcine gastric mucus gel ($n = 3$, bars = SD). (From Mortazavi and Smart, 1993a.)

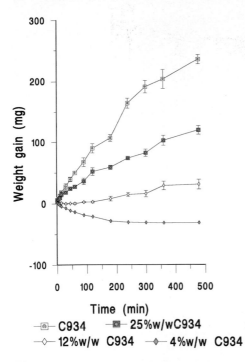

Figure 7 The weight gain of Carbopol 934 (C934) compacts and gels when placed into contact with a homogenized porcine gastric mucus gel ($n = 3$, bars = SD). (From Mortazavi and Smart, 1993a.)

was seen to form as a result of water movement from the mucus gel (Fig. 8). A darkened area after 15 minutes suggests the presence of a dehydrated region within the mucus gel. The same effect was observed when other mucoadhesive materials were tested.

Further evidence for substantial water movement was supplied by Jabbari et al. (1993), who observed a very large water peak at 1640 cm^{-1} after 6 minutes in the poly(acrylic acid) layer, indicating substantial water ingress.

Dehydrating a mucus gel increases its cohesive nature, and this was shown by Mortazavi and Smart (1993a). High-concentration gels were prepared by centrifugation, and tensiometer data and dynamic oscillatory rheology studies indicated a substantial increase in their adhesive behavior and resistance to elastic deformation (Figs. 9 and 10). Thus dehydration, which is a rapid process, substantially alters the physicochemical properties of a mucus gel, making it locally a more cohesive layer and promoting the retention of a delivery system. It is possible that some larger glycoprotein

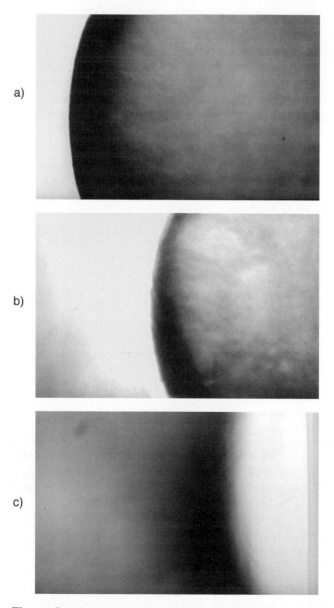

Figure 8   The interface between the edge of a 50-mg Carbopol 934 compact and a homogenized porcine gastric mucus gel after (a) 0, (b) 1, and (c) 15 min contact. (Magnification × 40.)

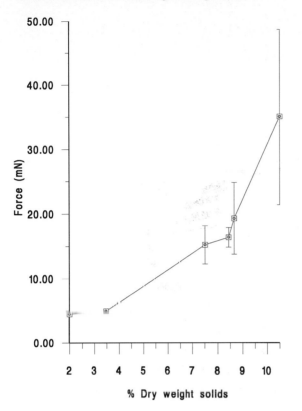

Figure 9 The force required to pull a disc and surface apart, when held together by homogenized porcine gastric mucus gels of varying water contents ($n = 4$, bars = SD). (From Mortazavi and Smart, 1993a.)

components would be carried with the flow of water into the mucoadhesive material, and some interpenetration could arise as a secondary event.

In case 3 mucoadhesion, if the dosage form is sufficiently large, dehydration effectively collapses the mucus layer, preventing any possibility of interpenetration of the gel networks. In these cases the affinity for water shown by these mucoadhesive polymers seems to generate an almost "suction-like" effect, and it has been shown to be much more difficult to remove test compacts on application of tensile stresses compared with the application of shear stresses (Smart and Johnson, 1996).

In case 2 and 4 mucoadhesion, involving fully hydrated dosage forms in which no further swelling can arise, the adhesive joint cannot be consol-

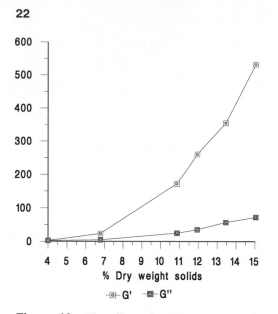

Figure 10 The effect of water content on the storage modulus and loss modulus of homogenized porcine gastric mucus gels ($n = 4$, bars = SD). (From Mortazavi and Smart, 1993a.)

idated by water movement. Surface energy effects (although interpenetration could also play a role in these circumstances) would be predicted to be the most important mechanism of adhesion.

IV. CONCLUSIONS

After the formation of intimate contact between a solid or partially hydrated mucoadhesive material and a mucus gel, a consolidation step is required to allow the formation of a strong stable mucoadhesive joint. Of the two theories of consolidation, the interpenetration theory is largely theoretical, with little supporting experimental evidence. The dehydration theory proposed in this chapter is theoretically plausible, the results can easily be observed under a microscope and have been clearly demonstrated experimentally. This theory, however, can apply only when a dosage form is brought into contact with a mucous membrane in a dry or partially hydrated state, and the adhesion of gels, fully swollen systems, or non–hydrogel-forming materials must be explained by other, probably surface energy, effects.

REFERENCES

Allen A, Cunliffe WJ, Pearson JP, Venables CW. 1990. J Intern Med 228(suppl 1): 83.

Chen JL, Cyr CN. 1970. In: Manly RS, ed. Adhesion in Biological Systems. New York, Academic Press, p 163.

Duchene D, Touchard F, Peppas NA. 1988. Drug Dev Ind Pharm 14:283.

Esposito P, Colombo I, Lovrecich M. 1994. Biomaterials 15:177.

Gu JM, Robinson JR, Leung SHS. 1988. Crit Rev Ther Drug Carrier Syst 5:21.

Hassan EE, Gallo JM. 1990. Pharm Res 7:491.

Helliwell M. 1993. Adv Drug Deliv Rev 11:221.

Jabbari E, Wisniewski N, Peppas NA. 1993. J Controlled Release 26:99.

Kalal J. 1983. In: Finch CA, ed. Chemistry and Technology of Water-Soluble Polymers. New York: Plenum, p 71.

Kerss S, Allen A, Garner A. 1982. Clin Sci 63:187.

Khare AR, Peppas NA, Massimo G, Columbo P. 1992. J Controlled Release 22:239.

Khokhlov AR, Starodubtzev SG, Vasilevskaya VV. 1993. Adv Polym Sci 109:123.

Lehr CM. 1994. Crit Rev Ther Drug Carrier Syst 11:119.

Lehr CM, Bowstra JA, Bodde HE, Junginger HE. 1992a. Pharm Res 9:70.

Lehr CM, Bowstra JA, Spies F, Onderwater J, van het Noordeinde J, Vermeij-Keers C, Van Munsteren CJ, Junginger HE. 1992b. J Controlled Release 18:249.

Madsen F, Eberth K, Smart JD. 1996. Pharm Sci 2:563.

Marriott C, Gregory NP. 1990. In: Lanaerts V, Gurny R, eds. Bioadhesive Drug Delivery Systems. Boca Raton, FL: CRC Press, p 1.

Mikos AG, Peppas NA. 1990. In: Lanaerts V, Gurny R, eds. Bioadhesive Drug Delivery Systems. Boca Raton, FL: CRC Press, p 25.

Mortazavi SA, Smart JD. 1993a. J Contr Rel 25:197.

Mortazavi SA, Smart JD. 1993b. J Pharm Pharmacol 45:1111.

Mortazavi SA, Smart JD. 1994. J Contr Rel 31:207.

Mortazavi SA, Smart JD. 1995. Int J Pharm 116:223.

Mortazavi SA, Carpenter BG, Smart JD. 1992. Int J Pharm 83:221.

Peppas NA, Buri PA. 1985. J Contr Rel 2:257.

Peppas NA, Korsmeyer RW. 1987. Hydrogels Med Pharm 3:109.

Refugo MF. 1976. In: Andrade JD, ed. Hydrogels for Medical and Related Applications. ACS Symposium Series 31. Washington, DC: American Chemical Society, p 37.

Rillosi M, Buckton G. 1995. Int J Pharm 117:75.

Silberberg-Bouhnik M, Ramon O, Ladyzhinski I, Mizrahi S, Cohen Y. 1995. J Polym Sci 33:2269.

Sonju T, Cristensen TB, Kornstad L, Rolla G. 1974. Caries Res 8:113.

Smart JD, Johnson ME. 1996. Eur J Pharm Sci. 4:S65.

Smart JD, Kellaway IW, Worthington HEC. 1984. J Pharm Pharmacol 36:295.

Tsibouklis J, Petty M, Song Y-P, Richardson R, Yarwood J, Petty MC, Feast WJ. 1991. J Mater Chem 1:819.

Voyutskii SS. 1963. Autoadhesion and Adhesion of High Polymers. New York: John Wiley & Sons/Interscience.

3

A Rheological Approach to Explain the Mucoadhesive Behavior of Polymer Hydrogels

Carla Marcella Caramella, Silvia Rossi, and Maria Cristina Bonferoni
University of Pavia, Pavia, Italy

I. INTRODUCTION

The task of dealing with the rheological aspects of mucoadhesion is rather intriguing, as is always the case when rheology is involved in explaining the behavior of a system. In fact, the rheological properties of a system, either liquid or semisolid, depend on so many factors (the chemical nature of the components, the state of the material, its derived physicochemical properties, the medium, the temperature, the previous history of the system, and the way the properties are measured, to cite only the well known ones) that it is difficult, on the one hand, to consider the rheological spectrum as a fingerprint of the mechanical properties of a given system and, on the other hand, to relate the rheological behavior to the functional characteristics of the system itself. In this regard, mucoadhesive hydrogels represent no exception.

In the context of the present chapter, the term hydrogel will be used in its broader sense, that is, a matrix formed by either a water-swellable material (usually a cross-linked polymer with limited swelling capacity) or a water-soluble material (usually a hydrophilic polymer that swells indefinitely and eventually undergoes complete dissolution). Most of the mucoadhesive polymers or materials belong to one of these two categories.

It must also be pointed out that mucoadhesion, so-called wet adhesion, invariably involves the presence of a hydrated gel phase. The mucoadhesive

25

hydrogel may be applied as such to the mucosal surface, or the hydrated gel phase is formed in situ upon hydration of a solid mucoadhesive system in contact with the mucus layer.

In previously published reviews on mucoadhesion (Gu et al., 1988; Kellaway, 1990; Helliwell, 1993), outstanding authors and experts have just started investigating the rheological aspects of the mucoadhesive phenomenon and have limited their considerations to no more than one or two paragraphs. We thank the editor for acknowledging the fact that this topic deserves a chapter in the present book and for asking us to discuss the rheological aspects of mucoadhesive hydrogels. The reasons for doing this are at least three:

1. At present, there is still a need for a better understanding of the mechanisms of mucoadhesion. Even though a few chapters of this book are devoted to the physicochemical principles of mucoadhesion and to understanding its underlying mechanisms, there is no unique explanation for the phenomenon and therefore any contribution would be welcome; rheological analysis may contribute to such an explanation.
2. The increasing interest in bioadhesive pharmaceutical dosage forms has stimulated the search for new mucoadhesive polymers and materials; therefore an understanding of the properties that are relevant to the interactions with mucus, including the rheological ones, is essential for a rational design of such polymers and materials.
3. A certain number of papers have been devoted to the rheological properties of mucoadhesive systems; these papers need to be reviewed and linked to each other.

In the present chapter, a knowledge of basic rheology is taken for granted.

II. RATIONALE FOR THE RHEOLOGICAL APPROACH TO EXPLAIN THE MUCOADHESIVE FUNCTION

In the past, many adhesive polymers were found empirically—cellulose derivatives, alginates, natural gums, polyacrylates, hyaluronic acid, scleroglucan, gelatin, pectin, pregelatinized and modified starches, etc.—and classified according to their adhesive performance. The characteristics of a polymer that are most relevant to its mucoadhesive properties have been determined by screening mucoadhesive polymers (Park and Robinson, 1985; Ch'ng et al., 1985; Gu et al., 1988; Duchene et al., 1988; Junginger, 1991).

First of all, the chemical nature must be considered; then the charge density, molecular weight, chain flexibility, and surface properties have to be examined. In particular, the presence of carboxylic groups favors the establishment of ionic and hydrogen bond interactions between polymer and mucin chains. Other functional groups, such as hydroxy, amino, and sulfate groups, are also likely to participate in the formation of chemical bonds. Charge density, that is, the spacing of charged groups (linked to the poly-electrolyte nature of the polymer), determines the possibility of ionic inter-action with mucin macromolecules. The molecular chain length (linked to molecular weight) determines the strength of mucoadhesive bonds related to chain entanglements; it is believed that an optimally high molecular weight corresponds to maximum adhesiveness. Molecular mobility and flexibility (expressed, for example, by glass transition temperature) also affect the abil-ity of the two macromolecular species to form entanglements. The surface properties (expressed by contact angle measurements) reflect the hydropho-bicity or polarity of the polymer, which in turn determines hydrogen bond formation.

The viscosity and the viscoelastic properties of polymers have also been considered (Anders and Merkle, 1989; Tamburic and Craig, 1995, 1996). Anders found that the viscosity grade of polymers does not neces-sarily predict the mucoadhesive properties of films. As in the case of mo-lecular weight, it may happen that the mucoadhesiveness increases with increasing viscosity until a maximum value, corresponding to optimal ad-hesion, is reached. On the other hand, Craig found that the viscoelastic nature of Carbopol gels is a good predictor of their adhesive properties.

In the design of mucoadhesive systems, the characterization of the mucoadhesive interface—that is, the understanding of the nature of the in-teraction between the polymer chains and the mucus chains—is as important as the characterization of the polymer. Various explanations have been pro-posed for the mucoadhesive phenomenon, the most common of which are the wetting-spreading theory and the interpenetration-interdiffusion theory (Peppas and Buri, 1985). The former analyzes the adhesive behavior of mucoadhesive systems in terms of their ability to spread over a biological surface and underlines the importance of the surface energy thermodynamics of mucus and the mucoadhesive systems in the formation of the adhesive bond. Accordingly, in the latter, the main physical mechanism of mucoad-hesion is the interdiffusion of the chains of polymer and mucus to a sufficient depth to create a semipermanent adhesive bond. Basically, as illustrated in Fig. 1, during the interpenetration process, the molecules of the mucoad-hesive and the glycoprotein network are brought in intimate contact, and,

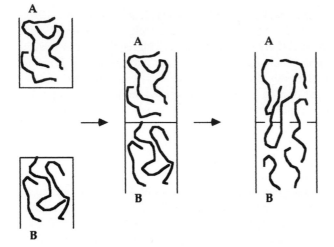

Figure 1 Chain interpenetration during bioadhesion of a polymer (A) with the glycoprotein network of mucus (B).

due to the concentration gradient, the mucoadhesive polymer chains penetrate into the mucus.

Complete agreement about the interpretation of the mucoadhesion phenomenon has not yet been reached. Accordingly, for most authors, both processes have to be taken into account to explain the phenomenon. Mucoadhesion is regarded as a two-step process (Duchene et al., 1988; Junginger, 1991). Initially, intimate surface contact between the mucoadhesive and the mucous substrate has to be established; in a consecutive step, both phases may interdiffuse or interpenetrate to a certain extent. The first step is considered to be an interfacial phenomenon influenced by the surface energy effects and the spreading process of both phases, the mucoadhesive hydrogel and the mucus. The second step involves chain entanglements due to the interpenetration of mucoadhesive and mucus chains. We definitely share the opinion of Lehr (1991), who stated that discriminating between spreading and interpenetration appears to be a purely semantic exercise. In fact, after the contact between the mucoadhesive and the mucous substrate, dynamic changes occur at the adhesive interface.

Some of the properties of the mucoadhesive interface that have been investigated, such as wettability for solid forms, spreadability for liquid forms, and interface energy, are more relevant to the wetting-spreading theory. Other properties, such as surface roughness, hydration, and swelling, are compatible with both the wetting-spreading and the interpenetration processes. The rheological properties of the mucoadhesive interface, in other words of the hydrated gel formed at the interface, instead should both influence and be influenced by the occurrence of the interpenetration step. In fact, during this process, mucoadhesive polymer and mucin macromolecules are interdiffused. Chain interlocking, conformational changes, and chemical interactions (such as hydrogen and van der Waals bonds), which occur between polymer and mucin chains, are likely to produce changes in the rheological behavior of the two macromolecular species.

The rheological approach to the explanation of mucoadhesion is also supported by the following observations and facts:

> The viscoelastic nature of mucus. The mucus present throughout the gastrointestinal tract acts both as a protective layer and as a lubricant. To serve this dual function, the mucus gel has particular physical characteristics, the most important of which is viscoelasticity. This property enables it to shield the mucosal surface from shear forces and to flow under the influence of peristalsis (Marriott and Gregory, 1990).
>
> The indirect evidence of the viscoelastic properties of the mucoadhesive interface reported by some authors. In particular, Peppas et al. (1987) analyzed the viscoelastic properties and the dynamic behavior of preswollen polyacrylic acid containing systems kept in contact with mucus and found that these properties were related to the bioadhesive properties. This approach was based on the evaluation of stress relaxation behavior as a measure of entanglements between polymer-mucin chains and of their ability to rearrange themselves to give permanent deformation under applied stress. In a more recent development of this concept, the time dependence of the viscoelastic term of the energy of fracture was pointed out and related to the irreversible deformation process of the bulk adhesive (Ponchel et al., 1991).

From a methodology point of view, the rheological approach involves the investigation of the changes in rheological properties that mucoadhesive polymers and hydrogels undergo when they are mixed with mucus (or mucins).

III. RHEOLOGICAL MEASUREMENTS USED TO EVALUATE POLYMER-MUCIN INTERACTIONS

A. Viscosity Measurements

Even though other authors had already published works on rheological synergism (Allen et al., 1986), Hassan and Gallo (1990) are considered to have pioneered the work on the rheological assessment of mucin-polymer bioadhesive bond strength. In fact, the authors observed that, when a putative mucoadhesive polymer and mucin were mixed together, there was a synergistic increase in viscosity.

Because the viscosity of a mucin dispersion is the net result of the resistance to flow exerted by individual chain segments, physical chain entanglements, and noncovalent intermolecular interactions, which are the same as the interactions involved in the process of mucoadhesion, Hassan suggested that the forces of interaction involved in a mucin-bioadhesive system could be evaluated by viscosity measurements. Both physical and chemical bonds in mucin-polymer mixtures cause changes in the shape or arrangement of macromolecules that are the basis for viscosity changes. Thus, the viscosity of a dispersion containing mucin and a bioadhesive system has to be considered as the result of the contribution of different components, such as the viscosities of bioadhesive polymer and mucin and the viscosity component due to bioadhesion (η_b). This was calculated by Hassan and Gallo according to the equation $\eta_b = \eta_t - \eta_m - \eta_p$, where η_t is the viscosity coefficient of the polymer-mucin system and η_m and η_p are the individual viscosity coefficients of mucin and the bioadhesive polymer, respectively. Moreover, the authors proposed an equation identical to the viscosity equation of Newtonian solutions for calculating the force of bioadhesion F (dyne/cm^2): $F = \eta_b \cdot$ shear rate.

These two parameters (η_b and F) give a direct estimate of the polymer-mucin interactions occurring in mucoadhesion. Using these parameters, different polymers (cationic, anionic, and neutral ones) in mixtures with porcine gastric mucin were evaluated for mucoadhesion potential in different hydration media (0.1 N HCl and 0.1 N acetate buffer, pH 5.5).

In Table 1, the η_b and F values obtained for some of the polymers investigated by Hassan and Gallo are reported. Different explanations were proposed by the authors for the results obtained, depending on the chemical nature and the molecular conformation of polymers.

In the case of the cationic polymer chitosan, high interaction parameters were observed both in hydrochloric acid and in acetate buffer. At a pH value of 1.0, where mucin carboxyl groups (from terminal sialic acids) are not ionized, the force of interaction could not be contributed to by ionic interactions, whereas at higher pH values, where sialic acid is in the anionic form, the combination of ionic interactions with other forces resulted in stronger bioadhesive bonds. Also anionic polyacrylic acid demonstrated a stronger interaction with mucin at pH 5.5 than at pH 1.0. Hassan explained this on the basis of the shape of polyacrylic acid molecules, which at pH 1.0 are not ionized and have a contracted form due to intramolecular hydrogen bonds, whereas at pH 5.5 they become ionized and have a stretched

Table 1 Component of Bioadhesion (η_b) and Force of Bioadhesion (F) Calculated for Mixtures of Different Polymers (0.5% w/v) with Mucin (15% w/v) in 0.1 N HCl and 0.1 N Acetate Buffer pH 5.5 (Mean Values ± SD, $n = 3$)

Polymer	0.1 N HCl		0.1 N acetate buffer pH 5.5	
	η_b (cps)[a]	F (dyne/cm^2)	η_b (cps)[a]	F (dyne/cm^2)
Chitosan	54.32 (±1.17)	5052 (±109)	70.63 (±1.54)	6596 (±140)
Polyacrylic acid	11.49 (±2.40)	1069 (±223)	15.03 (±2.03)	1398 (±189)
Dextran	6.24 (±1.17)	580 (±109)	8.54 (±1.71)	794 (±159)

[a]Calculated at 93 s^{-1} shear rate.
Adapted from Hassan and Gallo (1990).

cylindrical shape more able to penetrate the mucin network. A rheological interaction was also observed for a neutral polymer, dextran. The lower values of η_b and F observed for dextran with respect to the other polymers were attributed to polymer molecular branching that reduces interaction with mucin.

B. Dynamic Oscillatory Measurements

Almost simultaneously with Hassan and Gallo, Kerr et al. (1990) proposed the use of dynamic oscillatory measurements to evaluate the rheological interaction between a polymer and purified mucin.

Such measurements were expected to give useful information about the structure of the polymer-mucin network for two reasons that are linked to each other: (a) the viscoelastic nature of the mucus gel layer and the presumably viscoelastic nature of the mucoadhesive interface, already cited; and (b) the nondestructive nature of the dynamic oscillatory analysis. This kind of analysis is especially advised when dealing with gels, which have an internal tridimensional structure that may be destroyed by a classical rotational viscometry test. In fact, whereas in classical viscometry measurements, the sample is forced to flow and its structure can be partially destroyed, in viscoelastic tests, whenever they are performed in a linear viscoelastic region, polymer conformation and intermolecular bonds are stretched but not destroyed.

Kerr et al. investigated, at different pH values, the viscoelastic properties of glycoprotein (mucin) gels both alone and in mixtures with polyacrylic acids with different molecular weights. The addition of polyacrylic acid produced an increase in the viscoelastic parameters storage modulus

(G') and loss modulus (G''). The polymer with the highest molecular weight had the greatest effect on the elasticity of the glycoprotein gel. The strongest interactions were observed at the lowest pH examined (pH 4.0).

The usefulness of viscoelastic parameters in the evaluation of polymer-mucin interactions was also pointed out some time later by Mortazavi et al. (1992), who studied the effect of polyacrylic acid (Carbopol 934P) on the rheological properties of mucus, along with the influence of pH and temperature. A large increase in G' was observed in polymer-mucin mixtures in comparison with the values obtained when the mucus gel and the polyacrylic acid gel were evaluated separately at the same concentration. Such an increase was markedly affected by pH: it was minimal at pH 4.25, as evidenced in Fig. 2. The G'' values were much smaller and did not change significantly with pH. An increase in temperature in the range 5–45°C reduced the G' and G'' values but the mixtures retained their gel properties.

In subsequent work (Caramella et al., 1992, 1993), we wanted to investigate further the concept of rheological synergism. In particular, it was interesting to extend the dynamic oscillatory studies to mucoadhesive polymers other than the well-known polyacrylic acid and to investigate which rheological parameters were more suitable for characterizing the polymer-mucin interactions. To this end, three different mucoadhesive polymers—

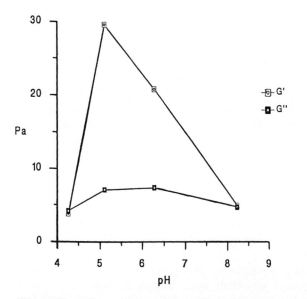

Figure 2   Effect of pH on the rheological behavior of PAA-mucus mixtures. (From Mortazavi et al., 1992.)

polyacrylic acid (Carbopol 934PH), sodium carboxymethylcellulose (NaCMC), and methacrylic acid methyl methacrylate copolymer (Eudispert)—were compared in their rheological interaction with mucin. The polymers examined were hydrated in 0.1 M phosphate buffer (pH 7.0). Preliminary rheological interaction experiments were carried out using solutions of the three polymers prepared at the same concentration. Because the polymer solutions had quite different viscosities, in order to better appreciate the extent of the rheological interaction with mucin, it was decided to prepare and test isoviscous solutions (i.e., comparable viscosities in a given shear rate range). The polymer concentrations were 0.96% (w/w) for Carbopol, 1.92% (w/w) for NaCMC, and 4.80% (w/w) for Eudispert. Mixtures of the polymers with a commercial purified submaxillary mucin were also prepared in the same medium and at the same polymer concentration as the polymer solutions. The mucin concentration was fixed at 4% (w/w).

A thorough rheological characterization of polymer and mucin solutions and of their mixtures was effected using a rotational rheometer (Bohlin CS Rhometer, Bohlin Instrument Division, Metric Group Ltd, Cirencester, UK). In particular, the apparent viscosity of the samples was measured at increasing shear rate ranging from 50 to 250 s^{-1}. Dynamic oscillatory tests were performed within the linear viscoelastic range. The viscoelastic parameters storage modulus (G') and loss modulus (G'') were measured at increasing frequency ranging from 0.1 to 4 Hz.

Figure 3 shows the apparent viscosity of the polymer solutions and their mixtures with mucin, measured at a representative value of shear rate (100 s^{-1}). For all three polymers a synergic increase in viscosity was observed when they were mixed with mucin: the viscosities of the mixtures were higher than the values calculated by simple addition of the viscosities of polymer and mucin alone. No significant differences were observed between the rheological synergisms of the three polymers, obtained, according to Hassan and Gallo (1990), as the difference between the measured and calculated viscosities.

Figure 4 shows the viscoelastic parameters G' and G'' measured at a representative value of frequency (1.0 Hz) for the polymer solutions and their mixtures with mucin. For all three polymers examined, an increase in both G' and G'' was observed in the presence of mucin. It must be mentioned that the mucin dispersion at the concentration employed (4% w/w) did not show any detectable viscoelastic properties; this means that the increase in viscoelastic parameters observed was definitely due to the polymer-mucin interactions. Although isoviscous, the three polymer solutions showed different values of both G' and G''; this made it difficult to compare directly the changes in viscoelastic properties after addition of mucin.

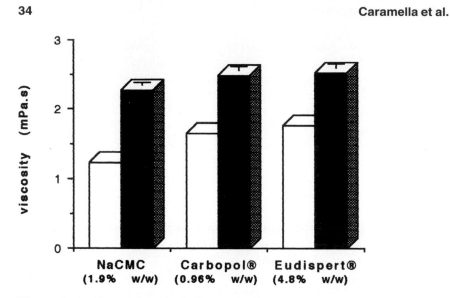

Figure 3 Evidence of rheological synergism between the three polymers examined and bovine submaxyllary mucin. (□) Viscosity of polymer-mucin mixture calculated as sum of the viscosities of polymer solution and mucin dispersion at 100 s^{-1} shear rate. (■) Viscosity of polymer-mucin mixture measured at 100 s^{-1} shear rate (mean values ± SD; $n = 3$).

Therefore, besides elastic and loss modulus, the loss tangent (tg δ), i.e., the ratio between G'' and G', was also considered. This parameter represents the balance between the viscous and elastic components of the rheological behavior. The more pronounced the elastic behavior with respect to the viscous behavior, the lower the loss tangent. Thus, tg δ was believed to be a suitable parameter for comparing the viscoelastic behavior of polymers with different G' and G'' values. Moreover, the more or less pronounced increase in storage modulus with respect to the loss modulus could be described by a change in tg δ. The tg δ variation, which was eventually observed when the polymer was mixed with mucin, conceivably expressed the extent to which elastic behavior changed with respect to viscous behavior.

Figure 5 shows the tg δ patterns (at increasing frequency) obtained for the polymer solutions and their mixtures with mucin. The three polymers examined showed different tg δ patterns. For NaCMC and Eudispert, which had poor initial elastic properties (high tg δ values), mixing with mucin caused a significant increase in the elastic properties, as evidenced by the decrease in tg δ. On the contrary, in the case of Carbopol, which was characterized by pronounced elastic properties (low tg δ values), a slight increase in tg δ was observed in the presence of mucin, indicating that the interaction

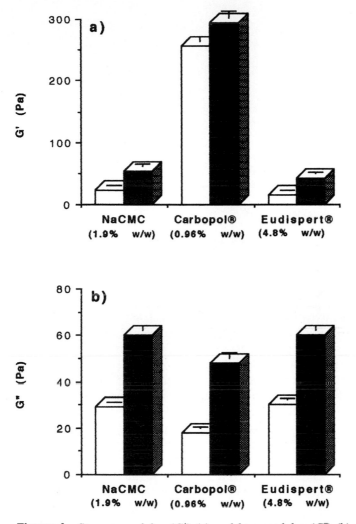

Figure 4   Storage modulus (G') (a) and loss modulus (G'') (b) measured at 1 Hz frequency for polymer solutions (□) and polymer-mucin mixtures (■) (mean values ± SD; $n = 3$).

with mucin did not change the elastic structure of the gel much and resulted in an increase in the viscous properties rather than in the elastic character of the polymer.

In conclusion, for all three polymers tested, a rheological interaction with mucin involving both viscosity and viscoelastic properties was ob-

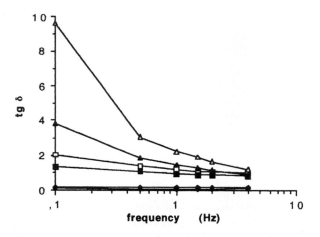

Figure 5 Loss tangent (tg δ) measured at increasing frequency values for polymer solutions (open symbols) and polymer-mucin mixtures (closed symbols). (Square) NaCMC; (rhombus) Carbopol; (triangle) Eudispert.

served. In particular, dynamic viscosity measurements provided a useful tool for differentiating the three polymers according to their interaction with mucin. In particular, the tg δ parameter was believed to provide a complete characterization of these rheological changes.

Mortazavi and Smart (1994) also compared the G' and G'' changes observed when different polymeric gels [based on Carbopol 934P, hydrox-ypropylcellulose (HPC), polyvinyl alcohol (PVA), and polybrene] were mixed with mucus. Different polymer concentrations were considered. The results are shown in Table 2.

As already observed in a previous paper (Mortazavi et al., 1992), mixing Carbopol 934P with homogenized pig gastric mucus produced gel strengthening as indicated by G' values greater than the sum of the G' values obtained when mucus and polymer were evaluated separately at the same concentration. Weaker gels were obtained with mixtures containing the non-ionic polymers HPC and PVA at the same concentration as Carbopol (5 mg g^{-1}). This suggests that PVA and HPC are less capable of establishing the interactions necessary for gelation. Because they do not contain a proton-donating carboxyl group, their ability to form hydrogen bonds with mucus glycoproteins is reduced. The cationic polybrene was found to produce little gel strengthening even at the highest concentration. Signs of gel breakdown were also observed, suggesting precipitation of an intermolecular complex resulting from electrostatic interactions.

The results so far illustrated provide definite experimental evidence of the rheological changes taking place when polymer-mucin mixtures are

Table 2 Comparative Rheological Assessment of Carbopol 934P, HPC, PVA, and Polybrene Mixtures with Mucus and Water at pH 6.2

Sample	G' (Pa)	G'' (Pa)
Mucus-C934 (5 mg g^{-1})	23.89	10.39
Water-C934 (5 mg g^{-1})	2.18	4.39
Mucus-HPC (5 mg g^{-1})	1.04	1.29
Water-HPC (5 mg g^{-1})	0.24	2.43
Mucus-HPC (20 mg g^{-1})	9.20	9.11
Water-HPC (20 mg g^{-1})	8.75	9.23
Mucus-PVA (5 mg g^{-1})	0.39	0.39
Water-PVA (5 mg g^{-1})	0.31	0.23
Mucus-PVA (100 mg g^{-1})	0.91	3.69
Water-PVA (100 mg g^{-1})	0.28	1.69
Mucus-polybrene (5 mg g^{-1})	0.58	0.50
Water-polybrene (5 mg g^{-1})	0.32	0.21
Mucus-polybrene (60 mg g^{-1})	0.21	0.29
Water-polybrene (60 mg g^{-1})	0.27	0.28
Mucus-water	0.43	0.42

From Mortazavi and Smart (1994).

formed, even though a comparison of the data obtained by various authors is rather difficult because of the different experimental conditions used (different techniques and apparatus, different types of mucin or mucus employed, different parameters measured, different test conditions).

Apart from this, it is still to be demonstrated whether the rheological interactions observed actually correspond to mechanical strengthening of the mucoadhesive interface that is formed when the polymer gel comes into contact with the mucus layer, which in turn should be linked to the adhesive performance (strength and duration of adhesion). Such a comparison and correlation between the rheological properties and mechanical resistance of the interface will be discussed in the following section.

Indeed, the correspondence between rheological interactions and adhesive performance was predicted by Hassan and Gallo (1990), who pointed out that the rheological interactions observed for the various polymers were in line with the results previously obtained by Park and Robinson (1984) using the fluorescent probe technique.

Also, Mortazavi and Smart (1994) found agreement between the rank order of G' values observed for the mixtures of Carbopol, HPC, and PVA and the adhesiveness of the same materials reported in other papers (Smart et al., 1984; Longer and Robinson, 1986). However, these comparisons were made for materials treated in different ways and tested under different experimental conditions.

IV. COMPARISON BETWEEN RHEOLOGICAL AND TENSILE TESTING

The methods most commonly employed for in vitro testing of mucoadhesive systems are based on measurement of the force necessary to detach a polymer layer from the mucus substrate. These methods are called peel, shear, or tensile testing methods, depending on the direction of the force applied. More often such methods enable us to calculate the adhesion work from the force versus displacement curves. In a typical tensile test, the breaking strength, which is measured when the polymer layer is detached from the mucus layer, can be attributed to the mechanical resistance of the mucoadhesive interface that is created when the two phases match each other.

One of the major disadvantages of such tests is the difficulty of distinguishing between mucoadhesive and cohesive forces. It all depends on the locus of the failure, which can occur within the mucous substrate or the polymer layer instead of occurring at the mucoadhesive interface. This is the main reason for the high intrinsic variability of tensile measurements. To overcome this drawback and to differentiate the mucoadhesive properties of different materials on a sound statistical basis, a high number of repetitions and a visual inspection of the polymer and the mucus layers after their separation are needed.

Given the widespread use of tensile testing techniques and taking into consideration the preceding warnings, it seemed interesting to understand whether the mechanical strengthening at the mucoadhesive interface was correlated with the rheological properties.

In the following, we will report the outcome of some of our research, in which tensile measurements have been systematically combined with rheological interaction measurements. This section will be divided into three parts. In the first part, the general relationships between tensile and rheological measurements for a series of miscellaneous mucoadhesive polymers will be reported. The other two parts will deal with the specific cases of sodium carboxymethylcellulose, chitosan hydrochloride, and modified starches.

A. General Relationships Between Rheological and Tensile Testing for a Series of Miscellaneous Polymers

In a pioneer work of ours (Caramella et al., 1994), we performed a thorough rheological characterization of some bioadhesive polymers and their mixtures with mucin and, in parallel, we measured the mucoadhesive performance of the same polymers and polymer mucin combinations with a tensile tester. The polymers examined included both natural gums, such as xanthan gum and scleroglucan; semisynthetic derivatives, such as sodium carboxymethylcellulose (NaCMC); and synthetic products, such as polyacrylic acid (Carbopol 934PH), methacrylic acid/methyl ester copolymer (Eudispert), and methylvinyl ether/maleic anhydride copolymer (Gantrez)—all claimed to possess mucoadhesive properties. Isoviscous polymer solutions in pH 7.0 phosphate buffer were prepared at the following concentrations: 2% (w/w) NaCMC, 3% (w/w) scleroglucan, 3% (w/w) xanthan gum, 1% (w/w) Carbopol, 5% (w/w) Eudispert, and 5.5% (w/w) Gantrez. The polymer solutions obtained were mixed with the exact amount of bovine submaxillary mucin (commercial sample) necessary to get a final mucin concentration of 4% (w/w). Also, 4% (w/w) mucin dispersion and polymer solutions having exactly the polymer concentration of the mixtures were prepared in pH 7.0 phosphate buffer.

Viscosity measurements and dynamic rheological analysis were performed on polymer-mucin mixtures, polymer solution, and mucin dispersion using a rotational rheometer (Bohlin CS rheometer). The rheological synergism parameter was calculated according to Hassan's method (Hassan and Gallo, 1990). From the values of the viscoelastic parameters G', G'' and the tg δ observed for the polymer solutions and mixtures with mucus, the following viscoelastic interaction parameters were calculated: $\Delta G'$ = difference between the storage modulus (G') of the polymer-mucin mixture and that of the polymer solution (blank); $\Delta G''$ = difference between the loss modulus (G'') of the polymer-mucin mixture and that of the polymer solution (blank); Δtg δ = difference between the loss tangent (tg δ) calculated for the polymer solution (blank) and that calculated for the polymer-mucin mixture.

The differential parameters $\Delta G'$ and $\Delta G''$ were also normalized with respect to the values of G' and G'' of the polymer solutions to take into account the initial differences in G' and in G'' of the various polymers. Since tg δ already represented a ratio (between G' and G''), Δtg δ was not normalized.

Tensile testing was performed on the same polymeric solutions and polymer-mucin mixtures. The equipment employed was a TA.XT2 Texture Analyzer (Stable Micro System, Haslemere, Surrey, UK). It basically con-a)

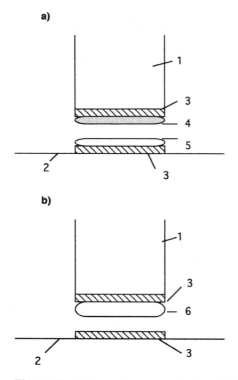

b)

Figure 6 Schematic representation of the test conditions in tensile testings: (a) interface method, (b) bulk method. (1) Upper (movable) probe; (2) platform; (3) filter paper disc; (4) polymer solution; (5) mucin dispersion (buffer in blank measurements); (6) polymer-mucin mixture (polymer alone in blank measurements).

sisted of a movable measuring probe and a fixed platform (Fig. 6). Samples were layered on filter paper discs, one fixed to the upper probe and the other to the lower platform. Test conditions chosen were as similar as possible to those that were realized during the rheological testing, so that we could provide comparable information. In particular, two methods were used. In the first, the "interface method" (Fig. 6a), 65 µL of the 4% (w/w) mucin dispersion was applied to the lower disc. After a 5-minute rest, 40 mg of the polymer solution was applied to the filter paper disc fixed to the probe, which was lowered in order to bring the two discs into contact with each other. A preload of 0.5 N was applied for 60 seconds (preload time); then the probe was raised to a constant speed of 0.1 mm/s. Blank measurements were also performed without mucin: instead of the mucin dispersion, 65 µL of phosphate buffer was applied to the lower disc.

In the second method, the "bulk method" (Fig. 6b), 50 mg of either the polymer-mucin mixture or the polymer solution alone (blank measurements) was applied to the upper disc; then the upper probe was lowered in order to bring the two discs into contact. As before, a preload of 0.5 N was applied for 60 seconds (preload time), and the probe was raised to a constant speed of 0.1 mm/s.

In both methods, the force versus displacement curves were recorded until complete detachment of the two surfaces occurred. Fifteen repetitions were performed on each sample.

The adhesion work was calculated as the area under the force-displacement curve (AUC). The differential parameter (ΔAUC) was also calculated by subtracting the AUC (mean value) obtained from blank measurements from the AUC (mean value) measured in the presence of mucin.

With regard to the interface method, for all the polymers considered, AUC significantly increased in presence of mucin with respect to the blanks. Therefore positive ΔAUC values were calculated for all the polymer-mucin combinations examined. These values were different from one polymer to another, meaning that, in the interface method, failure does not take place within the mucin layer, otherwise similar ΔAUC values would have been obtained for different mucoadhesive polymers (the cohesion of the mucin alone at the concentration employed being very low). On the other hand, failure could not have occurred within the polymer layer, otherwise there would have been no increase in AUC in the presence of mucin.

A significant increase in AUC in the presence of mucin was also observed in the bulk method, in which the polymer was directly mixed with mucin before tensile measurement.

This suggests that the work of adhesion (expressed by ΔAUC) is due not only to interface and cohesion phenomena (which might play a role only in the interface method) but also to the interpenetration between polymer and mucin chains.

The rheological parameters were systematically compared with those derived from tensile tests in order to establish whether a correlation existed between the two measurements. Linear regression analysis was performed between the rheological parameters of polymer-mucin mixtures (the absolute values G', G'', tgδ and the differential ones $\Delta\eta$, $\Delta G'$, $\Delta G''$, $\Delta G'/G'$, $\Delta G''/G''$, and Δtgδ) and the adhesion work obtained with the same polymer-mucin systems (AUC, ΔAUC). In most cases no direct correlation was found between rheological and tensile testing parameters.

A significant correlation was found only when the relationship between differential adhesion work and loss tangent variation was considered (Fig. 7). This suggested the relevance of this rheological parameter, which pro-

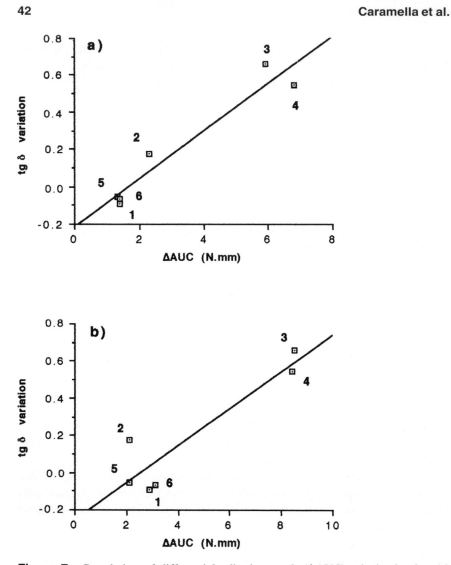

Figure 7 Correlation of differential adhesion work (ΔAUC), obtained using (a) interface method and (b) bulk method, and tg δ variation. (1) Carbopol; (2) NaCMC; (3) Gantrez; (4) Eudispert; (5) xanthan gum; (6) scleroglucan. (From Caramella et al., 1994.)

vided a more intrinsic characterization of the gel solutions with respect to the others, as already evidenced in the early studies reported in Sec. II.

The lack of a direct correlation between the rheological parameters and the adhesion work, found in most cases, was attributed to the diversity of the chemical nature of the polymers considered. In fact, because rheological measurements are very sensitive to the many different characteristics of the polymer chains and their behaviour when mixed with mucus, it was intrinsically difficult to use them to compare polymers of various chemical natures, which were likely to give different bonds and interactions with mucin and were differently affected by environmental conditions.

B. Relationships Between Rheological and Tensile Testing: The Cases of Sodium Carboxymethylcellulose and Chitosan Hydrochloride

In a subsequent study (Rossi et al., 1996) the correlation between rheological and tensile parameters of three viscosity grades (low, medium, and high) of the same polymer (sodium carboxymethylcellulose) was investigated. Isoviscous solutions of the three viscosity grades in distilled water were analyzed for rheological interaction with mucin and for work of adhesion by means of tensile testing. The following polymer concentrations were used: 1.92% (w/w) for high viscosity grade (h.v.), 3.84% (w/w) for medium viscosity grade (m.v.), and 7.68% (w/w) for low viscosity grade (l.v.). The mucin employed was a commercial porcine gastric mucin at 4% (w/w) concentration. The rheological interaction parameters $\Delta\eta/\eta$, $\Delta G'/G'$, $\Delta G''/G''$, and $\Delta tg\delta$ were calculated as previously described. Tensile testing was performed on the same polymeric solutions and polymer-mucin mixtures as previously described.

In Table 3 the rheological interaction parameters calculated at a representative value of shear rate ($100 \ s^{-1}$) and frequency (1 Hz) are reported. Similar results were obtained in the whole shear rate range ($50-250 \ s^{-1}$) and frequency range ($0.1-1.5$ Hz) considered.

All three grades showed positive rheological interaction with mucin. The same rank order between the three grades was found when rheological synergism $\Delta\eta/\eta$ and differential viscoelastic parameters $\Delta G'/G'$, $\Delta G''/G''$, and $\Delta tg\delta$ were considered: the low viscosity grade showed the highest interaction parameters, followed by the medium and high viscosity grades.

The samples examined were also subjected to stationary viscoelastic measurements ("creep test"): a rapid step in shear stress (chosen in the viscoelastic linear region) was applied to the sample and held constant for a predetermined period of time. The creep compliance (J) was recorded as a function of time (creep curve).

Table 3 Rheological Synergism $\Delta\eta|\eta$ and Viscoelastic Interaction Parameters $\Delta G'/G'$, $\Delta G''/G''$, and $\Delta tg\delta$ Calculated for the Three Different Viscosity Grades (High, Medium, and Low) of Sodium Carboxymethylcellulose (Mean Values $\pm$ SD, $n = 3$)

Polymer	$\Delta\eta/\eta^a$	$\Delta G'/G'^b$	$\Delta G''/G''^b$	$\Delta tg\delta^b$
l.v. grade (7.68% w/w)	1.33 ($\pm$0.175)	2.02 ($\pm$0.094)	1.16 ($\pm$0.140)	2.26 ($\pm$0.202)
m.v. grade (3.84% w/w)	0.90 ($\pm$0.077)	1.34 ($\pm$0.151)	0.90 ($\pm$0.081)	0.60 ($\pm$0.166)
h.v. grade (1.92% w/w)	0.37 ($\pm$0.058)	0.49 ($\pm$0.118)	0.37 ($\pm$0.089)	0.10 ($\pm$0.021)

[a]Calculated at 100 s^{-1} shear rate.
[b]Calculated at 1 Hz frequency.
Adapted from Rossi et al. (1996).

The creep curves were fitted to the following equation (Sherman, 1970): $J(t) = J_0 + J_m[1 - \exp(-t/\tau_m)] + t/\eta_N$, where J_0 = instantaneous elastic compliance (1/Pa), J_m = mean retarded elastic compliance (1/Pa), τ_m = mean retardation time (s), and η_N = residual viscosity (Pa s). Figure 8 shows the creep curves of the three polymer solutions and of their mixtures with mucin.

The low viscosity grade showed the highest compliance $J(t)$; because J represented the ratio between the strain and the stress applied, this result suggested a weaker structure for the low viscosity grade (which also had the lowest values of G' and G'') with respect to the other two grades. Moreover, the compliance values obtained for the mixtures of the three polymers with mucin were lower than those obtained for the related polymer solutions. This result indicated that the deformation of the polymeric network under the stress applied was smaller for the mixtures than for the polymer solutions. This was conceivably due to the strengthening of the polymeric network due to the physicochemical interactions between polymer and mucin chains.

In order to quantify the extent of such interactions, the J_m (mean retarded elastic compliance), τ_m (mean retardation time), and η_N (residual viscosity) values obtained by fitting creep curves were examined (insets in Fig. 8). The J_m values were lower for the polymer-mucin mixtures than for the

Figure 8 Creep profiles obtained for the polymer solutions (□) and the polymer-mucin mixtures (■) of the three viscosity grades of sodium carboxymethylcellulose: (a) low viscosity grade; (b) medium viscosity grade; (c) high viscosity grade. In the insets the best fit parameters (J_m, τ_m, and η_N) are reported (mean values $\pm$ SD; $n = 3$). (From Rossi et al., 1996.)

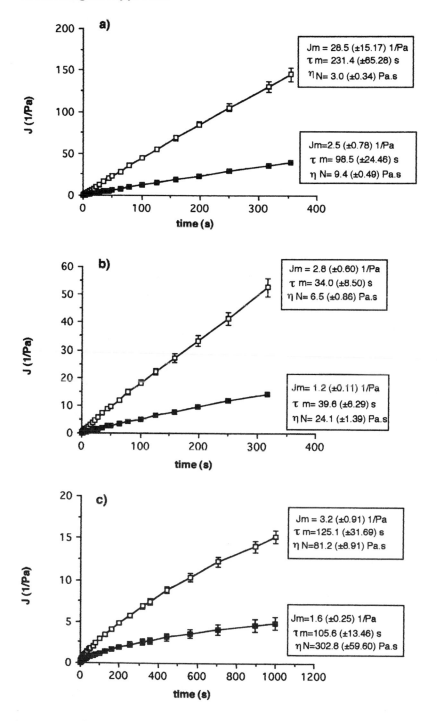

polymers alone. In particular, the decrease in J_m in the presence of mucin was more pronounced for the low viscosity grade than for the other two grades, indicating that a stronger interaction occurred for this grade. For all grades an increase in η_N was observed when the polymer was mixed with mucin. A significant variation in τ_m was observed only in the case of the low viscosity grade.

Table 4 shows the adhesion work measured with tensile testing. A significant increase in adhesion work was observed for all polymers in the presence of mucin. The interactions between mucin and polymer occurred both when a preequilibrated mixture of polymer and mucin was tested (bulk method) and when interpenetration between polymer and mucin layers was realized in situ (interface method). When the increase in adhesion work (ΔAUC) was considered, the same rank order as observed for rheological parameters was obtained: the low grade showed the highest value of ΔAUC, followed by the medium and high grades.

The relationship found between rheological and tensile parameters proved that for sodium carboxymethylcellulose, the strengthening of the polymer-mucin interface is linked to the rheological changes occurring when the polymer is mixed with mucin. In fact, the mucoadhesive performance of the polymers examined, in term of work of adhesion, is correctly predicted both by rheological synergism and by viscoelastic interaction parameters. The lower interaction of medium and high viscosity grades of sodium carboxymethylcellulose could be explained either by the lower concentration (at isoviscous conditions) or by the more entangled structure of polymer solutions; this hinders deep interpenetration between polymer and mucin molecules.

We subsequently performed an analogous study of three different viscosity grades (low, medium and high) of chitosan hydrochloride. Again, a relationship was found between rheological synergism and tensile testing parameters (Ferrari et al., 1997a).

In Fig. 9 the rheological synergism values calculated, as a function of shear rate, for the mixtures of the three viscosity grades of chitosan HCl with mucin are given. The mixtures considered contained 8% (w/w) partially purified pig gastric mucin and the following concentrations of chitosan HCl: 2.5% (w/w) for the high viscosity grade, 5.0% (w/w) for the medium viscosity grade, and 24% (w/w) for the low viscosity grade. The hydration medium employed was 0.1 M HCl. The rheological synergism was more pronounced for the low viscosity grade, followed by the medium and high grades, indicating that a stronger polymer-mucin interaction occurred in the case of the low viscosity grade.

Table 5 shows the maximum force of detachment (F_{max}) and the work of adhesion (AUC) obtained for the polymer-mucin mixtures and the cor-

Table 4 Adhesion Work (N mm) (Tensile Testing) of the Three Viscosity Grades of Sodium Carboxymethylcellulose (Mean ± SD, $n = 15$)[a]

	Interface method			Bulk method		
Polymer	Polymer (blank)	Polymer plus mucin	ΔAUC	Polymer (blank)	Polymer plus mucin	ΔAUC
l.v. grade	9.79 (±2.543)	17.52 (±2.908)	7.73	12.59 (±1.954)	20.68 (±2.012)	8.09
m.v. grade	11.88 (±3.849)	16.38 (±1.897)	4.50	12.96 (±1.160)	16.22 (±1.424)	3.26
h.v. grade	8.34 (±0.938)	10.94 (±1.054)	2.60	10.07 (±1.312)	11.46 (±2.640)	1.39

[a]Differences in adhesion work calculated in presence of mucin and in blank measurements were significant to Student's t-test ($P < .05$). From Rossi et al. (1996).

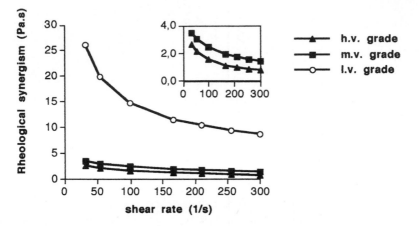

Figure 9 Rheological synergism values measured, at increasing shear rate, for the three viscosity grades of chitosan hydrochloride. In the inset the rheological synergism values of the medium and low viscosity grades are compared (mean values ± SE; $n = 3$). (From Ferrari et al., 1997a.)

responding polymer solutions (blanks), together with the normalized adhesion parameters ($\Delta F_{max}/F_{max}$ and $\Delta AUC/AUC$). The low grade showed the highest values for both normalized parameters, followed by the medium and high grades. This rank order is the same as that observed in the rheological synergism.

C. Relationships Between Rheological and Tensile Testing: The Case of Modified Starches

In other work of ours (Ferrari et al., 1997b), both rheological analysis and tensile testing were used to assess the possibility of inducing or improving mucoadhesive properties of starches by technological processes such as grinding or dry milling. Pregelatinized starches have been described as having mucoadhesive properties (Bottenberg et al., 1991). These are thermally modified starches in which the native structure of the grains has been destroyed by means of a technological process, such as spray drying (where the starch is first cooked in water and then spray dried in a hot chamber or tower), roll drying (where a watery starch paste is simultaneously cooked and dried on heated rolls), and drum drying (similar to the roll-drying process except that a thicker coating of starch paste is applied to heated rolls and the dried product is then ground to the desired particle size). In all cases starch gelatinization is carried out in a watery suspension.

Table 5 Force of Detachment (mN) and Work of Adhesion (µJ) of the Three Viscosity Grades of Chitosan Hydrochloride Hydrated in 0.1 M HCl (Mean Values ± SE; $n = 10$)[a]

Polymer	Force of detachment			Work of adhesion		
	Polymer (blank)	Polymer + mucin	$\Delta F_{max}/F_{max}$	Polymer (blank)	Polymer + mucin	ΔAUC/AUC
l.v. grade	667.6* (±26.35)	1345.9** (±46.01)	0.97	71.4* (±3.06)	182.5** (±3.32)	1.56
m.v. grade	784.8* (±35.07)	1154.9** (±52.34)	0.47	85.8* (±3.04)	170.2** (±4.84)	0.98
h.v. grade	1038.8* (±51.08)	1420.0** (±43.04)	0.37	101.3* (±5.64)	166.6** (±6.67)	0.64

[a]$P < .05$, Mann-Whitney test for comparison of two groups (* versus **).
From Ferrari et al. (1997a).

The induction of mucoadhesive properties in native starches via a simple process such as dry milling could be more advantageous from a technological point of view (because the process is very cheap and does not require expensive equipment) than a complicated procedure such as spray drying, roll drying, and drum drying.

We investigated two native starches (maize starch and waxy maize starch) and one pregelatinized waxy maize starch.

The waxy maize starch was subjected to a grinding process. The technological variables analyzed were the type of mill (from laboratory to pilot-plant scale) and the process conditions (milling time and cooling temperature of the grinding chamber). In particular, laboratory-scale experiments were performed with a high-energy ball mill apparatus (IGW, Giuliani, Torino, Italy) using different grinding times (3, 6, and 12 hours) and different cooling temperatures of the grinding chamber ($-50°C$ and $20°C$). As pilot-plant scale instruments, two different mills were used: a high-energy mill (VIBRO-Energy mill DM-3, Sweco, Los Angeles, CA) and a hammer mill with a horizontal shaft (CF, Bantham, Summit, NJ).

Native maize starch, pregelatinized starch, and waxy starch milled with the Sweco apparatus for 6 hours at room temperature were subjected to a spray-drying process in a laboratory plant apparatus (SD04, Labplant, Huddersfield, UK).

All the starches examined were hydrated in distilled water by gentle heating at about 70°C. The concentrations of the starch solutions were 18% (w/w) for pregelatinized starch and spray-dried pregelatinized starch, 7% (w/w) for waxy starch before and after milling, and 4.5% (w/w) for maize starch.

The hydrated starches were mixed with a commercial porcine gastric mucin. In all the mixtures the mucin concentration was 4% (w/w). The exact final concentration of the starches in the mixtures was calculated. Starch solutions having exactly the same concentration as in the mixtures were prepared in distilled water (blanks). A 4% (w/w) mucin dispersion in distilled water was also prepared.

For each starch-mucin mixture, the rheological synergism ($\Delta\eta$) was calculated according to Hassan's method (Hassan and Gallo, 1990). In order to compare the interaction properties of starches having different viscosities, the rheological synergism was normalized with respect to the viscosity of the starch solution alone. The starches were therefore compared on the basis of the normalized interaction parameter ($\Delta\eta/\eta$).

Tensile stress tests were performed with an apparatus set up in our laboratory. The mucoadhesive interface was formed by keeping in contact for 3 minutes under a preload of 800 mN two filter paper discs on which were applied 50 mg of starch solution, previously hydrated in distilled water to reach 25% (w/w) concentration, and 40 µL of a 4% (w/w) mucin dispersion or distilled water (blank measurements), respectively. Force of detachment versus displacement curves were recorded and the work of adhesion (AUC) was calculated. Ten repetitions were performed on each sample.

The differential parameter ΔAUC was calculated (as already described) as the difference between the work of adhesion (mean value) obtained in the presence of mucin and the work of adhesion (mean value) obtained for the blank, which is likely to measure both the surface tensional and the intrinsic cohesive properties of the starch sample. Because these properties were different for the various starches, the differential parameter was normalized with respect to the work of adhesion obtained from the blank measurements. The starches were therefore compared on the basis of the normalized interaction parameter (ΔAUC/AUC).

Figure 10 shows the normalized interaction parameters $\Delta\eta/\eta$ (calculated at 100 s^{-1}) and ΔAUC/AUC for the three initial starches (maize, waxy maize, and pregelatinized waxy).

For waxy and maize starches, both parameters showed negative values. This indicates that neither starch possessed mucoadhesive properties. On the contrary, the pregelatinized starch showed positive values for both normalized rheological synergism and work of adhesion. This confirms that it had mucoadhesive properties, as already observed by other authors (Bottenberg et al., 1991).

Positive values for both normalized interaction parameters were observed for waxy maize starch milled by means of a high-energy mill (Giu-

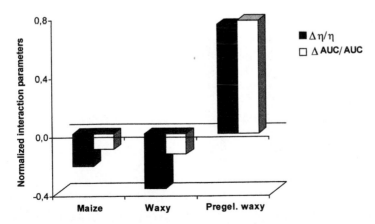

Figure 10 Normalized rheological synergism ($\Delta\eta/\eta$) and normalized work of adhesion (ΔAUC/AUC) of the two native starches (maize and waxy maize) and of pregelatinized waxy maize starch. (From Ferrari et al., 1997b.)

liani apparatus), as shown in Fig. 11. An increase in such parameters was also observed on increasing milling time. No significant influence of the cooling temperature of the grinding chamber was observed.

In Fig. 12 the results of rheological and tensile testing of waxy starch milled with two different types of pilot scale equipment (Sweco and Bantham apparatus) are compared with those of native waxy starch. In the case of the Sweco apparatus, the influence of milling time was also investigated. Waxy starch milled with the Sweco apparatus showed positive values for both rheological and tensile parameters. Such parameters increased with increasing milling time. On the contrary, the same starch milled with the Bantham apparatus did not show mucoadhesive properties. It can be stated that the grinding process induces mucoadhesive properties in a nonmucoadhesive starch only if grinding is effected in high-energy mills.

Figure 13 shows the influence of spray drying on the mucoadhesive properties of maize starch, pregelatinized starch, and waxy starch milled in a high-energy apparatus (Sweco). Whatever the material considered, the spray-drying process did not produce a significant modification of mucoadhesive properties.

As also found in previous studies, the results of rheological testing were in line with those of tensile testing: whenever the normalized rheological synergism was positive, positive values were also observed for the normalized work of adhesion; moreover, when the rheological synergism increased, an increase in the work of adhesion was observed.

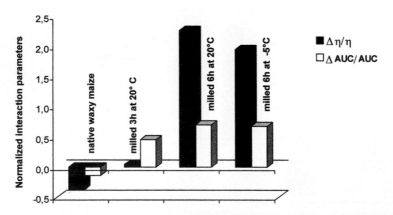

Figure 11   Normalized rheological synergism ($\Delta\eta/\eta$) and normalized work of adhesion (ΔAUC/AUC) of the native waxy maize starch and of the same starch after grinding by means of a Giuliani mill under different experimental conditions (chamber temperature and process time). (From Ferrari et al., 1997b.)

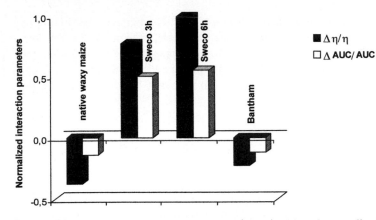

Figure 12 Normalized rheological synergism ($\Delta\eta/\eta$) and normalized work of adhesion (ΔAUC/AUC) of the native waxy maize starch and of the same starch after grinding by means of two different kinds of pilot scale equipment (Sweco and Bantham apparatus). (From Ferrari et al., 1997b.)

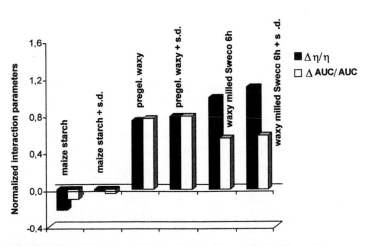

Figure 13 Influence of spray drying (+ s.d. in the figure) on normalized rheological synergism ($\Delta\eta/\eta$) and normalized work of adhesion (ΔAUC/AUC) of maize starch, pregelatinized waxy starch, and waxy starch milled with Sweco apparatus. (From Ferrari et al., 1997b.)

The results so far illustrated demonstrate that, whenever the rheological test is applied to homologous or similar materials, the mucoadhesive interaction parameters obtained from rheological measurements are somehow validated by the results of tensile testing conducted under appropriate conditions. These parameters therefore seem relevant to mucoadhesive performance.

V. RHEOLOGICAL INTERACTION: INFLUENCE OF TEST CONDITIONS

A criticism that is commonly raised about completely in vitro tests, such as the rheological approach described here, is that the test conditions—namely the hydration medium composition, pH and ionic strength, temperature, and polymer and mucin concentrations—greatly influence the rheological changes observed. For that reason, the test conditions chosen should mimic as closely as possible the in vivo situations envisaged, keeping in mind that, as happens for the other in vitro methods proposed (e.g., those based on tensile testing or on surface energy analysis), the relationship with in vivo behavior is not immediate but must be checked in every case.

Another criticism, perhaps the main one, of the rheological method is that it requires the use of mucus or mucin samples isolated from the mucus instead of mucous tissue. Processes of extraction, purification, and preservation of the glycoprotein fraction, which is assumed to be responsible for the interaction with polymers, can modify the native structure of mucin or spoil it with exogenous substances. Therefore the particulars of the mucin employed (origin, purification grade, and effect of further treatments such as freezing or freeze drying) are likely to be critical for the interaction.

Both freshly prepared and commercial mucin samples have been employed in the literature. Both types of mucin have advantages and drawbacks. In particular, in the case of fresh mucin, samples have to be used within a short time after preparation. Thus, batches are necessarily small and derive from a small number of animals, which causes batch-to-batch variability to be high. To counter this drawback, extraction and purification methods for fresh mucin are used and such mucin is not subjected to preservation processes, such as freezing and freeze-drying, which represent a critical factor in the use of commercial mucins. Commercial mucins show lower variability between batches and are ready to use.

To clarify partially the problems linked to mucin choice, we investigated the influence of mucin type on the rheological interaction of two well-known mucoadhesive polymers, sodium carboxymethylcellulose (NaCMC) and polyacrylic acid (Carbopol 934 PH) (PAA) (Rossi et al., 1992, 1995).

Table 6 Apparent Viscosity and Viscoelastic Moduli of 1.92% (w/w) NaCMC Solutions Prepared in Distilled Water, in pH 7.0 Phosphate Buffer, and in pH 4.5 Acetate Buffer (Mean Values ± SD, $n = 3$)

Medium	η (Pa s)[a]	G' (Pa)[b]	G'' (Pa)[b]
Distilled water	1.26 (±0.024)	17.3 (±3.78)	26.2 (±4.02)
pH 7.0 phosphate buffer	1.26 (±0.092)	18.5 (±3.43)	28.1 (±3.00)
pH 4.5 acetate buffer	1.16 (±0.114)	19.2 (±1.65)	28.0 (±2.82)

[a]Measured at 100 s^{-1} shear rate.
[b]Measured at 1 Hz frequency.
From Rossi et al. (1995).

Polymers were hydrated in media with different pH and ionic strength values to verify the influence of degree of ionization and ion composition on rheological parameters. The following hydration media were used: distilled water, 0.1 M phosphate buffer (H_3PO_4/NaOH) at pH 7.0 ($\mu = 0.136$), and 0.1 M acetate buffer (CH_3COOH/CH_3COONa) at pH 4.5 ($\mu = 0.044$). The polymer concentrations were 1.92% (w/w) for NaCMC and 0.96% (w/w) for PAA.

Tables 6 and 7 show the apparent viscosity (measured at 100 s^{-1} shear rate) and the viscoelastic moduli G' and G'' (measured at 1 Hz frequency) of NaCMC and PAA, hydrated in different media. The rheological properties of the two polymers were differently influenced by the hydration medium. Whereas the viscosity and viscoelastic behavior of NaCMC are not affected by the pH and ionic strength of the hydration medium (Table 6), in the case of PAA, both viscosity and viscoelastic moduli are higher in distilled water than in buffers (Table 7). This is due to the sensitivity, already described in

Table 7 Apparent Viscosity and Viscoelastic Moduli of 0.96% (w/w) PAA Solutions Prepared in Distilled Water, in pH 7.0 Phosphate Buffer, and in pH 4.5 Acetate Buffer (Mean Values ± SD, $n = 3$)

Medium	η (Pa s)[a]	G' (Pa)[a]	G'' (Pa)[b]
Distilled water	4.78 (±0.538)	496.7 (±17.05)	40.3 (±3.85)
pH 7.0 phosphate buffer	1.63 (±0.014)	254.0 (±2.83)	19.4 (±2.49)
pH 4.5 acetate buffer	0.93 (±0.083)	222.8 (±32.3)	10.9 (±2.70)

[a]Measured at 100 s^{-1} shear rate.
[b]Measured at 1 Hz frequency.
From Rossi et al. (1995).

the literature (Mortazavi et al., 1992; Mortazavi and Smart, 1994), of PAA
to the pH and ionic strength of the hydration medium.

Then PAA and NaCMC were mixed with three different commercial
mucins: a purified mucin from submaxillary glands (BSMG) and two brands
of crude mucin from pig stomach (PSI and PSII). The mixtures contained
1.92% (w/w) NaCMC or 0.96% (w/w) PAA and 4% (w/w) mucin. The
polymer-mucin rheological interaction was investigated.

Figure 14 shows the rheological synergism values ($\Delta\eta$) for the
NaCMC mixtures, hydrated in different media, with the three mucins under
examination. All the NaCMC mixtures examined showed positive rheological synergism values. Moreover, within each type of mucin, rheological
interaction was not significantly affected by the pH and ionic strength of the
hydration medium (at least in the ranges considered). Higher synergism was
observed with bovine submaxillary mucin (BSMG) than with porcine gastric
mucins (PSI and PSII) in all the hydration media employed. Similar results
were observed for the viscoelastic interaction parameters $\Delta G'$ and $\Delta G''$.

In the case of PAA, because the viscosity of the polymer solutions
varied markedly from one hydration medium to another, the normalized
parameter $\Delta\eta/\eta$ was considered in order to compare the results obtained
under different conditions. In Fig. 15 the normalized synergism values ($\Delta\eta/\eta$) calculated for the various PAA-mucin mixtures in different media are
given.

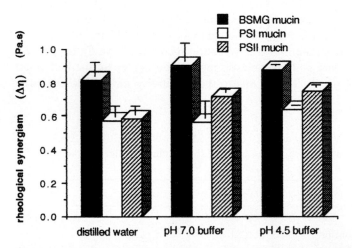

Figure 14 Rheological synergism values (measured at $100 \ s^{-1}$ shear rate) of
NaCMC mixtures with bovine submaxillary mucin BSMG, porcine gastric mucin
PSI, and porcine gastric mucin PSII in different hydration media (mean values $\pm$
SD, $n = 3$). (From Rossi et al., 1995.)

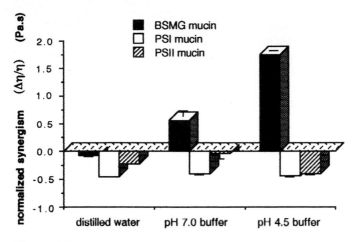

Figure 15 Normalized rheological synergism values (measured at 100 s^{-1} shear rate) of PAA mixtures with bovine submaxillary mucin BSMG, porcine gastric mucin PSI, and porcine gastric mucin PSII in different hydration media (mean values + SD, $n = 3$). (From Rossi et al., 1995.)

A lack of rheological interaction (indicated by negative values for $\Delta\eta/\eta$) was observed when PAA was mixed with porcine gastric mucins (PSI and PSII) in all the hydration media considered (Fig. 15). This can be explained by the presence of ions in commercial gastric mucins. These ions, whose presence is probably linked to mucin extraction and purification processes, may interact with PAA macromolecules and thus cause a breakdown in the gel network. PAA mixtures with BSMG mucin showed lack of interaction ($\Delta\eta < 0$) when the polymer was hydrated in distilled water and a positive value for rheological synergism in buffers. The lack of interaction observed in distilled water was probably due to the ions contained in the BSMG mucin, which break down the polymer network. The PAA solutions in buffers already showed lower viscosity values than the PAA solutions in distilled water. The addition of ions through mixture with mucin caused no further breakdown of the polymer network, thus allowing expression of the interaction of polymer and mucin macromolecules. Similar results were observed for the viscoelastic interaction parameters $\Delta G'$ and $\Delta G''$.

In the same work, one of the porcine gastric mucins (PSI) was subjected to purification and solubilization treatments such as dialysis, treatment with a solubilizing agent, and centrifugation. Dialysis aimed at removing ions from the mucin sample, and the other two treatments aimed at improving the solubilization of the glycoprotein fraction. The mucin concentration in the treated samples was assessed by a colorimetric method. The rheological interaction between the polymers and the treated mucin was then checked. For this purpose, polymer-mucin mixtures containing the same percentages of the two polymers as already cited and 4% (w/w) treated mucin were prepared.

In Fig. 16 the rheological synergism values observed for the mixtures (in pH 4.5 buffer) of the two polymers with the treated PSI mucin are compared with the values obtained for the mixtures with the untreated commercia! mucin. In the case of PAA, mixtures with PSI mucin showed a positive rheological interaction only when treatment included dialysis (B, C, E). This confirms the hypothesis that the ions contained in PSI mucin and removed by dialysis are indeed responsible for the decrease in viscosity observed for PAA mixtures. The increase in rheological synergism was more marked when the dialysis was accompanied by treatment with a solubilizing agent (C) or by centrifugation (E), which removed the insoluble glycoprotein fraction. PAA mixtures with centrifuged mucin (D) showed lack of rheological interaction, which was due to the ions still contained in the mucin sample.

NaCMC rheological synergism improved [with respect to the mixture with untreated mucin (A)] when the fraction of glycoprotein in solution increased [after treatment with SDS (C) or centrifugation (D,E)]. For

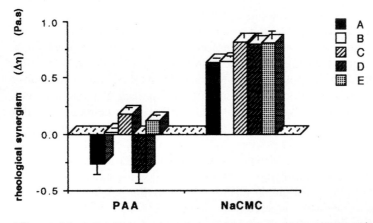

Figure 16 Rheological synergism values (measured at 100 s^{-1}) of PAA and NaCMC mixtures with porcine gastric mucin PSI: (A) untreated commercial mucin; (B) dialyzed mucin; (C) mucin treated with SDS and dialyzed; (D) centrifuged mucin; (E) centrifuged and dialyzed mucin (mean values $\pm$ SD, $n = 3$). (From Rossi et al., 1995.)

NaCMC, which is less sensitive to pH and ionic strength of the hydration medium, dialysis (B) did not yield any increase in synergism with respect to the value obtained for mixture with untreated mucin (A). Although thorough investigations and biochemical studies would be required to clarify factors responsible for the variability between commercial mucins, the results obtained in this work suggest that the rheological interaction with mucoadhesive polymers can be affected not only by the biochemical structure of glycoproteins but also by the way the sample is prepared (extracted, purified, stored).

In a study by Kocevar-Nared et al. (1997), differences were observed between the rheological properties of a commercial porcine gastric mucin and those of mucus obtained by scraping from a washed pig stomach. With respect to commercial mucin, the mucus was better hydrated but probably still contained components other than glycoproteins that affected the rheological properties.

Poor knowledge and standardization of the composition of glycoproteins of fresh samples are probably responsible for the discrepancies observed between some results reported in the literature. In particular, whereas Mortazavi et al. (1992) observed a gel breakdown (i.e., a decrease in viscoelastic moduli) in mixtures containing polyacrylic acid and homogenized porcine gastric mucus hydrated at (and below) pH 4.25, Kerr et al. (1990) did not report any similar decrease in viscoelastic moduli in mixtures containing polyacrylic acid and purified glycoproteins obtained from porcine stomach at low pH values. The decrease in viscoelastic moduli observed by Mortazavi et al. was attributed by the authors to the presence of other nonglycoprotein components in the mucus gel.

In subsequent work, Mortazavi et al. (1993a) compared the rheological behavior of polyacrylic acid mixtures with homogenized porcine stomach mucus and with the glycoprotein fraction obtained by extraction and purification. The results obtained with purified glycoprotein fraction were similar to those observed with homogenized mucus: a gel breakdown at lower pH values was also observed when purified glycoproteins were used. The mucus and the purified glycoprotein fraction were analyzed using sodium dodecyl sulfate (SDS)–polyacrylamide gel electrophoresis for the presence of glycoprotein and protein fractions. The homogenized mucus contained, besides glycoproteins, numerous other bands of carbohydrate residues, whereas the purified glycoprotein showed only traces of these smaller molecules. The electrophoresis, however, revealed the presence of small proteins, even in the purified glycoprotein fraction. The authors suggested that the gel breakdown observed with both mucus and glycoprotein fraction could be due to the presence of the low-molecular-weight proteins.

Given these results, further investigations are needed in order to obtain more standardized mucin samples with mechanical and rheological properties (when compared at the same glycoprotein content) more similar to those of the native substrate.

VI. CONCLUDING REMARKS

The rheological approach so far described for the evaluation of mucoadhesive hydrogels represents a completely in vitro test. Like all in vitro tests, it is of limited value as a predictor of in vivo performance. In fact, it is commonly accepted that the in vivo relevance of the mucoadhesive character of a polymer or system cannot be inferred from in vitro results but has to be confirmed through in vivo data.

However, the purpose of the present chapter is not to discuss the feasibility of in vitro–in vivo correlation of mucoadhesion tests but to justify their possible use in the preliminary screening of mucoadhesive polymers or in routine controls. Having said that, the results so far illustrated indicate that, whenever the rheological approach has been used for screening or evaluating a series of homologous or similar polymers, the same rank order of mucoadhesiveness observed with rheology has also been found with tensile testing.

Even though the conditions of the two in vitro tests are different (in tensile testing an interface between the polymer and the mucus layer is artificially created, more like what happens in vivo, whereas in rheological testing the two phases are already mixed together and no mucoadhesive interface exists), the rank order correlation found indicates that whenever the polymer-mucin interaction produces rheological synergism in a mixture, gel strengthening is also observed at the corresponding mucoadhesive interface.

Considering that tensile testing is the technique most commonly employed in screening procedures, the association established between the rheological and tensile tests provides a kind of validation of the rheological approach, even though it does not allow us to infer any extrapolation to an in situ or an in vivo situation.

Given these premises, why should we decide to use a rheological approach for screening or evaluating mucoadhesive polymers or gels and in what circumstances? What are the advantages and limitations of the rheological approach with respect to more classical tensile testing, and how can the two methods complement each other?

In choosing one or the other of the two techniques, we should take into consideration the following points: the applicability of the technique, the reproducibility of test conditions, the variability of measurements, and the information content.

Concerning applicability, the obvious difference between the two methods is that whereas tensile testing may be carried out on solid and semisolid (gel) systems, the rheological approach is applicable to semisolid (gel) and liquid systems, including those with low viscosities. In fact some authors (Albasini and Ludwig, 1995) have proposed the rheological approach for testing the mucoadhesiveness of ophthalmic solutions and their dilutions with lacrimal fluid. Similarly, we have applied the rheological method to evaluate the mucoadhesive properties of different viscosity grades of hyaluronic acid at low concentrations, in view of their possible use in rectal or ophthalmic formulations (Busetti et al., 1996). The same approach has been used for comparing the mucoadhesive properties of ophthalmic formulations based on hyaluronic acid, hydroxypropylmethylcellulose, and polyvinyl alcohol (Ronchi et al., 1997).

Concerning test conditions, these are generally rather critical when setting up a tensile test, in which many variables (sample layering or deposition on mucous substrate, type and quantity of hydration medium, hydration time, preload applied to establish smooth contact, preload time, rate of detachment at the interface, etc.) have to be optimized and fixed to obtain reliable and reproducible results. As a consequence, the intersample variability of the measurements of force of detachment is rather high and a reasonably high number of repetitions is needed to detect small significant differences between blank and sample measurements or between two sets of measurements.

In rheological testing, the number of variables to be set is smaller than in the case of tensile testing. In particular, because measurements are performed on already equilibrated mixtures, the composition, pH, and ionic strength of the environment, as well as the temperature, can be carefully controlled; the homogeneity of the polymer-mucin mixture can be checked; the physicochemical stability of the mixture itself can easily be verified throughout the measurements. As a consequence, the reproducibility of rheological measurements is fairly good and a smaller number of repetitions is generally needed than in the case of tensile measurements to achieve statistically significant data.

Concerning the information contents of the two methods, we can comment that the detachment force measured with tensile testing (and from which adhesion work is calculated) may not be as unambiguous as expected. In fact, because the mucoadhesive interface is not always well defined and the tensile fracture may occur either at the interface or inside the mucoadhesive layer, it is not always possible to distinguish between the adhesive and the cohesive components of the forces involved, which introduces a methodological complication in tensile measurements.

The rheological approach unambiguously measures the mechanical properties, that is, the resistance to flow and to deformation, of polymers and tells us how they change in the presence of mucin. However, it may be argued that it does not give any direct information about what actually occurs at the interface.

It can therefore be concluded that the information contents of the two methods (tensile and rheological) are different and complementary to each other. This does not exclude the fact that the two measurements can be related to each other in opportunely controlled conditions. It is also conceivable that the two methods contribute to a different extent to the explanation of the mucoadhesive phenomenon depending on which of the main mechanisms are involved, that is, depending on the type of system, the polymer, the step of adhesion, etc.

The last item we want to discuss also has to do with the information contents of rheological testing. Confirming the assumption that secondary bond (especially hydrogen bond) formation and physical entanglements following interpenetration are responsible for the mucoadhesive interactions, a few authors have used molecular spectroscopic methods to detect new bond formation in the presence of mucin. Moreover, some of them found a relationship between the results obtained with these techniques and the results obtained by dynamic oscillatory rheology, which is referred to as mechanical spectroscopy.

In particular, Kerr et al. (1990), using ^{13}C nuclear magnetic resonance (NMR) spectroscopy, reported interactions between the mucus glycoproteins and polyacrylic acid as a result of hydrogen bonding involving carboxylic acid groups of polyacrylic acid. Tobyn et al. (1992), using Fourier transform infrared spectroscopy, have also provided evidence of hydrogen bonding between the pig gastric mucus glycoproteins, reconstituted in USP simulated gastric fluid, and the test mucoadhesive. Jabbari et al. (1993) developed attenuated total reflection Fourier transform infrared spectroscopy (ATR-FTIR) to investigate chain interpenetration at a polyacrylic acid and mucin interface. The experimental results showed evidence in support of chain interpenetration between polymer and mucin. Mortazavi et al. (1993b) suggested the formation of hydrogen bonds between the terminal sugar residues on the mucus glycoprotein and the mucoadhesive polyacrylic acid, using infrared and ^{13}C-NMR. In subsequent work, Mortazavi (1995) investigated the nature of interactions (in particular hydrogen bonding) between the mucus gel and the model mucoadhesive polyacrylic acid (Carbopol 934P) using three separate techniques: dynamic oscillatory rheology, ^{13}C-NMR, and a tensiometer technique. He found that the addition of hydrogen bond breaking agents to mucus-polymer or model monosaccharide-polymer systems resulted in a reduction in viscoelastic properties of polymer-mucin mixtures, a positional change in the chemical shift of the polyacrylic acid signals, and a decrease in mucoadhesive strength in vitro.

The experimental evidence for new bond formation provided by spectroscopic methods and the correlation found between rheological and spectroscopic methods (Kerr et al., 1990; Mortazavi, 1995) further validated the rheological approach. In fact, even if spectroscopic methods may provide a direct estimate of the extent of hydrogen bond formation and/or disruption, the methods to be employed are generally demanding in terms of equipment needed and require careful data interpretation because of the spectroscopic complexity of mucin macromolecules and of the mucin-polymer interaction products.

Lacking molecular spectroscopy studies, mechanical spectroscopy, such as dynamic oscillation rheology, may provide a suitable alternative in investigating the extent of interaction between mucus and polymer macromolecules. In conclusion, an accurate study of the mechanical properties may represent a suitable surrogate for complete spectroscopic characterization.

REFERENCES

Albasini M, Ludwig A. 1995. Il Farmaco 50(9):633.

Allen A, Foster SNE, Pearson JP. 1986. Br J Pharmacol 87:126P.

Anders R, Merkle HP. 1989. Int J Pharm 49:231.

Bottenberg P, Cleymae R, De Muynck C, Remon JP, Coomans D, Michpotte Y, Slop D. 1991. J Pharm Pharmacol 43:457.

Busetti C, Orlandi R, Rossi S, Gazzaniga A, Caramella C. 1996. Valutazione "in vitro" delle proprietà mucoadesive di tre gradi di sodio ialuronato a diverso peso molecolare. Proceedings of the AFI-ADRITELF Symposium, Riccione, Italy, p 180.

Caramella C, Rossi S, Bonferoni MC, La Manna A. 1992. A rheometric method for the assessment of polymer-mucin interaction. Proceed. Intern. Symp. Control. Rel. Bioact. Mater., Orlando, Florida, pp. 90–91.

Caramella C, Rossi S, Bonferoni MC, La Manna A. 1993. Rheological characterization of some bioadhesive systems. Proceedings of the International Symposium on Controlled Release of Bioactive Materials, Washington, DC, pp 240–241.

Caramella C, Bonferoni MC, Rossi S, Ferrari F. 1994. Eur J Pharm Biopharm 40(4): 213.

Ch'ng HS, Park H, Kelly P, Robinson JR. 1985. J Pharm Sci 74(4):399.

Duchene D, Touchard F, Peppas NA. 1988. Drug Dev Ind Pharm 14(2&3):283.

Ferrari F, Rossi S, Bonferoni MC, Caramella C. 1997a. Il Farmaco 52(6–7):493.

Ferrari F, Rossi S, Martini A, Muggetti L, De Ponti R, Caramella C. 1997b. Eur J Pharm Sci 5:277.

Gu JM, Robinson JR, Leung SHS. 1988. CRC Crit Rev Ther Drug Carrier Syst 5: 21.

Hassan EE, Gallo JM. 1990. Pharm Res 7(5):491.

Helliwell M. 1993. Adv Drug Delivery Rev 11:221.

Kellaway IW. 1990. In vitro test methods for the measurement of mucoadhesion. In: Gurny R, Junginger HE, eds. Bioadhesion—Possibilities and Future Trends. Stuttgart: Wissenschftliche Verlagsgesellschaft, p 86.

Kerr LJ, Kellaway IW, Rowlands C, Parr GD. 1990. The influence of poly(acrylic) acids on the rheology of glycoprotein gels. Proceedings of the International Symposium on Controlled Release of Bioactive Materials, Reno, NV, pp 122–123.

Kocevar-Nared J, Kristl J, Smid-Korbar J. 1997. Biomaterials 18:677.

Jabbari E, Wisniewski N, Peppas NA. 1993. J Contr Rel 26:99.

Junginger HE. 1991. Pharm Ind 53(11):1056.

Lehr MC. 1991. Bioadhesive drug delivery systems for oral application. Doctoral thesis, Leiden.

Longer MA, Robinson JR. 1986. Pharm Int 7:114.

Marriott C, Gregory NP. 1990. Mucus physiology and pathology. In: Lenaerts L, Gurny R, eds. Bioadhesive Drug Delivery Systems. Boca Raton, FL: CRC Press, p 1.

Mortazavi SA. 1995. Int J Pharm 124:173.

Mortazavi SA, Carpenter BG, Smart JD. 1992. Int J Pharm 83:221.

Mortazavi SA, Carpenter BG, Smart JD. 1993a. Int J Pharm 94:195.

Mortazavi SA, Carpenter BG, Smart JD. 1993b. J Pharm Pharmacol 45(suppl):1141.

Mortazavi SA, Smart JD. 1994. J Pharm Pharmacol 46:86.

Park K, Robinson JR. 1984. Int J Pharm 19:107.

Park K, Robinson JR. 1985. J Contr Rel 2:47.

Peppas NA, Buri PA. 1985. J Contr Rel 2:257.

Peppas NA, Ponchel G, Duchene D. 1987. J Contr Rel 5:143.

Ponchel G, Lejoyeux F, Duchene D. 1991. Bioadhesion of poly(acrylic acid) containing-systems. Thermodynamical and rheological aspects. Proceedings of the International Symposium on Controlled Release of Bioactive Materials, Amsterdam, pp 111–112.

Ronchi C, Porziotta E, Mazza R, Rossi S, Caramella C. 1997. Valutazione delle proprietà bioadesive di vari polimeri in formulazioni oftalmiche. Proceedings of AFI-ISPE Symposium, Montecatini Terme, Italy, p 135.

Rossi S, Bonferoni MC, Bertoni M, Caramella C, La Manna A. 1992. "Do rheological measurements contribute to the characterization of mucoadhesive interface?", Proceedings of the 1st European Congress of Pharmaceutical Science, Amsterdam, p 74.

Rossi S, Bonferoni MC, Ferrari F, Bertoni M, Caramella C. 1996. Eur J Pharm Sci 4:189.

Rossi S, Bonferoni MC, Lippoli G, Bertoni M, Ferrari F, Caramella C, Conte U. 1995. Biomaterials 16:1073.
Sherman P. 1970. Industrial Rheology. London: Academic Press, p 15.
Smart JD, Kellaway IW, Worthington HEC. 1984. J Pharm Pharmacol 36:295.
Tamburic S, Craig QM. 1995. J Controlled Release 37:59.
Tamburic S, Craig QM. 1996. Pharm Res 13(2):279.
Tobyn MJ, Johnson JR, Gibson SAW. 1992. J Pharm Pharmacol 44(suppl):1048.

4

Functional Histology of Epithelia Relevant for Drug Delivery: Respiratory Tract, Digestive Tract, Eye, Skin, and Vagina

Udo Schumacher and Dietlind Schumacher
University of Hamburg, Hamburg, Germany

I. INTRODUCTION

In order to interact with a cellular receptor in a target organ, a drug that is not intravenously administered has to be absorbed and transported to the target organ, thereby passing through tissue barriers. These barriers can prevent or modify the uptake of this drug. Tissue barriers can even exist within an organ and thus prevent the uptake of the drug at the target cell, as is the case in the brain, where the blood-brain barrier prevents many drugs from entering the brain tissue, where the target cell population of neurons are located.

In principle, the following reactions can occur at these tissue barriers: (a) the tissue acts as a diffusion barrier, (b) the tissue can act as an enrichment for a specific drug, and (c) a metabolic modification of the drug by the cell can occur. All these processes are generally performed by epithelial cells or epithelia. Epithelia cover all our external and internal surfaces and show particular morphological and functional adaptations to their protective function, which depends on and varies according to their site. In all these different locations epithelia use, depending on whether the function is more protective as in the skin or more resorptive as in the intestine, the same subcellular building blocks, enabling them to perform these required functions.

II. WHAT IS AN EPITHELIUM?

Epithelia are one of the four basic tissues, the others being connective, nerve, and muscular tissue. A tissue is an aggregation of cells of the same type including the extracellular matrix produced by them. Organs in turn are composed of the different tissues, and in general epithelia confer the specific properties of a particular organ, while connective tissue gives the non–organ-specific support, including blood and nerve supply for the epithelia.

Epithelia are different from many other tissues because of their polarity. This means that two different regions of the cell can be distinguished: the apical and the basal one. This morphological distinction between apical and basal is also featured in functionally differing membrane domains, the

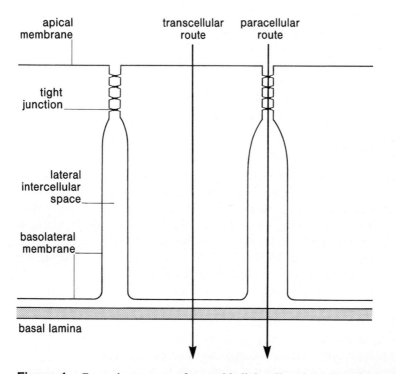

Figure 1   General structure of an epithelial cell and modes of transport across epithelia. The epithelial prototype cell rests on a basal lamina, and basal and apical membrane compartments can be distinguished. Differentiation between the two membrane compartments is possible due to the presence of tight junctions at the apical-basolateral border. (Modified from Widdicombe, 1994.)

apical and the basolateral membrane (Fig. 1). In epithelia that consist of only one layer of cells, the basally located region of the cell membrane is opposite the basal lamina, often also called basement membrane. It is typical for an epithelial cell to be found in an assembly with other epithelial cells, which then form sheets of epithelia, which are grouped and classified as outlined in Fig. 2.

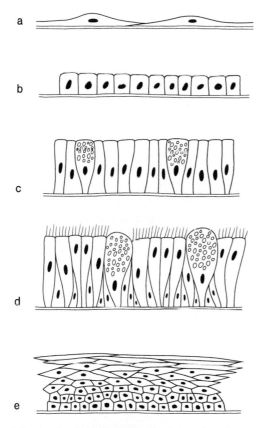

Figure 2 Principal arrangement of surface-covering epithelia into different forms of epithelial tissue. (a) Squamous epithelium; (b) cuboidal epithelium; (c) columnar epithelium with goblet cells characterized by their apical granules; (d) pseudostratified (respiratory) epithelium with goblet cells and cilia; (e) keratinized stratified squamous epithelium. Note that the classification of the epithelia is according to the shape (morphology) of the uppermost layer. Beneath this layer the morphology of the cells can be different. (Modified from Widdicombe, 1994.)

III. WHICH DIFFUSION BARRIERS OPERATE AT THE CELL AND TISSUE LEVEL?

The first contact a drug encounters with the cell is its membrane, which builds up the first barrier for a substance to pass (an exception is skin, where dead cells form the first contact zone). There are two possible choices for crossing the epithelial membrane: (a) using the route directly through the cell or (b) using a paracellular pathway slipping through tight junctions (see later).

A. Cell Membrane

The transport through the cell membrane is determined by its structure. The cell membrane consists of a lipid bilayer, in which the membrane proteins are dissolved (Fig. 3). The two sides of this lipid bilayer differ with respect to their composition: glycolipids are found on the extracytoplasmic side, whereas phospholipids are found only on the intracytoplasmic side. Lipids within the plane of the lipid film easily change their position, but transition from one plane to the other (flip-flop) is more difficult and special translocating enzymes are required to achieve this. Exceptions to this rule are cholesterol and its derivatives, which can easily change the planes of the lipid bilayer. This membrane asymmetry is of functional importance: the GM_1 ganglioside found on the apical surface of the intestinal epithelium is the binding partner for the cholera toxin. In addition, the carbohydrate residues of glycolipids and glycoproteins can serve as binding partners for lectins, carbohydrate binding proteins, and can therefore act as target structures in drug targeting.

B. Pathways Across Epithelia

Epithelia form the principal barriers between the environment and the inside of the body. Small lipophilic molecules can cross the lipid bilayer, but small hydrophilic molecules have to use channels or transmembrane transport systems. This transcellular route may be one mode of transport across an epithelium. An alternative route would be through tight junctions between two epithelial cells, the paracellular route. If substances can pass through these tight junctions, the junctions are often referred to as "leaky."

This mode of transport across epithelia, however, is not open to larger molecules, which cannot cross the membrane directly. They can be transported by vesicular transport. The simple form of this transport is called micropinocytosis. Macromolecules in the fluid above the apical membrane are incorporated in vesicles that are transported across the cytoplasm and

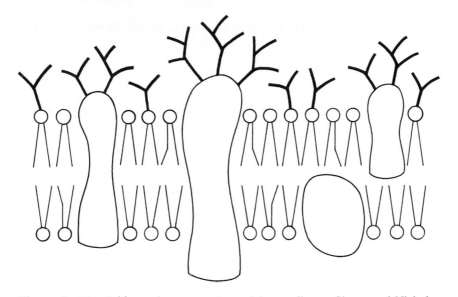

Figure 3 The fluid membrane mosaic model according to Singer and Nicholson. The membrane proteins are dissolved within the lipid bilayer. The lipids as well as the membrane proteins are asymmetrically distributed in the membrane. Phospholipids are located at the cytoplasmic sides of the membrane, while the carbohydrate residues of glycolipids face the extracytoplasmic side of the membrane. The integral membrane proteins, which cross the lipid bilayer, have a hydrophobic core that enables them to span the hydrophobic inside of the membrane and can carry carbohydrate residues on the extracytoplasmic side of the membrane. The surface coat of the cell is often referred to as the glycocalyx and can reach a height of about 1 μm in some covering epithelia.

are released with their content by fusion of the vesicle with the basolateral cell membrane. The extent of this transport system for intact macromolecules across the cell membrane may be rather limited, since most endocytotic vesicles are likely to end up in lysosomes, where the content of the vesicles is subject to digestion by acid hydrolases.

The other possible transport mechanism consists of endocytosis by clathrin-coated pits and vesicles, also called receptor-mediated endocytosis. This pathway, however, is open only to macromolecules that bind to a receptor at the apical cell surface. The receptor-ligand complexes cluster at the cell surface and are internalized. Again, they are subject to lysosomal degradation, but at least in theory some molecules might slip through and can be delivered to the basolateral surface. Transport of intact macromolecules with these mechanisms has been observed in M cells in Peyer's patches (see later).

C.   Cell Junctions

Epithelial tissues are characterized by the almost complete absence of extracellular matrix; hence neighboring epithelia are in close contact with each other. Several different junctions between epithelial cells can be distinguished functionally: (a) occluding junctions, (b) communicating junctions, and (c) anchoring junctions.

There is only one type of occluding junction, the tight junction (for a summary of the cell junctions see Fig. 4). It is located near the apical surface

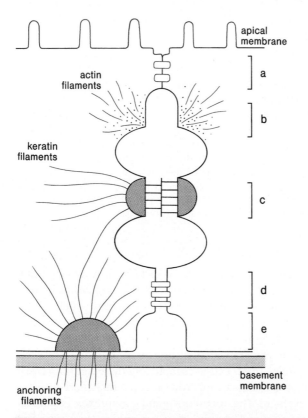

Figure 4   Principal arrangement of epithelial cell contacts: (a) zonula occludens (a row of tight junctions); (b) zonula adherens; (c) desmosome; (d) gap junction; (e) hemidesomosomes. (Modified from Widdicombe, 1994.)

of epithelial sheets, where it prevents even small molecules from leaking from one side of the sheet to the other. Hence it is of particular interest in transport processes. In addition, it separates apical from basolateral membrane domains, thus enabling epithelial cell polarity, including distribution of receptor and transport proteins in different cell membrane domains.

Among the communicating junctions, gap junctions are of interest in drug delivery. They consist of proteins that form tunnels between cells through which small molecules (up to about 500 daltons) can pass between two cells. If these molecules are ions, cells are electrically coupled. These junctions are of greater importance for transport processes between epithelial cells than for transport processes through cells.

Anchoring junctions link the cells mechanically. These attachment sites also act as focal adhesion points of intracellular fibers, either actin or intermediate (cytokeratin) filaments. The actin attachment sites encompass adhesion belts (cell-cell) or focal contacts (cell-matrix), and the cytokeratin attachment sites are desmosomes (cell-cell) or hemidesmosomes (cell-matrix). These functions are of less importance for drug delivery purposes.

D. Basal Lamina

Epithelia rest on a basal lamina, sometimes also called basement membrane, which is a feltlike mat composed of extracellular matrix molecules including type IV collagen, heparan sulfate proteoglycan, laminin, and entactin. Basal laminae are involved in the anchorage of epithelial cells to the underlying connective tissue and also influence cell metabolism and cell differentiation and act as a guiding structure for cell migration. Most important for drug delivery, they act as a highly selective filter.

E. Connective Tissue

The tissue below the basement membrane is called connective tissue. This tissue is of particular importance for drug delivery, because blood and lymphatic vessels, which are needed for drug transport, are located in the connective tissue. The epithelia do not contain these vessels, hence drugs have to cross through connective tissue. Whereas epithelia are characterized by the almost complete absence of intercellular spaces, connective tissue has abundance of extracellular matrix as its morphological hallmark. This matrix is produced by the principal cell of the connective tissue, the fibroblast. Two intricately interwoven components of the matrix can be distinguished: (a) the formed substances, the fibers (collagen, reticulin, elastic), and (b) the unformed substances, mainly glycoproteins and glycosaminoglycans. Because of their large carbohydrate component, the glycosaminoglycans form a hydrated gel and are important space fillers within the extracellular matrix. This hydrated gel is negatively charged and could well function as an ion exchange chromatography matrix within the tissue.

IV. THE RESPIRATORY TRACT

Two different but interlocking functions are performed by the respiratory system, namely transport of oxygen and carbon dioxide and gas exchange. The former has its morphological correlate in the conducting portions of the airways; the latter is morphologically represented in the respiratory area of the lungs, which makes up the bulk of the lung tissue.

A. Conducting Portions

The conducting portions consist of the nose, the pharynx, the larynx, the trachea, and the bronchial tree, which ends at the respiratory bronchioles, which have a respiratory function as well. These are continuous with the alveolar ducts and alveoli, which make up the respiratory portion.

Except for the parts of the airways that are shared with the pathway for food, all the surfaces are covered by pseudostratified ciliated epithelium. The epithelium is pseudostratified because all cells reach the basal lamina but the nuclei are stacked at different heights, giving at first sight the impression of a stratified epithelium. The principal cells found in this respiratory epithelium are the ciliated cell, the goblet cell, the brush cell, the basal cell, neuroendocrine cells, and migrating lymphocytes and granulocytes.

The *ciliated cell* makes up the bulk of the surface area of the airways. The apical membrane of the ciliated cells forms numerous microvilli, through which the cilia project into the lumen of the airways. The presence of microvilli is a morphological correlate for increased resorption of water, electrolytes, and other small molecules across the cell surface. Near the apical surface, tight junctions seal the luminal surface off. Below the tight junctions, neighboring cells are linked via desmosomes to give mechanical strength.

Interspersed between the ciliated cells are the *goblet cells*. They are characterized by an expanded apical region that is packed with mucigen granules. When the membrane of these granules fuses with the apical plasma membrane, the mucigen content of the granules is released onto the surface of the epithelium, where the mucin molecules, high-molecular-weight glycoproteins with a high percentage of carbohydrate residues, hydrate to form part of the mucus layer on top of the epithelium. The number of goblet cells decreases toward the smaller bronchi, and no goblet cells are found at the lower end of the airways at the level of the bronchioles.

The *brush cell* is a slender columnar cell that lacks cilia; its function(s) is essentially unknown, but speculations include the idea that brush cells are depleted goblet cells or that they represent intermediate stages in the differentiation between basal cells and ciliated epithelial cells. *Basal cells* do not reach the lumen of the airways, and some of them represent the stem cell population of the airway epithelium. They are of particular interest in repair and malignant transformation processes. *Neuroendocrine (Kulchitsky) cells* are rare in the bronchial epithelium and their functional role and participation in pathological processes are equally uninvestigated. *Lymphocytes* and *granulocytes* can be found in small numbers as migratory cells in the bronchial epithelium. Some of the granulocytes are special and resemble mast cells because they contain metachromatic granules; they are called globule leukocytes.

These cells make up the general cell population of the respiratory epithelium with the exception of the respiratory bronchioles, which contain the nonciliated bronchiolar cell, the *Clara cell*. Clara cells are the predominant cell population of airways about 1 mm in diameter and hence are located in the respiratory bronchioles of humans. The cell is characterized by an apical bulge of its cytoplasm, which can project above the other cells of the epithelium. It secrets a surfactant-like material and is rich in cytochrome P-450, thereby contributing to drug metabolism in the lung.

The other specialization of the airway epithelium is found in the olfactory region of the nose, which contains the receptor cells for smell.

All the preceding epithelial cells rest on a basal lamina, which is unusually thick in the respiratory tract. Loose connective tissue is located below this basal lamina. A delicate network of lymphatic vessels is distributed in this layer, which is also penetrated by blood vessels. Seromucous glands are located in this lamina propria and their secretions contribute considerably to the surface lining fluid of the airways.

B. Respiratory Portion

The respiratory portion consists of respiratory bronchioles (see earlier), alveolar ducts, and alveoli. The latter two are covered by type I and type II alveolar cells. The *type I* cell is also called the squamous alveolar cell; it covers about 90% of the surface area of the alveoli. Except for the area of the nucleus, the cell is just 0.2 μm thick and has a smooth surface. It is linked via occluding junctions to the neighboring cells, which can be either of the same type or type II cells. *Type II* cells, or great alveolar cells, are generally located in the angle between neighboring alveolar septa. They are thicker than type I cells, their apical surface is covered by small microvilli, and their cytoplasm contains numerous granules called lamellar bodies. The content of the granules, surfactant, a phospholipid- and protein-containing fluid that reduces the surface tension in the alveoli, is secreted by exocytosis. The diffusion barrier for oxygen and carbon dioxide consists of the surfactant overlying a type I or type II cell, the basal lamina of these cells, a small area of extracellular matrix, the basal lamina of the endothelium, and the endothelial cells. In many parts, no extracellular matrix is present between the two basal laminae, so they are fused. Macrophages are present within the alveoli.

C. Histophysiological Considerations for Drug Delivery

A few morphological and functional aspects of the respiratory system are of interest from the drug delivery point of view. The general morphology of the surface epithelium of the nose is outlined as above; however, the submucosa of the nose is characterized by a particularly rich vascular network with large venous plexuses. This transport capacity makes the nasal epithelium ideal for the delivery of small bioactive peptides, which are transported through this epithelium without apparent loss of function.

Another important aspect for drug delivery is the presence of mucociliary clearance. Mucins are high-molecular-weight glycoproteins that are well hydrated. Because some of the terminal carbohydrate residues of the mucins are negatively charged, they can serve as an absorptive matrix. By the constant beat of the cilia, which is directed toward the nearest orifice, inhaled substances that are trapped in the mucus are expelled. Because of the large gas exchange area (140 m^2) and the small diffusion barrier, the lung is ideally suited for drug delivery. If drugs have to be delivered to the respiratory area of the lung, however, it has to be borne in mind that the size of the inhaled particles is of critical importance in reaching these deep compartments of the lungs.

V. THE DIGESTIVE TRACT

The digestive tract is a hollow tube that consists of several layers. The principal layers are the same throughout the whole gastrointestinal tract; however, modifications according to the different functions of a particular organ are obvious. The main functions are food digestion and absorption of nutrients, and because these functions take place in or near the lumen, all major changes in the structure of the gastrointestinal tract are found in the layer that borders the lumen and is called the mucosa. The following layers are the submucosa, the tunica muscularis, and the serosa; the last one is absent in some parts of the gastrointestinal system. In the context of drug delivery, only the mucosa and submucosa are of interest, which will be considered further. The mucosa consists of (a) a surface epithelium, which rests on a basal lamina; (b) a small strip of loose connective tissue called the lamina propria, and (c) a thin layer of longitudinally oriented smooth muscle cells called the muscularis mucosae.

A. The Mouth and Esophagus

The oral cavity is the chamber in which food is mechanically fragmented by chewing and lubricated by saliva. This places a relatively high mechanical stress on the superficial epithelium, which is stratified squamous throughout and contains abundant tonofilaments. It does not keratinize, and no zonation such as that characteristic of the skin is seen. In contrast to the skin, the top layers contain condensed nuclei and seem to be viable. Occasionally, Langerhans cells, which are involved in antigen priming, are found within the epithelium. The underlying connective tissue, the lamina propria, differs in its thickness and degree of mobility from region to region. The gingiva differs from this epithelium and shows a stratum basale, stratum spinosum, stratum granulosum, and stratum corneum, which keratinizes.

From the drug delivery point of view, the thin nonkeratinized epithelium covering the floor of the mouth and under the tongue is more permeable than the other epithelia of the oral cavity. Nitroglycerin and other low-molecular-weight components can be administered in this way, thus avoiding first-pass effects in the liver. Mucins and digestive enzymes secreted from large and small salivary glands into the lumen of the oral cavity may also influence drug delivery, as does the contact time of the drug in the oral cavity.

The esophagus is a hollow tube connecting the pharynx with the stomach and is lined by the same type of stratified squamous epithelium as found in the oropharynx. Stratum germinativum, stratum spinosum, and stratum corneum with a few keratohyalin bodies can be distinguished. Esophageal glands add their secretions as lubricants into the lumen of the esophagus. Because of the short transit time of the food bolus, the esophagus is of little interest from the drug delivery point of view.

B. The Stomach

The main function of the stomach is to break down the ingested food enzymatically. This is achieved by digestive enzymes, which are secreted into the lumen of the stomach. The digestive juice inside the stomach has a pH of about 1. This low pH serves two main functions: (a) it destroys bacteria and other pathogens that have been swallowed and (b) it acts as an activator of the digestive enzymes.

Because of the aggressive nature of the stomach juice, several protective mechanisms have to be employed by the stomach to protect itself and the surrounding structures of the gastrointestinal tract from autodigestion. Both the entrance region of the stomach, the cardiac region, and the exit region, the pylorus, differ in their fine structure from the main part of the stomach, the fundus and body. The surface epithelium of the stomach is the same throughout. The mucosa of the stomach is folded; the height of the folds depends on the degree of filling. The surface lining epithelium consists of a single layer of columnar cells that have a few microvilli on their apical surface. The apical membrane including the microvilli is covered by a conspicuous glycocalyx on the luminal surface. Above this layer, the cells are covered by a blanket of mucus, which is part of the protective system against the aggressive luminal content. The apical portion of the surface lining epithelium is rich in mucigen granules. The cells are linked together by tight junctions and desmosomes.

The surface of the epithelium is regularly pierced by the openings of the gastric glands. The infoldings of the surface where the gastric glands terminate are called the gastric pits. The gastric glands of the body and fundus region contain different cell populations, which are responsible for the regeneration of the epithelium, the production of digestive enzymes, and the secretion of hydrochloric acid. Because only protective and not digestive functions have to be carried out in the cardia and pylorus, their glands are different. Hence these glands consist mainly of mucin-producing cells.

From a drug delivery point of view, the stomach is of little interest because its morphological specializations are geared toward protection and not absorption. The presence of mucins, which can serve as absorbing matrices, and the acidic pH combined with proteolytic enzymes make this region rather hostile for drug delivery purposes.

C. The Small Intestine

The main function of the small intestine is the absorption of nutrients. In order to achieve this, an increased surface area is created at the gross anatomic level by the formation of mucosal folds, the plicae circulares; at the tissue level by the formation of crypts and villi; and at the ultrastructural level by the formation of microvilli. To facilitate absorption, the gut is lined by a single layer of columnar epithelium. Mainly three types of cells are found in the epithelium: enterocytes (absorptive cells), goblet cells, and enteroendocrine cells.

The cells specialized for absorption are the *enterocytes*. Their apical surface consists of densely packed microvilli, which are covered by a well-developed glycocalyx. The cells are linked together by tight junctions just below the apical surface, followed by desmosomes, which form a continuous band around the whole cell. This junctional complex functions not only to seal the lumen off from the inside of the body but also to provide the separation between apical and basolateral membrane domains. The separation is necessary because the apical domain contains several digestive enzymes, such as leucine aminopeptidase, sucrase, lactase, and maltase, which break down dietary components into building blocks such as hexoses. These building blocks (monosaccharides, amino acids) are transported across the membrane by carrier proteins, which are also located in the apical membrane. The building blocks cross the cells and are transported away by the capillaries and veins, which ultimately drain into the portal vein, so all blood drained from the small intestine passes through the liver first.

Fat is absorbed in a different way. Free fatty acids and monoglycerides (products of lipid digestion by pancreatic lipase in the intestinal lumen) combine with bile salts to form micelles, which diffuse across the lipid bilayer of the plasma membrane. Both are resynthesized in the smooth endoplasmic reticulum to triglycerides, which are subsequently transported to the Golgi apparatus. Here the lipids are converted into chylomicrons, complex glycolipoprotein complexes, which are released into the lateral cleft between the lower two thirds of the enterocytes. These chylomicrons are transported via the lymphatics to the thoracic duct, where they enter the systemic circulation.

The *goblet cells* are mucin-producing unicellular glands of the intestine that are intercalated between the enterocytes. The goblet cells are integrated into the tissue formation via tight junctions and desmosomes in the same way as enterocytes. The mucin released by the goblet cells forms part of the mucus blanket covering the epithelia.

Scattered between these two cell populations are *enteroendocrine cells*. These cells release hormones, which can modify the local environment and can influence the activity of the bowel movements. In the initial part of the small intestine of some species, including humans and rodents, *Paneth cells* are located at the base of the crypts. They secrete lysozyme, and their number is dependent on the bacterial colonization of the gut.

A particular morphological specialization of the gut immune system is located in the terminal ileum, Peyer's patches. They consist of aggregated lymphatic follicles, which bulge the covering epithelium into the gut lumen. This specialized dome epithelium contains *M cells*, which contain lymphocytes in their basal pockets. These cells are specialized for processing of luminal antigens and are involved in the afferent part of the intestinal immune response. M cells can also be found in the epithelium covering single follicles that are scattered throughout the intestinal tract. Their precise role in priming of the immune response, however, needs to be clarified, in particular how many of its functions can also be performed by normal enterocytes.

These cell populations rest on a basement membrane, followed by a small layer of connective tissue. The connective tissue is particularly well developed in the core of the villi, where large lymphatic and blood vessels are located to facilitate the transport of the ingested nutrients. Below the thin layer of connective tissue at the base of the crypts, the final layer of the mucosa, a thin layer of smooth muscle fibers called the lamina muscularis mucosae borders the submucosa, which consists of loose connective tissue. The submucosa is similar in structure in the small and large intestine except for the duodenum, where Brunner's glands are located in the submucosa. These glands deliver a mucin via a specialized duct to the base of the crypt. The other layers of the gut wall are not of particular interest with respect to drug delivery and will not be described in detail.

D.   The Large Intestine

The main function of the colon is absorption of water and electrolytes. The principal cell populations, namely enterocytes, goblet cells, and enteroendocrine cells, are the same as in the small intestine. The main difference between large intestine and small intestine at the tissue level is the absence of villi in the former. At the cellular level, the main difference is the ratio of enterocytes to goblet cells, which is shifted toward the goblet cells in the colon.

E.   Histophysiological Considerations for Drug Delivery

The large surfaces of the small and large intestine are ideal for drug delivery, and oral application of drugs is widely used. New drugs and vaccines that target particular cell populations such as the M cells are under development. Although the enterocytes seem to be of a uniform character, this is not so. A gradient from crypt to villus and from duodenum to rectum in the distribution of membrane receptors and mucin composition exists and can influence absorptive processes. In addition, the gut content in terms of both diet and microbial flora can modify the gut structure and hence its absorptive properties.

The transport of particulate matter across the gut mucosa has gained renewed interest, as has the role of M cells in the gut immune system. In addition to absorption of small molecules through membrane carriers and channels, absorption of larger molecules through endocytosis is theoretically possible. This can happen as micropinocytosis or via clathrin-coated pits and vesicles (particularly in M cells). The extent to which these mechanisms contribute to uptake of macromolecules in the gut and its biological significance are still unclear.

VI. THE EYE

The parts of the eye that are of interest for drug delivery purposes are the epithelium covering the anterior surface of the cornea and the epithelia of the inner surface of the eyelids and their accessory structures. The stratified squamous corneal epithelium is about six layers thick. The cells in the outer layer are linked together by zonulae occludentes, and their apical membranes form a system of apical ridges or microplicae. The epithelia rest on a fibrillar lamina, Bowman's membrane, which forms the border to the stroma below. The cornea itself is avascular, and this lack of blood vessels explains why no graft rejection occurs after corneal transplantation into an allogeneic host. Hence, the cornea is not of interest for systemic drug delivery.

The outer surface of the eyelids is covered a stratified squamous epithelium, which is modified at the margin of the eyelid to form the palpebral conjunctiva, which is continuous with the bulbar conjunctiva. The recesses between the two conjunctivae are called the superior and inferior fornices, respectively. The conjunctival sac is the space between the eyelids and is continuous with these fornices. In comparison with the outer layer of the eyelids, the layers of the conjunctival epithelium become fewer and superficial cells become low cuboidal. Interspersed in this epithelium are goblet cells, which produce a mucin that is of great importance in maintaining the tear film over the eyes. Both the lacrimal glands located in the upper lateral side of the eyeballs and small accessory lacrimal glands called tarsal lacrimal glands secrete into the conjunctival sac. On the medial side of each eyelid, the tear fluid accumulates temporarily in the lacrimal lake, which is drained via the nasolacrimal duct into the nose. The epithelium of the sac and duct is a tall columnar pseudostratified epithelium.

Local drug delivery is of prime importance in a number of ocular diseases such as glaucoma. In addition, devices have been constructed to deliver drugs over extended periods of time to allow even systemic drug delivery at this site. The challenge to drug delivery at this site is the continuous washing of the absorptive surfaces by the tear fluid. Enhanced attachment of drugs to the resorptive surfaces would be of great benefit. This has limitations, however, because the continuous production of a covering tear film is necessary for proper functioning of the eye.

VII. THE SKIN

Because of its large surface area, the skin is one of the largest organs of the body. Its main functions are protection against desiccation and injury and an important role in thermoregulation and water balance. Two principal layers are distinguished: the epidermis, which is the surface layer, and the dermis or corium, which is a subadjacent connective tissue layer.

The principal cell of the skin is the keratinocyte. It is stacked in several layers to form the keratinized squamous stratified epithelium. The layers are stratum basale, stratum spinosum, stratum granulosum, stratum lucidum, and stratum corneum. The keratinocytes of the basal layer rest on a basal lamina, which is the border to the dermis below the epithelial layer. The cells are cuboidal to low columnar and mitosis is regularly encountered in this layer (hence sometimes stratum germinativum: the transit time to the top layer is 20–30 days). The cells in the stratum spinosum are densely interdigitated and tightly connected with each other by desmosomes. Cytokeratin filaments end in these desmosomes. Lamellar membrane-bound granules are also present in the cytoplasm. Three to five layers of keratinocytes form the stratum granulosum. Its name is derived from the presence of keratohyalin granules in the cytoplasm of the cells in this layer. They contain a lipid-rich substance that is not delineated from the surrounding cytoplasm by a limiting membrane, hence cytokeratin bundles can be incorporated in their periphery. The granules are exocytosed and their content forms a coating of the cell membranes. Because of this lipid-rich coating, the extracellular space increases from less than 1% in the lower layers of the epithelium to 5–30% of the tissue volume in this layer. This waterproof lipid layer is the major diffusion barrier in the skin.

The stratum lucidum is usually identifiable in the thicker regions of the skin. It consists of about five rows of very flat cells, which are lightly stained. The nuclei start to degenerate in the outer part of this layer. The stratum corneum consists of several layers of flattened, dead cells that desquamate in the top layers. The thickness of these layers in general, but particularly in the stratum corneum, varies considerably according to the mechanical stress placed on them. Hence the epithelium on the sole of the feet has the thickest layer and the skin is much thinner in the areas of the eyebrows. Another feature is the presence of hairs, which also varies considerably between the different regions. In contrast, scattered Langerhans cells, which are part of the accessory cells of the immune system and are involved in the priming of T cells, are present throughout the whole epidermis. Melanocytes and Merkel cells are also present in the skin but have no apparent role in drug delivery and hence are not considered further.

From the drug delivery point of view, water-soluble macromolecules cannot traverse the lipid barrier in the stratum granulosum; principally lipids and lipophilic substances can cross this barrier.

VIII. THE VAGINA

Typical stratified squamous epithelium with an underlying lamina propria makes up the mucosa of the vagina. The height of the epithelium varies according to the phases of the menstrual cycle: about 45 layers in the follicular phase and 30 in the luteal phase. The superficial layer undergoes little keratinization but contains a few keratohyalin granules. Glycogen deposits are found in these cells in midcycle. Glycogen from exfoliated cells is broken down by the bacterial flora, thus creating a low pH in the lumen of the vagina, which has antimicrobial properties. These glycogen deposits are influenced by estrogens; hence the pH of the vaginal fluid is lowest at midcycle. Administration of estrogen leads to an increased amount of glycogen in the epithelium, which leads to a decrease of the intravaginal pH and may thus be a measure to influence it. Lamellar intercellular lipid deposits and the formation of tight junctions contribute to the functionally observable barrier for water-soluble macromolecules. Langerhans cells are present below this permeability barrier.

REFERENCES

Alberts B, Bray D, Lewis J, Raff M, Roberts K, Watson JD. 1994. Molecular Biology of the Cell. 3rd ed. New York: Garland.

Fawcett DW, 1994. Bloom and Fawcett, a Textbook of Histology. 12th ed. New York: Chapman & Hall.

Weiss L. 1988. Cell and Tissue Biology. 6th ed. Baltimore: Urban & Schwarzenberg.

Widdicombe JH. 1994. Structure, growth and repair of epithelia. In: Goldie R. Immunopharmacology of Epithelial Barriers. London: Academic Press, pp 1–17.

5
Biochemical and Functional Aspects of Mucus and Mucin-Type Glycoproteins

Barry James Campbell
University of Liverpool, Liverpool, Merseyside, England

I. INTRODUCTION

Throughout the body, the internal epithelial cell surfaces are lined by an important protective and lubricating gel of "sticky" viscous mucus, secreted from both nonspecialized and specialized (goblet) epithelial cells. As well as forming a diffusion barrier between the luminal substances (including ions, nutrients, proteases, drugs, and toxins) and the cell surface, mucus functions to bind bacteria, parasites, and viruses and may play important roles by interacting with and modulating the immune response, inflammation, and tumorigenesis. The physical and chemical properties of mucus had been poorly understood until recently, but rapid progress has been made in the last 5 years in this state-of-the-art fast-moving research field. This chapter will focus on the biochemical and functional aspects of the mucus barrier, chiefly a function of the high-molecular-weight mucus glycoproteins called mucins, including current knowledge of the genomic organization of mucins, the structural properties of the apomucin core peptides, control of mucin glycosylation, the mechanisms of secretion, and the role of mucins as recognition molecules relevant to epithelial disease.

II. MUCUS COMPOSITION

The gel-forming properties of mucus are due to the presence of mucins, giant glycoproteins with a relative molecular mass range of $1-40 \times 10^6$

85

daltons. Mucins possess a linear protein core, typically of high serine and threonine content, that is heavily glycosylated by oligosaccharide side chains that contain blood group structures. These oligosaccharide chains account for 50–80% of the dry weight of mucins and are initiated by N-acetylgalactosamine that is O-linked to serine or threonine in the protein core. Mucins should not be confused with mucopolysaccharides, an obsolete term used to describe tissue glycosaminoglycans such as chondroitin, dermatan, heparan, and keratan, which contain iduronic or glucoronic acid.

Mucins may be classified into two classes, membrane-bound and secretory forms. The *membrane-bound mucins* possess a hydrophobic membrane-spanning domain, are attached to cell surfaces, and may play important roles by modulating, for example, immune response, inflammation, and tumorigenesis (Varki, 1993; Springer, 1994; Hounsell et al., 1997). Evidence has suggested that the increased expression of cell surface mucins, such as the sialomucin MUC1 (episialin), may also affect cell-cell and cell-matrix interactions, functioning to shield epithelial cell surface receptors and antigens in certain pathological conditions (Hilkens et al., 1995; Wesseling et al., 1996). A study has identified a novel 230-kDa membrane-bound mucin-like glycoprotein (gp230) distinct from MUC1 and CD44 in oral and cervical squamous epithelia, with loss of expression occurring with progression to neoplasia (Nielsen et al., 1997b).

Secretory mucins are secreted from both mucosal absorptive epithelial cells and specialized goblet cells. They constitute the major component of mucus gels of the gastrointestinal, ocular, respiratory, and urogenital surfaces. The "bottle brush" arrangement of oligosaccharides around the protein core, as described by Lamont (1992), allows the mucin to bind large amounts of water, resulting in a gel that expands rapidly after secretion into the intestinal lumen. This gel not only constitutes a physical barrier and lubricant but also generates a protective diffusion barrier for the underlying epithelium. This effect is particularly important in the stomach, where the low pH also increases mucus viscosity at the lumenal surface (Bhaskar et al., 1991). Mucus also acts as a free radical scavenger, partly as a consequence of its ability to bind lipids (Gong et al., 1991). Important constituents secreted in mucus include growth factors, such as epidermal growth factor (EGF) (Wright et al., 1990) and trefoil peptides (Wright et al., 1993), both secreted by the specialized mucus-secreting cells found as an adaptive phenomenon adjacent to areas of ulcerated mucosal tissue that may be important in maintaining the barrier function of the mucosal surface, enhancing cell migration, and facilitating healing after injury. The trefoil peptides have also been speculated to play a role in aggregation of mucins, acting as a putative link peptide (Sands and Podolsky 1996). Other mucus constituents include secretory immunoglobulin A (IgA), lysozyme, lactoferrin, α_1-antitrypsin, dialyzable salts, and N-glycosylated glycoproteins. A 60-kDa stress (heat shock) protein is also secreted with mucus, functioning as a chaperone molecule associating with mucin and possibly aiding in its synthesis and/or secretion (Winrow et al., 1993). A mucin-associated protease inhibitor has been described and sequenced (Van-Seuningen et al., 1989) and may well have an important role in protecting the epithelium.

The physical and chemical structure of mucins themselves is becoming better understood; at least nine different protein core (apomucin) sequences have been identified. The genes are differentially expressed in different tissues and in different disease states but all share the common feature of containing tandem repeats of DNA sequence, which lead to tandem repetition of amino acid motifs. These tandemly repeated domains can constitute 50% or more of the mucin polypeptide. They vary in sequence and in length but typically have high content of serine and threonine, which are the sites for O-glycosylation. Genetic variation in the number of tandem repeats is a common feature of mucins and may be of functional significance in mucosal disease.

The O-linked chains are always initiated by N-acetylgalactosamine, α-linked onto serine or threonine, but further extension of the oligosaccharide chain (up to 15 or more carbohydrates) is characterized by enormous variation in monosaccharide composition (i.e., glucose, galactose, N-acetylglucosamine, N-acetylgalactosamine, and fucose), branching, linkage, substitution by ester sulfate or sialic acid (N-acetylneuraminic acid), or O-acetylation of the sialic acids themselves. This leads to much greater polydiversity than would be achievable with similar numbers of amino acid residues. Fully glycosylated, each mucin chain with a relative molecular mass range of $1-4 \times 10^6$ daltons probably contains 150 or more O-linked oligosaccharides and thereafter can assemble into structures of relative molecular mass range $5-40 \times 10^6$ daltons (Carlstedt et al., 1993; Sheehan et al., 1996). Indeed, most of the mucin core is so protected by glycosylation that it is resistant to protease attack, but mucins vary considerably in the size of subunits that are released by protease digestion. In contrast to other cellular or plasma glycoproteins, in which the majority of oligosaccharides are initiated by N-acetylglucosamine N-linked to asparagine residues, mucins contain only a few N-linked oligosaccharides, added before O-glycosylation can proceed (Strous and Dekker 1992), but they are particularly present in the cysteine-rich N- and C-terminal nonrepetitive domains that flank the tandem repeat sequences (Gum et al., 1992; Bobek et al., 1993).

A. Similarities Between Epithelial Mucins and "Endothelial" Mucin-Like Glycoproteins

Not all O-glycosylated glycoproteins are mucins. Mucin-like glycoproteins (proteins with 20–55% proline, serine, and threonine composition and 40–

80% of their mass as O-linked oligosaccharides) have also been found in nonepithelial tissues that are not barrier tissues. They can act both as a selective barrier protecting the cell and in the adhesion cascade initiating the process of inflammation, serving as ligands for selectins, and are involved in lymphocyte trafficking (Shimizu and Shaw, 1993; van Klinken et al., 1995). These transmembrane or membrane-associated mucin-like glycoproteins include CD34, CD45, CD96 (TACTILE), glycosylation-dependent cell adhesion molecule 1 (GlyCAM-1), mucosal addressin cellular adhesion molecule 1 (MAdCAM-1), and P-selectin glycoprotein ligand (PSGL-1). Another membrane-associated mucin-like molecule, leukosialin (CD43) present on leukocytes, has been reported to be synthesized and secreted into the medium by a colon cancer cell line (Baeckstrom et al., 1995). Unlike the epithelial mucins, the endothelial/leukocyte mucins possess little or no tandem repeat sequence structure in their proline-, serine-, and threonine-rich domains (with the exception perhaps of CD43 and PSGL-1); however, they do share similar biochemical properties with the epithelial mucins, a function of their high degree of O-glycosylation.

III. EPITHELIAL MUCIN GENES, APOMUCIN STRUCTURE, AND TISSUE EXPRESSION

To date, nine different human epithelial mucin (*MUC*) genes have been identified, each of which contains distinct tandemly repeated sequences that encode (apo)mucin core polypeptides (MUC). MUC1 is a membrane-bound epithelial mucin, whereas MUC 2 to 8 are secretory gel-forming mucins (Table 1). However, only three mucin complementary DNAs (cDNAs) have been fully sequenced, these being *MUC1*, 2, and 7.

A. MUC1: A Membrane-Bound Epithelial Mucin Unable to Form a Gel in Solution

The *MUC1* gene, the first to be discovered, was localized to chromosome 1 within the region 1q21-24 (Swallow et al., 1987). *MUC1* encodes a membrane-bound polymorphic epithelial mucin (PEM/PUM) that is 300–600 kDa in its fully glycosylated form, with an extracellular domain containing 20 to 80 (depending on the allele) 20-amino-acid tandem repeat peptide sequences rich in potential O-glycosylation sites (serine and threonine residues; see Table 1), a short 31-amino-acid transmembrane domain, and a 69-amino-acid cytoplasmic "tail" that may interact with cytoskeletal actin filaments (Gendler et al., 1990; Lan et al., 1990). MUC*1* is highly expressed on apical membranes of the lactating mammary gland, pancreas, bronchus, salivary gland, prostate, and uterus (Gendler and Spicer, 1995). *MUC1* is only sparsely expressed in the small intestine and colonic epithelium, although higher expression in the colon can be detected by immunohistochemistry following α-fucosidase treatment to reveal the antibody epitope (Bara et al., 1993).

B. MUC2 AND MUC3: The Major Secreted Intestinal Mucins

Both genes encode secreted intestinal mucins (Gum et al., 1989, 1990) and are localized to chromosomes 11p15 and 7q22, respectively (Griffiths et al., 1990; Fox et al., 1992). Whereas *MUC2* has been completely sequenced, only a partial sequence of *MUC3* is known. MUC2 possesses two different tandem repeat motifs, one 23 amino acid residues in length (approximately 100 in total), rich in threonine (potential O-glycosylation sites), and a smaller 7 to 40 (mean 16) amino acid imperfectly conserved tandem repeat, toward the apomucin N-terminus (Toribara et al., 1991). MUC2 also possesses unique cysteine-rich subdomains located both upstream and downstream of its central repetitive region, and these are likely to be involved in the polymerization of MUC2 into biopolymers via intramolecular S—S bonds, as well as the expected linear end-to-end biopolymers (Gum et al., 1992). MUC2 is expressed in jejunum, duodenum, ileum, and colon and to a lesser extent in the gallbladder and bronchus (Gambús et al., 1993).

MUC3, which contains tandem repeat peptides of 17 amino acids (Gum et al., 1990) and has an EGF-like motif in its C-terminal cysteine-rich domain, may also not polymerize although the entire sequence is not known. *MUC3* is expressed in jejunum, ileum, colon, gallbladder, and pancreas (Gambús et al., 1993; Audie et al., 1993; Balague et al., 1995). The lack of MUC2 and MUC3 in the normal stomach may indicate that the major secreted gastric mucin is structurally adapted for resistance against acid, while the presence of high levels of MUC3 in gallbladder epithelium and intestine perhaps indicates that this mucin is more protective against bile salts. A study demonstrated that MUC2 and MUC3 show differences in their intracellular localization (MUC2 in intestinal goblet cells and MUC3 in both goblets and absorptive cells lacking secretory granules) and provides evidence for the existence of a maturational gradient for MUC3 but not MUC2, reflecting possible functional differences between the two (Chang et al., 1994a).

C. MUC4: A Tracheobronchial Secretory Mucin

MUC4 cDNA cloned from tracheobronchial mucosae contains 39 sequence repeats of 48 base pairs each sequence encoding a serine- and threonine-rich tandem repeat domain (Porchet et al., 1991) and the gene is localized to chromosome 3 (Gross et al., 1992) (see Table 1). *MUC4* is found to be strongly expressed in the bronchus, prostate, and endocervix (particularly in the luteal phase of the ovulatory cycle). *MUC4* is also expressed in all regions of the gastrointestinal tract except for the gallbladder and submaxillary glands (Porchet et al., 1995; Ogata et al., 1992).

D.   MUC5(AC and B): Two Major Secretory Mucins of the Tracheobronchus, Salivary Glands, and Stomach

MUC5 cDNAs were originally cloned from a tracheobronchial cDNA library and divided into three distinct groups of nonoverlapping clones, namely *MUC5A*, *B*, and *C* (Aubert et al., 1991). The data now indicate that *MUC5A* and *MUC5C* are part of the same gene, now called *MUC5AC*, which is distinct from *MUC5B* (Guyonnet-Duperat et al., 1995), with both having been mapped to chromosome 11p15 (Nyugen et al., 1990; Dufosse et al., 1994).

Sequence data have demonstrated that MUC5*B* contains four super-repeats of 528 amino acids each comprising 11 repeats of the irregular tandem repeat sequence of 29 amino acid residues (the most common amino acid sequence is shown in Table 1) with a cysteine-rich subdomain similar to those of MUC5AC and MUC2 (Desseyn et al., 1997). The superrepeat present in *MUC5B* is the largest ever determined in a mucin gene, and the central exon is the largest reported for a vertebrate gene. *MUC5B* is strongly expressed in the bronchus, endocervix, pancreas, and the acinar cells of the salivary glands (Audie et al., 1993; Balague et al., 1995; Neilsen et al., 1997). Two glycoforms of the MUC5B mucin have been demonstrated in human respiratory mucus with evidence for a cysteine-rich sequence repeated within the molecule (Thornton et al., 1997).

MUC5AC is characterized by 24-base-pair tandem repeat sequences, each encoding the eight-amino-acid peptide motif (see Table 1), and is expressed predominantly in the trachea, the gastric (fundic and antral) epithelial mucosae, and also the uterine endocervix (Audie et al., 1993; Porchet et al., 1995). MUC5AC has 3′ end cystiene-rich clusters and a similar carboxyl terminus domain apparently homologous to pre-pro von Willebrand factor and MUC2 (Klomp et al., 1995; Lesuffleur et al., 1995).

E.   MUC6: A Major Secretory Mucin of the Stomach

MUC6 is one of two gastric mucin cDNAs described by Toribara et al. (1993). The other gene was previously described in tracheal mucosae (i.e., *MUC5AC*). The tandem repeat unit of MUC6 consists of 169 amino acid

Table 1 Characteristics of Human Mucins[a]

Mucin	High-level expression	Chromosome	Tandem repeat (amino acids)	Gel forming	vWF-like domains[a]
MUC1 (PEM)	Breast Pancreas	1q21-24	**GSTAPPAHGVTSAPDTRPAP** (20)	No	No
MUC2	Intestine Tracheobronchus	11p15	**PTTTPITTTTTVTPTPTPTGTQT** (23) **PPTTTPSPPTTTTTF** (imperfect repeat, average 16)	Yes	Yes
MUC3	Intestine Gallbladder Pancreas	7q22	**HSTPSFTSSITTTETTS** (17)	Yes	?
MUC4	Tracheobronchus Colon Uterine endocervix	3q29	**TSSASTGHATPLPVTD** (16)	Yes	?
MUC5A/C	Tracheobronchial Stomach Uterine endocervix Ocular	11p15	**TTSTTSAP** (8)	Yes	Yes
MUC5B	Tracheobronchial Salivary (MG1)	11p15	**SSTPGTAHTLTVLTTTATTPTATGSTATP** (29)	Yes	Yes
MUC6	Stomach Gallbladder	11p15	**SPFSSTGPMTATSFQ**TTTYPTPSHPQTTLPTHVPP FSTSLVTPSTGTVITFTHAQMATSASIHSTPTGTIP PPTTLKATGSTHTAPMTPTTSGTSQAHSSFSTAK TSTSHTHTSSTHHPEVTPTSTTTTPNPTSTGTSTPV AHTTSATSSRLPTPFTTHSPPTGS (169)	Yes	Yes
MUC7	Salivary (MG2)	4q21.2	**TTAAPPTPSATTPAP-SSSAPPE** (23)	No	No
MUC8	Tracheobronchus Reproductive tract	12q24.3	TSCPRPLQEGTPGSRAAHALSR RVHELPTSSPGGDTCF (41) TSCPRPLQEGTRV (13)	Yes	?

[a]Possible O-glycosylation sites of the apomucin tandem repeat amino acid sequences are shown in bold type. Single-letter symbol abbreviations for amino acids are used. vWF, von Willebrand factor; PEM, polymorphic epithelial mucin.

residues and possesses at least three cysteine-rich domains similar to those seen in MUC5AC. High expression of *MUC6* has been demonstrated in the gastric glands, the gallbladder (Toribara et al., 1993), Brunner's glands of the duodenum, and seminal vesicles (Bartman et al., 1997). Weak expression was noted in the ileum, the colon (Toribara et al., 1993), and the endocervical and basal endometrial glands (Bartman et al., 1997). *MUC6* is expressed by 18–20 weeks of gestation, and the tissue distribution suggests that its primary function is protection of vulnerable epithelial surfaces from damaging agents such as gastric acid, bile, and proteases (Bartman et al., 1997). MUC6 and MUC5AC represent major secretory mucins in the stomach and are localized to distinct cell types (Ho et al., 1995a; De Bolos et al., 1995). MUC6 is localized to neck mucous cells of the fundus and antrum and may be a soluble mucin functioning to protect the glands locally from digestion or may be secreted to bind bacteria to be eliminated from the stomach. In contrast, MUC5AC is expressed in and released from the surface mucous cells of the fundic and antral gland and may protect against gastric luminal HCl and peptic digestion.

F.   MUC7: A Salivary Gland Mucin

Submandibular saliva and sublingual saliva contain two distinct mucin forms, a high-molecular-weight mucin (MG1) and a low-molecular-weight mucin (MG2), with differential expression in mucous and serous cells, respectively (Nielsen et al., 1996). *MUC7* cDNA was cloned by Bobek et al. (1993) from submandibular mucosae, is localized to chromosome 4, and encodes MG2. The messenger RNA (mRNA) encodes a 377-amino-acid polypeptide that contains six 23-amino-acid tandem repeats with 54 potential O-glycosylation sites. *MUC7* appears to be restricted to the sublingual and submandibular salivary glands with absent expression in the parotid glands. A study has identified the gene encoding MG1 as the tracheobronchial mucin gene *MUC5B* and found that *MUC5B* mRNA is expressed in all mucous cells of salivary glands (Nielsen et al., 1997a). This confirms previous studies that detected *MUC5B* expression in sublingual gland by in situ hybridization (Audie et al., 1993).

G.   MUC8: A Tracheobronchial and Reproductive
Tract Mucin

MUC8, with a tracheobronchial cDNA of 941 base pairs with imperfect 41-nucleotide tandem repeats, encodes a unique polypeptide with two consensus repeats of 13 amino acids and 41 amino acids (see Table 1) (Shankar et al. 1994). The C-terminal domain of *MUC8* has been cloned and the gene mapped to chromosome 12 (12q24.3) (Shankar et al., 1997). *MUC8* is highly expressed in both tracheobronchial and reproductive tract mucosae (D'Cruz et al., 1996).

H. Organization, Regulation, and Polymorphism

The location of *MUC2*, *5AC*, *5B*, and *6* on the short arm of chromosome 11 in the region 11p15.3-5 suggests a functional clustering of secretory mucin genes with implications for possible evolutionary linkage and also possible coregulation (Guyonnet-Duperat et al., 1995). The gene order was determined to be *MUC6*, *2*, *5AC*, and *5B*, and this order on the map corresponds to the relative order of their expression along the anterior-posterior axis of the body, suggesting a possible functional significance of the gene order (Desseyn et al., 1997; Pigny et al., 1996). Information on the organization of the mucin genes at chromosome 11p15 has identified an additional putative mucin gene (*MUCX*) at this site, located upstream of *MUC2* and with an expression pattern similar to that of *MUC5B* and *MUC6*. Whether this gene is *MUC6* or a novel mucin is as yet unclear (Velcich et al., 1997).

Overall, very little is known of the regulation of mucin genes. The promoter region of *MUC1* has been characterized to some extent and a sequence called E-MUC1 (-84 to -74 bp) appears to determine tissue-specific expression (Kovarik et al., 1993). Shirotani et al. (1994) have also shown a responsive mucin element in the promoter region of *MUC1* that binds to a soluble nuclear protein capable of stimulating the production of MUC1 by human colon carcinoma cells. Other regulatory regions have been identified, although their function in mucin gene expression is currently unknown (Gendler & Spicer, 1995). Similarly, the promoter region of *MUC2* has been characterized (Gum et al., 1994) and a study has demonstrated the influence of DNA methylation as a possible regulatory mechanism for *MUC2* gene expression, with increased methylation in the promoter region concomitant with the decrease of *MUC2* mRNA expression (Hanski et al., 1997).

Polymorphism of mucins is a common finding that has been demonstrated at the level of both DNA and mRNA and is a feature consistent with the size heterogeneity of biochemically purified mucins even under the most stringent purification conditions. Mucin transcripts (mRNAs) also tend to be large (up to 16 kb), and polydisperse transcripts observed for all mucin apoprotein genes cloned, except MUC*1* (Verma and Davidson, 1994), are generally thought to arise from allelic variations in the number of encoded tandem repeats (Swallow et al., 1987; Toribara et al., 1991). However, there is evidence that intact apomucin transcripts from a given allele are more typically monodisperse but very susceptible to degradation. Thus, the pronounced polydispersity of mature mucin molecules is more likely to be caused primarily by highly variable glycosylation (Baeckstrom & Hansson, 1996).

IV. BIOSYNTHESIS OF MUCIN O-LINKED CARBOHYDRATE SIDE-CHAINS

After synthesis of the protein core in the ribosomes, followed by leading sequence cleavage and translocation into the lumen of the endoplasmic reticulum, posttranslational modifications of mucin (including glycosylation) take place and continue as the molecule migrates from the cis to the trans side of the Golgi apparatus (Gleeson et al., 1994). Mucin glycosylation is controlled by an extensive family of glycosyltransferases, type II transmembrane enzymes, that are responsible for catalysis of the addition of monosaccharide units to an oligosaccharide chain or the apomucin polypeptide core. Many of the relevant glycosyltransferases have been cloned and sequenced (Field and Wainwright 1995; Hardiuin-Lepers et al., 1995) and demonstrate a very high degree of specificity for not only the nucleotide sugar donor but also the oligosaccharide acceptor molecule monosaccharide composition and glycosidic linkage. All glycosyltransferases demonstrate structural similarities, having an N-terminal cytoplasmic tail, a hydrophobic transmembrane sequence (thought to act as a signal anchor), a "stem" region that is prone to proteolysis, and perhaps the most important domain, that which determines enzyme function (i.e., the lumenally orientated C-terminal catalytic domain). Breakdown of the stem region allows release of the soluble, active catalytic domain (Colley, 1997).

Each mucin probably contains 30–40 different oligosaccharide structures, but it is not yet clear what determines their sequence. At present it seems more likely to depend on the relative proportions of the different glycosyltransferases rather than on the type of mucin core protein sequence. As yet, little is known about how glycosyltransferase expression may be regulated, but some studies have demonstrated control by both intracellular and extracellular factors including cytokines (Piller et al., 1998), hormones, second messengers, and nutritional factors (Biol et al., 1992; Li et al., 1995).

A. Initiation of Mucin O-Linked Glycosylation

The initiation of mucin-type O-glycosylation is different from that of N-glycosylation, which involves the synthesis of a lipid-linked oligosaccharide and the transfer of the whole oligosaccharide to the polypeptide chain. O-glycosylation is initiated by binding of *N*-acetylgalactosamine (GalNAc) as its uridine diphosphate nucleotide derivative (UDP-GalNAc) to the oxygen of the serine or threonine residues (i.e., an O-glycosidic bond) by the action of a glycosyltransferase, UDP-GalNAc:polypeptide α-*N*-acetylgalactosaminyltransferase (GalNAc-transferase) (Babczinski, 1980). The addition of the GalNAc to Ser or Thr residues can occur in the endoplasmic reticulum, a transitional compartment, or in the cis-Golgi, depending on cell type or differentiation status. It is now known that the initiation of O-glycosylation is controlled by a family of polypeptide GalNAc transferases encoded by at least four distinct and highly homologous human gene sequences (designated GalNAcT1 to GalNAcT4) (Clausen and Bennett 1996) (Table 2). More GalNAc transferases probably exist, with nine or more candidate isoforms already cloned or partially cloned (Marth, 1996). Multiple GalNAc transferases suggest a complexity in initiation of O-glycosylation, with different GalNAc-transferases showing differentially regulated expression in different organs and distinct but overlapping peptide sequence acceptor specificities.

Further O-glycosylation then involves the step-by-step addition of monosaccharide units to the already attached GalNAc residue. As O-linked glycosylation is a posttranslational event taking place in the cis-Golgi compartment (Roth et al., 1994) following N-glycosylation, folding, and oligomerization (Asker et al., 1995), acceptor motifs must be exposed on the glycoprotein surface to be accessible to a GalNAc-transferase. In contrast to N-linked glycosylation, no simple consensus acceptor sequence has been identified for mucin-type O-linked glycosylation and little is known of the actual distribution of the carbohydrate in these glycosylation clusters or whether this distribution along the peptide core is random or whether a given Ser or Thr residue is preferentially O-glycosylated over others. The acceptor sequence patterns are highly dependent on the amino acid sequences flanking the serine and threonine. Mucin protein cores typically have a high proline content, which is thought to prevent α-helix formation, thus leaving the molecule in an expanded conformation that facilitates the high degree of glycosylation. A study using a synthetic peptide known as TAP25, which is based on the tandem repeat sequence of the mucin MUC*1*, showed the importance of surrounding proline residues to a putative O-glycosylation site, and nearby charged amino acids such as aspartic acid did not allow the glycosylation of a free threonine residue (Stadie et al., 1995). One transferase can form both GalNAcα-*O*-Ser and GalNAcα-*O*-Thr, although the amino acid sequence adjacent to the serine and threonine residues markedly influences their formation (Wang et al., 1993). In support of this, Elhammer et al. (1993), showed that the GalNac transferase species from bovine colostrum can transfer GalNAc to both serine and threonine but that threonine is glycosylated in preference to serine, with an activity that is 35 times greater.

Table 2 Initiation of Polypeptide O-Glycosylation: A Family of Human GalNAc $\rightarrow$ Serine/Threonine Transferases

GalNAc-transferase	Chromosomal localization	High tissue expression	References
GalNAcT1	18q12.1	Ubiquitous	Meurer et al., 1995; White et al., 1995 Takai et al., 1997; Clausen and Bennett, 1996
GalNAcT2	1q4.1-2	Muscle, pancreas, heart, ovary, colon, and liver	White et al., 1995; Clausen and Bennett, 1996
GalNAcT3	2q24-31	Pancreas	Bennett et al., 1996
		Testis	Clausen and Bennett, 1996
GalNAcT4	12q21.3-22	Ubiquitous (low level)	Clausen and Bennett, 1996

Gerken and colleagues (1997) have provided the first detailed analysis of the glycosylation pattern of a mucin tandem repeat sequence (81-amino-acid repeat from porcine submaxillary mucin) and suggested that mucin glycosylation is modulated by peptide sequence but not entirely as expected from existing O-glycosylation prediction algorithms. All sites could be glycosylated, although the serine shows a wider range of glycosylation (30–100%) than the threonine (70–90%) residues.

B. Mucin Oligomerization

Following initiation of O-glycosylation, the chain can be terminated by the addition of N-acetylneuraminic acid (NeuAc/sialic acid), forming the tumor-associated sialyl-Tn antigen, NeuAcα2–6GalNAcα-O-Ser/Thr, or more typically can undergo oligomerization, i.e., be progressively elongated by further addition of alternating N-acetylglucosamine (GlcNAc) and galactose (Gal). Core region and terminal glycan biosynthesis of O-glycoproteins proceeds in an organ-specific and differentiation-dependent manner. This is determined by the varying activities and specificities of the competing glycosyltransferases and their compartment-specific localization. The resultant oligosaccharide side chains can be considered as comprising core, backbone, and peripheral regions.

At least eight types of mucin core regions have been identified (Table 3) (Hounsell et al., 1996). Cores 1, 2, and 3 are common to many serum, cell membrane, and mucin glycoproteins (Hounsell & Feizi 1982), whereas other core structures have a more restricted organ distribution. Core 4 has been found in mucins of several species including humans (Breg et al., 1988). Cores 5 and 6 have been identified in human meconium (Hounsell et al., 1985; Feeney et al., 1986). Core 7 has as yet been demonstrated only in bovine submaxillary mucin (Chai et al., 1992), whereas core 8 has been identified in human respiratory mucin (van Halbeek et al., 1994). Galβ1–6GalNAc, the sequence previously reported to be a novel core structure in gastric mucins, has since been demonstrated to be absent from this source (Hanisch et al., 1993).

Further glycosylation is probably controlled at least to some extent by the structure and glycosylation of the peptide core of the mucins; for example, Galβ1–3→GalNAc glycoprotein transferase, active in O-glycan core 1 synthesis, shows a preference for GalNAcα-O-Thr over GalNAcα-O-Ser, but the presence of Galβ1–3GalNAcα side chains adjacent to GalNAc-Thr reduces activity (Granovsky et al., 1994; Brockhausen et al., 1996). Extended glycosylation of the core oligosaccharide structures by addition of galactose (Gal) and N-acetylglucosamine (GlcNAc) residues occurs on both branches at C-6 and C-3 of GalNAc-O-Ser/Thr in the Golgi complex. Both

Table 3 The O-Linked Oligosaccharide Region Sequences of Epithelial Mucins and Cell-Surface Glycoproteins

Periphery	Backbone	Core (→Ser/Thr)	
	Galβ1&3/4→R		
[A] GalNAcα1–3→R	Type 1 Galβ1–3GlcNAcβ1–3Galβ→R	Galβ1–3GalNAcα-O-Ser/Thr	Core 1
[B] Galα1–3→R	Type 2 Galβ1–4GlcNAcβ1–3Galβ→R	Galβ1–3[GlcNAcβ1–6]GalNAcα-O-Ser/Thr	Core 2
[H] Fucα1–2Galβ1–4/3GlcNAcβ→R	Type 3 Galβ1–3GalNAcα1–3Galβ→R	GlcNAcβ1–3GalNAcα-O-Ser/Thr	Core 3
[Leᵃ] Galβ1–3[Fucα1–4]GlcNAcβ→R	Type 4 Galβ1–3GalNAcβ1–3Galβ→R	GlcNAcβ1–3[GlcNAcβ1–6]GalNAcα-O-Ser/Thr	Core 4
[Leˣ] Galβ1–4[Fucα1–3]GlcNAcβ→R	Galβ1–4GlcNAcβ1–6Galβ→R	GalNAcα1–3GalNAcα-O-Ser/Thr	Core 5
[Leᵇ] Fucα1–2Galβ1–3[Fucα1–4]GlcNAcβ→R	GlcNAcβ1–3[GlcNAcβ1–6]Galβ→R	GlcNAcβ1–6GalNAcα-O-Ser/Thr	Core 6
[Leʸ] Fucα1–2Galβ1–4[Fucα1–3]GlcNAcβ→R		GalNAcα1–6GalNAcα-O-Ser/Thr	Core 7
		Galα1–3GalNAcα-O-Ser/Thr	Core 8

Abbreviations: Gal, D-galactose; GlcNAc, N-acetyl-D-glucosamine; GalNAc, N-acetyl-D-galactosamine; Fuc, L-fucose; Ser, serine; Thr, threonine.

repeating Galβ1–3GlcNAc (type 1) and alternating Galβ1–3GlcNAc (type 1)/Galβ1–4GlcNAc (type 2) sequences are commonly found, giving rise to several different linear backbone structures (Hounsell et al., 1989) (see Table 3). Type 2 sequences can also be linked via a GlcNAcβ1–6Gal bond either in a straight chain or as a branch GlcNAcβ1–3[GlcNAcβ1–6]Gal (Hounsell et al., 1988, 1989). Less frequently, core structures can be elongated with a type 3 chain or a type 4 chain containing GalNAc residues in either an α- or β-linked conformation (Hakamori, 1989) (see Table 3).

The chains are finally terminated or terminally branched by an α-linked sugar residue, either sialic acid, *N*-acetylgalactosamine (blood group A), or galactose (blood group B), in association with fucose [blood group H(O)] on the subterminal galactose [depending on the presence or absence of the fucosyltransferase that is encoded by the secretor (*Se*) and *H* genes] (Hounsell and Feizi, 1982; Oriol et al., 1992). Most, arguably all, of the mucin oligosaccharide structures are blood group antigens, with Lewis (Le), I, TF, Tn, Forssman, as well as ABO antigens commonly being expressed.

C. Mucin Sulfation

Sulfation of mucin O-linked oligosaccharides occurs particularly in colonic and respiratory mucins and conveys important functional properties. There is increasing evidence that this sulfation imparts a strong negative charge, which probably influences the physical properties of the mucus (Forstner and Forstner, 1994). In addition, studies by our own group and others have shown that these sulfate moieties significantly increase the resistance of the normal colonic mucus to degradation by enteric bacterial enzymes (Tsai et al., 1992, 1995; Corfield et al., 1993; Roberton et al., 1993), perhaps reflecting a protective adaptation to an infected environment.

The addition of O-sulfate esters to epithelial mucin-type oligosaccharides (linked to galactose, GalNAc, or GlcNAc) is controlled by sulfotransferases catalyzing the transfer of sulfonate moieties from the donor molecule 3′-phosphoadenosine-5′-phosphosulfate (PAPS) within the trans-Golgi network as a late step in glycoprotein synthesis (Carter et al., 1988). An increased variation in oligosaccharide structure, function, and antigenicity is obtained with sulfation, which can include sulfate at C-3 of terminal galactose (Gal) residues or Gal linked β1–4 to *N*-acetylglucosamine (GlcNAc) in core 2, at C-6 of the terminal or internal Gal, at C-6 of GlcNAc in core 2, and at C-4 of terminal galactose or terminal GalNAc linked β1–4 to GlcNAc (Lamblin et al., 1991; Yuen et al., 1992; Lo-Guidice et al., 1994, 1997; Hooper et al., 1995; Karlsson et al., 1996). However, it is unknown just how many sulfotransferases are involved in the synthesis of sulfated O-linked oligosaccharides, and likewise it is unclear what mechanisms regulate

their activity. It seems likely that in intestinal epithelial cells, the same sulfotransferases may be responsible for sulfation of O-linked oligosaccharides
on cell surface glycoproteins and also for sulfation of O-linked oligosaccharides on secreted mucins. As yet none of the relevant sulfotransferases
have been cloned, although activity for mucin-type oligosaccharides has
been demonstrated in normal and malignant human mucosal tissues (Lo-
Guidice et al., 1995; Vavasseur et al., 1994; Chandrasekaran et al., 1997).

D. Mucin Polymerization

Fully glycosylated, gel-forming mucin monomers undergo an end-to-end
intramolecular assembly process (macromolecular multimerisation) to form
dimers and possible tetramers, perhaps within maturing storage granules
(Sheehan et al., 1995). MUC2 possesses unique cysteine-rich subdomains
located both upstream and downstream of its central repetitive region (Gum
et al., 1992), and the sequence homology observed with these domains of
MUC2 to four D-domains of pre-pro von Willebrand factor (Gum et al.,
1994; Kim et al., 1996), an endothelial cell large multisubunit glycoprotein that is well characterized with regard to the formation of tetramers via
N—N and then C—C disulfide bonding and then further polymerization for
constitutive cellular secretion (Voorberg et al., 1991), suggests that these
homologous regions may be important in the processing of MUC2. Studies
do suggest that dimerization of both human MUC2 and porcine submaxillary
mucin monomers precedes polymerization analogous to von Willebrand factor polymerization (Asker et al., 1995; Perez-Vilar et al., 1996). Indeed,
physical and electron microscopic evidence suggests that mucin glycopeptide subunits are linked end to end via intramolecular disulfide (S—S)
bonds, so forming an elongated thread structure to generate structures with
a relative molecular mass of $5-40 \times 10^6$ daltons (Carlstedt et al., 1993;
Sheehan et al., 1986, 1995). Although the entire sequence is not yet known,
MUC3 is very different from the structure of MUC2 and it is not yet known
whether MUC3 can undergo polymerization. It is clear, however, that MUC3
does possess a cysteine-rich carboxyl terminal domain that also contains an
EGF-like motif (Gum et al., 1990). MUC5AC and MUC6 may undergo a
scheme of mucin multimerization similar to that suggested for MUC2, as
both possess cysteine-rich clusters and a carboxyl terminus domain apparently homologous to von Willebrand factor and MUC2 (Klomp et al., 1995;
Toribara et al., 1997). Two glycoforms of the MUC5B mucin have been
demonstrated in human respiratory mucus, also with evidence for a cysteine-
rich sequence repeated within the molecule (Thornton et al., 1997). In contrast, MUC1, the membrane-bound epithelial mucin, does not have a high

cysteine content and does not form these characteristic disulfide-dependent polymers (Gendler et al., 1991). Similarly, MUC7, which contains only two cysteine residues in its entire sequence, found in the amino terminal region only six residues apart (Bobek et al., 1993), is also highly unlikely to undergo polymerization given the absence of cysteine residues in its carboxyl terminus.

Mucus synthesis, generally measured in vitro by incorporation of radiolabeled N-acetyl-D-glucosamine or D-glucosamine, can be increased by butyrate (Finnie et al., 1995), carbenoxolone, corticosteroids (prednisolone and hydrocortisone), and nicotine (Finnie et al., 1996; Phillips et al., 1997). Gastric mucin synthesis has been shown to be inhibited by the corticosteroid dexamethasone (Turner et al., 1997). Lumenal factors in the gut such as bacterial N-formylated chemotactic peptides [e.g. formylmethionyl-leucyl-phenylalanine (fMLP)] (Campbell et al., 1997) and dietary lectins such as peanut agglutinin (PNA) (Ryder et al., 1994a) have also been shown to enhance mucus synthesis in cultured colonic cell lines and explants from patients with normal and diseased colonic epithelium.

V. MUCUS SECRETION

The mucosa consists of absorptive epithelial cells and specialized mucus-secreting goblet cells, both of which secrete mucus, but the goblet cell provides the main source of mucus to protect and lubricate the epithelial surface. Goblet cells can begin their lives in the lower crypt in one of two ways, arising from either multipotent basal crypt stem cells or poorly differentiated lower crypt oligomucous cells. Then they migrate up the crypts, mature, and are sloughed into the lumen over 2–3 days. As they migrate they mature, undergoing morphological change and acquiring an organized array of microtubules and intermediate filaments (theca), which separate granule from cytoplasm and give the goblet cell its typical shape (Radwan et al., 1990). In addition, changes in mucin type can occur (for example, the distal colonic lower crypt goblet cells contain predominantly sulfomucins, whereas upper crypt goblets contain fewer sulfated mucins) (Lapertosa et al., 1984).

Following completion of O-glycosylation in the Golgi network, mature mucins collect at the nodular dilations found at the end of the trans-Golgi stack and subsequently bud from the trans-Golgi tubules to form condensing secretory granules. The precise mechanisms targeting mucins to these granules have yet to be discovered, although the signal for storage of secretory proteins such as renin and growth hormone is generally contained within the primary protein sequence (Trahair et al., 1989; Chu et al., 1990). It is therefore likely that the primary sequence of mucins may also contain a signal directing them to storage granules and that mutation or alternative splicing of mucins might result in the redirection of the secretory product into the nonstorage, constitutive secretory pathway. Stored mucins and other mucus constituents are gradually concentrated as the secretory granules mature, as observed by an increase in electron density (Sandoz et al., 1985), although just why mucins appear to be targeted almost exclusively to condensing granules when a nonstorage baseline secretory route would be simpler and less costly to the cell still remains a mystery. These condensing granules are most numerous near the trans-Golgi network, not being found near the plasma membrane or in the main granule mass of the goblet cell. Within the condensing granules, a high intragranular calcium ion concentration is correlated with condensation of the mucins, probably via the phenomenon of "charge shielding," which may act to facilitate mucin apolar interaction and compaction and in association with other granule factors perhaps aids mucin polymer-polymer affinity (Verdugo, 1990). Upon maturation, the granules aggregate as a compact mass at the apical membrane and await secretion (Forstner, 1995).

Two well-established methods of goblet cell mucus secretion have been described, and our own group has preliminary evidence that there may be a third type of mucus secretion in inflamed tissue.

A.   Slow Baseline Secretion

Slow baseline secretion plays a key role in mucin secretion and involves the steady release of single mucus secretory granules via a vesicular constitutive pathway of conventional but immediate exocytosis in which no storage takes place (i.e., the intermittent fusion of a single mucous granule membrane and the apical plasma membrane). Slow baseline exocytosis is nonregulated, requiring no signal other than contact with the plasma membrane, although one study has shown that arachidonic acid partially inhibits baseline secretion (Yedger et al., 1992), and this probably ensures continual replenishment of the mucosal surface mucus layer. Following processing in the Golgi, mucin granules are guided to the lumenal cell surface by a vertical assembly of microtubules, which takes approximately 4 hours (Specian and Olivier, 1991). Interaction with the microtubules is essential for their orderly translocation, as depolymerization by nocodazole inhibits and disorients this constitutive baseline secretion (Olivier and Specian, 1991). Actin filaments at the cell apex may act as a functional barrier to secretion because depolymerization of actin by cytochalasin D and dihydrocytochalasin B accelerates baseline mucus granular secretion (Olivier and Specian, 1990).

B. Rapid Mucus Secretion

Rapid mucus secretion occurring in response to secretagogue stimuli and providing protection to a threatened epithelial surface involves fusion of multiple storage granules and the apical membrane, resulting in secretion of the mucin granule mass together with cytoplasm and excess apical membrane (i.e., regulated compound exocytosis). Rapid expansion of mucin molecules takes place on exposure to the extracellular fluid and may be a phenomenon of charge repulsion of the polyanionic chains once the intragranule Ca^{2+} charge shielding is negated (Verdugo, 1990). The empty or nearly empty goblet then gradually refills over a period of $1-2$ hours (Forstner, 1995). This process of compound exocytosis is not inhibited by drugs that disrupt the actin filaments of the cellular cytoskeleton but may be mediated by a rise in intracellular Ca^{2+} (Olivier and Specian, 1990).

C. Whole Goblet Expulsion: A Novel Mechanism of Secretion

Our own group has preliminary evidence that there may be a third type of mucus secretion in inflamed tissue. This involves expulsion of entire goblets from the goblet cells, leaving behind an apparently healthy intact cell not morphologically recognizable as a goblet cell (Sadek et al., 1994). A later study has demonstrated an absence of tubulin staining around the extruded goblet, implying that the microtubule basket was intact within the cell, probably in the apical segment of the cell (Leiper et al., 1997). This phenomenon is in keeping with the observation that there is an apparent selective loss of goblet cells relative to nongoblet epithelial cells in experimental colitis (Kaftan and Wright, 1989). The mechanism of this type of goblet expulsion is as yet unclear, but its probable specificity for inflamed tissue suggests that it might involve leukocyte components as secretagogues.

D. Mucus Secretagogues

Mucin-secreting cells respond to multiple stimuli and possess a wide variety of receptors coupled to an assortment of intracellular pathways, and precisely how signals are supplied to the plasma membrane and secretory granule to elicit exocytosis is now beginning to be elucidated. Induction of rapid mucus secretion has been demonstrated in response to anaphylaxis (Lake et al., 1980), mucosal irritants (such as mustard oil, alcohol, hypertonic saline, triglycerides, and bile acids), mechanical trauma, histamine (Neutra et al., 1982), and stress (Castagliulo et al., 1997). Mucin secretion can also be activated by bacterial enterotoxins release by *Vibrio cholerae* and *Esche-*

richia coli from small intestine and colon (Moon et al., 1971; Forstner et al., 1981; Chadee et al., 1991). A study has demonstrated that cholera toxin activates mucin secretion via the intracellular cyclic AMP (cAMP) pathway and is partially inhibited by a protein kinase A inhibitor and the microtubule inhibitor colchicine (Epple et al., 1997). The release of mucin secretion from human colonic cells induced by the protozoan *Entamoeba histolytica* is dependent on contact and protein kinase C activation (Keller et al., 1992).

Cholinergic agonists (e.g., acetylcholine) and cholinomimetic drugs (such as pilocarpine and carbachol) stimulate mucin secretion in small intestine mucin-producing cells (Phillips, 1992), trachea (Fung et al., 1992), stomach (Seidler and Pfeiffer, 1991), and colonic goblet cell lines (McCool et al., 1994; Epple et al., 1997). Muscarinic receptors have been identified on both villus and crypt cells of small intestine and colon (Rimele et al., 1981; Wahawaisan et al., 1983) and on colonic cell lines (Kopp et al., 1989). Muscarinic agonists stimulate mucin secretion by releasing inositol 1,4,5-trisphosphate (IP_3) and elevating intracellular concentrations of Ca^{2+} (Fleming et al., 1992; Seidler and Pfieffer, 1991). Agents that elevate intracellular calcium (ionophores), activate protein kinase C (diacylglycerol, phorbol esters), and elevate intracellular cAMP [forskolin, 3-isobutyl-1-methylxanthine (IBMX)] all stimulate mucin secretion in tracheal goblet cells (Steel and Hanrahan, 1997) and various colonic cell lines (McCool et al., 1990; Yedger et al., 1992; Hong et al., 1997a). Removal of extracellular Ca^{2+} and depletion of intracellular Ca^{2+} does not prevent forskolin-stimulated mucus secretion in gastric mucus cells (Seidler and Sewing, 1989) and colonic cell lines (Forstner et al., 1994); however, changes in extracellular Ca^{2+} may play a role in regulating production of cervical mucus (Gorodeski et al., 1997). In the airways, the inhibition of the anion secretion response to cholinergic stimulation leads to mucus accumulation of the gland ducts, resembling early cystic fibrosis (Inglis et al., 1997).

Other agonists including tetragastrin (Komuro et al., 1992), EGF (Kelly and Hunter, 1990), nitric oxide, and cyclic GMP (Brown et al., 1993) all stimulate gastric mucus secretion. Gastric mucus becomes more acidic, particularly due to sulfation, in response to histamine H_2 antagonists and proton pump inhibition (Matsumoto et al., 1992). Vasoactive intestinal polypeptide (VIP) stimulated secretion of mucin in the T84 colonic cell line (McCool et al., 1990) but not in colonic goblet cell line HT29-C1.16E even though VIP receptors were present (Laberthe et al., 1989). However, carbachol-induced secretion in HT29-C1.16E cells was strongly potentiated by VIP, proving these receptors functionally active in the control of mucin secretion (Laberthe et al., 1989). "Cross talk" probably occurs between the cAMP pathway stimulated by VIP and the Ca^{2+} pathway stimulated by neurotensin or carbachol. The combined action of carbachol and VIP requires extracellular calcium (Bou-Hanna et al., 1994). In addition to VIP, other neuropeptides such as neurotensin and neuromedin N have been shown to stimulate secretion in colonic cells, both via the same shared receptor (Augeron et al., 1992). Neurotensin stimulation was preceded by a rise in intracellular Ca^{2+} without an increase in cAMP, suggesting that receptor binding may activate phospholipase C.

In addition, a number of other agents have been shown to stimulate mucin secretion in intestinal cells and explants, including ATP (Merlin et al., 1994), immunoglobulins, and interleukin 1 (IL-1) (Cohan et al., 1991). Secretion with IL-1 and a macrophage-derived secretagogue (Sperber et al., 1993) suggests a link between the immune response and mucus hypersecretion, which may explain the mucus depletion that is a feature of most forms of mucosal inflammation. Overexpression of IL-4 has also been shown to induce mucin hypersecretion and mucin gene expression (Temann et al., 1997). A dose-related increase in mucin secretion in response to stimulated neutrophils was demonstrated in the colonic goblet cell line HT29-MTX and was completed inhibited by a specific inhibitor of human elastase, illustrating that human neutrophil elastase may be a potent mediator of mucin secretion in the inflamed colon (Milton et al., 1996; Leiper et al., 1997).

VI. ABERRANT MUCINS IN EPITHELIAL DISEASE

Although the synthesis and structure of the mucin core proteins are becoming better understood, very little is known about whether the structure of the oligosaccharides is determined by the mucin core sequence or alternatively by differential expression of the relevant glycosyltransferases. Pathological modifications of the biochemical structure of the mucosal barrier epithelial mucins will undoubtedly affect their normal gel-forming properties and the rate of mucus degradation and therefore directly influence the viability of the defensive barrier. Exposure of apomucin backbone structures of the mucin glycoproteins occurs in epithelial malignancies and can be due to abnormal glycosylation during biosynthesis. Dysregulation of tissue and cell-specific expression of mucin genes can also occur in epithelial cancers, but whether differential regulation of mucin genes affects the behavior of the tumor and results in the altered glycosylation commonly seen in these tumors is currently unknown.

A. Altered Mucin Gene and (Apo)mucin Core Expression

Increased MUC1 mucin peptide immunoreactivity correlating with increased *MUC* mRNA was first demonstrated in breast carcinomas and metastatic lesions (Gendler et al., 1990). More recently, levels of MUC1 and MUC2 apomucins have been demonstrated to be highly elevated in mucinous carcinoma of the breast, whereas invasive duct carcinomas of the breast express only increased levels MUC1 (Yonezawa et al., 1995). In addition, MUC1 overexpression may play a crucial role in the process of blood-borne metastases in breast cancer by inhibiting adhesion of the tumor cells, allowing escape from immune surveillance (Wesseling et al., 1996) and then interaction with endothelial adhesion molecules such as intracellular adhesion molecule 1 (ICAM-1) (Regimbald et al., 1996). MUC1 is also seen to be increased in ovarian and prostate cancer (Ho et al., 1993).

In the colon, an increased risk for malignant transformation is associated with increased expression of MUC1, 2, and 3 epitopes in hyperplastic polyps and adenocarcinoma of all histological subtypes (Nakamori et al., 1994; Ho et al., 1996); comparable *MUC* mRNA levels are decreased (Ho et al., 1996; Weiss et al., 1996). The highest expression of MUC2 and MUC3 epitopes has been observed in mucinous (colloid) cancers, and this correlates with decreased survival of patients (Ho et al., 1993). Evidence suggests that the increased expression of MUC2 epitopes in colon cancer is due to incomplete glycosylation of the MUC2 apomucin rather than an increase in MUC2 synthesis (Hong et al., 1997b). Increased *MUC4* and *MUC5AC* mRNA has been observed in colonic cancer tissue (Ogata et al., 1992; Ho et al., 1993) and an increase in *MUC5AC* mRNA has been seen in rectosigmoid villous adenomas (Buisine et al., 1996). In inflammatory intestinal tissues (ulcerative colitis and Crohn's colitis), *MUC2* mRNA levels and biosynthesis appear unchanged (Weiss et al., 1996; Tygat et al., 1996a, 1996b). In contrast, MUC3 has been shown to be markedly decreased or absent in the glands of severely inflamed colitic epithelium, suggesting that lack of *MUC3* expression may have an important role in the pathogenesis of colitis (Chang et al., 1994b). Another study, however, demonstrated conflicting results with no such change in *MUC3* mRNA expression regardless of whether the mucosa manifested active or quiescent inflammation (Weiss et al., 1996). More recently, *MUC* 3, 4, and 5B have been shown to be reduced in healthy and involved ileal mucosa of patients with Crohn's disease, suggesting a primary defect in expression of these genes (Buisine et al., 1997; Lucas et al., 1997).

Whereas normal stomach lacks *MUC2* and 3, both genes are highly expressed in gastric intestinal metaplasia and carcinoma, with a reciprocal decrease in *MUC5* and 6 expression. Gastric cancers of all types have demonstrated increased MUC5 and MUC6 epitope expression correlating with decreased in mRNA expression (Ho et al., 1995a, 1995b). Increased MUC1 immunoreactivity is also seen in most adenocarcinomas of stomach (Ho et al., 1993), and evidence suggests that individuals with small *MUC1* alleles are more susceptible to the risk of gastric carcinoma (Carvalho et al., 1997).

In contrast to normal pancreatic tissue, mucinous hyperplasias and pancreatic cancers show an increase in *MUC3*, *MUC4*, and *MUC5AC* gene expression (Balague et al., 1995). Immunohistochemical studies have also demonstrated increased levels of MUC3 and MUC5/6, while MUC1 apomucin remains positive and MUC2 remains almost negative during neoplastic transformation of the pancreas (Terada et al., 1996).

Similar abnormalities of mucins have also been described in epithelial diseases of the airways. In particular, mucus-secreting lung adenocarcinomas exhibit increased *MUC1* and *MUC4* mRNA levels (Seregni et al., 1996), with similar pattern of expression for well-differentiated cancers including increased *MUC3* mRNA (Nguyen et al., 1996). Changes in mucin gene expression are less well characterized in hypersecretory lung diseases such as chronic bronchitis and cystic fibrosis. However, evidence is now beginning to improve our understanding of the pathogenesis of cystic fibrosis, including demonstration that up-regulation of *MUC2* gene expression occurs in nasal epithelium of patients with cystic fibrosis (Li et al., 1997a) and that *Pseudomonas aeruginosa* infection of cystic fibrosis airway epithelium leads to mucus overproduction and transcriptional up-regulation of *MUC2* via a tyrosine kinase–dependent pathway (Li et al., 1997b).

At present, it is not known whether the apomucins MUC7 and MUC8 show altered or differential expression in a tumor-specific manner as has been demonstrated for other *MUC* gene products.

B. Altered Mucin Glycosylation in Malignancy and Premalignant Disease

In many cases, blood group carbohydrate antigens normally expressed during fetal development but absent in adult epithelium return as oncofetal antigens during carcinogenesis. Sometimes blood group antigens may be expressed that are incompatible with the blood group of the host (Yuan et al., 1985). Glycosyltransferase activities for blood groups A and B have been shown to be reduced, whereas up-regulation of α-1,2-fucosyltransferase responsible for the synthesis of the H-antigen has been observed in colon and gastric tumors (see Kim et al., 1996). Some of these cancer-associated blood group antigens, such as sialyl-Le^a and sialyl-Le^x (Ogata et al., 1995), are present not only on secreted mucus but also on the cell surface. They serve as ligands for the endothelial cell adhesion molecules, such as the selectins (Mannori et al., 1995), and may play an important role in the adhesion of cancer cells to vascular endothelium (Takada et al., 1993) before extravasation of cancer cells from the capillaries (Fukuda, 1996). Advanced primary colorectal carcinomas and liver metastases also show loss of sulfomucin, increased sialomucin expression, and increased sialyl-dimeric-Le^x antigen expression. This extended Le^x antigen seems to function as an ectopic adhesion ligand that promotes metastatic tumor cell implantation, and its high expression is correlated with a poor prognosis (Hoff et al., 1989; Matsushita et al., 1991). Le^x and Le^y antigens (positional isomers of Le^a and Le^b, but formed by the action of an α-1,3-fucosyltransferase rather than an α-1,4-fucosyltransferase) minimally expressed in normal colonic epithelia are both increased in colonic cancer and adenomatous polyps (Kim et al., 1986; Itzkowitz et al., 1986). Sialylated Le^a antigen is another oncofetal-associated antigen observed in colon cancers and is also expressed in the dysplastic areas of inflammatory bowel tissue. Often these structures are extended by chain elongation, internal fucosylation, and sialylation, particularly for type 1 backbone chains (Galβ1–3GlcNAcβ1–3Galβ→R), increasing expression of extended sialyl-Le^x or sialyl-Le^a. However, when α1,3/1,4-fucosyltransferase activity was compared in normal and cancerous tissues, no apparent differences were noted (Yang et al., 1994; Hanski et al., 1996). In addition, extended type 2 chains, polylactosamine backbone structures (blood group i antigen, Galβ1–4GlcNAcβ1–3Galβ→R), are synthesized preferentially by premalignant and malignant colonocytes (Miyake et al., 1989) as a result of increased β1→3GlcNAc-transferase activity (Vavasseur et al., 1995).

The commonest epithelial disease–related mucin oligosaccharide epitopes in primary adenocarcinomas (including breast, lung, ovary, colon, and most other gastrointestinal tumors) are short "incomplete" structures, such as the core 1 Thomsen-Friedenreich (TFα; Galβ1–3GalNAcα-O-Ser/Thr) antigen, or prematurely terminated structures such as Tn (GalNAcα-O-Ser/Thr) and sialosyl-Tn (NeuAcα2–6GalNAcα-O-Ser/Thr) antigens (Schmitt et al., 1995; Terada and Nakanuma, 1996; Hounsell et al., 1997). Sialosyl-Tn is increasingly expressed during the dysplasia-carcinoma sequence of the colorectum and has been associated with advanced disease, poor prognosis (Itzkowitz et al., 1990; Siddiki et al., 1993), and increased malignant potential of adenomatous polyps (Itzkowitz et al., 1992). Immunotherapy with tumor antigens (such as sialosyl-Tn, Tn, and TF), in order to enhance immune recognition, has been explored as an approach to cancer therapy (O'Boyle et al., 1992; Taylor-Papadimitriou et al., 1993; Springer et al., 1995). Furthermore, mucins bearing sialosyl-Tn have been shown to mediate inhibition of natural killer cell cytotoxicity, thereby providing immune escape for cancer cells (Ogata et al., 1996). Several mucin glycosylation alterations seen in colorectal cancer have also been demonstrated in inflammatory bowel disease (Rhodes, 1997). A common end result of the glycosylation alterations in both ulcerative colitis and Crohn's disease is increased expression of the oncofetal Thomsen-Friedenreich (TF) antigen

(Campbell et al., 1995). It is also worth noting that secreted conjunctival mucins have been shown to possess a unique oligosaccharide pattern containing Tn and sialosyl-Tn structures (Berry et al., 1996), indicating normal roles in human ocular mucins for these antigens, which are disease markers in other tissues.

The mechanism of increased expression of shorter or prematurely terminated oligosaccharide side chains (such as sialyl-TF, TF, Tn, and sialosyl-Tn) in premalignant and malignant epithelial tissue may involve altered glycosyltransferase expression. Evidence to support this has come from studies of colorectal and breast carcinoma cell lines. Enzyme activity and mRNA studies have demonstrated changes in the profile of glycosyltransferases related to aberrant glycosylation of MUC1 seen in breast cancer, with loss of the core 2 branching enzyme (β1,6 GlcNAc-transferase) and increased α-2,3-sialyltransferase activity resulting in an increase in sialylated core 1 structures being found in the MUC1 glycans (Brockhausen et al., 1995; Brockhausen 1997; Whitehouse et al., 1997). Data also suggest that in addition to shorter oligosaccharide side chains on MUC1 in breast carcinomas, fewer sites may be glycosylated (Lloyd et al., 1996). In human colon carcinoma and adenoma cell lines, decreased activity of the core 3 GlcNAcβ1$\rightarrow$3GalNAc-transferase is associated with increased expression of the TF (core 1) and Tn antigens in colon cancer mucins (Vavasseur et al., 1994, 1995; King et al., 1994). However, currently very little is known about the biochemical mechanisms regulating the relevant glycosyltransferase changes responsible for the altered expression of carbohydrate antigens in cancer cells.

As in colorectal cancer and metastases (Bresalier et al., 1996), overall mucin sialylation is increased in inflammatory bowel disease (Parker et al., 1995). In addition, loss of O-acetylation of sialomucin has been demonstrated in ulcerative colitis as well as in colonic adenomas and carcinomas (Jass and Smith, 1992; Milton et al., 1993; Ogata et al., 1995). During neoplastic transformation, the loss of O-acetylation may account for the increased expression of antigenic epitopes of sialosyl-Tn, sialyl-Le[x], and sialyl-Le[a] observed in colon cancer (Ogata et al., 1995), and a more recent study has demonstrated that decreased sialyl-Le[x] O-acetylation on mucins correlates with colorectal carcinoma progression (Mann et al., 1997). Whether increased expression and activity of O-acetyltransferase occur in cancerous tissues and just how far the sialic acid de-O-acetylation facilitates the tumor cell capacity to metastasize are as yet undetermined. It is worth noting, however, that studies have demonstrated a genetic inheritance of O-acetylation (Fuller et al., 1990). Another member of the sialic acid family, N-glycolylneuraminic acid (NeuGc), until recently not thought to occur on human epithelial mucins and cell surface glycoproteins, has been demonstrated to occur on breast carcinoma MUC1 mucin as the novel core-type sialyl-T(TF) antigen, Galβ1–3{NeuGcα2–6}GalNAcα (Devine et al., 1991; Hanisch et al., 1996).

C. Altered Mucin Sulfation

Aberrant sulfation of epithelial mucins and cell surface glycoconjugates may play an important pathophysiological role by altering mucus function, cell-cell interactions, and cell-pathogen interactions. In the gut, increased sulfation is seen in gastric intestinal metaplasia (Jass and Filipe, 1981) often associated with *Helicobacter pylori* colonization (Craanen et al., 1992) and in the neocolonic epithelium that forms in the human ileum after surgical pouch reconstruction (Corfield et al., 1992), presumably as a response to bacterial colonization. In contrast, the expression of sulfomucins is decreased in colorectal cancer progression (Vavasseur et al., 1994; Matsushita et al., 1995), and studies by ourselves and others have shown that mucosal samples from the colons of European patients with inflammatory bowel disease (ulcerative colitis) incorporate less sulfate into their mucins than those of non–inflammatory bowel disease control subjects (Raouf et al., 1992; Corfield et al., 1996), supporting earlier histochemical evidence of reduced mucin sulfation (Filipe, 1979). Another study has also shown that although they have similar glycosylation changes, South Asians with colitis, unlike their European counterparts, do not have reduced sulfomucin (Probert et al., 1995), and it has been speculated that this might be related to their apparent low risk for colitis-associated colon cancer. The molecular basis of the defect in sulfomucin has not yet been established, although it is generally assumed that this reduced mucin sulfation may be secondary to the disease in parallel with the glycosylation changes that have been demonstrated. However, a pattern is beginning to emerge that makes more plausible the hypothesis that altered structure of O-linked oligosaccharides might be a primary genetic factor in inflammatory bowel disease, with reduced oligosaccharide sulfation as the prime suspect for this genetic factor in Europeans, analogous to secretor (*Se*) gene-mediated alterations in mucosal fucosylation (Rhodes, 1996). This hypothesis is supported by the studies showing an altered ion exchange profile of colonic mucins from unaffected monozygous twins of ulcerative colitis patients (Tysk et al., 1991). In addition, studies from Inka Brockhausen's group have shown the core 1 TF antigen to be a major site for ester sulfation in the colonic epithelium (Vavasseur et al., 1994; Kuhns et al., 1995) and this raises the interesting possibility that reduced activity of the core 1 Galβ1–3GalNAcα-sulfotransferase could provide an alternative explanation for the increased TF antigen expression seen in ulcerative colitis (Campbell et al., 1995) and also for the decrease in mucin sulfation (Raouf et al., 1992). Our own studies have demonstrated that core 1 structure can be revealed in normal colon by mild acid hydrolysis (Campbell et al., 1995) but neither significantly by sialidase nor fucosidase treatment (Campbell, unpublished observations), and this would be in keeping with its concealment by ester sulfation.

In studies outside the gastrointestinal tract, particularly of breast and airway disease, alterations in mucin sulfation have also been demonstrated. In breast cancer cells, mucin sulfotransferase activity is reduced compared with normal mammary cells (Brockhausen, 1997). In cystic fibrosis lung disease, studies have demonstrated increased rates of mucin synthesis in nasal mucosa of cystic fibrosis with higher levels of sulfation (Frates et al., 1983; Cheng et al., 1989). It has been postulated that the increased sulfomucin seen in cystic fibrosis might be the result of altered pH in intracellular organelles including the Golgi network, because the cystic fibrosis gene product is a chloride ion channel (Riordan et al., 1989).

It is also worth noting that biological roles as ligands for cell adhesion molecules [sulfated Lewis blood group structures are the preferred ligands for the selectins (Yuen et al., 1992; Green et al., 1995)] have been described for sulfated oligosaccharides of the type predicted to occur on colonic and respiratory mucins (Crottet et al., 1996), suggesting that soluble mucus oligosaccharides may be more directly involved in cell–cell and cell–matrix interactions, leukocyte trafficking, and epithelial malignancy.

VII. LECTIN-MUCUS INTERACTIONS

Lectins (of plant, mammalian, and microorganism origin) are carbohydrate-binding proteins or glycoproteins of nonimmune origin that possess high specific affinity for carbohydrates and are sensitive to conformation of the glycan chain. As such, lectins have been extensively used as histochemical probes to demonstrate changes in epithelial oligosaccharide structure in malignant and premalignant disease states, particularly but not exclusively in the colon (Lance and Lev, 1991). The normal adult colon expresses receptors for wheat germ agglutinin (WGA), which binds to GlcNAc and NeuAc (sialic acid), and receptors for concanavalin A (Con A) (mannose and glucose binding) are present in the proximal but not distal colon. *Ulex europaeus* (UEA1) (fucose binding) and peanut agglutinin (PNA) [Galβ1–3GalNAcα (TF) binding] show little or no binding in the normal colon; however, in malignant and hyperplastic colonic epithelia binding is greatly increased (Jacobs and Huber, 1985; Rhodes et al., 1986, 1988; Boland,

1988). Despite these observations and the known presence of lectin-containing foods in the diet, it was not previously known whether this abnormal glycoconjugate expression had functional implications. However, studies have proposed that increased core 1 TF antigen expression, not only on intestinal mucins but also on O-linked epithelial cell surface glycoproteins, may allow interaction with dietary lectins, such as PNA. This lectin, like many others, is highly resistant to cooking or digestion and is present in active form in the colonic lumen after ingestion (Pusztai et al., 1990). Interaction with such lectins when TF antigen is expressed has the potential to have marked effects on epithelial proliferation both in vitro (Ryder et al., 1992, 1994a, 1994b; Yu et al., 1993) and in vivo (Ryder et al., 1998), which may be quantitatively as important as interaction with growth factors in determining the increased proliferation that occurs in hyperplasia and as a prelude to malignant change (Rhodes et al., 1996).

More recently, lectins have attracted interest as bioadhesive agents because of their high specificity for membrane-bound receptors on epithelial cells (to be discussed in later chapters). Briefly, lectins that recognize cell surface carbohydrates have been used as agents mediating drug transport across the intestinal epithelium by carbohydrate-mediated endocytosis (Lehr and Lee, 1993; Naisbett and Woodley, 1994) or lectin-conjugated liposome (proteoliposome) targeting (Jones, 1994). Indeed, lectins have even been used for gene vector transfer to airway epithelial cells in which the transfected cell can express and may therefore be of significant relevance to future gene therapy of cystic fibrotic airway epithelial cells (Yin and Cheng, 1996). In addition, lectins detected in animals and microorganisms have been shown to be involved in adhesion of symbiotic and pathogenic bacteria to host mucosae, in specific adhesion of tumor cells to organ cells in metastatic spread, and in certain interactions with the immune system (see Beuth et al., 1995). Thus, a greater understanding of the specific changes in mucus-secreting cell and absorptive epithelial cell surface glycoconjugates, their alterations in epithelial disease states, and their lectin-binding capacity may in the future allow not only modulation of growth of the hyperplastic and neoplastic epithelia but also inhibition of both the metastatic process and bacterial adhesion by blocking with lectins specific for appropriate oligosaccharides or glycoconjugates.

REFERENCES

Asker N, Baeckstrom D, Axelsson MA, Carlstedt I, Hansson GC. 1995. The human MUC2 mucin apoprotein appears to dimerise before O-glycosylation and shares epitopes with the insoluble mucin of rat small intestine. Biochem J 308: 873–880.

Aubert JP, Porchet N, Crepin M, Duterque-Coquillaud M, Vergnes G, Mazzuca M, Debuire B, Petitprez D, Degand P. 1991. Evidence for different human tracheobronchial mucin peptides deduced from nucleotide cDNA sequences. Am J Respir Cell Mol Biol 5:178–185.

Audie JP, Janin A, Porchet N, Copin MC, Gosselin B, Aubert JP. 1993. Expression of human mucin genes in respiratory, digestive and reproductive tracts ascertained by in situ hybridisation. J Histochem Cytochem 41:1479–1485.

Augeron C, Voisin T, Laboisse CL. 1992. Neurotensin and neuromedin N stimulate mucin output from human goblet cells (Cl.16E) via neurotensin receptors. Am J Physiol 262:G470–G476.

Babczinski P. 1980. Evidence against the participation of lipid intermediates in the in vitro biosynthesis of serine (threonine) N-acetyl D galactosamine linkages in sub-maxillary mucin. FEBS Lett 117:207–211.

Baeckstrom D, Hansson GC. 1996. The transcripts of the apomucin genes *MUC2*, *MUC4*, and *MUC5AC* are large and appear as distinct bands. Glycoconj J 13: 833–837.

Baeckstrom D, Zhang K, Asker N, Ruetsch UEM, Hansson GC. 1995. Expression of the leukocyte-associated sialoglycoprotein CD43 by a colon carcinoma cell line. J Biol Chem 270:13688–13692.

Balague C, Audie J-P, Porchet N, Real FX. 1995. In situ hybridisation shows distinct patterns of mucin gene expression in normal, benign and malignant pancreas tissues. Gastroenterology 109:953–964.

Bara J, Imberty A, Perez S, Imai K, Yachi A, Oriol R. 1993. A fucose residue can mask the MUC1 epitopes in normal and cancerous gastric mucosae. Int J Cancer 54:607–613.

Bartman AE, Toribara NW, Aubert JP, Neihans GA, Kim YS, Crabtree JE, Ho SB. 1997. *MUC6* secretory mucin gene expression in tissues exposed to endogenous mucosal damaging agents (abstr). Gastroenterology 112:A861.

Bennett EP, Hassan H, Clausen H. 1996. cDNA cloning and expression of a novel human UDP-GalNAc N-acetylgalactosaminyltransferase, GalNAc-T3. J Biol Chem 271:17006–17012.

Berry M, Ellingham RB, Corfield AP. 1996. Polydispersity of normal human conjuctival mucins. Invest Ophthalmol Vis Sci 37:2559–2571.

Beuth J, Ko HL, Pulverer G, Uhlenbruck G, Pichlmaier H. 1995. Importance of lectins for the prevention of bacterial infections and cancer metastasis: Glycopinion mini-review. Glycoconj J 12:1–6.

Bhaskar KR, Gong DH, Bansil R, Pajevic S, Hamilton JA, Turner BS, LaMont JT. 1991. Profound increase in viscosity and aggregation of pig gastric mucin at low pH. Am J Physiol 261:G827–832.

Biol MC, Martin A, Louisot P. 1992. Nutritional and developmental regulation of glycosylation processes in digestive organs. Biochimie 74(1):13–24.

Bobek LA, Tsai H, Biesbrock AR, Levine MJ. 1993. Molecular cloning, sequence, and specificity of expression of the gene encoding the low molecular weight human salivary mucin (*MUC7*). J Biol Chem 268:20563–20569.

Boland CR. 1988. Lectin histochemistry in colorectal polyps. Prog Clin Biol Res 279:277–287.

Bou-Hanna C, Berthon B, Combettes L, Claret M, Laboisse CL. 1994. Role of calcium in carbachol- and neurotensin-induced mucin exocytosis in a human colonic goblet cell line and cross talk with the cAMP pathway. Biochem J 299:579–585.

Breg J, van Halbeek H, Vliengenthart JFG, Klein A, Lamblin G, Roussel P. 1988. Primary structure of neutral oligosaccharides derived from respiratory mucus glycoproteins of a patient with bronchiectasis, determined by combination of 500-MHz ^{1}H-NMR spectroscopy and quantitative sugar analysis. 2. Structure of 19 oligosaccharides having the GlcNAcβ1–3GalNAcol core (type 3) or the GlcNAcβ1–3(GlcNAcβ1–6)GalNAcol core (type 4). Eur J Biochem 171: 643–654.

Bresalier RS, Ho SB, Schoeppner HL, Kim YS, Sleisenger MH, Brodt P, Byrd JC. 1996. Enhanced sialylation of mucin-associated carbohydrate structures in human colon cancer metastasis. Gastroenterology 110:1354–1367.

Brockhausen I. 1997. Biosynthesis and functions of O-glycans and regulation of mucin antigen expression in cancer. Biochem Soc Trans 25:871–874.

Brockhausen I, Toki D, Brockhausen J, Peters S, Bielfeldt T, Kleen A, Paulsen H, Meldel M, Hagen F, Tabak LA. 1996. Specificity of O-glycosylation by bovine colostrum UDP-GalNAc:polypeptide α-N-acetylgalactosaminyltransferase using synthetic glycopeptide substrates. Glycoconj J 13:849–856.

Brockhausen I, Yang JM, Burchell J, Whitehouse C, Taylor-Papadimitriou J. 1995. Mechanisms underlying aberrant glycosylation of MUC1 mucin in breast cancer cells. Eur J Biochem 233:607–617.

Brown JF, Keates AC, Hanson PJ, Whittle BJR. 1993. Nitric oxide generators and cGMP stimulate mucus secretion by rat gastric mucosal cells. Am J Physiol 265:G418–422.

Buisine MP, Desreumaux P, Gambiez L, Aubert JP, Colombel JF, Porchet N. 1997. Expression of MUC3, MUC4 and MUC5B mRNA is decreased in the ileal mucosa in Crohn's disease (abstr). Gastroenterology 112:A943.

Buisine MP, Janin A, Manunoury V, Audie JP, Delescaut MP, Copin ML, Colobel JF, Degand P, Aubert JP, Porchet N. 1996. Aberrant expression of a human mucin gene (MUC5AC) in rectosigmoid villous adenomas. Gastroenterology 110:84–91.

Campbell BJ, Finnie IA, Hounsell EF, Rhodes JM. 1995. Direct demonstration of increased expression of Thomsen-Friedenreich (TF) antigen in colonic adenocarcinoma and ulcerative colitis mucin and its concealment in normal mucin. J Clin Invest 95:571–576.

Campbell BJ, Jenkinson MD, Yu LG, Rhodes JM. 1997. Presence of a fMLP-receptor, stimulation of mucin synthesis and release of IL-8 by fMLP in human mucus-secreting gastrointestinal epithelial cell-lines (abstr). Gastroenterology 112:A543.

Carlstedt I, Hermann A, Karlson HH, Sheehan JK, Fransson LA, Hansson GC. 1993. Characterisation of two different glycosylated domains from the insoluble mucin complex of rat small intestine. J Biol Chem 268:18771–18781.

Carter SR, Slomiany A, Gwozdzinski K, Liau YH, Slomiany BL. 1988. Enzymatic sulfation of mucus glycoprotein in gastric mucosa. J Biol Chem 263:11977–11984.

Carvalho F, Seruca R, David L, Amorim A, Seixas M, Bennet E, Clausen H, Sobrinho-Simoes M. 1997. *MUC1* gene polymorphism and gastric cancer—an epidemiological study. Glycoconj J 14:107–111.

Castagliulo I, LaMont JT, Qiu B, Fleming SM, Bhaskar KR, Nikulasson ST, Kornetsky C, Pothoulakis C. 1997. Acute stress causes mucin release from rat colon: Role of corticotropin releasing factor and mast cells. Am J Physiol 271: G884–892.

Chadee K, Keller K, Forstner J, Innes DJ, Ravdin JI. 1991. Mucin and non-mucin secretagogue activity of *Entamoeba histolytica* and cholera toxin in rat colon. Gastroenterology 100:986–997.

Chai W, Housell EF, Cashmore GC, Rosankiewicz JR, Bauer CJ, Feeney J, Feizi T, Lawson A. 1992. Neutral oligosaccharides of bovine submaxillary mucin: A combined mass spectrometry and ^{1}H-NMR study. Eur J Biochem 203:257–268.

Chandrasekaran EV, Jain RK, Vig R, Matta KL. 1997. The enzymatic sulfation of glycoprotein carbohydrate units: Blood group T-hapten specific and two other distinct Gal:3-*O*-sulfotransferases as evident from specificities and kinetics and the influence of sulfate and fucose residues occurring in the carbohydrate chain on C-3 sulfation of terminal Gal. Glycobiology 7:753–768.

Chang SK, Dohrman AF, Basbaum CB, Ho SB, Tsuda T, Toribara NW, Gum JR, Kim YS. 1994a. Localization of mucin (MUC2 and MUC3) messenger RNA and peptide expression in human normal intestine and colon cancer. Gastroenterology 107:28–36.

Chang SK, Park ES, Chung WS, Son SW, Park SM, Gum JR, Kim YS. 1994b. Expression of MUC2, MUC3 apomucins and mRNA in ulcerative colitis. Gastroenterology 106:662.

Cheng PW, Boat TR, Cranfill K, Yankaskas JR, Boucher RC. 1989. Increased sulfation of glycoconjugates by cultured nasal epithelial cells from patients with cystic fibrosis. J Clin Invest 84:68–72.

Chu WN, Baxter JD, Reudelhuer TL. 1990. A targeting sequence for dense secretory granules resides in the active renin protein moiety of human preprorenin. Mol Endocrinol 90:1905–1913.

Clausen H, Bennett EP. 1996. A family of UDP-GalNAc:polypeptide *N*-acetylgalactosaminyl-transferases control the initiation of mucin-type O-linked glycosylation. Glycobiology 6:635–646.

Cohan VL, Scott AL, Dinarello CA, Prendergast RA. Interleukin-1 is a mucus secretagogue. Cell Immunol 136:425–434.

Colley KJ. 1997. Golgi localization of glycosyltransferases: More questions than answers. Glycobiology 7(1):1–13.

Corfield AP, Myerscough N, Bradfield N, Do Amaral-Corfield C, Gough M, Clamp JR, Durdey P, Warren B, Bartolo DCC, King K, Williams JM. 1996. Colonic mucins in ulcerative colitis: Evidence for loss of sulphation. Glycoconj J 13: 809–822.

Corfield AP, Wagner SA, O'Donnell LJD, Durdey P, Mountford RA, Clamp JR. 1993. The roles of enteric bacterial sialidase, sialate O-acetyl esterase and glycosulfatase in the degradation of human colinic mucin Glycoconj J 10: 72–81.

Corfield AP, Warren B, Bartolo DCC, Wagner SA, Clamp JR. 1992. Mucin changes in ileoanal pouches monitored by metabolic labelling and histochemistry. Br J Surg 79:1209–1212.

Craanen ME, Blok P, Dekker W, Ferwerda J, Tygat GNJ. 1992. Subtypes of intestinal metaplasia and *Helicobacter pylori*. Gut 33:597–600.

Crottet P, Kim YJ, Varki A. 1996. Subset of sialylated, sulfated mucins of diverse origins are recognised by L-selectin. Lack of evidence for unique oligosaccharide sequences mediating binding. Glycobiology 6:191–208.

D'Cruz OJ, Dunn TS, Pichan P, Hass GG, Sachdev GP. 1996. Antigenic cross-reactivity of human tracheal mucin with human sperm and trophoblasts correlates with the expression of mucin 8 gene messenger RNA in reproductive tissues. Fertil Steril 66:316–326.

De Bolos C, Garrido M, Real FX. 1995. MUC6 apomucin shows a distinct normal tissue distribution that correlates with Lewis antigen expression in the human stomach. Gastroenterology 109:723–734.

Desseyn JL, Guyonnet-Duperat V, Porchet N, Aubert JP, Laine A. 1997. Human mucin gene *MUC5B*, the 10.7kb large central exon encodes various alternative subdomains resulting in a super repeat. Structural evidence for a 11p15.5 gene family. J Biol Chem 272:3168–3178.

Devine PL, Clark BA, Birrell GW, Layton GT, Ward BG, Alewood PF, McKenzie IFC. 1991. The breast tumour–associated epitope defined by monclonal antibody 3E1.2 is an O-linked mucin carbohydrate containing N-glycolylneuraminic acid. Cancer Res 51:5826–5836.

Dufosse J, Porchet N, Audie JP, Guyonnet-Duperat V, Laine A, Van Seuningen I, Marrakchi S, Degand P, Aubert JP. 1993. Degenerate 87 base pair tandem repeats create hydrophilic/hydrophobic alternating domains in human peptide mucins mapped to 11p15. Biochem J 293:329–337.

Elhammer AP, Poorman RA, Brown E, Maggiora LL, Hoogerheide JG, Kezdy JF. 1993. The specificity of UDP-GalNAc:polypeptide N-acetyl-galactosaminyl-transferase as inferred from a database of in vivo substrates and from the in vitro glycosylation of proteins and peptides. J Biol Chem 268:10029–10038.

Epple HJ, Kreusel KM, Hanski C, Schulke JD, Riecken EO, Fromm M. 1997. Differential stimulation of intestinal mucin secretion by cholera toxin and carbachol. Pflugers Arch 433:638–647.

Feeney J, Frienkel TA, Hounsell EF. 1986. Complete 1-H-NMR assignments for two core-region oligosaccharides of human meconium glycoproteins, using 1D and 2D methods at 500 MHz. Carbohydr Res 152:63–72.

Field MC, Wainwright LJ. 1995. Molecular cloning of eukaryotic glycoprotein and glycolipid glycosyltransferases: A survey. Glycobiology 5:463–472.

Filipe I. 1979. Mucins in the human gastrointestinal epithelium: A review. Invest Cell Pathol 2:195–216.

Finnie IA, Campbell BJ, Taylor BA, Milton JD, Sadek SK, Yu L-G, Rhodes JM. 1996. Stimulation of colonic mucin synthesis by corticosteroids and nicotine. Clin Sci 91:359–364.

Finnie IA, Dwarakanath AD, Taylor BA, Rhodes JM. 1995. Colonic mucin synthesis is increased by sodium butyrate. Gut 36:93–99.

Fleming N, Maellow L, Bhullar D. 1992. Regulation of the cAMP signal transduction pathway by protein kinase C in rat submandibular cells. Eur J Physiol 421:82–90.

Forstner G. 1995. Signal transduction, packaging and secretion of mucins. Annu Rev Physiol 57:858–605.

Forstner G, Zhang Y, McCool D, Forstner J. 1994. Regulation of mucin secretion by T84 adenocarcinoma cells by forskolin: Relationship to Ca^{2+} and PKC. Am J Physiol 266:G1096–G1102.

Forstner JF, Forstner G. 1994. Gastrointestinal mucus. In: Johnson LR, ed. Physiology of the Gastrointestinal Tract. New York: Raven Press, pp 1255–1283.

Forstner JF, Roomi NW, Fahim REF, Forstner GG. 1981. Cholera toxin stimulates secretion of immunoreactive intestinal mucin. Am J Physiol 240:G10–G16.

Fox MF, Lahbib F, Pratt W, Attwood J, Gum JR, Kim YS, Swallow DM. 1992. Regional localisation of the intestinal mucin gene MUC3 to chromosome 7q22. Ann Hum Genet 56:281.

Frates RC, Kaizu TT, Last JA. 1983. Mucus glycoproteins secreted by respiratory epithelial tissue from cystic fibrosis patients. Pediatr Res 17:30–34.

Fukuda M. 1996. Possible roles of tumour-associated carbohydrate antigens. Cancer Res 56:2237–2244.

Fuller CE, Davies RP, Williams GT, Williams ED. 1990. Crypt restricted heterogeneity of goblet cell mucus glycoprotein in histologically normal human colonic mucosa: A potential marker of somatic mutation. Br J Cancer 61:382–384.

Fung DCK, Beacock DJ, Richardson PS. 1992. Vagal control of mucus glyconjugate secretion into the feline trachea. J Physiol (Lond) 453:435–447.

Gambús G, De Bolos C, Andreu D, Franci C, Egea G, Real FX. 1993. Detection of the MUC2 apomucin tandem repeat with a mouse monoclonal antibody. Gastroenterology 104:93–102.

Gendler SJ, Lancaster CA, Taylor-Papadimitriou J, Duhig T, Peat N, Burchell J, Pemberton L, Lalani EN, Wilson D. 1990. Molecular cloning and expression of the human tumour-associated polymorphic epithelial mucin (PEM). J Biol Chem 265:15286–15293.

Gendler SJ, Spicer AP. 1995. Epithelial mucin genes. Annu Rev Physiol 57:607–634.

Gendler SJ, Spicer AP, Lalani EN, Duhig T, Peat N. 1991. Structure and biology of a carcinoma-associated mucin, MUC1. Am Rev Respir Dis 144:S42–47.

Gerken TA, Owens CL, Pasumarthy M. 1997. Determination of the site specific O-glycosylation pattern of the porcine submaxillary mucin tandem repeat glycopeptide. J Biol Chem 272:9709–9719.

Gleeson PA, Teasdale RD, Burke J. 1994. Targeting of proteins to the Golgi apparatus. Glycoconj J 11:381–394.

Gong DH, Turner B, Bhaskar KR, Lamont JT. 1991. Lipid binding to gastric mucin: Protective effect against oxygen free radicals. Am J Physiol 259:G681–686.

Gorodeski GT, Jin W, Hopfer U. 1997. Extracellular Ca^{2+} directly regulates tight junctional permeability in the human cervical cell line CaSki. Am J Physiol 272:C511–524.

Granovsky M, Bielfeldt T, Peters S, Paulson H, Medal M, Brockhausen J, Brockhausen I. 1994. UDP galactose:glycoprotein-N-acetylgalactosamine 3-beta-D-galactosyltransferase activity synthesising O-glycan core 1 is controlled by the amino acid sequence and glycosylation of glycopeptide substrates. Eur J Biochem 221:1039–1046.

Green PJ, Yuen CT, Childs RA, Chai W, et al. 1995. Further studies on the binding specificity of the leukocyte adhesion molecule L-selectin, towards oligosaccharides—suggesting a link between the selectin- and the integrin-mediated lymphocyte adhesion systems. Glycobiology 5:29–38.

Griffiths B, Matthews DJ, West L, Attwood J, Povey S, Swallow DM, Gum JR, Kim YS. 1990. Assignment of the polymorphic intestinal mucin gene (*MUC2*) to chromosome 11p15. Ann Hum Genet 54:277–285.

Gross MS, Guyonnet-Duperat V, Porchet N, Bernheim A, Aubert JP, Nguyen VC. 1992. Mucin 4 (*MUC4*) gene: Regional assignment (3q29) and RFLP analysis. Ann Genet 35:21–26.

Gum JR, Byrd JC, Hicks JW, Toribara NW, Lamport DTA, Siddiki B, Kim YS. 1989. Molecular cloning of human intestinal cDNAs. Sequence analysis and evidence for genetic polymorphism. J Biol Chem 264:6480–6487.

Gum JR, Hicks JW, Swallow DM, Lagace RL, Byrd JC, Lamport DTA, Siddiki B, Kim YS. 1990. Molecular cloning of cDNAs derived from a novel human intestinal mucin gene. Biochem Biophys Res Commun 171:407–415.

Gum JR, Hicks JW, Toribara NW, Rothe EM, Lagace RE, Kim YS. 1992. The human MUC2 intestinal mucin has cysteine-rich subdomains located both upstream and downstream of its central repetitive region. J Biol Chem 267:21375–21383.

Gum JR, Hicks JW, Toribara NW, Siddiki B, Kim YS. 1994. Molecular cloning of human intestinal mucin (MUC2) cDNA. Identification of the amino-terminus and overall sequence similarity to prepro-von Willebrand factor. J Biol Chem 269:2440–2446.

Guyonnet-Duperat V, Audie JP, Debailleul V, Laine A, Buisine MP, Galiegue-Zouitina S, Pigny P, Degand P, Aubert JP, Porchet N. 1995. Characterisation of the human mucin gene *MUC5AC*: A consensus cysteine rich domain for 11p15 mucin genes? Biochem J 305:211–219.

Hakamori S. 1989. Aberrant glycosylation in tumors and tumor-associated carbohydrate antigens. Adv Cancer Res 52:718–723.

Hanisch F-G, Chai W, Rosankiewicz, Lawson AM, Stoll, Fiezi T. 1993. Core-typing of O-linked glycans from human gastric mucins: Lack of evidence for the occurrence of the core sequence Gal1-6GalNAc. Eur J Biochem 217:645–655.

Hanisch F-G, Stadie TRE, Peter-Katalinic J. 1996. MUC1 glycoforms in breast cancer cell-line T47D as a model for carcinoma-associated alterations of O-glycosylation. Eur J Biochem 326:318–327.

Hanski C, Klussmann E, Wang J, Bohm C, Ogorek D, Hanski ML, Kruger-Krasagakes S, Eberle J, Schmitt-Graff A, Fiecken EO. 1996. Fucosyltransferase III and sialyl-Le(x) expression correlate in cultured colon carcinoma cells but not in colon carcinoma tissue. Glycoconj J 13:727–733.

Hanski C, Riede E, Gratchev A, Foss HD, Bohm C, Klussmann E, Hummel M, Mann B, Buhr HG, Stein H, Kim YS, Gum J, Riecken EO. 1997. *MUC2* gene suppression in human colorectal carcinomas and their metastases: In vitro evidence of modulatory role of DNA methylation. Lab Invest 77(6):685–695.

Hardiuin-Lepers A, Recchi M-A, Delannoy P. 1995. 1994, the year of sialyltransferases. Glycobiology 5:741–758.

Hilkens J, Wesseling J, Vos HL, Storm J, Boer B, Van der Valk SW, Maas MCE. 1995. Involvement of the cell surface–bound mucin, episialin/MUC1, in progression of human carcinomas. Biochem Soc Trans 23:822–826.

Ho SB, Ewing SL, Montgomery CK, Kim YS. 1996. Altered mucin core peptide immunoreactivity in the colon polyp-carcinoma sequence. Oncol Res 8(2): 53–61.

Ho SB, Neihans GA, Lyftogt C, Yan PS, Cherwitz DL, Gum ET, Dahiya R, Kum YS. 1993. Heterogeneity of mucin gene expression in normal and neoplastic tissues. Cancer Res 53:641–651.

Ho SB, Roberton AM, Shekels LL, Lyftogt CT, Niehans GA, Toribara NW. 1995a. Expression of gastric mucin cDNA and localisation of mucin gene expression. Gastroenterology 109:735–747.

Ho SB, Shekels LL, Toribara NW, Kim YS, Lyftogt CT, Cherwitz DL, Niehans GA. 1995b. Mucin gene expression in normal, preneoplastic and neoplastic human gastric epithelium. Cancer Res 55:2681–2690.

Hoff SD, Matsushita Y, Ota DM, Cleary KR, Yamori T, Hakamori S, Irimura I. 1989. Increased expression of sialyl-dimeric Le^x antigen in advanced primary colorectal carcinomas and liver metastases. Cancer Res. 49:6883–6888.

Hong DH, Forstner JF, Forstner GG. 1997a. Protein kinase C-epsilon is the likely mediator of mucin exocytosis in human colonic cell lines. Am J Physiol 272: G31–37.

Hong JC, Kwan J, Gum JR, Sleisenger MH, Kim YS. 1997b. Altered glycosylation and processing of MUC2 apomucin in colon cancer (abstr). Gastroenterology 112:A579.

Hooper LV, Hindsgaul O, Baenziger JU. 1995. Purification and characterisation of the GalNAc-4-sulfontransferase responsible for sulphation of GalNAcβ1–4GlcNAc–bearing oligosaccharides. J Biol Chem 270:16327–16332.

Hounsell EF, Davies MJ, Renouf DV. 1996. O-linked protein glycosylation structure and function. Glycoconj J 13(1):19–26.

Hounsell EF, Feizi T. 1982. Gastrointestinal mucins, structures and antigenicities of their carbohydrate chains in health and disease. Med Biol 60:227–236.

Hounsell EF, Lawson AM, Feeney J, Cashmore GC, Kane DP, Stoll M, Feizi T. 1988. Identification of a novel oligosaccharide backbone structure with a galactose residue monosubstituted at C-6 in human foetal gastrointestinal mucins. Biochem J 256:397–401.

Hounsell EF, Lawson AM, Feeney J, Gooi HC, Pickering N, Stoll MS, Lui SC, Feizi T. 1985. Structural analysis of the O-glycosidically linked core-region oligosaccharides of human meconium glycoproteins which express oncofetal antigens. Eur J Biochem 148:367–377.

Hounsell EF, Lawson AM, Stoll M, Kane DP, Cashmore GC, Carruthers RA, Feeney J, Feizi T. 1989. Characterisation by mass spectrometry and 500-MHz proton nuclear magnetic resonance spectroscopy of penta- and hexa-saccharide chains of human foetal gastrointestinal mucins (meconium glycoproteins). Eur J Biochem 186:597–610.

Hounsell EF, Young M, Davies MJ. 1997. Glycoprotein changes in tumours: A renaissance in clinical applications. Clin Sci 98:287–293.

Hutton DA, Pearson JP, Allen A, Foster SNE. 1990. Mucolysis of the colonic mucus barrier by faecal proteinases: Inhibition by interacting polyacrylate. Clin Sci 78:265–271.

Inglis SK, Corboz MR, Taylor AE, Ballard ST. 1997. Effect of anion transport inhibition on mucus secretion by airway submucosal glands. Am J Physiol 272:L372–377.

Itzkowitz SH, Bloom EJ, Kokal WA, Modin G, Hakamori S, Kim YS. 1990. Sialosyl Tn: A novel mucin antigen associated with poor prognosis in colorectal cancer patients. Cancer 66:1960–1966.

Itzkowitz SH, Bloom EJ, Lau TS, Kim YS. 1992. Mucin associated Tn and sialosyl Tn antigen expression in colorectal polyps. Gut 33:518–523.

Itzkowitz SH, Yuan M, Fukushi Y, Palekar A, Phelps PC, Shamsuddin AM, Kim YS. 1986. Lewis^x and sialylated Lewis^x-related antigen expression in human malignant and non-malignant colonic tissues. Cancer Res 46:2627–2632.

Itzkowitz SH, Yuan M, Montgomery CK, Kjeldsen T, Kakahashi HK, Bigbee WL, Kim YS. 1989. Expression of Tn, sialosyl Tn and T antigens in human colon cancer. Cancer Res. 49:197–204.

Jacobs LR, Huber PW. 1985. Regional distribution and alterations of lectin binding to colorectal mucin in mucosal biopsies from controls and subjects with inflammatory bowel disease. J Clin Invest 75:112–118.

Jass JR, Filipe MI. 1981. The mucin profile of normal gastric mucosa, intestinal metaplasia and its variants and gastric carcinoma. Histochem J 13:931–939.

Jass JR, Smith M. 1992. Sialic acid and epithelial differentiation in colorectal polyps and cancer; a morphological, mucin and lectin histochemical study. Pathology 24:233–242.

Jones MN. 1994. Carbohydrate-mediated liposomal targeting and drug delivery. Adv Drug Delivery Res 13:215–250.

Kaftan SM, Wright NA. 1989. Studies on the mechanisms of mucous cell depletion in experimental colitis. J Pathol 159:75–85.

Karlsson NG, Johansson ME, Asker N, Karlsson H, Gendler SJ, Carlstedt I, Hansson GC. 1996. Molecular characterisation of the large heavily glycosylated domain glycopeptide from the rat small intestinal Muc2 mucin. Glycoconj J 13:823–831.

Keller K, Olivier M, Chadee K. 1992. The fast release of mucin secretion from human colonic cells induced by *Entamoeba histolytica* is dependent on contact and protein kinase C activation. Arch Med Res 23:217–221.

Kelly SM, Hunter JO, 1990. Epidermal growth factor stimulates synthesis and secretion of mucus glycoproteins in human gastric mucosa. Clin Sci 79:425–427.

Kim YS, Gum J, Brockhausen I. 1996. Mucin glycoproteins in neoplasia. Glycoconj J 13:693–707.

Kim YS, Yuan M, Itzkovitz SH, Sun Q, Kaizu T, Pelakar A, et al. 1986. Expression of Le^y and extended Le^y blood group–related antigens in human malignant, premalignant and non-malignant colonic tissues. Cancer Res 46:5985–5992.

King M-J, Chan A, Roe R, Warren BF, Dell A, Morris HR, Bartolo DCC, Durdey P, Corfield AP. 1994. Two different glycosyltransferase defects that result in GalNAcα-*O*-peptide (Tn) expression. Glycobiology 4(3):267–279.

Klomp LW, van Rens L, Strous GJ. 1995. Cloning and analysis of human gastric mucin cDNA reveals two types of conserved cysteine rich domains. Biochem J 308:831–838.

Kopp R, Lambrecht G, Mutschler E, Moser U, Tacke R, Pfeiffer A. 1989. HT29 colon carcinoma cells contain muscarinic M3 receptors coupled to phosphoinositide metabolism. Eur J Pharmacol 172:397–405.

Kovarik A, Peat N, Wilson D, Gendler SJ, Taylor-Papadimitriou J. 1993. Analysis of the tissue-specific promoter of the *MUC1* gene. J Biol Chem 268:9917–9926.

Kuhns W, Jain RK, Matta KL, Paulsen H, Baker MA, Geyer R, Brockhausen I. 1995. Characterisation of a novel mucin sulphotransferase activity synthesising sulphated O-glycan 1,3-sulphate-Galβ1–3GalNAcα-R. Glycobiology 5:689–697.

Kumoro Y, Ishihara K, Ohara S, Saigenji K, Hotta K. 1992. Effects of tetragastrin on mucus glycoprotein in rat gastric mucosal protection. Gastroenterol Jpn 27:597–603.

Laberthe MA, Augeron C, Rouyer-Fessard C, Roumagnac I, Maoret JJ, Grasset E, Laboisse C. 1989. Functional VIP receptors in the human mucin-secreting colonic epithelial cell-line CL.16E. Am J Physiol 256:G443–G450.

Lake AM, Bloch KJ, Sinclair KJ, Walker W. 1980. Anaphylactic release of intestinal goblet cell mucous. Immunology 39:173–178.

Lamblin G, Rahmoune H, Wieruszeski JM, Lhermitte M, Strecker G, Roussel P. 1991. Structure of two sulphated oligosaccharides from respiratory mucins of a patient suffering from cystic fibrosis. Biochem J 275:199–206.

Lamont JT. 1992. Mucus: The front line of intestinal mucosal defence. Ann N Y Acad Sci 664:190–201.

Lan MS, Batra SK, Qi WN, Metzar RS, Hollingsworth MA. 1990. Cloning and sequencing of a human pancreatic tumor mucin cDNA. J Biol Chem 265:15294–15299.

Lance P, Lev R. 1991. Colonic oligosaccharide structures deduced from lectin-binding studies before and after desialylation. Hum Pathol 22:307–312.

Lapertosa G, Fulcheri E, Acquarone M, Filipe MI. 1984. Mucin profiles in the mucosa adjacent to large bowel nonadenocarcinoma neoplasias. Histopathology 8:805–811.

Lehr CM, Lee VHL. 1993. Binding and transport of some bioadhesive plant lectins across Caco-2 cell monolayers. Pharm Res 12:1796–1799.

Leiper K, Helliwell TR, Sadek SK, Milton JD, Yu LG, Democratis J, Rhodes JM. 1997. What is goblet cell depletion (abstr)? Gut 41(suppl 3):A114, P149A.

Lesuffleur T, Roche, F, Hill AS, Lacasa M, Fox M, Swallow DM, Zweibaum A, Real FX. 1995. Characterisation of a mucin cDNA clone isolated from HT29 mucus secreting cells. The 3′ end of MUC5AC? J Biol Chem 270:13665–13673.

Li D, Wang D, Majumdar S, Jany B, Durham SR, Cottrell J, Caplen N, Geddes DM, Alton EW, Jeffery PK. 1997a. Localisation and up-regulation of mucin (MUC2) gene expression in nasal biopsies of patients with cystic fibrosis. J Pathol 181:305–310.

Li JD, Dohrman AF, Gallup M, Miyata S, Gum JR, Kim YS, Nadel JA, Prince A, Basbaum CB. 1997b. Transcriptional activation of mucin by Pseudomonas aeruginosa lipopolysaccharide in the pathogenesis of cystic fibrosis lung disease. Proc Natl Acad Sci U S A 94:967–972.

Li M, Andersen V, Lance P. 1995. Expression and regulation of glycosyltransferases for N-glycosyl oligosaccharides of fresh human surgical and murine tissue and cell lines. Clin Sci 89:397–404.

Lloyd K, Burchell J, Kudryashov V, Yin BWT, Taylor-Papadimitriou J. 1996. Comparison of O-linked chains in MUC1 mucin from normal breast epithelial cell-lines and breast carcinoma cell-lines. Demonstration of simpler and fewer glycan chains in tumour cells. J Biol Chem 271:33325–33334.

Lo-Guidice J-M, Herz H, Lamblin G, Plancke Y, Roussel P, Llhermitte M. 1997. Structures of sulfated oligosaccharides isolated from the respiratory mucins of a non-secretor (O,Le^{a+b-}) patient suffering from chronic bronchitis. Glycoconj J 14:113–125.

Lo-Guidice J-M, Perini JM, Lafitte JJ, Decourouble MP, Roussel P, Lamblin G. 1995. Characterisation of a sulfotransferase from human airways responsible for the 3-O-sulfation of terminal galactose in N-acetyllactosamine–containing mucin carbohydrate chains. J Biol Chem 270:27544–27550.

Lo-Guidice J-M, Wieruszeski JM, Lemoine J, Verbert A, Roussel P, Lamblin G. 1994. Sialylation and sulfation of the carbohydrate chains in respiratory mucin from a patient with cystic fibrosis. J Biol Chem 269:18794–18813.

Lucas WB, Keates AC, Kelly CP, Offiner GD, Nunes DP. 1997. Expression of mucin and trefoil protein genes in inflammatory bowel disease (abstr). Gastroenterology 112:A1029.

Mann B, Klussman E, Vandamme-Feldhaus V, Hanski ML, Riecken EO, Buhr HJ, Schauer R, Hanski C. 1997. The decrease of sialyl-Le^x O-acetylation correlates with colorectal carcinoma progression (abstr). Gastroenterology 112:A609.

Mannori G, Crottet P, Cecconi O, Hanasaki K, Aruffo A, Nelson RM, Varki A, Bevilacqua MP. 1995. Differential colon cancer cell adhesion to E, P, and L selectin: Role of mucin-type glycoprotein. Cancer Res 55:4425–4431.

Marth JD. 1996. Complexity in O-linked oligosaccharide biosynthesis engendered by multiple polypeptide *N*-acetylgalactosaminyltransferases. Glycobiology 6: 701–706.

Matsumoto A, Asada S, Okumura Y, Takiuchi H, Hirata I, Ohshiba S. 1992. Effects of anti-acid secretory agents on various types of gastric mucus. J Clin Gastroenterol 14:S94–97.

Matsushita Y, Nakamori S, Seftor EA, Hendrix MJ, Irimura T. 1991. Human colon carcinoma cells with invasive capacity obtained by selection for sialyl-dimeric Le^x antigen. Exp Cell Res 196:20–25.

Matsushita Y, Yamamoto N, Shirahama H, Tanaka S, Yonezawa S, Yamori T, Irimua T, Sato E. 1995. Expression of sulfomucin in normal mucosae, colorectal adenocarcinomas and metastases. Jpn J Cancer Res 86:1060–1061.

McCool DJ, Marcon MA, Forstner JF, Forstner GG. 1990. The T84 human adenocarcinoma cell line produces mucin in culture and releases it in response to various secretagogues. Biochem J 267:491–500.

McCool DJ, Forstner JF, Forstner GG. 1994. Synthesis and secretion of mucin by the human colonic tumour cell line LS180. Biochem J 302:111–118.

Merlin D, Augeron C, Tien XY, Guo X, Laboisse CL, Hopfer U. 1994. ATP-stimulated electrolyte and mucin secretion in the human intestinal goblet cell-line HT29-CL.16E. J Membr Biol 137:137–149.

Meurer JA, Naylor JM, Baker CA, Thomsen DR, Homa FL, Elhammer AP. 1995. cDNA cloning, expression and chromosomal localisation of a human UDP-GalNAc:polypeptide, *N*-acetylgalactosaminyltransferase. J Biochem 118:568–574.

Milton JD, Eccleston D, Parker N, Raouf A, Cubbin C, Hoffman J, Hart CA, Rhodes JM. 1993. Distribution of O-acetylated sialomucin in the normal and diseases gastrointestinal tract shown by a new monoclonal antibody. J Clin Pathol. 46: 323–329.

Milton JD, Yu LG, Democratis J, Rhodes JM. 1996. Neutrophil elastase is a secretagogue for goblet differentiated human colon cancer cells (HT29-MTX abstr). Gut 38:A637.

Miyake M, Kohno N, Nudelman ED, Hakomori SI. 1989. Human IgG3 monoclonal antibody directed to an unbranched repeated type 2 chain Galβ1–4GlcNAcβ1 3Galβ1–4GlcNAcβ1–3Galβ-R which is highly expressed in colonic and hepatocellular carcinoma. Cancer Res 49:5689–5695.

Moon HW, Whipp DC, Baetz AL. 1971. The comparative effects of enterotoxins from *Escherichia coli* and *Vibrio cholerae* on rabbit and swine small intestine. Lab Invest 25:133–140.

Naisbett B, Woodley J. 1994. The potential use of tomato lectin for oral drug delivery. 2. Mechanism of uptake in vitro. Int J Pharm 110:127–136.

Nakamori S, Ota DM, Cleary KR, Shirotani K, Irimura T. 1994. MUC1 mucin expression as a marker of progression and metastasis of human colorectal carcinoma. Gastroenterology 106:353–361.

Neutra MR, O'Malley LJ, Specian RD. 1982. Regulation of intestinal goblet cell secretion. II. A survey of potential secretagogues. Am J Physiol 242:G380–387.

Nguyen PL, Niehans GA, Chewitz DL, Kim YS, Ho SB. 1996. Membrane-bound (*MUC1*) and secretory *MUC2, MUC3, MUC4* mucin gene expression in human lung cancer. Tumour Biol 17(3):176–192.

Nguyen VC, Aubert JP, Gross MS, Porchet N, Degand P, Frezal J. 1990. Assignment of human tracheobronchial mucin gene(s) to 11p15 and a tracheobronchial mucin-related sequence to chromosome 13. Hum Genet 86:167–172.

Nielsen PA, Bennett EP, Wandall H, Therkildsen MH, Hannibal J, Clausen H. 1997a. Identification of a major human high-molecular weight salivary mucin (MG1) as tracheobronchial mucin MUC5B. Glycobiology 7:413–419.

Nielsen PA, Mandel U, Therkildsen MH, Clausen H. 1996. Differential expression of human high-molecular weight salivary mucin (MG1) and low-molecular weight salivary mucin (MG2). J Dent Res 75:1820–1826.

Nielsen PA, Mandel U, Therkildsen MH, Ravn V, David L, Reiss CA, Wandall HH, Dabelsteen E, Clausen H. 1997b. Loss of a novel mucin-like epithelial glycoprotein in oral and cervical squamous cell carcinomas. Cancer Res 57:634–640.

O'Boyle KP, Zamore R, Adluri S, Cohen A, Kemeny N, Welt S, Lloyd KO, Oettgen HF, Old LJ, Livingston PO. 1992. Immunisation of colorectal cancer patients with modified ovine submaxillary gland mucin and adjuvants induces IgM and IgG antibodies to sialylated Tn. Cancer Res 52:5663–5667.

Ogata S, Ho I, Chen A, Dubois D, Maklansky J, Singhal A, Hakamori S, Itzkowitz SH. 1995. Tumour-associated sialylated antigens are constitutively expressed in normal human colonic mucosa. Cancer Res 55:1869–1874.

Ogata S, Maimonis PJ, Itzkowitz SH. 1996. Mucins bearing the cancer-associated sialosyl-Tn antigen mediate inhibition of natural killer cell cytotoxicity. Cancer Res 52:4741–4746.

Ogata S, Uehara H, Chen A, Itzkowitz SH. 1992. Mucin gene expression in colonic tissues and cell lines. Cancer Res 52:5971–5978.

Olivier MG, Specian RD. 1990. Cytoskeleton of intestinal goblet cells: Role of actin filaments in baseline secretion. Am J Physiol 259:G991–G997.

Olivier MG, Specian RD. 1991. Cytoskeleton of intestinal goblet cells: Role of microtubules in baseline secretion. Am J Physiol 260:G850–G857.

Oriol R, Mollicone R, Coullin P, Dalix M-A, Candelier J-J. Genetic regulation of the expression of ABH and Lewis antigens in tissues. APMIS Suppl 27:28–38.

Parker N, Tsai HH, Ryder SD, Raouf AH, Rhodes JM. 1995. Increased rate of sialylation of colonic mucin by cultured ulcerative colitis mucosal explants. Digestion 56:52–56.

Perez-Vilar J, Eckhardt AE, Hill RL. 1996. Porcine submaxillary mucin forms disulfide-bonded dimers between its carboxyl-terminal domains. J Biol Chem 271:9845–9850.

Phillips JA, Yu LG, Rhodes JM, Campbell BJ. 1997. Differing effects of corticosteroids on mucin synthesis by colorectal and gastric cell-lines (abstr). Gut 41(suppl 3):A255, P954A.

Phillips TE. 1992. Both crypt and villus intestinal goblet cells secrete mucin in response to cholinergic stimulation. Am J Physiol 262:G327–G331.

Pigny P, Guyonnet-Duperat V, Hill AS, Pratt WS, Galiegue-Zouitina S, d'Hooge MC, Laine A, Van-Seuningen I, Degand P, Gum JR, Kim YS, Swallow DM, Aubert JP, Porchet N. 1996. Human mucin genes assigned to 11p15.5: Identification and organisation of a cluster of genes. Genomics 38(3):340–352.

Piller F, Piller V, Fox RI, Fukuda M. 1988. Human T lymphocyte activation is associated with changes in O-glycan biosynthesis. J Biol Chem 263:15146–15150.

Porchet N, Cong NV, Dufosse J, Audie J, Guyonnet-Duperat V, Gross MS, Denis C, Degand P, Bernheim A, Aubert JP. 1991. Molecular cloning and chromosomal localisation of a novel human tracheo-bronchial mucin cDNA containing tandemly repeated sequences of 48 base pairs. Biochem Biophys Res Commun 175:514–422.

Porchet N, Pigny P, Buisine MP, Debailleul V, Degand P, Laine A, Aubert JP. 1995. Human mucin genes: Genomic organisation and expression of MUC4, MUC5AC, and MUC5B. Biochem Soc Trans 23:800–805.

Probert CSJ, Warren BF, Perry T, Mackay EH, Mayberry JF, Corfield AP. 1995. South Asian and European colitics show characteristic differences in colonic mucus glycoprotein type and turnover. Gut 36:696–702.

Pusztai A, Ewen SWB, Grant G, Peumans WJ, Van Damme EJM, Rubio L, Bardocz S. 1990. The relationship between survival and binding of plant lectins during small intestinal passage and their effectiveness as growth factors. Digestion 46(suppl 2):308–316.

Radwan KA, Oliver MG, Specian RD. 1990. Cytoarchitectural reorganisation of rabbit colonic goblet cells during baseline secretion. Anat Rec 189:365–376.

Raouf AH, Tsai HH, Parker N, Hoffman J, Walker RJ, Rhodes JM. 1992. Sulphation of colonic and rectal mucin in inflammatory bowel disease: Reduced sulphation of rectal mucus in ulcerative colitis. Clin Sci 83:623–626.

Regimbald LH, Pilarski LM, Longnecker BM, Reddish MA, Zimmermann G, Hugh JC. 1996. The breast mucin MUC1 as a novel adhesion ligand for endothelial intercellular adhesion molecule 1 in breast cancer. Cancer Res 56:4244–4249.

Rhodes JM. 1996. A unifying hypothesis for inflammatory bowel disease and associated colon cancer: Sticking the pieces together with sugar. Lancet 3487:40–44.

Rhodes JM. 1997. Mucins and inflammatory bowel disease. Q J Med 90:79–82.

Rhodes JM, Black RR, Savage A. 1986. Glycoprotein abnormalities in colonic carcinoma, adenoma and hyperplastic polyps shown by lectin peroxidase histochemistry. J Clin Pathol 39:1331–1334.

Rhodes JM, Black RR, Savage A. 1988. Altered lectin binding by colonic epithelial glycoconjugates in ulcerative colitis and Crohn's disease. Dig Dis Sci 33:1359–1363.

Rimele TJ, O'Doriso MS, Gaginella T. 1981. Evidence of muscarinic receptors on rat colonic epithelial cells: Binding of ^{3}H-quinuclidinyl benzilate. J Pharmacol Exp Ther 218:426–434.

Riordan JR, Lap-Chee Tsui, Collins FS, et al. 1989. Identification of the cystic fibrosis gene. Science 245:1059–1079.

Roberton AM, McKenzie CG, Sharfe N, Stubbs LB. 1993. A glycosulphatase that removes sulphate from mucus glycoprotein. Biochem J 239:683–689.

Roth J, Wang Y, Eckhardt AE, Hill RL. 1994. Subcellular localization of the UDP-N-acetyl-D-galactoamine:polypeptide N-acetylgalactosaminyltransferase mediated O-glycosylation reaction in the submaxillary gland. Proc Natl Acad Sci U S A 91:8935–8939.

Ryder SD, Jacnya MR, Levi AJ, Rizzi PM, Rhodes JM. 1998. Eating peanuts increases rectal proliferation in individuals with mucosal expression of the peanut lectin receptor. Gastroenterology 114:44–49.

Ryder SD, Parker N, Eccleston D, Haqqani MT, Rhodes JM. 1994a. Peanut lectin stimulates proliferation in colonic explants from patients with ulcerative colitis and colonic polyps. Gastroenterology 106:117–124.

Ryder SD, Smith JA, Rhodes EGH, Rhodes JM. 1994b. Proliferative response of HT29 and Caco2 human colorectal cancer cells to a panel of lectins. Gastroenterology 106:85–93.

Ryder SD, Smith JA, Rhodes JM. 1992. Peanut lectin is a mitogen for normal human colonic epithelium and HT colorectal cancer cells. J Natl Cancer Inst 84:1410–1416.

Sadek SK, Finnie IA, O'Dowd GM, Rhodes JM. 1994. Effect of formyl-methionyl-leucyl-phenylalanine on mucus secretion in the normal human colon: A novel mechanism of mucus secretion (abstr). Clin Sci 86:33.

Sandoz D, Nicolas G, Laine M. 1985. Two mucous cell types revisited after quick-freezing and cryosubstitution. Biol Chem 54:79–88.

Sands BE, Podolsky DK. 1996. The trefoil family of peptides. Annu Rev Physiol 58:253–273.

Scawen M, Allen A. (1977). The action of proteolytic enzymes on the glycoprotein from pig gastric mucus. Biochem J 163:363–368.

Schmitt FC, Figuerido P, Lacerda M. 1995. Simple mucin-type carbohydrate antigens (T, sialosyl T, Tn and sialosyl Tn) in breast carcinogenesis. Virchows Arch 427:251–258.

Seidler U, Pfeiffer A. 1991. Inositol phosphate formation and $[Ca^{2+}]i$ in secretagogue-stimulated rabbit gastric mucous cells. Am J Physiol 260:G133–141.

Seidler U, Sewing K. 1989. Ca^{2+}-dependent and -independent secretagogue action on gastric mucus secretion in rabbit mucosal explants. Am J Physiol 256:G739–746.

Seregni E, Botti C, Lombardo C, Cantoni A, Bogni A, Cataldo I, Bombardieri E. 1996. Pattern of mucin gene expression in normal and neoplastic lung tissues. Anticancer Res 16(4B):2209–2213.

Shankar V, Gilmore MS, Elkins RC, Sachdev GP. 1994. A novel human airways cDNA encodes a protein with unique tandem-repeat organisation. Biochem J 300:295–298.

Shankar V, Pichan P, Eddy RL, Tonk V, Nowak N, Sait SN, Shows TB, Shultz RE, Gotway G, Elkins RC, Gilmore MS, Sachdev GP. 1997. Chromosomal localisation of a human mucin gene (*MUC8*) and cloning of the cDNA corresponding to the carboxy terminus. Am J Respir Cell Mol Biol 16:232–241.

Sheehan J, Thornton DJ, Howard M, Carlstedt I, Corfield AP, Paraskeva C. 1996. Biosynthesis of the MUC2 mucin: Evidence for a slow assembly of fully glycosylated units. Biochem J 315:1055–1060.

Sheehan JK, Hanski C, Corfield AP, Paraskeva C, Thornton DJ. 1995. Mucin biosynthesis and macromolecular assembly. Biochem Soc Trans 23:819–821.

Sheehad JK, Oates K, Carlstedt I. 1986. Electron microscopy of cervical, gastric and bronchial glycoproteins. Biochem J 239:147.

Shimizu Y, Shaw S. 1993. Mucins in the mainstream. Nature 366:630–631.

Shirotani K, Taylor-Papadimitriou J, Gendler SJ. 1994. Transcriptional regulation of *MUC1* gene in colon carcinoma cells by a soluble factor: Identification of a regulatory element. J Biol Chem 269:15030–15035.

Siddiki B, Ho JJL, Huang J, Byrd JC, Lau E, Yuan M, Kim YS. 1993. Monoclonal antibody directed against colon cancer mucin has high specificity for malignancy. Int J Cancer 54:467–474.

Specian RD, Olivier MG. 1991. Functional biology of intestinal goblet cells. Am J Physiol 260:G183–193.

Sperber K, Ogata S, Sylvester C, Aisenberg J, Chen A, Mayer L, Itzkowitz S. 1993. A novel human macrophage-derived intestinal mucin secretagogue: Implications for the pathogenesis of inflammatory bowel disease. Gastroenterology 104:1302–1309.

Springer GF, Desai PR, Spencer BED, Tegemeyer H, Calrstedt SC, Scanlon EF. 1995. T/Tn antigen vaccine is effective and safe in preventing recurrence of advanced breast carcinoma. Cancer Detect Prev 19(4):173–182.

Springer TA. 1994. Traffic signals for lymphocyte recirculation and leukocyte emigration: The multi-step paradigm. Cell 76:301–314.

Stadie RE, Chai W, Lawson AM, Byfield PGH, Hanisch F-G. 1995. Studies on the order and site specificity of GalNAc transfer to MUC1 tandem repeats by UDP-GalNAc:polypeptide *N*-acetylgalactosaminyltransferase from milk or mammary carcinoma cells. Eur J Biochem 229:140–147.

Steel DM, Hanrahan JW. 1997. Muscarinic-induced mucin secretion and intracellular signaling by hamster tracheal goblet cells. Am J Physiol 272:L230–237.

Strous GJ, Dekker J. 1992. Mucin type glycoproteins. Crit Rev Biochem Mol Biol 27(1/2):57–92.

Swallow DB, Gendler S, Griffiths B, Kearney A, Povey S, Sheer D, Palmer RW, Taylor-Papadimitriou J. 1987. The hypervariable gene locus PUM, which codes for the tumour associated epithelial mucins is located on chromosome 1, within the region 1q21-24. Ann Hum Genet 51:289–294.

Takada A, Ohmori K, Yoneda T, Tsuyuoaka K, Hasegawa A, Kiso M, Kannagi R. 1993. Contribution of carbohydrate antigens sialyl Lewis[a] and sialyl Lewis[x] to adhesion of human cancer cells to vascular endothelium. Cancer Res 53:354–361.

Takai S, Hinoda Y, Adachi T, Imai K, Oshima M. 1997. A human UDP-GalNAc; polypeptide *N*-acetylgalactosaminyltransferase type 1 gene is located at chromosomal region 18q12.1. Hum Genet 99:293–294.

Taylor-Papadimitriou J, D'Souza B, Burchell J, Kyprianou N, Berdichevsky F. 1993. The role of tumour-associated antigens in the biology and immunotherapy of breast cancer. Ann N Y Acad Sci 698:31–47.

Temann UA, Prasad B, Gallup MW, Gasbaum C, Ho SB, Flavell RA, Rankin JA. 1997. A novel role for murine IL4 in vivo: Induction of *MUC5AC* gene expression and mucin hypersecretion. Am J Respir Cell Mol Biol 16:471–478.

Terada T, Nakamura Y. 1996. Expression of mucin carbohydrate antigens (T, Tn and sialyl Tn) and MUC-1 gene product in intraductal papillary-mucinous neoplasm of the pancreas. Am J Clin Pathol 105:613–620.

Terada T, Ohta T, Sasaki M, Nakanuma Y, Kim YS. 1996. Expression of MUC apomucins in normal pancreas and pancreatic tumours. J Pathol 180(2):160–165.

Thornton DJ, Howard M, Khan N, Sheehan JK. 1997. Identification of two glycoforms of the MUC5B mucin in human respiratory mucus. Evidence for a cysteine-rich sequence repeated within the molecule. J Biol Chem 272: 9561–9566.

Toribara NW, Gum JR, Culhane PJ, Legace RE, Hicks JW, Petersen GM, Kim YS. 1991. MUC2 human small intestinal mucin gene structure: Repeated arrays and polymorphism. J Clin Invest 88:1005–1013.

Toribara NW, Ho SB, Gum E, Gum JR, Lau P, Kim YS. 1997. The carboxyl terminal sequence of the human secretory mucin gene *MUC6*: Analysis of the primary amino-acid sequence. J Biol Chem 272:16398–16403.

Toribara NW, Roberton AM, Ho SB, Kuo WL, Gum ET, Gum JR, Byrd JC, Siddiki B, Kim YS. 1993. Human gastric mucin: Identification of unique species by expression cloning. J Biol Chem 268:5879–5885.

Trahair JF, Neutra MR, Gordon J. 1989. Use of transgenic mice to study routing of secretory proteins in intestinal epithelial cells. Analysis of human growth hormone compartmentalisation as a function of cell type and differentiation. J Cell Biol 109:3231–3242.

Tsai HH, Dwarakanath AD, Hart CA, Milton JD, Rhodes JM. 1995. Increased faecal mucin sulphatase activity in ulcerative colitis: A potential target for treatment. gut 36:570–576.

Tsai HH, Sunderland D, Gibson G, Hart CA, Rhodes JM. 1992. A novel mucin sulphatase from human faeces: Its identification, purification and characterisation. Clin Sci 82:447–454.

Turner BS, Bhaskar KR, LaMont JT. 1997. Corticosteroid and retinoid mediation of mucin secretion and gene expression in pig stomach mucosal explant culture (abstr). Gastroenterology 112:A316.

Tygat KM, Opdam FJ, Einerhand AW, Buller HA, Dekker J. 1996a. MUC2 is the prominent colonic mucin expressed in ulcerative colitis. Gut 38:554–563.

Tygat KM, van der Wal JW, Einerhand AW, Buller HA, Dekker J. 1996b. Quantitative analysis of MUC2 synthesis in ulcerative colitis. Biochem Biophys Res Commun 224:397–405.

Tysk C, Riedsel H, Lindberg E, Panzini B, Podolsky D, Jarnerot G. 1991. Colonic glycoproteins in monozygomatic twins with inflammatory bowel disease. Gastroenterology 100:419–423.

van Halbeek H, Strang AM, Lhermitte M, Rahmoune H, Lamblin G, Roussel P. 1994. Structures of monosialyl oligosaccharides isolated from the respiratory secretions of an on-secretor (O, Lea+b−) patient suffering from chronic bronchitis. Characterisation of a novel type of mucin carbohydrate core structure. Glycobiology 42(2):203–219.

van Klinken BJW, Dekker J, Buller HA, Einerhand AW. 1995. Mucin gene structure and expression: Protection vs. adhesion. Am J Physiol 269:G613–G627.

Van-Seuningen I, Davril M, Hayem A. 1989. Evidence for the tight binding of human mucus proteinase inhibitor to highly glycosylated macromolecules in sputum. Biol Chem Hoppe-Seyler 370:749–755.

Varki A. 1993. Biological roles of oligosaccharides. Glycobiology 3:97–130.

Vavasseur F, Dole K, Yang J, Matta KL, Myerscough N, Corfield A, Paraskeva C, Brockhausen I. 1994. O-glycan biosynthesis in human colorectal adenoma cells during progression to cancer. Eur J Biochem 222:415–424.

Vavasseur F, Yang JM, Dole K, Paulsen H, Brockhausen I. 1995. Synthesis of O-glycan core 3: Characterisation of UDP-GlcNAc:GalNAc-R beta 3-N-acetylglucosaminyltransferase activity from colonic mucosal tissues and lack of activity in human colon cancer cell-lines. Glycobiology 5(3):351–357.

Velcich A, Palumbo L, Selleri L, Evans G, Augenlicht L. 1997. Organisation and regulatory aspects of human intestinal mucin gene (MUC2) locus. J Biol Chem 272:7968–7976.

Verdugo P. 1990. Goblet cell secretion and mucogenesis. Annu Rev Physiol 52:157–176.

Verma M, Davidson EA. 1994. Mucin genes, structure, expression and regulation. Glycoconj J 11:172–179.

Voorberg J, Fontijn R, Calafat J, Janseen H, Mourik JA, Pannekok H. 1991. Assembly and routing of von Willebrand factor variants: The requirements for disulphide-linked dimerisation reside within the carboxy-terminal 151 amino acids. J Cell Biol 113:195–205.

Wahawaisan R, Wallace LJ, Gaginella TS, 1983. Muscarinic receptors exist on ileal crypt and villus cells of the rat. Fed Proc 42:761.

Wang Y, Agrwal N, Eckhardt AE, Stevens RD, Hill RL. 1993. The acceptor specificity of porcine submaxillary UDP-GalNAc:polypeptide N-acetylgalactosaminyltransferase is dependent on the amino acid sequence adjacent to serine and threonine residues. J Biol Chem 268:22979–22983.

Weiss AA, Babyatsky MW, Ogata S, Chen A, Itzkowitz SH. 1996. Expression of MUC2 and MUC3 mRNA in human normal, malignant and inflammatory intestinal tissues. J Histochem Cytochem 44:1161–1166.

Wessling J, Van der Valk SW, Hilkens J. 1996. Mechanism for the inhibition of E-cadherin—mediated cell-cell adhesion by the membrane associated mucin episialin/MUC1. Mol Biol Cell 7:565–577.

White T, Bennett EP, Takio K, Sorensen T, Bonding N, Clausen J. 1995. Purification and cDNA cloning of a human UDP-N-acetyl-alpha-D-galactosamine:polypeptide N-acetylgalactosminyltransferase. J Biol Chem 270:24156–24165.

Whitehouse C, Burchell J, Gschmeissner S, Brockhausen I, Lloyd KO, Taylor-Papadimitriou J. 1997. A transfected sialyltransferase that is elevated in breast cancer and localizes to the medial/trans Golgi apparatus inhibits the development of core-2-based O-glycans. J Cell Biol 137:1229–1241.

Winrow VR, Mojdehi GM, Ryder SD, Rhodes JM, Blake DR, Rampton DS. 1993. Stress proteins in colorectal mucosa: Enhanced expression in ulcerative colitis. Dig Dis Sci 38:1994–2000.

Wright NA, Pike C, Elia G. 1990. Induction of a novel epidermal growth factor–secreting cell lineage by mucosal ulceration in human gastrointestinal stem cells. Nature 343:82–85.

Wright NA, Poulson R, Stamp GWH, Van Norden S, Sarraf C, Elia G, et al. 1993. Trefoil peptide expression in gastrointestinal cells in inflammatory bowel disease. Gastroenterology 104:12–20.

Yang JM, Byrd JC, Siddiki BB, Chung YS, Okuno M, Sowa M, Kim YS, Matta KL, Brockhausen I. 1994. Alteration of O-glycan biosynthesis in human colon cancer tissue. Glycobiology 4:873–884.

Yedger S, Eidelman O, Malden E, Roberts D, Etcheberrigaray R, Goping G, Fox C, Pollard HB. 1992. Cyclic AMP–independent secretion of mucin by SW 1116 human colon carcinoma cells. Biochem J 283:421–426.

Yin W, Cheng PW. 1996. Lectin-mediated gene transfer into airways epithelial cells (abstr). Glycoconj J 12(4):426 (S7).

Yonezawa S, Nomoto M, Matsukita S, Xing PX, McKenzie IFC, Hilkens J, Kim YS, Sato E. 1995. Expression of *MUC2* gene product in mucinous carcinoma of breast—comparison with invasive ductal carcinoma. Acta Histochem Cytochem 28(3):239–246.

Yu L, Fernig DG, Smith JA, Milton JD, Rhodes JM. 1993. Reversible inhibition of proliferation of epithelial cell lines by *Agaricus bisporus* (edible mushroom) lectin. Cancer Res 53:4627–4632.

Yuan M, Itzkowitz SH, Palekar A, Shamsuddin AM, Phelps PC, Trump BF, Kim YS. 1985. Distribution of blood group antigens A, B, H, Lewis a and Lewis b in human normal, fetal and malignant colonic tissue. Cancer Res 45:4499–4511.

Yuen CT, Lawson AM, Chai W, Larkin M, Stoll MS, Stuart AC, Sullivan FX, Ahern TJ, Feizi T. 1992. Novel sulfated ligands for the cell adhesion molecule E-selectin revealed by the neoglycolipid technology among O-linked oligosaccharides on an ovarian cystadenoma glycoprotein. Biochemistry 31:9126–9131.

6

Adaptation of a Microbalance to Measure Bioadhesive Properties of Microspheres

Donald E. Chickering III
Acusphere, Inc., Cambridge, Massachusetts

Camilla A. Santos and Edith Mathiowitz
Brown University, Providence, Rhode Island

I. INTRODUCTION

Identification and selection of bioadhesive materials are the first steps in developing a bioadhesive drug delivery system. The most direct method for determining the adhesive properties of polymeric materials is by physical measurement using mechanical testing equipment (1–5). Unfortunately, most devices that are readily available to measure tensile forces are not designed to measure *microscopic* interactions such as those that may occur between microparticles and biological tissues. Furthermore, environmental conditions that can significantly affect the performance of bioadhesives, such as hydration, temperature, and pH, have not typically been well controlled.

By modifying the operation of a contact angle microbalance, a simple and reproducible method was developed for measuring bioadhesive properties of polymer microspheres. While utilizing a miniature tissue chamber to maintain specific physiologic conditions, in vivo interactions were mimicked in a controlled in vitro setting. The system was designed to measure 11 parameters from load versus deformation curves generated with each experiment, including fracture strength, deformation to failure, and tensile work. By analyzing such properties, it is possible to identify materials that have particular affinity for specific tissues.

Using this system, a wide variety of polymeric microspheres, including thermoplastics, hydrogels, degradables, and nondegradables, were screened for their bioadhesive potential (6,7). Polymers known to be rich in carboxylic acid moieties, such as poly(fumaric-co-sebacic anhydride), produced the strongest adhesive interactions with rat intestinal mucosa. Identification and classification of such compounds are important steps in generating a database of materials that can be utilized for the development of bioadhesive drug delivery systems.

In this chapter we report on the microbalance method utilized to screen bioadhesive polymer candidates. In addition, we present a study analyzing the effects of copolymer ratio on bioadhesive potential. Clear differences can be measured between microspheres made of polyanhydrides with various proportions of fumaric acid to sebacic acid.

II. MATERIALS AND METHODS

A. The Microbalance

A Cahn Dynamic Contact Angle Analyzer (model DCA-322, CAHN Instruments, Cerritos, CA) was modified to perform adhesive microforce measurements. Although this specialized piece of equipment is designed for measuring contact angles and surface tensions using the Wilhelmy plate technique, it is essentially a glorified microbalance. The DCA-322 system includes a balance stand assembly, an IBM-compatible computer, and a printer. The microbalance unit consists of stationary sample and tare loops and a stepper motor-powered z-translation stage. The balance has a sensitivity rated at 1×10^{-5} mN but can be operated with samples weighing as much as 3.0 g. The speed of the stage is variable and can be operated over a range of 20 to 264 μm/s.

In its designed application, a glass microscope coverslip is suspended from the sample loop and balanced by applying weight to a pan hanging from the tare loop. A small beaker of liquid is placed on the stage several millimeters below the coverslip. Using the DCA software, the stage is programmed to rise to a particular height at a given stepper motor speed, pause for a set interval, and then return down to its original position at the same rate that it traveled up. During this sequence, force measurements are sent from the microbalance to the computer. The software uses this information to determine the interfacial tension and dynamic contact angle between the coverslip and the liquid.

To develop an automated, reproducible method for measuring bioadhesive properties between polymer microspheres and biological substrates, it was necessary to modify the operation of the balance and stage. Instead of rising to a preset level, the stage had to rise until the balance measured a preset compressive load between the biadhesive and the tissue. To achieve this goal, the microbalance's controller had to be reworked.

The standard IBM-compatible computer system was replaced with an Apple Macintosh II computer. The computer and microbalance were interfaced through the computer's modem port. An external power supply (DC Power Supply 1630, BK Precision, Chicago, IL) was required to provide a constant -10 V dummy signal to the balance, forcing it to sense that it was still connected to the parent computer. This connection was hard-wired through the RS-232 port on the balance.

CAHN Instruments was gracious enough to supply the command codes essential to reprogramming the microbalance operation. LabView II software was used to write a user-friendly, menu-driven package to perform tensile experiments automatically, with easily adjustable graphical settings to control stage speed, compressive load, and time of adhesion. After each run, the software generates graphs of load versus stage position and load versus time. In addition, 11 parameters are calculated automatically: (a) compressive deformation, (b) peak compressive load, (c) compressive work, (d) yield point, (e) deformation to yield, (f) returned work, (g) peak tensile load, (h) deformation to peak tensile load, (i) fracture strength, (j) deformation to failure, and (k) tensile work.

B. The Tissue Chamber

A small, temperature-controlled tissue chamber was constructed to maintain physiological temperature and pH during adhesive force measurements (Fig. 1). The chamber was fabricated of Plexiglas and consists of a 3-mL tissue cell heated by a water jacket. Temperature control was achieved by using a Fisher Scientific Isotemp refrigerated circulator (model 9000) and a ther-

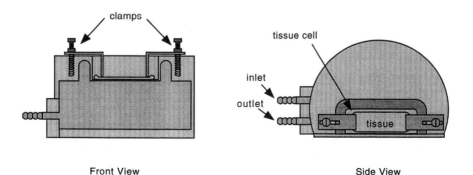

Figure 1 Tissue chamber showing tissue clamps and inlet and outlet flow ports for temperature control.

mocouple. A microtip pH probe (model MI-710, Microelectrodes, London-derry, NH) was used to monitor pH simultaneously. Two stainless steel clamps with thumb screws were used to secure tissue samples to the bottom of the tissue cell.

C. Microsphere Sample Preparation

A series of poly(fumaric-co-sebacic acid) polymers [P(FA:SA)] were synthesized (8) according to the methods of Domb et al. (9–12). Fumaric acid (FA) and sebacic diacid (SA) monomers (Sigma Chemical Company, St. Louis, MO) were purified by recrystallization from 95% ethanol or 100% ethanol, respectively, and drying with vacuum in a $CaCl_2$-containing desiccator. Prepolymers (FAPP and SAPP) were produced by refluxing purified monomer mixtures in acetic anhydride, evaporating 80–90% at 50–60°C (Buchi RE-111 Rotavapor and 461 Water Bath system), and recrystallizing at −20°C. Precipitates were filtered, dissolved in toluene with heat, precipitated overnight at room temperature, and then further recrystallized in an ice bath for 10 hours. Precipitates were filtered, washed (in 200 mL of petroleum ether or a 1:1 volume ratio of ethyl and petroleum ether, for FAPP or SAPP, respectively), and dried under vacuum. For polymerization, prepolymers were melted under reduced pressure and periodically purged with nitrogen gas to remove acetic anhydride. The resulting viscous polymer was cooled, dissolved in methylene chloride, filtered, precipitated in hexane, dried, and stored at −20°C. Molecular weight was determined with gel permeation chromatography (GPC) to be between 5000 and 15,000, depending on the batch. Ratios studied included 10:90, 20:80, 30:70, 50:50, 60:40, and 70:30. Microspheres were prepared by the hot melt technique (13). Microspheres were sieved to a size range of 710–850 μm.

Rigid pins were required to attach the microspheres to the microbalance sample loop. A melting technique was used to mount thermoplastic microspheres to the tips of fine iron wires (280 μm in diameter, Leeds & Northrup Company, Philadelphia, PA). Although this technique could alter the surface morphology of some materials, scanning electron microscope (SEM) analysis (not reported here) showed that the structure of the microspheres remained unchanged on the hemisphere intended for analysis. The nonloaded ends of the wires were attached to a sample clip and suspended in the microbalance enclosure.

D. Tissue Collection and Handling

Rat jejunal tissue was used in all experiments. This tissue represents a significant fraction (by length) of the overall gastrointestinal tract and is therefore a good representative of the target tissue for orally administered bioadhesive drug delivery systems.

Unfasted, male, VAF, CD rats (250–350 g) were anesthetized by intraperitoneal (IP) injection of nembutal (pentobarbital sodium, 55 mg/kg) and perfused transcardially with ice-cold saline (0.9% NaCl). Approximately 10–15 cm of jejunum was carefully excised from the abdominal cavity. The tissue segment was flushed with Dulbecco's phosphate-buffered saline containing 100 mg/dL glucose (DPBSG) and stored in DPBSG at 4°C for no more than 3 hours.

E. Microbalance Operation

The basic operation of the system is quite simple: apply a load between a microsphere and tissue, and then measure the force required to fracture any adhesive interactions. However, certain steps must be performed and particular parameters must be set in the software before the measurements can be made. Typically, the system was initialized and set up 30 minutes prior to collecting data. This allowed time for the tissue chamber to reach its operating temperature of 37°C. Next, a freshly excised section of tissue was clamped in the buffer-filled chamber (pH 7.4), and a mounted microsphere was hung from the sample loop. In the setup window of the software, the stage speed, applied load, and adhesion time were set to 50.2 μm/s, 25 mg, and 7 minutes, respectively. The diameter of each microsphere was also entered for calculations of force per surface area.

After initiating the run cycle, the stage rose at 50.2 μm/s until the applied load between microsphere and tissue reached the preprogrammed set point of 25 mg. The stage was held stationary at this position for the duration of adhesion time (7 minutes). To fracture the adhesive interactions, the stage was then lowered to its initial position (Fig. 2). Following this sequence of events, certain bioadhesive parameters were calculated and graphs of force versus position and time were plotted. A minimum of 4 (maximum of 23) samples were run for each polymer. In every case, a new microsphere and a fresh tissue site were used for the bioadhesive test.

III. RESULTS AND DISCUSSION

A. Interpretation of the Microbalance Recordings

The plotted output from the microbalance software is unique in that it displays both the compressive and tensile portions of the experiment. Figure 3 shows a typical load versus deformation graph, where the position and mass are represented by the horizontal and vertical axes, respectively. Compres-

Figure 2 Photograph of tissue chamber showing microsphere at the moment of separation from intestinal segment.

sive forces are represented in the negative direction, while tensile or adhesive forces are positive. It should be noted that the position axis represents the position of the stage, not the microsphere, because the microsphere remains stationary throughout the experiment and the stage, supporting the tissue sample, is moved up or down.

Initially, the microsphere is suspended above the tissue segment at zero position and zero load. The horizontal line of zero force is the distance that the stage is raised before contact is made between the tissue and the microsphere. The point at which the microsphere initially makes contact with the tissue is represented by position A. The segment from point A to point B is the compressive portion of the experiment. Point B represents the maximum compressive load. This load can be varied and will affect the degree of penetration into the tissue. The distance from point A to point B is the compressive deformation. At position B the stage is held stationary. The microsphere and tissue are held in compression to favor adhesion.

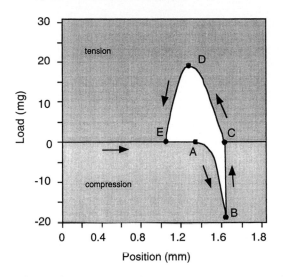

Figure 3 Typical load versus deformation curve indicating direction of tracing.

As the stage is lowered and the compressive load is released, the tensile region of the experiment begins. Segment BC represents downward motion of the stage to the point of 0 mg applied force (point C). The slope of the linear portion of the curve following point C indicates the stiffness of the adhesive interaction. During this time, the force increases as a function of stage position, and it is assumed that the contact area between the sphere and the mucus remains constant. The yield point, which indicates the first sign of adhesive failure, is determined by the point of intersection of the two lines tangent to the curve during and after the initial linear region of the tensile curve. The distance from B to this point is referred to as the deformation to yield.

Peak load is represented by point D on the curve. Fracture stress was calculated by dividing the peak load by the projected surface area of the microsphere. It was assumed, based on visual observations made during the experiments, that the surface area in contact with the tissue remained constant from the point of zero force until the peak load was reached. The distance from point B to point D is known as the deformation to peak load.

The final segment (DE) of the tensile curve indicates a period of partial detachment between the biological surface and polymeric device. During this phase, the contact area decreased as separation occurred. In some instances this region was elongated and marked with several small peaks and valleys. We believe that in such cases the adhesion was the result of many small individual bonds. As the load increased, some bonds were stretched and the curve rose until one or more of the bonds failed. The curve then dropped momentarily until the load was transferred to other bonds. This cycle continued until finally no adhesions remained, resulting in total detachment. Figure 4 schematically depicts the disruption of the adhesive union where several focal attachments may be responsible for the adhesive bond strength.

The last point (E) is the point of complete detachment of the sphere from the mucosal layer and can also be termed the point of failure. The distance from point B to point E is known as the deformation to failure. Materials exhibiting large deformations to failure produced adhesive interactions that were more compliant and less susceptible to disruption from outside forces.

Three different work values can be obtained from the graphs. Compressive work is the work done in joining the two surfaces and is calculated as the area between the compressive region of the curve and the zero load line. Returned work is the work done by the tissue on the microsphere when the compressive load is released. This component is the area bounded by the negative region of the tensile curve and the zero load line. Tensile work is the work done in separating the two surfaces and is the area between the positive portion of the tensile curve and the zero load line.

In order to compute fracture strength (F_m/A_0) of a bioadhesive bond, the contact area has to be determined. The area of contact between the microsphere and mucosa was estimated to be the projected surface area of the spherical cap defined by the depth of penetration of the microsphere

Figure 4 Schematic diagram showing the progression of bioadhesive fracture.

below the surface level of the tissue. This area can be calculated using the following relation:

$$\text{Area} = A_0 = \pi R^2 - \pi (R - a)^2$$

where R is the microsphere radius and a is the depth of penetration. The depth of penetration, a, was visually estimated to be approximately equal to the microsphere radius for microspheres between 710 and 850 μm. Thus, $A_0 = \pi R^2$. Fracture strengths were obtained by normalizing peak load by the projected area of this cap.

B. System Limitations

The microbalance-based system has several limitations: (a) microspheres smaller than 300 μm are difficult to mount, (b) overshoot of the applied load must be accounted for, (c) very small and very large applied loads are difficult to control, and (d) changes in applied load over time due to tissue relaxation or contraction are not accounted for once the stage has stopped moving.

First, because of the mounting method, it is difficult to attach microspheres smaller than 300 μm to the microbalance. A more sophisticated method using micropipettes and a micropositioner (14) could possibly be employed to mount microspheres as small as 10 μm. However, estimations of contact area for fracture strength calculations would be difficult.

Second, because the sampling rate of the microbalance is 1 Hz, stage motion commands can be sent only once every second. Therefore, as the microbalance reaches the applied load set point, there is a 1-second delay between the time when the software commands the stage to stop moving and when the motion actually ceases. This can result in overshoot of the applied load in the range of 5–50%. For moderate applied loads (i.e., 10–50 mg), the overshoot is consistent, easily adjusted for, and rarely greater than 20%.

Third, very small and very large applied loads are difficult to control because of the response time of the microbalance and the nature of the motor controlling the stage movement. The stepper motor supplied with the DCA-322 moves in incremental jumps (as opposed to smooth motion). When an attempt is made to apply very small loads, a small amount of overshoot can be greater than 100% of the intended value. When very large loads are attempted, the slope of the load versus deformation curve is so steep that the overshoot is always very large. Therefore, moderate forces (10–50 mg) are most easily attained.

Fourth, in many cases the applied load changes while the stage is not in motion. This occurs due to relaxation or contraction of the smooth muscle in the tissue. Because of unsuitable response time and relatively large incremental movements of the stepper motor, accurate adjustments cannot be made in the stage position to account for these small dynamic effects.

C.   Analysis of Polymer Microspheres

Despite the aforementioned drawbacks and limitations, bioadhesive measurement made using this system are quite reproducible. More important, the sensitivity is great enough to show statistical differences between materials.

Table 1 gives a comparative analysis of three key bioadhesive parameters for the series of polymer microspheres analyzed. Fracture strength is the maximum force per surface area required to break the adhesive bond. This parameter is perhaps the most frequently measured and reported descriptor of bioadhesive materials. Deformation to failure is the distance required to move the stage before complete separation occurs. This parameter is dependent on both the material stiffness and the intensity or strength of the adhesion. Work of adhesion is a function of both the fracture strength and the deformation to failure. Therefore, it tends to be the strongest indicator of bioadhesive potential.

Fracture strengths were obtained by dividing the peak load by the projected contact area of adhesion, which for microspheres between 700 and 850 μm was estimated to be approximately 0.00487 cm^2. Polymers composed of a majority of fumaric acid (i.e., $\geq 50\%$ FA) produced the highest fracture strengths (Fig. 5).

Analysis of deformation to failure also showed that copolymers rich in fumaric acid produced the greatest adhesive interactions (Fig. 6). The adhesive bonds formed with the tissue were resilient enough to elongate the

Table 1   Comparison of Fracture Strength, Deformation to Failure, and Work of Adhesion

Polymer	Fracture strength (mN/cm^2)	Deformation to failure (mm)	Work of adhesion (nJ)
P(FA:SA) 10:90	31.6 ± 5.6	1.13 ± 0.20	83.0 ± 12.8
P(FA:SA) 20:80	17.5 ± 1.8	0.72 ± 0.09	33.0 ± 5.4
P(FA:SA) 30:70	24.3 ± 4.8	0.84 ± 0.15	66.9 ± 19.0
P(FA:SA) 50:50	58.2 ± 4.6	1.80 ± 0.14	315.3 ± 43.6
P(FA:SA) 60:40	67.4 ± 6.8	2.00 ± 0.17	315.0 ± 43.9
P(FA:SA) 70:30	99.7 ± 17.4	1.95 ± 0.20	453.2 ± 47.7

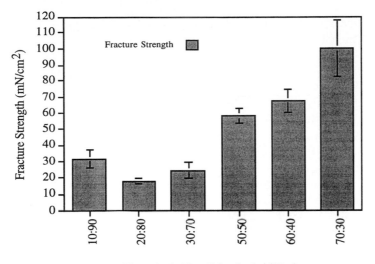
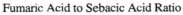

Figure 5 Fracture strength versus FA:SA ratio.

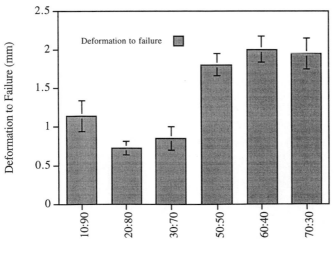
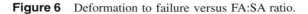

Figure 6 Deformation to failure versus FA:SA ratio.

bioadhesive-tissue composite to a length nearly twice that for polymers made mainly of sebacic acid.

As could be expected, tensile work was very large for the polymers that displayed high fracture strengths and large deformations to failure (Fig. 7). P(FA:SA) 50:50, 60:40, and 70:30 produced mean tensile works that were 3.5 to 13.5 times greater than those of the other polymers studied. The interaction formed between these materials and rat jejunal tissue was notable. In fact, in many instances the bond strength was greater than the intratissue strength, resulting in fracture within the tissue rather than at the tissue-bioadhesive interface.

It is interesting to note the range of bioadhesive interactions of the poly(fumaric-co-sebacic anhydride) copolymers with rat jejunal tissue. The copolymers low in FA content [such as P(FA:SA) 10:90 and 20:80] display lower adhesive forces than the copolymers with high FA content, such as P(FA:SA) 60:40 and 70:30. The microbalance technique provides a simple method of ranking materials for bioadhesive potential, even when there are only slight differences in their chemical structure.

Using other bioadhesion assays, similar trends among the P(FA:SA) copolymers were observed (15). As seen in Fig. 8, there is a strong correlation between the results obtained with the microbalance technique and the

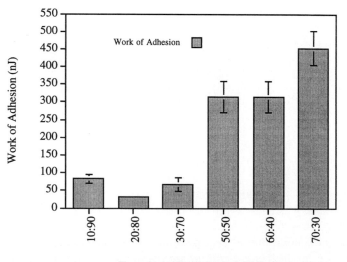

Figure 7   Work of adhesion versus FA:SA ratio.

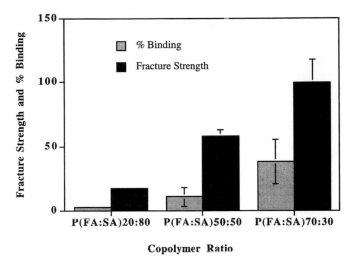

Figure 8 Correlation between data obtained by microbalance measurements (fracture strength) and by the everted intestinal sac technique (% binding).

perhaps more common everted intestinal sac method. In the everted sac technique (16, 17), microspheres are allowed to incubate in a physiologic buffer with an everted segment of intestinal tissue. The percentage of microsphere that adheres to the tissue is an indicator of the bioadhesive potential. Again, the P(FA:SA) ratios high in FA content yield the most bioadhesive measurements in both assays.

Several factors of this phenomenon have been explored. One hypothesis is that the crystallinity may play a role, with the most crystalline copolymers displaying the least bioadhesive interactions with mucus or mucosa. Differential scanning calorimetry and X-ray diffraction measurements (18) have revealed high crystallinity at both ends of the copolymer ratio spectrum, with 10:90 and 20:80 ratios being more crystalline than 50:50, 60:40, and 70:30 ratios. Analysis of molecular weight data (Table 2) obtained by gel permeation chromatography, by the method of Pekarek et al. (19), reveals lower molecular weights for the copolymer ratios high in FA content, suggesting that molecular weight may also play a role in bioadhesion. In addition, water contact angle measurements made on melt cast films of the copolymers show decreasing hydrophobicity with increasing FA content, another factor that may play a significant role in the attraction between the polyanhydride and mucosal tissue.

Table 2   Comparison of Molecular Weight and Contact Angle Data

Copolymer ratio	Weight average Mw (Da)	Number average Mn (Da)	Water contact angle
10:90	16,000	6,900	71 ± 10°
20:80	11,700	4,700	70 ± 12°
30:70	10,000	4,300	78 ± 4°
50:50	6,800	2,100	79 ± 7°
60:40	6,200	2,100	46 ± 1°
70:30	4,000	1,500	36 ± 9°

IV. CONCLUSIONS

In summary, although the microtensiometer described in this chapter has some limitations, it offers several advantages over previous devices. The setup enables determination of bioadhesive forces between single microspheres and biological tissues. The technique allows measurement of bioadhesive properties of candidate materials in a spherical form, which in many cases is the intended geometry for drug delivery devices. The use of a tissue chamber mimics physiological temperature and pH conditions of a living animal and maintains viable tissue segments for testing. Because experiments are conducted in an aqueous environment, problems in distinguishing surface tension forces at an air-liquid interface from forces at the microsphere-mucus interface are eliminated.

A LabView-based software program was developed to convert the operation of the contact angle analyzer into a bioadhesive tensiometer. Advantages of the program include ease of operation, elimination of operator-dependent variables, precise control of the microbalance, and automatic data calculation and graphing. By analyzing the output, it is possible to identify materials with strong bioadhesive properties. Specific parameters obtained from the experiments that we believe are key predictors of bioadhesive potential include (a) fracture strength, (b) deformation to failure, and (c) tensile work. Correlation has been observed between the results obtained with this technique and those obtained using other methods of bioadhesion evaluation, but the differences between materials may be more pronounced with the microbalance measurements. Analyses of materials using this or similar instrumentation will surely provide valuable information for the development of future microsphere-based, bioadhesive drug delivery systems.

REFERENCES

1. Park K, Park H. Test methods of bioadhesion. In: Lenaerts V, Gurny R, eds. Bioadhesive Drug Delivery Systems. Boca Raton, FL: CRC Press, 1990, pp 43–66.
2. Jacques Y, Buri P. Optimization of an ex-vivo method of bioadhesion quantification (abstr). Proceedings of the 18th International Symposium on Controlled Release of Bioactive Materials, Controlled Release Society, Amsterdam, 1991.
3. Smart JD, Kellaway IW. In vitro techniques for measuring mucoadhesion. J Pharm Pharmacol 34:70P, 1982.
4. Gu JM, Robinson JR, Leung SHS. Binding of acrylic polymers to mucin/epithelial surfaces: Structure-property relationships. CRC Crit Rev Ther Drug Carrier Syst 5:21–67, 1988.
5. Ponchel G, Touchard F, Duchêne D, Peppas N. Bioadhesive analysis of controlled-release systems. I. Fracture and interpenetration analysis in poly(acrylic acid)-containing systems. J Controlled Release 5:129–141, 1987.
6. Chickering DE, Jacob JS, Panol G, Mathiowitz E. A tensile technique to evaluate the interaction of bioadhesive microspheres with intestinal mucosa (abstr). Proceedings of the 19th International Symposium on Controlled Release of Bioactive Materials, Controlled Release Society, Orlando, FL, 1992.
7. Chickering DE, Mathiowitz E. Bioadhesive microspheres: I. A novel electrobalance-based method to study adhesive interactions between individual microspheres and intestinal mucosa. J Controlled Release 34:251–262, 1995.
8. Chickering DE, Jacob JS, Mathiowitz E. Bioadhesive microspheres: II. Characterization and evaluation of bioadhesion involving hard, bioerodible polymers and soft tissue. Reactive Polym 25:189–206, 1995.
9. Domb A, Langer R. Polyanhydrides. I. Preparation of high molecular weight polyanhydrides. J Polym Sci 25:3373–3386, 1987.
10. Domb AJ, Gallardo FC, Langer R. Polyanhydrides. 3. Polyanhydrides based on aliphatic-aromatic diacides. Macromolecules 22:3200–3204, 1989.
11. Domb A, Mathiowitz E, Ron E, Giannos S, Langer R. Polyanhydrides. IV. Unsaturated polymers composed of fumaric acid. J Polym Sci 29:571–579, 1991.
12. Leong KW, Brott BC, Langer R. Bioerodible polyanhydrides as drug-carrier matrices. I: Characterization, degradation and release characteristics. J Biomed Mater Res 19:941–955, 1985.
13. Mathiowitz E, Langer R. Polyanhydride microspheres as drug carriers. I. Hot melt microencapsulation. J Controlled Release 5:13–22, 1987.
14. Guilford WH, Gore RW. A novel remote-sensing isometric force transducer for micromechanics studies. Am J Physiol 263:C700–707, 1992.
15. Santos C, Jacob J, Freedman B, Press D, Harnpicharnchai P, Mathiowitz E. Correlation of two bioadhesion assays. Proceedings of the 24th International symposium on Controlled Release of Bioactive Materials, Controlled Release Society, Stockholm, 1997.

16. Wilson TH, Wiseman G. The use of sacs of inverted small intestine for the study of the transference of substances from the mucosal to the serosal surface. J Physiol (Lond) 123:116–125, 1954.

17. Jacob J, Santos C, Carino G, Chickering D, Mathiowitz E. An in vitro bioassay for quantification of bioadhesion of polymer microspheres to mucosal epithelium (abstr). Proceedings of the 22nd International Symposium on Controlled Release of Bioactive Materials, Controlled Release Society, Seattle, 1995.

18. Mathiowitz E, Ron E, Mathiowitz G, Langer R. Morphological characterization of bioerodible polymers. 1. Crystallinity of polyanhydride copolymers. Macromolecules 23:3212–3218, 1990.

19. Pekarek KJ, Dryud MJ, Ferrer K, Jong YS, Mathiowitz E. In vitro and in vivo degradation of double-walled polymer microspheres. J Controlled Release 40: 169–178, 1996.

7
Novel Magnetic Technique to Measure Bioadhesion

Benjamin A. Hertzog and Edith Mathiowitz
Brown University, Providence, Rhode Island

I. INTRODUCTION

In the course of designing a new bioadhesive drug delivery system (BDDS), our laboratory employs a number of qualitative and quantitative techniques for predicting in vivo performance. We have developed two specialized measurement techniques to aid in the evaluation of bioadhesive microspheres. The first, the CAHN machine, is a sensitive electrobalance, which is described in detail in the previous chapter. The CAHN machine was developed and first described in 1995 (Chickering and Mathiowitz, 1995). We have been using the CAHN machine for a number of years, and we have gained a wealth of knowledge and understanding about bioadhesive interactions from tensile experiments. The CAHN machine is a powerful tool, but inherent in its measurement technique are a few limitations that make it better suited for large microspheres (diameter > 300 mm) that have been adhered to tissue in vitro. In an effort to expand our understanding of the bioadhesive processes of smaller microspheres, we have developed a second evaluation tool that we call the electromagnetic force transducer or EMFT. Both the CAHN machine and the EMFT were designed to provide quantitative tensile information about bioadhesive interactions, but the ways in which these machines measure tensile forces are very different. Consequently, the capabilities, limitations, and even the preferred uses of each technique vary.

The EMFT is a remote sensing instrument that uses a calibrated electromagnetic to detach a magnetic-loaded polymer microsphere from a tissue sample (Fig. 1). The adhesive force experienced between the tissue and

147

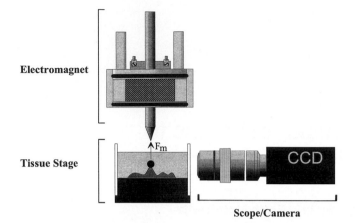

Figure 1   The working element of the EMFT. The tissue sample is mounted in a special chamber with the microsphere positioned directly under the magnet tip. The stage is slowly moved away from the tip, and the video camera is used to detect motion of the sphere. As the sphere moves away from the magnet tip, the control system increases the magnet current accordingly. The change in magnetic field strength results in a force (F_m) that pulls the magnetic sphere back into its original position. This process is repeated until the sphere is pulled free from the tissue.

polymer is proportional to the current through the electromagnet. No physical attachment is required between the force transducer and the microsphere, and therefore the system can perform bioadhesive measurements on small microspheres that have been implanted and incubated in vivo.

Currently, the EMFT and the CAHN machine are used to screen polymers for bioadhesive tendencies toward specific tissue types. This information will be used to develop new bioadhesive drug delivery systems. The bioadhesive properties of the polymer can be used to increase gastrointestinal (GI) transit time, target specific diseased tissues, or simply enhance the intimate contact between polymer and target tissue to optimize the efficiency of drug delivery.

II. BACKGROUND

A. Bioadhesive Delivery Systems

Controlled-release drug delivery systems are an important aspect of many pharmaceutical therapies. The development of polymer-based controlled-release systems (CRSs) during the past decade has led to new products that allow more efficient delivery of pharmaceuticals than standard, noncontrolled drug administration. Local delivery has become a popular option in circumstances in which the target tissue for the drug is relatively small and centralized. Local delivery devices make it possible to maintain local therapeutic levels of a pharmaceutical agent that is otherwise toxic if delivered at the same concentration systemically. In addition, expensive agents and experimental drugs that are available in very limited supply can be delivered directly to the site of action in much smaller doses than would be required to achieve therapeutic levels systemically.

In the effort to enhance the performance of CRSs, it has been recognized that more intimate contact of the device and target tissue enhances drug uptake. This knowledge has sparked growing interest in what has come to be known as bioadhesive drug delivery systems. These devices may have delivery characteristics similar to those of CRSs, but they are specifically designed to increase adhesion between the device and the specific biologic substrate targeted for drug uptake.

Previous research has demonstrated that certain polymers exhibit adhesive tendencies toward specific tissue types (Chickering et al., 1997), and these properties have since been used to develop new bioadhesive drug delivery systems. The adhesive properties are used to target specific diseased tissues, increase residence time, or enhance the intimate contact between the drug delivery system and the target tissue.

Experimental data support the notion that increasing bioadhesion favorably affects duration of contact and intimacy of contact between the BDDS and target tissue. Research focusing on the use of BDDSs has produced two significant correlations between in vitro bioadhesion measurements and the performance of GI delivery of drugs using polymer microspheres (Chickering and Mathiowitz, 1995). (a) Microspheres that demonstrated stronger bioadhesion in the in vitro tests also demonstrated greater residency in the GI tract than microspheres of lesser bioadhesion. (b) Certain drugs delivered via more strongly bioadhesive polymer formulations showed increased bioavailability compared with less adhesive formulations. These results are extremely important because they verify that through proper selection of polymers and manufacturing or processing techniques it is possible to develop more efficient drug delivery systems that utilize the bioadhesive of polymers (Mathiowitz et al., 1997).

B. Measuring Bioadhesion

Efforts are being made to increase understanding of the mechanisms of bioadhesive interactions between the polymers and substrate. This work depends on data gathering for categorizing and cataloging the parameters that affect bioadhesive properties of polymer-based delivery systems. Key to the effort is the development of a rigorous, standardized testing device that allows researchers to gather a wide range of data on bioadhesive interactions.

Historically, the major difficulty in the design of bioadhesive delivery systems has been in quantifying the adhesive interaction between the delivery device and the target tissue. A number of techniques have been designed to measure the bulk adhesion of large polymer samples to biological substrates, and still others were developed to give qualitative information, but few are capable of sensitive tensiometric measurements of small microspheres under physiologic conditions (Chen and Cyr, 1970; Ch'ng et al., 1985; Duchene et al., 1988; Gu et al., 1988; Gurny et al., 1984; Ishda et al., 1981; Lehr et al., 1990, 1992, 1993; Mikos et al., 1991; Mikos and Peppas, 1983, 1986, 1990; Park and Robinson, 1985; Park and Park, 1990; Smart and Kellaway, 1982; Smart et al., 1984; Teng and Ho, 1987). In 1995, when we first introduced a new concept in the evaluation of bioadhesive delivery systems, we reported on a novel tensiometric bioadhesion testing system based on a CAHN series 300 dynamic contact angle measuring system (ATI CAHN Inc., Boston, MA) (Chickering et al., 1995). The CAHN system, described in the previous chapter, improved greatly upon previous bioadhesion testing systems, primarily because it provides a technique that can measure direct tensile forces in near-physiologic conditions with accurate device geometries for microspheres.

The CAHN system has provided a wealth of new information about bioadhesive interactions, and it has become an extremely valuable tool in the development of BDDSs. Unfortunately, we seem to have reached several limits of this device: (a) The microsphere must be physically attached to the sample loop. Practically speaking, this prohibits use of microspheres with diameters less than approximately 400 μm. This is fine for most oral drug delivery systems, but injectable systems and systems targeting cellular uptake necessitate much smaller microspheres. (b) In addition to the size limitation, it is nearly impossible to perform adhesion experiments on microspheres that were implanted and incubated in vivo because it is difficult to attach the metal wire to a sphere that is already in contact with tissue. (c) The CAHN hardware imposes on the ability to control the compressive load precisely or record high-frequency adhesion events because the electrobalance and motorized stage electronics can be interrogated by the computer only at 1.0-second intervals.

For these reasons, other novel techniques for evaluating bioadhesive interactions between polymer microspheres and viable tissue were investigated. The electromagnetic force transducer, described in this chapter, represents the culmination of these efforts. The EMFT is a remote-sensing bioadhesive testing system based on an electromagnetic force measurement technique described by Guilford and Gore (1992). To the best of our knowledge, no other system has been reported that can match the EMFT's unique ability to record remotely and simultaneously force information and high-magnification video images of bioadhesive interactions at near-physiologic conditions.

III.   EXPERIMENTAL DESIGN AND METHODS

A.   The Electromagnetic Force Transducer

The preliminary design for the new EMFT evolved from an immediate need to overcome the shortcomings of earlier bioadhesion testing systems. The following design objectives are not only to enhance the data collection capabilities and remove shortcomings of the CAHN hardware but also to develop a system with increased accuracy, functionality, and potential usefulness.

1.   Design Objectives

(a) The device should be remote sensing; that is, there should be no direct physical connection between the load generating mechanism and the specimen under investigation. This would allow the use of very small microspheres. (b) The feedback loop should be as fast and efficient as possible to capture high-frequency adhesion events as the microsphere is pulled free of the tissue. (c) The tests can be conducted on spheres administered in vivo that have not been artificially compressed into the tissue substrate. The effects of the compressive work cycle are thus eliminated, giving a clearer picture of the forces involved in adhesion. (d) The system should be capable of recording video images of the experiment in progress. (e) The system should be automated to the extent that it is easy and convenient to use. (f) The computer interface should be user friendly and intuitive as well as providing a potential foundation for future upgrades.

Based on these design goals, a preliminary design for the EMFT was developed and constructed. The resultant system is controlled through a graphical computer interface that has complete control over all aspects of the new system: video analysis, motion control, feedback loop, power supplies, data collection, data analysis, and data presentation. Functionally, the new system operates as shown in Fig. 2.

2.   Electromagnetic Force Measurement

The EMFT measures adhesive forces by monitoring the magnetic force required to exactly oppose the force of a bioadhesive interaction. To test a microsphere, it must first be attached to the sample of tissue. An electro-

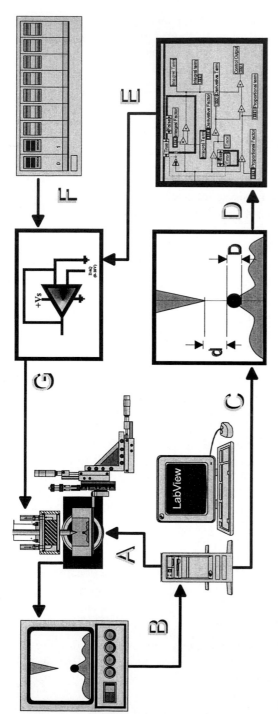

Figure 2   The process flow in a typical bioadhesion experiment. Before an experiment can be run, a sample of tissue is placed in the tissue chamber and held at physiologic conditions. The motion control system is used to locate and center a microsphere in the video field and under the tip of the electromagnet. When the experiment is started, the computer sends a signal to the motorized stage, causing it to descend at a predetermined velocity (A). Simultaneously, the video analysis system starts capturing and analyzing images (B). Using the captured video frames, the computer calculates d, the microsphere-electromagnet separation, and D, the diameter of the microsphere (C). The microsphere diameter is determined once and is used for converting electromagnet current into calibrated force. The tip-sphere separation, d is calculated continuously. As the tissue sample moves down and d increases slightly, the computer sends an error signal to the PID controller (D). The PID controller estimates the correction needed to pull the microsphere up and return d to its original value. The power for the electromagnet is supplied by the GPIB power supply (F), but the power is regulated by a high-power amplifier circuit (G). The amplifier circuit acts as a voltage-controlled current source and is driven by the output signal from the PID controller (E). This control loop operates continuously until the microsphere is pulled free from the tissue while the computer simultaneously records magnet power, time, and position data. After the experiment, a calibration data set can be used to convert the magnet power to force values, and the computer will automatically calculate pertinent adhesion parameters (e.g., fracture strength, peak tensile load, work of detachment).

magnet is activated, producing a magnetic field, and the force on the microsphere, and thus the tissue, is given by

$$F_m = \chi V_m H(\nabla H) \tag{1}$$

where V_m is the volume of the microsphere, χ is the microsphere magnetic susceptibility, H is the magnetic field strength, and ∇H is the gradient of the magnetic field. For a given configuration, the variation in H and ∇H is determined by the construction of the electromagnet and the magnet current, which are properties of the instrument design. As a result, the range of measurable forces is determined by the product (χV_m). Thus, the magnitude of the susceptibility term can be adjusted accordingly, depending on the properties and quantity of the magnetic material used in the microspheres.

The position of the microsphere is used as a feedback signal to make adjustments to the applied magnetic force. The descending tissue sample displaces the sphere, and the control system adjusts the current through the electromagnet, altering $H(\nabla H)$ until the sphere returns to its original position. By measuring current through the electromagnet and comparing it with a calibration data set, the force being applied to the sphere is calculated.

3. System Hardware

The EMFT was designed, from the beginning, as a microtensiometric device. Each element of the hardware was chosen to simplify the construction and programming of the completed system as well as to optimize both performance and flexibility. Many different components were integrated to construct a flexible, extensible testing device. There are six major components of the EMFT hardware: microscope, computer, video analysis system, electromagnet control system, motion control system, and the tissue chamber. The motion control, video analysis, and electromagnet feedback systems are controlled by a 200-MHz Pentium Pro–based computer (Dimension XPS-Pro200, Dell Computer, Austin, TX) through an interface developed in LabVIEW (National Instruments, Austin, TX) and are organized according to Fig. 3.

The EMFT is built around a Mitutoyo model FS-110T microscope with a transmitted light stand (MTI Corp., Aurora, IL) that has been mounted on its back to provide a vertical focus plane. Mitutoyo long-working-distance bright-field objectives are used to provide a sufficient working distance for placement of the tissue chamber directly in front of the objective lens. Tissue samples are mounted in a small delrin chamber designed to maintain physiologic conditions (temperature and pH). The chamber has optical-quality glass windows to allow passage of light and observation of the adhesion

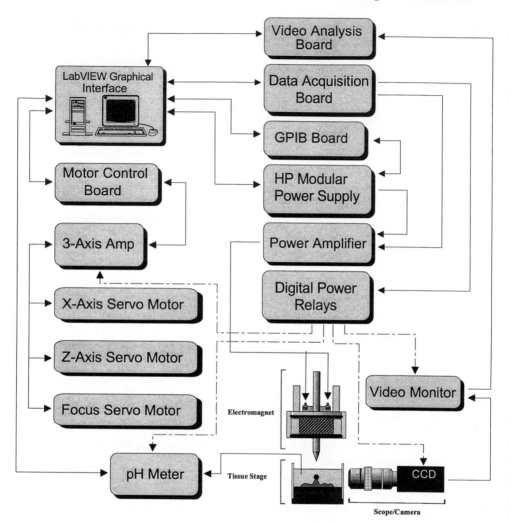

Figure 3 Information and data flow in the EMFT. The EMFT hardware has been integrated with the LabVIEW interface via the motor-control, video analysis, data acquisition, and GPIB boards installed in the computer. Power to auxiliary equipment is controlled using the digital output lines of the data acquisition board and a series of AC-DC power relays.

event with the microscope. The tissue is held in place with two aluminum clips, and the chamber is filled with approximately 10 mL of buffered saline.

The tissue stage is mounted on a custom three-axis motion control system built with high-resolution servo stages (National Aperture, Salem, NH). The servomotors are controlled by the computer interface using an external, four-axis servo amplifier and a control board installed in the computer (MC3-SA and MC3-SIO, nuLogic, Windham, NH).

Magnetic force is generated by an electromagnet mounted on the microscope vertically above the tissue chamber (Fig. 1). The number of turns of wire per length of coil and the current through the wire determine the strength of the spatially varying magnetic field. At a given point in space, force experienced by a magnetic microsphere is adjusted by varying the current through the magnet coil. A number of different magnets have been constructed for the force transducer, but all of the magnets used for preliminary testing were of similar design (Fig. 4). The magnet was wound with 1391 turns of 30 AWG Heavy Armored Poly-Thermaleze magnet wire (Belden Wire and Cable, Richmond, IN) on an aluminum spool. The magnet core was made from a length of 0.25-inch HyMu "80" alloy rod stock (Carpenter Technology, Reading, PA). The tip of the rod was machined to a 15° point and then carefully sharpened with a high-speed diamond grinding wheel. The entire magnet spool is jacketed with a clear acrylic sleeve sealed with two O-rings. Cold water is circulated around the magnet windings via

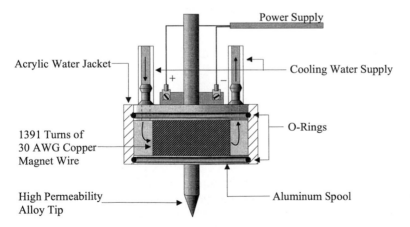

Figure 4 The electromagnet used in the EMFT. On an aluminum spool, 1391 turns of 30 AWG copper magnet wire were wound. The coil has a DC resistance of 31.3 ohms and an inductance of 9 mH. The core of the magnet was machined from high-permeability alloy designed to minimize hysteresis and residual magnetism. The magnet is water jacketed to permit efficient heat dissipation.

two ports on the top surface of the spool. The water circulation helps regulate the temperature of the magnet, thus minimizing heat transfer to the tissue sample and enabling the magnet to operate at much higher currents by dissipating excess heat.

The magnet core material was chosen specifically to maximize the field strength and minimize residual magnetism due to hysteresis (hysteresis is described briefly later in this chapter). Even with the use of the HyMu alloy, the magnet tip exhibits a certain degree of residual magnetism. Between experiments the tip must be demagnetized by driving the magnet with a decaying (90% per half-cycle) bipolar sinusoidal current as described by Shahvarooghi and Moses (1994). This demagnetization routine ensures that the net attractive force between the magnet tip and a sphere will be zero when the power to the magnet is turned off.

When an experiment is started, the graphical interface prompts the user to position a microsphere directly under the sharpened tip of the electromagnet as shown in Fig. 1. The sample image is captured by a Hitachi Denshi model KP-M1U black-and-white CCD camera and displayed on a video monitor at all times throughout the experiment. After the sphere has been positioned properly under the magnet tip, the computer captures an image from the video camera using a PCI frame-grabber card (Data-Raptor 4M PCI, Bit Flow, Woburn, MA) and displays it on the computer screen. Next, the user is prompted to select a point at the very top of the microsphere using the computer's mouse. The video analysis system begins to scan continuously a vertical line of pixels containing the selected point. A video analysis algorithm uses the scanned pixels to determine the position of the sample sphere as well as the distance between the microsphere and magnet tip. The algorithm has been developed using CONCEPT V.i (Graftek France, Mirmande, France), a library of powerful LabVIEW image analysis functions.

After the computer has calculated the position of the microsphere, the tissue chamber is slowly moved down, away from the magnet tip. While this is happening, the video analysis is continuously calculating sphere position. Sphere position is used as an error signal by the electromagnet feedback system, which includes the PID controller, power supply, power amplifier, and electromagnet. As the sphere is displaced, the PID routine calculates a driving signal based on the magnitude, velocity, and duration of sphere displacement. The driving signal is used to adjust the output of the power amplifier and consists of an analog output voltage from -5.0 to $+5.0$ V. The power amplifier is a voltage-controlled current source that regulates the current flow from the power supply (Hewlett-Packard 66000A modular supply with 66104A module) to the electromagnet based on the driving signal input. The operational status of the power supply is controlled through the computer's GPIB board (GPIB TnT+, National Instruments). The new driving signal is used either to increase or to decrease the current through the electromagnet and, consequently, increase or decrease the magnetic force experienced by the microsphere. In effect, the electromagnet feedback system is used to hold the microsphere in a fixed position as the tissue sample is pulled free.

As the tissue sample slowly descends away from the magnet, the control loop (capturing the sample image, calculating sphere position, determining a driving signal, and adjusting the magnet current) operates continuously until the sphere is pulled free from the tissue. At this point the power supply is disabled, the tissue stage is returned to its original position, and video capture ceases. Finally, the computer prompts the user either to display the raw data or to convert it to a force versus displacement graph using the system calibration data.

B. Magnetic Microspheres

Successful operation of the EMFT relies on the ability to manufacture, sort, and characterize magnetic microspheres made with a wide range of bioadhesive polymers. Various methods have been utilized to manufacture magnetic microspheres. In a previous publication, we have reviewed many manufacturing techniques used to produce magnetic bioadhesive microspheres for use with the EMFT (Hertzog et al., 1997).

A number of factors, specific to the operation of the EMFT, were considered in the design and characterization of magnetic microspheres. The first consideration regards the differences between the calibration sphere and the experimental sphere. To calculate forces from calibration data accurately, the product of volume (V) and magnetic susceptibility (χ) must be known or calculated for each microsphere. It is important that spheres are manufactured in a way that ensures homogeneity of spheres from the same batch. Therefore, we can assume that all the spheres from any given batch will have similar magnetic susceptibilities. It can be seen from Eq. (1) that for a given magnetic field ($H\nabla H$ = constant) the force experienced by the microsphere is proportional to the product of microsphere volume and magnetic susceptibility. Consequently, the EMFT can account for differences between the calibration sphere and the test microspheres simply by utilizing the features of the video analysis system to calculate microsphere volume and scaling the calibration curve accordingly.

The bioadhesive forces produced by a polymer microsphere are a result of physical and chemical interactions at the tissue-polymer interface. Consequently, the second major manufacturing consideration for EMFT spheres is the effect of the magnetic material on bioadhesive tendencies of the polymer. Ideally, to best simulate the behavior of nonmagnetic polymer microspheres, the magnetic material should have no effect on the bioadhesive properties of the polymer. Because the bioadhesive properties are thought to be entirely a function of the surface properties of the microspheres, the best way to minimize the effects of the magnetic material is to concentrate it in the core, thus avoiding surface contact. We have developed various techniques for localizing magnetic material away from the surface of the sphere, and we have successfully manufactured single- and double-wall spheres with no apparent magnetic material on the surface (Fig. 5) (Hertzog et al., 1997).

1.   Magnetic Materials and Considerations

The choice of magnetic materials used to manufacture microspheres for the EMFT may drastically affect the performance of the adhesion measurement system. Many materials respond to magnetic fields, but to produce attractive forces of magnitude great enough to overcome the bioadhesive forces we are limited primarily to ferromagnetic and superparamagnetic materials. The responses of paramagnetic and ferromagnetic materials in a magnetic field are very different. Figure 6 shows the magnetization curves for typical para- and ferromagnetic materials. When a paramagnetic material is introduced into a magnetic field, the elementary atomic dipoles begin to line up with the field. The number of dipoles that line up with the external field is proportional to the strength of the external field and the amount of kinetic energy (temperature) in the sample. Figure 6A shows the typical response of a paramagnetic material. Ferromagnetic materials exhibit a phenomenon called exchange coupling that helps to align the dipoles and overcome the tendency of thermal energy to randomize their direction.

There are two primary differences between the two types of magnetism, at least as far as the EMFT is concerned. First, when a ferromagnetic material is introduced into a magnetic field and the magnetic field is subsequently removed, the magnetization curve does not exactly retrace itself (Fig. 6B). This property is called *hysteresis*, and the end effect is that even after the external field is removed the material retains some residual magnetism. Paramagnetic materials, on the other hand, do not experience hysteresis, and their magnetization goes to zero when the external field is removed. The second main difference is in the magnitude of the magnetization induced in each type of material. The amount of magnetization induced in ferromagnetic materials tends to be much greater than that of paramagnetic materials. Therefore, a microsphere loaded with a certain mass of ferromagnetic material can generally produce much larger attractive forces than a sphere loaded with an equal mass of paramagnetic material.

In many of our spheres we have used an iron oxide powder purchased from Fisher Scientific (Fair Lawn, NJ). It is a combination of ferrous (FeO)

Figure 5 Magnetically loaded polymer microspheres used in the EMFT for bioadhesion testing. These spheres were manufactured using techniques that produce spheres with surfaces devoid of magnetic material. Consequently, the adhesive forces measured by the EMFT are a result of the properties of the polymer, exclusive of the effects of the magnetic material. (A) Double-wall sphere with poly(L-lactic acid) (molecular weight 24,000) surface and a core of poly(styrene) (MW 50,000) loaded with 25% Fe_3O_4. (B) Single-wall poly(styrene) sphere (MW 50,000) loaded with 30% Fe_3O_4.

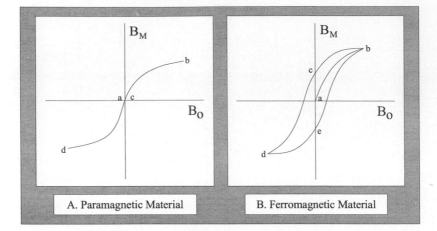

Figure 6   Generalized magnetization curves for (A) paramagnetic and (B) ferromagnetic materials. Unlike paramagnetic materials, whose magnetization curves retrace themselves (ABCD) under the influence of an alternating magnetic field, ferromagnetic materials exhibit a hysteresis loop (ABCDEB). Hysteresis results because the orientations of magnetic domains in a ferromagnetic material are not completely reversible. The end result is that ferromagnetic materials can have residual magnetism even after the external magnetizing field has been removed ($B_0 = 0$).

and ferric (Fe_2O_3) oxides and is supplied as a fine, dry, black powder that exhibits bulk paramagnetic properties. It has an average particle size of approximately 1.0 μm, but in most cases the powder was ground in a mortar and pestle to reduce the particle size further to approximately 200–300 nm. At 200–300 nm the Fisher powder is too large to distribute homogeneously in very small microspheres, and we have resorted to using 10-nm Fe_3O_4 (magnetite) particles isolated from samples of ferrofluids. Ferrofluids are suspensions of very small magnetic particles in various carriers. We isolated the magnetite particles by freezing and lyophilizing water-based ferrofluids (EMG 1111, EMG 807, Ferrofluidics Corp., Nashua, NH) and then loading the magnetite into spheres as a dry powder. The 10-nm magnetite particles exhibit superparamagnetic behavior. In cases in which the bioadhesive forces are expected to be high, we load spheres with ferromagnetic material such as pure iron powder (Aldrich Chemical Co., Milwaukee, WI).

Because of the paramagnetic nature of the iron oxides, it is desirable to use them when possible. Although ferromagnetic spheres can produce larger forces, the hysteresis of their magnetic material can cause problems.

For instance, as a microsphere is pulled off a tissue sample, other spheres in the same vicinity will experience a small magnetic field from the magnet tip. If these spheres are ferromagnetic, each may end up with a net magnetization of its own, and the spheres may consequently be attracted to each other. This net attraction can make running repetitive experiments with ferromagnetic spheres problematic. In addition, in calibration procedures for ferromagnetic spheres the magnetic hysteresis must be taken into consideration.

C. Calibration

Because of manufacturing variability, batch-to-batch differences between magnetic microspheres require that the EMFT is calibrated for each type of microsphere used. The system is calibrated with a few representative microspheres from each batch, and then the calibration is used to convert electromagnet current to force. Two techniques have been used to calibrate the EMFT, depending on the size of the microspheres being used and/or the range of forces to be measured. Both techniques are described in detail by Guilford and Gore (1992). The first technique uses glass microfilaments as small springs based on a technique described by Yoneda (1960), and the second technique measures the terminal velocity of the sphere in a solution of known viscosity.

Glass micropipettes are hot pulled to form a glass filament with a diameter of approximately 20 μm. Pipettes can be pulled with varying lengths and widths to adjust carefully the magnitude of the forces produced. In this way, the pipette can be designed specifically to calibrate the EMFT to a certain range of forces. As an example, longer, thinner pipettes (smaller bending constants) are used for small or less adhesive spheres, whereas shorter, thicker filaments are needed to produce the forces experienced by large, very adhesive spheres.

To calculate the bending constant, the glass filament is first mounted on a micrometer-controlled translation stage with the very tip of the filament resting against the sample loop of a sensitive electrobalance. The pipette is then stepped down in known increments and the force on the electrobalance is recorded. Figure 7 shows a number of pipette force curves from an EMFT calibration. Because the total distance the pipettes are bent (~1 mm) equates to such a small percentage of their overall length (~10–25 mm), the force versus displacement curves are almost perfectly linear. The slope of the force curve is calculated and recorded as the bending constant for the pipette (K_b). Next, the calibration sphere is mounted at the tip of the glass filament with epoxy, and the entire filament is mounted on the vertical translation stage of the EMFT with the sphere centered directly under the magnet tip (Fig.

Pipette Force Curves

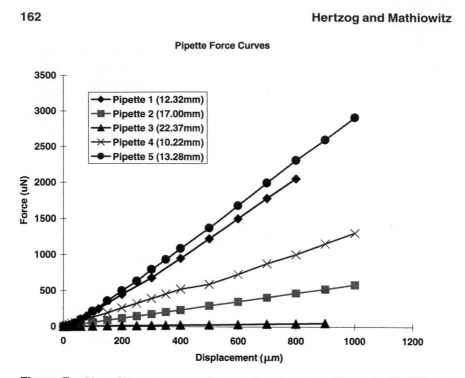

Figure 7 Glass filaments are used as small springs to calibrate the EMFT with relatively large spheres ($D > 50$ μm). The force curves of the glass filaments are very linear, and consequently the force produced at the tip of a pipette is simply equal to the produce of displacement and the bending coefficient (K_b, the slope of the force curve).

8A). The motion control system automatically moves the microfilament a known distance from the magnet tip, while the control system uses the electromagnet to hold the sphere in its original position. Consequently, the stage position is equal to the magnitude of filament bending as long as the sphere stays in its original position. Finally, using these data and the bending constant of the filament, a calibration curve can be constructed that relates magnet current to force for the given device geometries (magnet-sphere separation and sphere diameter).

The viscous fluid calibration technique works well with smaller spheres that are difficult to mount on the end of the glass filaments. A special temperature-controlled chamber has been developed to hold the viscous fluid. The calibration sphere is suspended in the fluid and the motion control

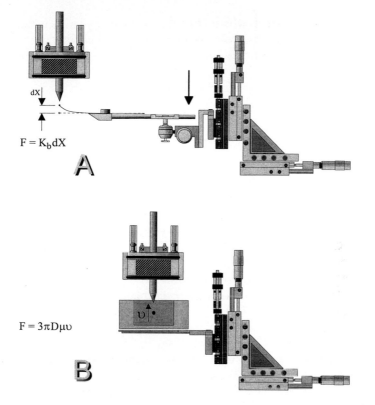

dX

$F = K_b dX$

A

$F = 3\pi D\mu\upsilon$

B

Figure 8 Two techniques are used to calibrate the EMFT. (A) Large spheres are calibrated using glass microfilaments with known bending constants (K_h). (B) Small spheres are calibrated by pulling them through a solution of known viscosity (μ) and using Stokes' law to calculate the force acting on the sphere based on its terminal velocity (v) and diameter (D).

system is used to focus on the sphere and position it directly under the magnet tip. A fixed current is supplied to the magnet and the video analysis system is used to measure the velocity of the sphere as it passes through a point at a predetermined distance from the magnet tip (Fig. 8B).

$$F = 3\pi D\mu v \tag{2}$$

Stokes' law [Eq. (2)] is then used to calculate the force acting on the sphere at the given magnet current, where D is the sphere diameter, μ is the fluid viscosity, and v is the terminal velocity of the sphere. This process is repeated for a range of magnet currents to produce a force versus current calibration curve.

IV. RESULTS AND DISCUSSION

A. Performance

The sensitivity of the EMFT is a difficult parameter to define because it is different for practically every microsphere tested. From Eq. (1) we see that the force experienced by the microsphere is equal to the product of magnetic field strength H, field gradient ∇H, microsphere volume V_m, and magnetic susceptibility χ. If we define the absolute sensitivity of the EMFT as the smallest measurable change in force, then we must look at how the EMFT measures force. The EMFT does not measure force directly; rather it measures the magnet current required to counteract the bioadhesive forces experienced by the microsphere. Therefore the measured value is really magnet current. The current to the electromagnet is regulated by the power amplifier that responds linearly to changes in the driving signal. The driving signal is a ± 5.0 V analog signal from the EMFT's 12-bit data acquisition card. Therefore, the smallest unit that the EMFT can record is a 1-bit change in the least significant bit (LSB). A change in the LSB for a 12-bit signal is 1 part in 4096 (2^{12}). Because the signal is 10.0 V full scale, the LSB would equate to a 2.44 mV signal change. From a calibration curve constructed from force versus driving signal data (Fig. 9) we can then calculate the absolute sensitivity of the EMFT by determining the change in force for a 2.44-mV change in driving signal.

Sensitivity depends on a number of other factors, including magnetic susceptibility of the microsphere, microsphere diameter, and the tip-sphere gap. The EMFT is much more sensitive when using small paramagnetic spheres than it is with large ferromagnetic ones. As a reference, the absolute sensitivity of the EMFT is approximately 134 pN for a 50-μm polystyrene sphere loaded with 30% (by weight) Fe_3O_4 and positioned 100 μm from the magnet tip.

Keep in mind that this is an absolute sensitivity, and the actual resolution of the EMFT is affected to some extent by extraneous mechanical and electrical noise and the limit to which displacement of the sphere can be measured. A vibration isolation system has been incorporated in the base of the EMFT to help limit mechanical noise, and we run many of the elec-

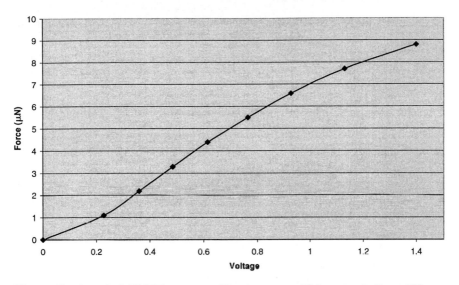

Figure 9 A typical EMFT system calibration curve. This curve is for a 200-μm poly(styrene) microsphere loaded with 30% Fe_3O_4 and located 100 μm from the magnet tip. This curve has been plotted against the control voltage to the power amplifier to demonstrate that the absolute sensitivity of the EMFT is equal to the product of the slope (force versus volts) and the smallest adjustment that can be made to the control voltage (0.00244 V).

tronic devices from 12-V DC automotive batteries to minimize ripple currents and other electrical noise. The EMFT's ability to measure microsphere movement is affected by the resolution of the video signal and the magnification of the microscope. We are experimenting with a number of different edge detection and imaging algorithms to quantify microsphere position quickly and accurately. These techniques are capable of tracking both subpixel and macro movements of the microsphere.

The video microscopy, image analysis, high-resolution motion control, and force measurement capabilities of the EMFT enable the system to perform a wide range of adhesion and cellular studies. Samples can be accurately positioned with a resolution 0.124 μm and a velocity range of 0–4000 μm/s. The three-dimensional motorized sample stage is well suited to repetitive and timed tasks.

B. Microsphere Adhesion

The EMFT was designed to overcome some of the limitations of the CAHN.
The primary advantage of the EMFT over the CAHN system is that it is
remote sensing and no physical attachment is required between the transducer and the microsphere. The EMFT makes it possible to take accurate
tensile measurements on very small microspheres that have been implanted
in vivo and then excised (along with the host tissue) for measurement. As
a result, the microspheres experience normal biological processes that may
drastically effect bioadhesive interactions. We have used the EMFT to manipulate spheres as small as 10 μm, although it is conceivable that we can
handle much smaller spheres.

Spheres can also be adhered to tissue in vitro if needed. We have used
a modified everted-sac technique to adhere microspheres to rat intestinal
tissue that had already been excised from the host animal (Fig. 10) (Jacob

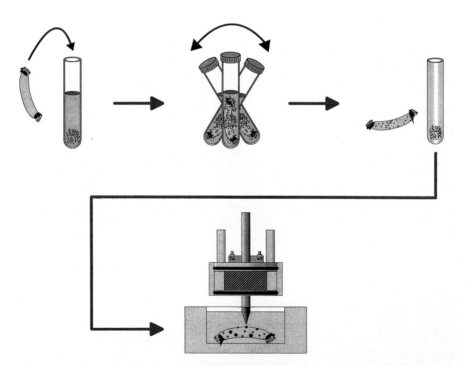

Figure 10   Microspheres can be adhered to tissue in vitro using the everted-sac
technique. Microspheres are agitated with an everted sac and PBSG in a sealed
container for 15 minutes at room temperature. The sac is then placed in the tissue
chamber of the EMFT and spheres are pulled from its surface.

et al., 1995). For the everted-sac technique, the small intestine is excised from sacrificed rats and flushed with approximately 10 mL of cold phosphate-buffered saline, pH 7.2, containing 200 mg/dL glucose (PBSG). The intestine is then cut into segments approximately 3 cm in length. The segments are everted (lumen side out) with a stainless steel rod and washed with PBSG to remove any remaining gut contents. Ligatures are used to occlude each end of the segment, and the resulting sac is filled with PBSG. Next, the sac is placed in a sealed vial with 10–100 mg of microspheres and 5 mL of cold PBSG. The vial is then agitated for 15 minutes at room temperature. Finally, the sac is removed from the sealed vial and loaded into the tissue chamber of the EMFT.

It is often impractical to sacrifice an animal every time the EMFT needs to be tested. Consequently, for developmental purposes, a simulated bioadhesive surface is used in place of living tissue. Generally, a thin layer of high-viscosity silicone vacuum grease is spread on a glass slide, and magnetic spheres are simply sprinkled on the surface of the grease. This simulated "bioadhesive surface" is then mounted on the sample stage of the EMFT. The grease layer exhibits adhesive forces that are in the same range as those typically observed with living tissue, and the mechanical properties of the grease are much more predictable than those of mucus. Figure 11 shows the adhesion curve for a 250-μm magnetic microsphere pulled from a simulated bioadhesive surface.

C. Cell Adhesion Experiments

A new project has been started in an attempt to use the EMFT to evaluate the adhesion between cultured cells and different man-made and bioartificial substrates. The aim of this study is to identify materials, substrates, coatings, and various surface properties that result in cellular adhesion and may ultimately lead to the design of specific surface treatments that enhance cell ingrowth into porous biomaterials. The advantages of testing cellular adhesion using the EMFT are that specific cell types can be tested under controlled conditions, precise forces can be measured, and the entire adhesion event can be observed and recorded using the microscope and video analysis capabilities of the EMFT.

We have been investigating the adhesion of neonatal rat calvarial osteoblasts (RCO-Bs) via integrin binding on inorganic substrates coated with a variety of peptide sequences (RGDC, CG, RADC, and YIGSR). For the EMFT to measure cellular adhesion, the cells must first be made magnetic. This is accomplished by letting the cells phagocytose polymer nanoparticles loaded with Fe_3O_4. The nanoparticles are made from poly(caprolactone) (molecular weight 32,000) using a phase-inversion nanoparticle (PIN) technique

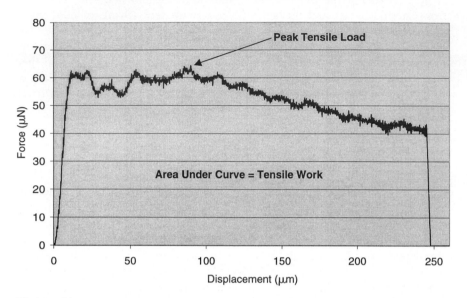

Adhesion Curve: 250 μm Magnetic Sphere on Silicone Grease

Figure 11   Force versus displacement curve of a 250-μm magnetic sphere pulled from a simulated bioadhesive surface (silicone vacuum grease) using the EMFT. Simulated tissue surfaces are often used because it is impractical to sacrifice animals to provide tissue samples for developmental purposes.

(Mathiowitz et al., 1997). The nanoparticles are loaded with 30% (by weight) 10-nm Fe_3O_4 particles, isolated from ferrofluids, and a small amount of rhodamine to make them fluorescent. The nanoparticles can be manufactured in various size ranges, but the spheres that seem to work best for our study have a number average size of approximately 3 μm as determined with a Coulter LS-230 particle size analyzer. The cells are plated in each well of a 12-well tissue culture plate at about 15,000 cells per well and are allowed to incubate for 24 hours before the magnetic particles are added (~1 mg per well).

The cells are incubated for 24 hours with the magnetic nanoparticles. Then, to determine that the cells have indeed engulfed the microparticles, the cells are trypsinized and viewed under a microscope. The osteoblasts that have phagocytosed the microparticles can clearly be seen. After the phagocytosis has been viewed, the cells are plated on sterilized 18-mm glass plates with peptide coatings and allowed to mature for 24 hours. Before the plates are placed in the sample chamber of the EMFT, each plate is rinsed with Hanks' balanced salt solution (HBSS) to remove any free magnetic particles.

Figure 12 demonstrates how the magnetic cells are positioned under the tip of the electromagnet. The image analysis and motion control systems are used to select and focus on a single cell. The cell is then positioned directly under and at a known distance from the magnet tip. The sample is slowly moved down and the magnet current is increased to pull the cell off the tissue. This cell is then placed on a fresh glass slide so that it can be observed later under a separate microscope, where we determine the magnetic loading of the cell.

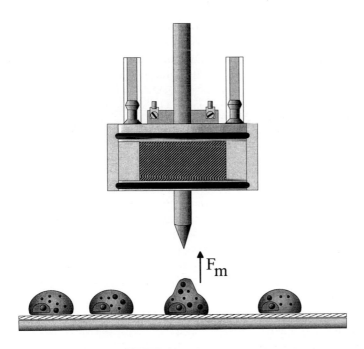

F_m

▨ Test Substrate

Figure 12 In addition to measuring the adhesion of polymer microspheres, the EMFT is used to quantify the adhesion of single osteoblasts to bioartificial substrates. Neonatal rat calvarial osteoblasts (RCOBs) are made magnetic by letting them phagocytose poly(caprolactone) (PCL) nanoparticles loaded with 30% Fe_3O_4. The cells are then pulled off of the substrate, and the force of adhesion is calculated.

To calculate the force of detachment from the magnet current data, we use a calibration curve for poly(caprolactone) spheres loaded with 30% Fe_3O_4. In bioadhesion tests using single microspheres we can simply calculate the volume of the sphere to know how much magnetic material it contains. Unfortunately, there seems to be a great deal of uptake variability between cells, so we are forced to view each cell with a microscope under ultraviolet light and count the number and size of the magnetic particles. Counting individual particles has proved difficult, so we have developed a technique using video analysis to estimate the volume of the magnetic material from a digitized image of each cell.

D. Future Perspectives

The EMFT has just progressed to a working state. Most of our efforts are now focused on streamlining the software, specifically the sample loop. We are making changes that will improve the sensitivity and usefulness of the device. One area that could use some improvement is the video feedback loop. The CCD video camera used in the EMFT greatly limits the bandwidth of force measurements. The sampling rate of the EMFT is ultimately limited by the 60-Hz refresh rate of the video system. Although the software can sample the video signal at speeds up to 500 Hz, the computer resamples the same image many times before the video image is refreshed. In addition, the main bottleneck in the software loop is the image acquisition. If we could eliminate the image acquisition portion of the feedback loop, the EMFT could run at much higher speeds.

With these two points in mind, we have started designing a new optical feedback loop in which we have added a silicon photodiode. The CCD camera will still be used for most functions such as sphere selection, auto-focus, dimensional measurement, and sample positioning, but the photodiode will be used to monitor sphere position during an experiment. The photodiode has a bandwidth of roughly 1.6 kHz, and by eliminating the image acquisition portion of the software feedback loop we should be able to increase drastically the overall bandwidth of the EMFT.

We plan to use the EMFT to reproduce some of the bioadhesion experiments performed using the CAHN tensiometric device (Chickering, et al., 1995, 1997; Chickering and Mathiowitz, 1995). These benchmark experiments will be used both to validate the new magnetic machine and to draw correlations between the two systems. Then the EMFT will be used to test a wide range of polymer microsphere delivery systems that were either difficult or impossible to test with the CAHN. We also plan to use the EMFT to research the mechanisms of bioadhesion and, possibly, the mechanisms and forces involved in the cellular uptake of microsphere delivery systems.

V. CONCLUSIONS

We have developed a powerful tool that should greatly increase the knowledge and understanding of bioadhesive interactions. The EMFT is on the frontier of the biomaterials testing. It offers a technique for accurately quantifying adhesive interactions between polymer microspheres and living tissue under near-physiologic conditions. The data collected from the EMFT will enable us to develop bioadhesive delivery systems to meet specific needs, target specific tissues, reduce cost, and improve patients' compliance. It is hoped that one day the bioadhesion experiments performed with the EMFT will lead to the development of a new generation of oral and injectable controlled-release delivery systems based on the specific bioadhesive properties of polymers.

REFERENCES

Chen JL, Cyr GN. 1970. Compositions producing adhesion through hydration. In: Manly RS, ed. Adhesion in Biological Systems. New York: Academic Press, pp 163–181.

Chickering DE, Harris WP, Mathiowitz E. 1995. A microtensiometer for the analysis of bioadhesive microspheres. Biomed Instrum Technol 29:501–512.

Chickering DE, Jacob JS, Desai TA, Harrison M, Harris WP, Morrell CN, Chaturvedi P, Mathiowitz E. 1997. Bioadhesive microspheres: III. An in vivo transit and bioavailability study of drug-loaded alginate and poly(fumaric-co-sebacic anhydride) microspheres. J Controlled Release 48:35–46.

Chickering DE, Mathiowitz E. 1995. Bioadhesive microspheres: I. A novel electrobalance-based method to study adhesive interactions between individual microspheres and intestinal mucosa. J Controlled Release 34:251–261.

Ch'ng HS, Park H, Kelly P, Robinson JR. 1985. Bioadhesive polymers as platforms for oral controlled drug delivery. ii: Synthesis and evaluation of some swelling, water-insoluble bioadhesive polymers. J Pharm Sci 74:399–405.

Duchene D, Touchard F, Peppas NA. 1988. Pharmaceutical and medical aspects of bioadhesive systems for drug administration. Drug Dev Ind Pharm 14:283–318.

Gu JM, Robinson JR, Leung SHS. 1988. Binding of acrylic polymers to mucin/epithelial surfaces: Structure property relationships. Crit Rev Ther Drug Carrier Syst 5:21–67.

Guilford WH, Gore RW. 1992. A novel remote-sensing isometric force transducer for micromechanics studies. Am J Physiol. (Cell Physiol 32) 263:C700–C707.

Gurny R, Meyer JM, Pappas NA. 1984. Bioadhesive intraoral release systems: Design, testing, and analysis. Biomaterials 5:336–340.

Hertzog B, Mottl T, Yim D, Mathiowitz E. 1997. Microspheres for use in a novel electromagnetic bioadhesion testing system. In: Schutt UHW, Teller J, Zborowski M, eds. Scientific and Clinical Applications of Magnetic Carriers. New York: Plenum, pp 77–92.

Ishda M, Machida Y, Nambu N, Nagai T. 1981. New mucosal dosage form of insulin, Chem Pharm Bull 29:810–816.

Jacob J, Santos C, Carino G, Chickering D, Mathiowitz E. 1995. An in vitro bioassay for quantification of bioadhesion of polymer microspheres to mucosal epithelium. International Symposium on Controlled Release of Bioactive Materials, pp 312–313.

Lehr C-M, Bodde HE, Bouwstra JA, Junginger HE. 1993. A surface energy analysis of mucoadhesion. II: Prediction of mucoadhesive performance by spreading coefficients. Eur J Pharm Sci 1:19–30.

Lehr C-M, Bouwstra JA, Schacht EH, Junginger HE. 1992. In vitro evaluation of mucoadhesive properties of chitosan and some other natural polymers. Int J Pharm 78:43–48.

Lehr C-M, Bouwstra JA, Tukker JJ, Junginger HE. 1990. Intestinal transit of bioadhesive microspheres in an in situ loop in the rat—a comparative study with copolymers and blends based on poly(acrylic acid). J Controlled Release 13:51–62.

Mathiowitz E, Jacob JS, Jong YS, Carino GP, Chickering DE, Chaturvedi P, Santos CA, Vijayaraghavan K, Montgomery S, Bassett M, Morrell C. 1997. Biologically erodable microspheres as potential oral drug delivery systems. Nature 386:410–414.

Mikos AG, Mathiowitz E, Langer R, Peppas NA. 1991. Interaction of polymer microspheres with mucin gels as a means for characterizing polymer retention on mucus. J Colloid Interface Sci 143:367–373.

Mikos AG, Peppas NA. 1983. Systems for controlled release of drugs. V. Bioadhesive systems. STP Pharm 2:705–716.

Mikos AG, Peppas NA. 1986. Comparison of experimental technique for the measurement of bioadhesive forces of polymeric materials with soft tissues. 13th International Symposium on Controlled Release of Bioactive Materials, p 97.

Mikos AG, Peppas NA. 1990. Bioadhesive analysis of controlled release systems. IV. An experimental method for testing the adhesion of microparticles with mucus. J Controlled Release 12:31–37.

Park H, Robinson JR. 1985. Physico-chemical properties of water insoluble polymers important to mucin/epithelial adhesion. J Controlled Release 2:47–57.

Park K, Park H. 1990. Test methods of bioadhesion. In: Gurny VLR, ed. Bioadhesive Drug Delivery Systems. Boca Raton, FL: CRC Press, pp 43–46.

Shahvarooghi A, Moses AJ. 1994. High-speed computerized dc magnetization and demagnetization of mild steel. J Magn Magn Mater 133:386–389.

Smart JD, Kellaway IW. 1982. In vitro techniques for measuring mucoadhesion. J Pharm Pharmacol 34:70P.

Smart JD, Kellaway IW, Worthington HEC. 1984. An in vitro investigation of mucosa-adhesive materials for use in controlled drug delivery. J Pharm Pharmacol 36:295–299.

Teng CLC, Ho NFL. 1987. Mechanistic studies in the simultaneous flow and adsorption of polymer-coated latex particles on intestinal mucus. J Controlled Release 6:133–149.

Yoneda M. 1960. Force exerted by a single cilium of *Mytilus edulis*. J Exp Biol 37:461–469.

8

Principles of Skin Adhesion and Methods for Measuring Adhesion of Transdermal Systems

Michael Horstmann, Walter Müller, and Bodo Asmussen
LTS Lohmann-Therapie Systeme GmbH, Andernach, Germany

I. (BIO-) ADHESION OF TRANSDERMAL SYSTEMS: FUNDAMENTALS

The conceptual understanding of bioadhesion also includes adhesion to the cornified skin as a main pharmaceutical application. The general mechanism and theory of bioadhesion have been discussed in detail in Chapter 1. However, because of the major differences between skin adhesion and the predominantly hydrophilic concepts of mucosal adhesion, a short reference to the mechanisms and theories involved should be helpful.

Transdermal patches, better known as transdermal therapeutic systems (TTSs), are pharmaceutical sustained-release devices that operate in a state firmly attached to human skin (Asmussen, 1991; Cleary, 1991; Ranade, 1991). Consequently, besides the delivery characteristics, adhesion properties are important attributes of this form of application (Ponchel, 1996). A transdermal adhesive has to provide firm, soft contact with skin but also has to allow the patch to be removed easily with only minor skin trauma.

Adhesion can be described as the formation of a new mechanical joint between surfaces of the adherent (here the skin) and the adhesive composition (Fig. 1).

There are several theories of adhesion and underlying forces that appear to act independently. Because of the magnitude of the materials and conditions involved, different degrees of their contribution are noted.

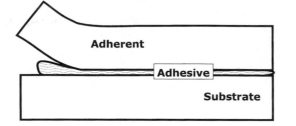

Figure 1 Principle of adhesive joint formation.

Very generally, an adhesive has to present both high adhesive (pre-dominantly surface-to-surface) *and* cohesive forces, which are necessary for mechanical transfer of the force of the newly generated bond between ad-herent and substrate. In contrast, in the state of bond generation, easy flow and wettability of the surfaces are preconditions.

Surface structure. Generally, a high surface area increases adhesive bonding. Rough and porous surfaces provide easier and stronger bonds than even ones, partly because of an increase in the effective area.

Bonding forces. A necessity for adhesion is high physicochemical interaction between the adhesive and the substrate surface. The forces involved follow general miscibility and wettability rules and include all types of molecular bonding forces. It is therefore conceivable that polar surfaces should be glued with polar adhesives and non-polar surfaces (such as polyolefins) with nonpolar adhesives. Generally, the adhesion force increases with polarity of both the adherent and adhesive.

The contact angle may be used to characterize the surface en-ergy and thus to predict adhesive performance.

Viscoelastic behavior of pressure-sensitive adhesives (PSAs). According-ing to Wetzel's (1957) theory, rubber-resin blends achieve their high tack through a two-phase structure in which a resin-dominated dis-persed phase (droplets of submicrometer size) is distributed in a polymer-dominated continuous phase. Intrinsically adhesive poly-mers such as polyacrylates behave more uniformly. Independent of their specific molecular structure, PSAs as polymer-based systems have both elastic ("rubber band–like") and viscous ("honey-like") properties.

Diffusion and interpenetration. This mechanism, in which polymeric surface structures partly dissolve and interact in the form of diffusing

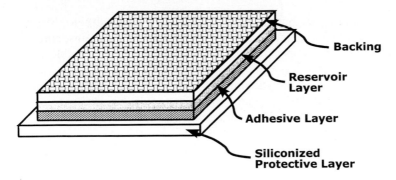

Figure 2 Construction of a transdermal therapeutic system (example).

polymer chain ends, is important in both mucous absorption and artificial polymer bonding. A clear border between the bonded structures thus disappears in this approach. We can also understand this highly effective mechanism to provide just a surface increase on a molecular level.

Most adhesives are used in a molten, dissolved, or non–cross-linked state in order to wet the surfaces and provide their bonds after cooling to room temperature, after evaporation of solvent, or after a cross-linking reaction.

Pressure-sensitive adhesives are adhesives of a specific physical form that are important in skin adhesion of transdermal systems essentially do not change their shape and adhesive force before and after bonding. Adhesives of this type can be regarded as viscous liquids that instantly provide tack to a suitable surface after short mechanical contact.

In the many published and marketed transdermal systems, at least one layer—the one that ultimately provides the skin contact—has to consist of a PSA in order to provide firm skin contact (Fig. 2).

II. SPECIAL ASPECTS OF ADHESION TO SKIN

In contrast to mucous surfaces, human skin does not provide a good substrate for hydrophilic ("classical") bioadhesives. Similarly, sticky skin adhesives do often not adhere much to mucosa. To give a practical example, chewing gum (e.g., a collapsing "bubble") provides nasty adhesion to facial skin but is fully dehesive to mucosa.

The lipophilicity of human skin resides only in the 20- to 40-μm-thick horny layer, which contains about 50% keratin-like proteins, approximately 20% lipids, and approximately 30% water and water-soluble compounds.

The lipid fraction of the stratum corneum is predominantly based on about 77% neutral lipids (triglycerides, free fatty acids, sterols, wax esters, squalene, alkanes) and about 20% sphingolipids, mainly ceramides, polar lipids, and cholesterol sulfate (Elias, 1983).

At medium air humidity and low temperature, the skin surface is basically dominated by lipids. Again, day-to-day experience teaches that pure water does not wet clean human skin well. In contrast, ethanol and petroleum ether spread evenly upon skin contact. With this common understanding, the success of adhesives with high or medium lipophilicity is conceivable. Upon skin contact, a good pressure-sensitive adhesive will instantly interact with surface lipids, provide mechanical flow into the "valleys" of the rough surface, and finally adhere directly to the keratin backbone.

Low-molecular-weight compounds generally provide a challenge for skin adhesion. On the one hand, water is generated by both passive diffusive transport ("perspiratio insensibilis," approximately 0.5 L/day on a total skin surface of about 1.8 m^2) and climate-dependent active sweating (an additional 0.5–1 L/d). Water has limited solubility in most adhesives (approximately 0.04% to several percent w/w) and may spread by the force of excretion between adhesive and skin surface.

Skin lipids are soluble to a higher extent but if not soaked up quickly may provide a film with decreased mechanical cohesion. Adhesives with a high capacity for lipids and speed in dissolving them will integrate these compounds as plasticizers and have a high likelihood of intense and long-lasting skin contact.

Sebum production is decreased with occlusion, and this nonocclusive adhesive systems are more likely to become softer with wearing time.

A. Stripping Effect

Upon detachment, most adhesives take up the upper stratum corneum layer ("stripping effect") after the first seconds of skin contact (experimentally, one can remove the major part of the stratum corneum layer by layer in this way; see Fig. 3). This is why reapplication of detached transdermal tapes is rarely possible. With the nitrazin yellow test (e.g., Gall, 1979), it is possible to detect minor fissures in the skin after stripping and study the resistance to alkali addition.

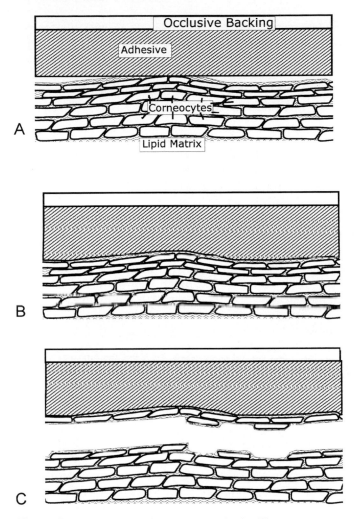

Figure 3 Detachment of a patch (''stripping'').

The stratum corneum layers are renewed completely about every 30 days. Hence, specifically after longer application times (1 week or more), the adhesion force on skin is based primarily on adhesion between single stratum corneum layers and thus dependent on individual skin behavior. With time, the contact between cornified keratocytes becomes loose and desquamation would have occurred if the skin had not been covered with a patch. Cholesterol sulfate is probably a key compound responsible for co-hesion between keratocytes in the epidermal lipid matrix (Elias, 1983).

III. BASIC PSA INGREDIENTS SUITABLE FOR USE IN TRANSDERMAL THERAPEUTIC SYSTEMS

Rarely, the adhesives used in transdermal delivery are just a uniform poly-mer. Because of specific requirements of drug delivery (permeability) for the active ingredient and the need to provide reservoir capacity for phar-maceutical vehicles, an adjustment to general skin PSAs is generally necessary.

The backbone polymer typically creates the cohesive basis of an ad-hesive. Cohesion is achieved mainly by high-molecular-weight molecules and a high level of entanglement, e.g., by branching or by chemical cross-linking of polymer chains, which leads to an overall gel-like network. Mo-lecular or micellar dissolved resin structures finally allow wetting and thus contact of surfaces.

Depending on this function in the skin adhesive layer, ingredients can be categorized as follows:

A. Backbone Polymers

The function of backbone polymers is to provide a cohesive network phys-icochemically compatible with both skin surface and other ingredients. Main examples are polyacrylates, polyisobutylene, polystyrene-polydiene block copolymers, natural or synthetic rubber, silicone adhesives, polyvinyl ethers, and vinyl acetate copolymers.

Polyacrylates or "acrylic adhesives" are widely used transdermal in-gredients because of their inherent pressure-sensitive behavior (without the addition of tackifiers), which they share only with polyether adhesives to this extent; their hypoallergenicity; their good oxidation stability; and their tolerance of both lipids and moisture (Lalla et al., 1994). For transdermal patches, in addition, their adequate solubility or diffusivity for active ingre-dients is a highly recognized property.

The polymers are available as solutions (in solvent mixtures contain-ing, e.g., alcohols, ethyl acetate, or even petroleum ether) and aqueous emul-sions. The advantage of emulsions is their generally higher molecular weight, but the necessary content of emulsifiers provides an undesirable softener load for adhesives prepared in this way. In addition, adhesives prepared from aqueous dispersions never completely lose their dispersion history and tend to disintegrate upon contact with water.

Many different monomers have been used for polyacrylate (always *co*polymers!) synthesis, but the main ingredients are 2-ethylhexyl acrylate, butyl acrylate, and isooctyl acrylate (Fig. 4). A small percentage of acrylic acid provides some polarity and the possibility of cross-linking by metallic ions or chelates. Many other cross-linking mechanisms have been reported in the patent literature.

Polyisobutylene (PIB) is used as a homopolymer with a widely varying chain length. PIBs with high molecular weights are rubber-like solid masses, whereas low-molecular-weight PIBs are more or less viscous liquids. By mixing PIBs of extremely high, medium, and low molecular weights (or mineral oil), a very inert adhesive can be prepared, but usually PIBs require the addition of tackifying resins. PIBs are usually processed in solutions of hydrocarbons and to a lesser extent in melt extrusions.

Styrene-isoprene (or butadiene)-styrene (SIS) block copolymers form a specific polymer backbone enabling the adhesive to cross-link physically in aromatic domains. The limited compatibility of both the styrene (inner) phase and the aliphatic (outer) phase with resins and other ingredients requires careful selection. SIS is very often used as a basic ingredient for hot-melt processing formulations because of the thermic reversibility of the physical cross-link.

Rubber (cis-polyisoprene) was the key ingredient of wound patches in the late 19th century and is still used today. Products of natural origin are, however, now replaced by high-quality synthetic types without a major heavy metal or polypeptide load.

B. Silicone Adhesives

Silicone pressure-sensitive adhesives (Fig. 5) are generally condensation products of polydimethyl (or polydimethyldiphenyl) siloxane ("polymer" component) with a resin-like three-dimensional silicate structure end capped to a large extent with trimethylsiloxy groups ("resin"). Polymer and resin react via still free silanol groups. Different molecular weights and different proportions of polymer and resin components allow adjustment for specific needs (Pfister et al., 1992; Toddywala et al., 1991). Generally, the higher the resin content, the lower the tack, but cohesive strength is improved. Further improvement of cohesion may be accomplished by the addition of cross-linkers, e.g., peroxides or metal salts.

The catalytic action of aminic drugs with free silanol groups (approximately 600–700 ppm) in transdermal systems often led to unexpected additional cross-linking during patch storage, which resulted in loss of tack. Modern, "amine-compatible" silicone PSAs undergo a final additional end capping with trimethylsilyl groups.

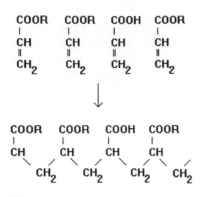

Figure 4 Polyacrylates.

Silicone adhesives are considered physiologically safe but expensive. Their adhesive behavior is soft and gentle but never reaches the strong bonding effect of rubber-based adhesives. Most silicone polymers may be reapplied to skin without major loss of tack, which is interpreted in terms of their apparent absence of a stripping effect. Because of their low solubility but high diffusivity for most active ingredients and their inertness, silicones provide a reasonable source of transdermal adhesives.

C. Resins

The function of resin derivatives in adhesive formulations is usually to increase tack, which is accomplished by both increasing fluidity and improving the wetting behavior at the skin surface. The expression "tackifying" resin

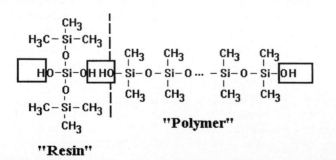

Figure 5 Silicone adhesives.

is more a technical than a structural term, and consequently several chemically very different "resinaceous" entities (lipophilic, brittle solids, or viscous fluids) are included.

Rosin derivatives, the largest group of resins for transdermal use, are based on *colophony*, a mixture of structurally related unsaturated resin acids (e.g., abietic acid); other neutral substances such as fatty acid esters, hydrocarbons, and terpene alcohols have also been isolated. Colophony, a resinous residue obtained on distilling oil of turpentine, is produced primarily from wood resin of *Pinus* species, but roots and tall oil also provide a source.

Colophony, as specified in the British Pharmacopoeia, is a resinous brittle mass that softens at about 73–80°C and has a density of about 1.07 g/mL. Moreover, it is subject to oxidative degradation and is a weak contact sensitizer. The main components of colophony, about 90% in the case of balsam colophony, are resin acids. The resin acids themselves are mixtures of tricyclic monobasic C_{20} hydrocarbon acids that differ only in the position and number of double bonds and side chains.

Hydrogenation (e.g., to di- or tetrahydroabietic acid) and/or esterification results in a high degree of stability to hydrolysis of the reaction products. Glycine, glycerin, or pentaerythrol esters resulting from hydrogenation of colophony (rosin) are used as components in paints, adhesives (especially in hot-melt adhesives that have to be resistant to heat and oxygen), surface coatings, print dyes, packaging (including food packaging), and chewing gum. This group of resins is also a preferred source for tackifiers in medical adhesives.

Hydrocarbon resins, in contrast to rosin derivatives, are true polymers obtained primarily from petroleum chemistry. In this group, *Polyterpene tackifiers* are polymerization products of β-pinene or other wood turpentine monomers. Furthermore, aliphatic (styrene or indene derived) or aliphatic (oligomerization products of the C_5 fraction) hydrocarbon resins are very commonly used to modify PSAs but are less important in medical applications.

D. Plasticizers

Plasticizers are convenient products for increasing tackiness and adjusting compatibility and wetting of surfaces. Examples are phthalates (e.g., dibutyl phthalate), mineral oil, citric acid esters, and glycerol esters.

For transdermal patches and medical tapes, other oily components of cosmetic and pharmaceutical origin (lanolin, oleyl oleate, medium-chain alcohols are also used). Technically, many compounds that are added in order to increase the permeability of the skin for the active ingredient ("permea-

tion enhancers'') also fall in this group, as do many active ingredients themselves (e.g., nicotine, nitroglycerin).

E. Fillers

Traditionally, rubber-based PSAs have been filled with finely dispersed inorganic substances such as zinc oxide, chalk, and silica. By addition of such
compounds, the consistency of the adhesive products can be adjusted and
light stability may be improved. With the introduction of acrylic adhesives,
the use of fillers has decreased, especially in dermal applications.

F. Cross-Linkers

There is a wide variety of chemical cross-linking in adhesives that are used
predominantly in technical applications. Cross-linking is used in acrylic adhesives in order to control cohesion without impairing the diffusion and tack
behavior too much. Chelates of titanium and aluminum, e.g., with acetyl
acetone, are examples of this group of compounds. Their addition to the
polymer rarely exceeds 1%.

G. Antioxidants

Antioxidants are typically already contained in raw materials containing unsaturated groups, such as rubber. Sometimes the adhesive formulation requires addition of extra antioxidants to protect the polymers or the active
ingredients. Typical concentrations are 0.5% of solids or lower.

H. Solvents

Solvents are—besides traces—not contained in the final product but are
necessary to create a uniform, spreadable mixture of the different excipients.
Petroleum ether fractions, other hydrocarbons, ethyl acetate, isopropanol,
ethanol, butanone-2 and acetone are typical examples. Their content in the
adhesive solution is typically in the range of 40 to 70% (w/w).

IV. HISTORY AND GENERAL CONCEPTS OF
MEDICAL ADHESIVES FOR TRANSDERMAL
THERAPEUTIC SYSTEMS

"Emplastra" in the pharmaceutical sense of the 15th to 19th centuries were
quite different in their concept (Aiache, 1984) from "patches" of today.

Many formulations of that time contributed complicated mixtures of plant, animal, and mineral origin and were even not primarily devoted to the treatment of external diseases but followed more a general (systemic) approach to medication. This concept reminds one of systemically acting transdermal systems. "Emplastrum lithargyri DAB6," a basic mixture of peanut oil, yellow lead oxide, and lardlead oxide, was a late example that survived into the 20th century but was finally used primarily as an ointment constituent.

Emplastra were, however, not very sticky and required other means of mechanical fixation. "Collemplastra," *adhesive* patches based on rubber, were introduced around 1870 by Seabury & Johnson in New York. These products also served purely fixation purposes and still contained plant products such as starch or medical herb root powders.

The "white patches" invented by Beiersdorf & Unna in 1901 (Leukoplast) were based on rubber, rosin, zinc oxide, and lanolin. These were already modern and stable textile medical (fixation) patches produced in a solvent-based coating process. In this way, patches containing, e.g., salicylic acid or topical antirheumatic ingredients were produced on an industrial basis and are still on the market today. Patches with water-soluble adhesives ("Emplastrum adhaesivum anglicum") were invented in parallel by the end of the 19th century but were—because of poor skin adhesion—not successful. Isinglass was the key ingredient.

Because of some skin irritation and contact allergy risks of the early resin types contained, after World War II polyacrylic ("hypoallergenic") adhesive masses replaced rubber-based adhesives for skin contact in most applications.

It is thus not surprising that the reinvention of transdermal delivery in the 1970s (e.g., the membrane-reservoir concept of Zaffaroni) appears to have been uncoupled from the older (and, primarily in Japan and Europe, still existing) experience with both adhesives and transdermal medication. Early transdermal patches of that time were generally bulky and rigid and retained much active ingredient after use—their adhesive behavior was to be improved. Since 1985, a kind of synthesis of modern adhesive patch technology with suitable drug candidates has occurred, providing a range of consumer-friendly, thin, and flexible patches that do their work reliably as pharmaceutical forms of application.

Because of worldwide regulatory constraints, the introduction of new adhesive technologies, especially involving new polymers, is restricted today as a consequence of expensive requirements for toxicological, clinical, and analytical qualification.

Typical recent examples of drug-carrying adhesive masses are shown in Table 1.

Table 1 Examples of Pressure-Sensitive Adhesives for Transdermal Systems

Type	Ingredients (example) (in order of decreasing content)	Reference	Advantage, application
Polyisobutylene adhesive	Polyisobutylene Mineral oil	US 4,031,894 Urquhart et al., 1976 Alza	No resin Low interaction In-line adhesive
Acrylic adhesive	Acrylic copolymer with [(2-alkoxy-2-phenyl-2-benzoyl)ethyl]acrylate	US 4,144,157 Guse et al., 1975 Beiersdorf	Very cohesive and softenizer-resistant medical adhesive
"Ion pair"	Cationic methacrylate Glycerol Fatty acids Adipic acid	US 5,133,970 Petereit & Roth, 1990 Röhm GmbH	Completely water soluble
Silicone adhesive	Silicone resin Silicone gum Ester plasticizer Alkylaryl siloxane extender 2,4-Dichlorobenzoyl peroxide (cross-linker)	US 3,929,704 Horning, 1975 General Electric	Example of progressed silicone adhesives allowing high softenizer addition
Polyacrylate with high softenizer content	Polyacrylate adhesive Polymethacrylate Softenizer	US 5,306,503 Müller et al., 1991 LTS	Shear-stable matrix to allow high softenizer addition

Thermoplastic elastomer adhesive	Oil/fatty acid Resin SIS copolymer	US 4,455,146 Noda et al., 1985 Hisamitsu	Adhesive with elastomeric behavior Hot-melt processing
Water-based acrylic adhesive	Polyacrylate latex	US 4,564,010 Coughlan & And, 1984	Processing without organic solvents
Polyisobutylene/resin adhesive	Ester resin Polyisobutylene Paraffin oil Rosin	US 4,623,346 von Bittera et al., 1984 Bayer AG	Typical PIB adhesive
Amine-resistant silicone adhesive	Silicone resin Silicone gum	US 4,655,767 Woodard & Metevia, 1985 Dow Corning	Amine (active ingredient) resistant compositions
Resin/SIS rubber composition	Resin Liquid paraffin Liquid polybutene rubber SIS copolymer Antioxidant	US 4,963,361 Kawazi, 1988 Teikoku Seiyaku	Low-molecular-weight rubber used as emollient
Basic acrylic copolymer	Acrylic copolymer with basic comonomers	US 5,458,885 Müller et al., 1994 LTS, Sandoz	Improvement of delivery for basic active ingredients

The objectives of transdermal formulators are smooth but safe adhesion to human skin, strong anchorage to backing, easy detachment of protective foil, no oozing out of oily or aqueous ingredients, and limited "cold flow" of the adhesive mass.

A short exemplary list of typical failures of adhesive behavior will help in understanding formulation mistakes and their circumvention:

A.   Adhesion to Pouch (Inner Side) During Storage

Upon slight (but sustained) pressure, the polymeric gel behaves like a viscous fluid and adhesive tends to ooze out with time beyond the boundaries of the backing material (Fig. 6). Because of this adverse property, the patch sticks to the inner side of the pouchstock, and it may be impossible to take the patch out of the pouch without destroying it.

Circumvention: Besides mechanical possibilities (e.g., embossing the surrounding release liner with spacers), cohesive forces in the active matrix should be strengthened. This may be accomplished by using a higher molecular weight polymer, by cross-linking, by mixing polymers of different polarities and molecular weights (Ko et al., 1995), or by increasing the proportion of high-melting resin.

B.   Formation of Adhesive Residue on Skin ("Dark Rings")

The patch sticks well on the skin, but after detachment textile fibers from clothing tend to mark the patch's former shape on the skin, indicating that residual adhesive is left. As in the previous case, cold flow due to low cohesive strength is the basic reason for this. In addition, skin sebum further dilutes the adhesive, especially in the area surrounding the patch. The effect is worsened when less occlusive backings and adhesives are used, and adhesive residue may be found on the complete application area.

Circumvention: Increase polymer cohesion as pointed out before. Use occlusive backing materials. Decrease the thickness of the adhesive layer.

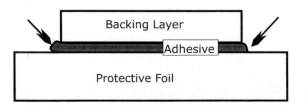

Figure 6   "Oozing out" of adhesive in transdermal patches due to cold flow.

C. Adhesion Failure of Backing Layer

Upon detachment, only the piece of backing layer is detached and the adhesive matrix still adheres to the skin. This kind of failure is easily prevented by performing peel adhesion tests with patches and by critical observation. If—even after storage—no failure of adhesion to the backing is noticed in the adhesion test, bad performance in vivo is very unlikely. In particular, backing materials with low polarity provide low anchorage.

Circumvention: If the adhesive cannot be modified to have lower polarity itself, the use of primers or corona treatment may help.

D. Protective Foil Difficult to Detach

This failure may have manifold causes: inhomogeneous coating of the release liner surface with adhesive; too high polarity of the protective liner surface in comparison with the adhesive; and formation of, e.g., microcrystalline anchorage points by resin, drug, or other solid component.

E. Poor Adhesion to Skin

The major reasons for poor skin adhesion are

> Overall adhesive polarity too high (rarely too low)
> Use of too high molecular weight polymers
> Too much cross-linker
> Excessive sweating due to the agent's own pharmacodynamic effect
> Tackifier content too low or high adhesive viscosity
> Change of adhesive properties because of diffusive loss of active ingredient
> Backing layer too stiff
> Patch area too large

Very often, excipients with a positive effect on drug transport interact badly with the polymeric adhesive structure. For example, the adhesive force of acrylic adhesives is markedly reduced after addition of polyethylene glycol (Dittgen and Adam, 1991). The interaction of adhesives with moisture in general provides a specific challenge that may be met by using polymers with high solubility of water (Schultz et al., 1995).

V. TEST METHODS FOR ADHESION

Many test methods are available for technical adhesives, but it is recommended that the beginner concentrate on a few significant and most widely used ones.

For the *development* of transdermal systems, most methods—especially those concentrating on tack and peel adhesion evaluation—do not provide adequate models for adhesion on human skin because there are major differences between artificial testing surfaces and the human stratum corneum surface. At present, no artificial skin model is available for in vitro testing. Even if such a model skin should be established in the future, we have to face the huge variation of human skin surface properties, which is primarily genetically based.

For routine control and stability testing of transdermal therapeutic systems, however, surface (e.g., steel) peel adhesion tests and the determination of the release force of the protective foil ("liner release") are usually done for new products and are required by some authorities for marketing applications.

Care should, however, be taken in the interpretation of results. Very often, slight (e.g., storage-dependent) modifications of the polymer interaction with other ingredients (active ingredient, other plasticizers, tackifying resins) result in marked changes of adhesion to steel that are not likely to be seen on living human skin. Also, results of the liner release test may vary with storage because of migration of small molecules, subtle resin crystallization, and polymer chain rearrangement. These changes are relevant only if they impair the ability of a patient to remove the release liner completely before use.

General descriptions of the basic principles of the most common tests and research methods for evaluation of pressure-sensitive adhesives are given next. They refer to test tack, adhesion or shear force, and combinations of these attributes.

A. Rolling Ball Tack

A solid ball (e.g., stainless steel, 11 mm in diameter*; see Fig. 7) is rolled down an inclined track passively and afterwards rolls out on a horizontal piece of the adhesive test structure. The rollout distance is inversely related to tack. By definition of the ball size and material and the ramp geometry, various combinations of speed and impressing force can be defined.

B. Surface Tension

As with other solid polymer surfaces, the surface tension is best tested by applying droplets of (not dissolving) fluids to the free surface and measuring

---

*Method 6. In: Test Methods for Pressure-Sensitive Adhesives. 6th ed. Itasca, IL: Pressure-Sensitive Tape Council.

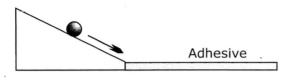

Figure 7 Rolling ball test.

contact angles. The "critical surface tension" of a surface can then be approximated by extrapolating the values found to a zero contact angle (Zisman, 1964; Dann, 1970). The lower the critical surface tension of the surface to be adhered to ("adherent," here the skin), the smaller the surface tension of the adhesive has to be.

C. Probe Tack

The most direct task measurement is the contact sensation of the finger of the formulator upon gentle fingertip contact with the adhesive surface. With the well-known "Polyken" Probe Tack tester (Hammond, 1963), (adjustable) contact force is variably applied on a defined contact area of a probe material (e.g., polyethylene) to the adhesive surface and the surfaces are then separated at a predefined rate. The tack is considered to be the force measured.

D. Peel Adhesion Test

In this test procedure (e.g., ASTM, PSTC, AFERA), the complete transdermal patch is adhered to a test surface, e.g., stainless steel, and then detached at a fixed rate (Fig. 8). The detachment force is recorded and taken as a measure of the adhesive force. The test is used with many variations of speed, type of substrate, and angle. A 90° angle is less sensitive to backing material failure and is preferred with transdermal systems. Peel adhesion is also used in a pharmacopoeial test (Pharmacopoeia Europaea).

The disadvantage of most in vitro adhesion tests, and especially of this one, is that they provide a description of a specific (unrealistic) type of failure of adhesion, more than a description of adhesion at work on the final biological surface, the skin.

Peel adhesion tests are performed technically in a very similar way.

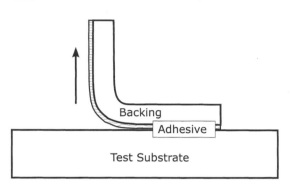

Figure 8   Peel adhesion test.

E.   Viscoelastic Properties

Dynamic mechanical analysis is a very special test that is used predominantly as a research tool for formulators and not so much for quality control. Adhesives have to provide both elastic and viscous properties in a balanced way.

For example, by application of sinusoidal oscillations to an adhesive sample firmly adhered between parallel plates, one can observe typical patterns in the time course of stress and strain. The main derived parameters are the *storage modulus*, a material constant dependent on the elastic behavior; the *loss modulus*, a material constant dependent on the viscous behavior; and *tan delta*, which is the ratio of the loss and storage moduli.

These parameters are dependent on the material, temperature, and frequency of oscillations. Most polymers, including adhesives, become more viscous than elastic with increasing temperature. At the glass transition temperature, a maximum of tan delta is usually found. Good adhesives provide a reduced storage modulus at a low shear frequency because of their ability to flow significantly.

With viscoelastic measurements it is possible to study in detail the compatibility of resin-oil and polymer mixtures.

As with general-purpose adhesives, with transdermal matrices an increase in adhesion on metal surfaces is noted with reduced polymer polarity (Toddywala et al., 1991). The author's notion, however, that the addition of minute amounts of steroid leads to a drastic decrease in adhesiveness in vitro appears unlikely to be true in the same way in vivo.

VI. IN VIVO ASSESSMENT OF ADHESION

Researchers with experience of adhesion both in vitro and in vivo are well aware of the limitations of tack and peel adhesion tests for extrapolation to in vivo conditions. In fact, the attribute of adhesion appears to be more difficult to assess than even the diffusional behavior of the active ingredient.

Until now, the specific interaction of skin in vivo with patches has not been mimicked completely by in vitro models, which would have to comply with the complicated active and passive moisture loss (e.g., see Hedenstrom et al., 1995), sebum production, desquamation, and other effects. Early comparison of results with the situation in vivo is therefore mandatory.

Spencer et al. (1990) have addressed this question by applying tensile testing techniques to living human skin. Even after 7 days of wear, they found peel adhesion values of 169 ± 100 and 152 ± 85 g/in with two different acrylic adhesives in seven volunteers. With increased moisture from

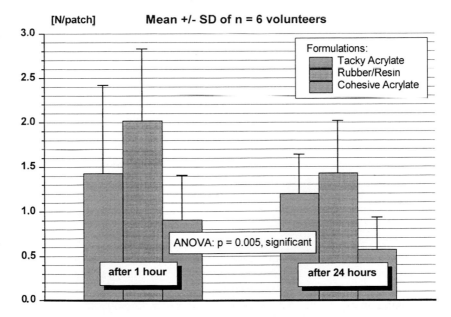

Figure 9 Adhesion in vivo: contribution of adhesive quality.

perspiration, values of 66 g/in and lower were measured after 24 hours. They concluded that an interval of 100–500 g/in reflected an adequate level of adhesion allowing firm fixation but gentle release.

The speed of removal of a tape from skin indicates (in the low-speed region) a change from a cohesive failure to an adhesive failure mode. The impact of this phenomenon on the formulation of skin adhesives is further evaluated by Schiraldi (1990).

Lücker et al. (P. W. Lücker et al., unpublished data, 1997) compared three differently adhesive placebo formulations in six healthy volunteers and found the method to differentiate between formulation (Fig. 9). However, the large individual differences between volunteers should not be forgotten. It is clear that adhesion to different human skin sources in vivo undergoes a higher degree of variation here than slight formulation improvements (Fig. 10). Nevertheless, this kind of experiment is the key to identifying in a bias-free way the optimal adhesive behavior in a limited number of formulation candidates. Ideally, such tests are combined with a pharmacokinetic study.

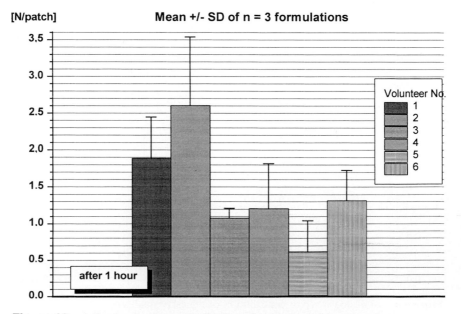

Figure 10   Adhesion in vivo: contribution of individual skin properties.

REFERENCES

Aiache JM. 1984. Historique des emplâtres. Bull Tech Gattefossé 77:9–17.

Asmussen B. 1991. Transdermal therapeutic systems—actual state and future developments. Methods Find Exp Clin Pharmacol 13:343–351.

Cleary GW. 1991. Transdermal drug delivery. Cosm Toil 106:97–109.

Dann JR. 1970. J Colloid Interface Sci 32:302–331.

Dittgen M, Adam H. 1991. Einfluß von Polyethylenglycol in der Haftschicht von Zweischicht-Transdermalpflastern auf die Liberation eines β-Blockers. Pharmazie 46:223–224.

Elias PM. 1983. Epidermal lipids, barrier function, and desquamation. J Invest Dermatol 80:044s–049s.

Gall H. 1979. Funktionsprüfungen der Hornschicht mit dem Indikator Nitrazingelb: Modifizierung der Alkaliresistenzbestimmung und Permeabilitätsprobe auf rekonvaleszenter Haut. Arbeitsmed Soz Med Prav Med 2/79:42–44.

Hammond FJ Jr. 1963. Polyken Probe Tack Tester. ASTM Special Publication 360, 1963.

Hedenstrom JC, Wick SM, Godbey KJ. 1995. The influence of adhesive, coating weight and backing on the adhesion to skin characteristics and MVTR of selected pressure sensitive adhesives utilized within transdermal drug delivery systems. Pharm Res (Annu Meet Am Assoc Pharm Sci 1995) 12:S-177.

Ko CU, Wilking SL, Birdsall J. 1995. Pressure sensitive adhesive property optimizations for the transdermal drug delivery systems. Pharm Res (Annu Meet Am Assoc Pharm Sci 1995) 12:S-143.

Lalla JK, Bapat VR, Malshe VC. Acrylate adhesives for transdermal therapeutic systems. Indian J Pharm Sci 56(1):5–9.

Pfister WR, Woodard JT, Grigoras S. 1992. Developing drug-compatible adhesives for transdermal drug delivery devices. Pharm Technol 16(1):42–83.

Ponchel G. 1996. Bioadhésion des formes dermiques. In: Seiller M, Martini M-C, eds. Formes pharmaceutiques pour application locale. Paris: Tec&Doc, pp 367–394.

Ranade VV. 1991. Drug delivery systems. 6. Transdermal drug delivery. J Clin Pharmacol 31:401–418.

Schiraldi MT. 1990. Peel adhesion of tapes from skin. Tappi Polymers, Laminations and Coatings Conference, Proceedings, Boston, pp 63–70.

Schultz HJ, Hart JR, Wick SM. 1995. Water uptake by adhesives and the associated affect on adhesion to skin characteristics. Pharm Res (Annu Meet Am Assoc Pharm Sci 1995) 12:S-177.

Spencer TS, Smith SE, Conjeevaram S. 1990. Adhesive interactions between polymers and skin in transdermal delivery systems. Polym Mater Sci 63:337–340.

Toddywala RD, Ulman K, Walters P, Chien YW. 1991. Effect of physicochemical properties of adhesive on the release, skin permeation and adhesiveness of adhesive-type transdermal drug delivery systems (a-TDD) containing silicone-based pressure-sensitive adhesives. Int J Pharm 76:77–89.

Wetzel FH. 1957. Rubber Age 82:291–295.

Zisman WA. 1964. In Contact Angle, Wettability and Adhesion. ACS Advances in Chemistry Series No. 1. Washington, DC: American Chemical Society.

9
Force Microscopy of Cells to Measure Bioadhesion

E. zur Mühlen, P. Koschinski, and S. Gehring
TopoMetrix GmbH, Darmstadt, Germany

Robert Ros and Louis Tiefenauer
Paul Scherrer Institute, Villigen, Switzerland

Eleonore Haltner, Claus-Michael Lehr, and Uwe Hartmann
Saarland University, Saarbrücken, Germany

Falk Schwesinger and A. Plückthun
University of Zurich, Zurich, Switzerland

I. INTRODUCTION

The direct study of molecular interactions with microscopes has been limited so far by the resolution limits of the wavelengths and the operating conditions. The traditionally used optical microscope provides a theoretical resolution of about half the wavelength of light, which is not sufficient to study effects on a molecular level. For investigations of biomolecules on this level, a field of view of some square micrometers with nanometer resolution is required. Electron microscopes, which provide much higher magnification than optical microscopes, suffer from the environmental conditions. The samples are studied in a vacuum chamber, and the samples usually have to be coated with a metal film to avoid charging by the electron beam. These conditions are far from the "real" conditions, e.g., a buffered solution (1).

A quite young technique, scanning probe microscopy (SPM), overcomes these restrictions and offers much more than the conventional microscopic techniques (2,3). SPM can be utilized under any environmental conditions, in air, under liquids, and in a vacuum. It can achieve magnifications as high as $10^9\times$, which makes it possible to visualize single atoms and atomic defects, and it measures a three-dimensional image of the surface. Moreover, other physical properties such as friction and viscoelastic, magnetic, electrical, optical, and thermal properties can be investigated simultaneously with the topography. All types and sizes of conducting and nonconducting samples can be handled with this nondestructive technique. Because of these advantages, SPM is used in material investigations, physics, semiconductor development and quality control, chemistry, and polymer studies and is gaining increasing popularity for biological applications.

SPM was invented by Binnig and Rohrer in the early 1980s (4). They presented the first scanning tunneling microscope (STM) and determined the atomic structure of a silicon surface. In 1986 they received the Nobel Prize for their work. The drawback that STM handles only conducting surfaces has been overcome with the scanning force microscope (SFM) also called the atomic force microscope (AFM), which was presented in 1986 (5). Scanning force microscopy is the technique mostly used today within the family of scanning probe techniques. SPMs are much more than magnification tools. Molecular interactions and forces can be measured and used to modify surfaces and molecules. SPM opens the way for nanotechnology, which will be one of the key technologies in the next century.

The goal of this chapter is to present the basic ideas of SPM and to demonstrate the function of the SFM as a force measurement tool for bioadhesion studies. In the biological area, SFM and also STM are used for high-resolution studies under "real" conditions and dynamical studies of cells, viruses, proteins, and DNA. Some references are given in Sec. IV for review purposes. The requirements for the applications of SPM for biological systems are discussed and results of adhesion measurements are presented.

II. SCANNING FORCE MICROSCOPY AND RELATED TECHNIQUES

A. Concepts of Scanning Probe Microscopy

In a scanning probe microscope, a sharp probing tip is brought either in contact with or in the immediate vicinity of a surface. The image is acquired by stirring the tip parallel to the sample surface, or alternatively the sample surface parallel to the tip, while acquiring surface properties in a rectangular array of coordinates. The actual surface properties probed by such a microscope are dependent mainly on the mechanism chosen to control the tip-to-surface distance. For example, the STM probes the density of electronic surface states by acquiring the current between a metal tip and a conductive surface. More important in biological applications, as nonconductive surfaces can be probed as well, the AFM or SFM is sensitive to the forces acting between tip and surface. In such a microscope the tip is attached to a soft cantilever. The bending of the cantilever thus provides information about the forces exchanged between tip and surface. Thus, the AFM is the most suitable tool for studying bioadhesion forces, and this chapter will focus on this SPM mode.

A further SPM technique, scanning near-field optical microscopy (SNOM), combines the measurements of surface forces with the acquisition of optical properties, e.g., luminescence, reflectivity, transparency, and even Raman signature. A high spatial resolution of typically 50 nm is achieved by using an illuminated glass fiber as the probing tip. As there exists a great variety of surface properties and thus SPM techniques, a discussion of SNOM and other SPM techniques lies beyond the scope of this chapter and the reader is referred to Refs. 6 and 7.

B. Contact AFM: Topography Imaging

1. The Tip-Cantilever System

The heart of any scanning force microscope is the tip-cantilever system. An image of such a system is displayed in Fig. 1, which shows the front part of the cantilever with a pyramid with a height of 4 μm pointing downward. The cantilever and pyramid are made from silicon nitride. A 3-μm-long tube of carbon is attached. The probing tip at the end of the tube remains sharp within the display limit; higher resolution images reveal that the tip end has a spherical shape with diameters typically ranging between 5 and 10 nm. The cantilever as displayed is a standard cantilever commercially available for contact imaging, i.e., imaging in the range of repulsive interatomic forces. It divides into two arms to provide lateral stiffness, each arm being about 220 μm in length and several micrometers thick. This geometry yields spring constants as low as $k = 0.032$ nN/nm. As a result, already small changes in force, e.g., a 0.1-nN change, a value below typical values for repulsive interatomic forces, leads to bending of a free tip-cantilever system of about 3 nm, a distance large compared with atomic corrugations. Thus, if such a tip is brought into immediate contact with a surface, the tip can generally be stirred across the surface, following surface features without destroying or otherwise altering the sample surface.

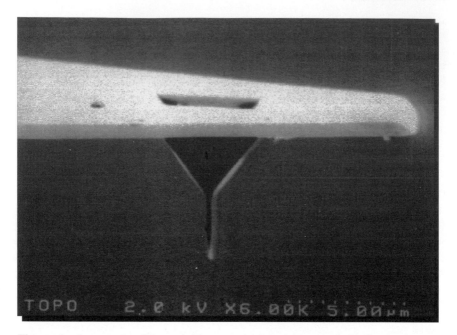

Figure 1 Image of a silicon nitride pyramidal tip with an attached carbon nanotip.

2. Typical Concept of a Scanning Force Microscope

To generate an image in contact mode AFM, the tip-cantilever system is brought into contact with the surface; i.e., tip and sample exchange inter-atomic repulsive forces. As the spring constant is small, the tip follows topographic features as the sample is moved parallel to the sample surface. Changes in topography can then be monitored by detecting the bending of the cantilever. This can be a very delicate task, considering that changes in cantilever deflection may be as small as atomic corrugations (<0.1 nm). In the development of atomic force microscopes, several approaches were used to measure the bending. Because of its reliability, versatility, and ease of use, the optical position-sensitive method is by far the most popular approach and thus will be discussed further.

A schematic setup of an atomic force microscope based on the optical position-sensitive method is displayed in Fig. 2 (8). Cantilever deflection changes are monitored by acquiring the position of a laser beam reflected from the cantilever. The laser beam is generated by a laser diode placed above the cantilever. The beam is aligned on the free end of the cantilever, which is mounted with a tilt of a few degrees downward relative to the horizontal plane. The beam is reflected from the cantilever at twice this angle

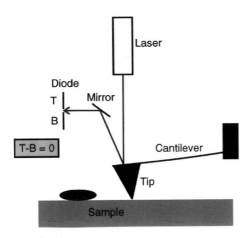

Figure 2 Basic components of a force microscope (not to scale). A tip is held by a soft cantilever and touches the sample surface. The vertical tip position is acquired by an optical deflection method: a laser beam is reflected from the cantilever, and the position of the reflected beam spot in a photodiode is monitored as the photocurrent difference between the top and bottom of the photodiode. Prior to data acquisition, the instrument is aligned to yield a previously chosen value for the photocurrent difference (e.g., T − B = 0). (From Ref. 8.)

relative to the incoming beam and propagated via a mirror into a position-sensitive photodiode. Positional sensitivity is gained by dividing the diode into four quadrants, two placed at the top (labeled T in Fig. 2) and two at the bottom (labeled B). As the tip touches the surface, repulsive interatomic forces bend the tip upward. Mirror and photodiode alignment is chosen in such a way that the reflected laser beam then rests in the vicinity of the center of the four quadrants. The exact location of the beam corresponds to the force acting between tip and sample, and the approach of the tip to the surface is stopped at a preset value, called the setpoint (chosen to be zero for simplicity in Fig. 2). As the sample is moved laterally relative to the tip, changes in topography cause a shift of the laser spot at the photodiode, which is electronically accessible as a change in the photocurrent difference between top and bottom quadrants relative to the set point.

For topographic imaging, the photocurrent difference serves as the input signal for a feedback loop, which controls the bending of the cantilever. In a first processing step the difference Δ between the actual photocurrent difference value and the set point is calculated. From this difference the loop tries to estimate an "improved" cantilever-to-sample position. The sample is raised or lowered to this new position, and again the photocurrent difference value is taken. The loop is passed repeatedly until the level of the set point is reached again. Height information can then be directly obtained by monitoring the total change of the vertical sample position while scanning the sample laterally. Thus, if a fast response of the feedback loop to topographic changes is provided, the cantilever deflection and consequently the force acting between tip and sample are kept at a constant value. A topographic image is then simply obtained by acquiring the total vertical change in sample position. The information is mapped in a square array of coordinates in the sample plane and stored in a computer with from 200×200 to 1000×1000 points per image. The resolution of the image is then simply determined by the distance between neighboring coordinates, as long as physical resolution boundaries are not surpassed. For display purposes the acquired height values are translated into a linear gray scale. Typically, dark gray values represent depressed regions of the sample and bright gray values are assigned to higher areas.

Two main factors determine the physical resolution limit. First, the resolution limit is given by the interaction volume between tip and sample, which will be discussed in the next section. Second, the positioning accuracy of the tip relative to the sample imposes a technical limit on the best possible resolution. Of course, it is impossible to reach a positioning accuracy on or even near an atomic scale with classical mechanical components. A more suitable approach had been found in the use of piezoceramic elements. Their movement is based on the fact that they alter their shapes if placed in an electric field. As an example, a thin ceramic tube with a grounded electrode on the inside and an electrode on the outside will change in length as a function of the charge on the outer electrode. If the outer electrode is divided along the long axis of the tube, motion perpendicular to the tube axis can be generated by applying opposite charges to the electrodes. The tube expands at one side and contracts at the opposite side, resulting in bending of the tube. Thus, if four equal electrodes are placed around a tube, it is possible to bend the tube in any direction perpendicular to its long axis. Finally, the length of the tube changes if a further bias voltage is applied to the inner electrode. Thus, if a sample is placed on one end of such a piezoceramic tube, it can be positioned laterally and perpendicularly relative to its surface. In general, the positioning accuracy of these mechanical motors, called scanners, can reach values as low as 0.01 nm and thus surpasses even atomic accuracy.

For large scan ranges, however, the use of tube scanners is disadvantageous. To generate a large scan area, tube scanners have to be quite long, which leads to mechanical instabilities. For this reason, scanners built from three different linear independently working piezo stacks, called tripod scanners, are favored for imaging areas larger than 10×10 μm. In this design two stacks are responsible for lateral motion and the third piezo stack works perpendicularly to the raster plane. A further advantage is that it is relatively easy to add control mechanisms to the piezo movements. Nonlinear components of the piezoceramic response to the applied voltage, such as hysteresis, are a well-known problem for large displacements. This is negligible for small imaging areas but causes large errors in length measurements on larger scales. These uncertainties can be overcome by hardware control of the piezo length during the scan process. For example, strain gages can be glued to the piezo stacks. These strain gages have negligible hysteresis and thus give precise feedback of the actual piezo lengths. The signal from these gages is processed by a feedback loop in the control electronics and used to correct piezo artifacts.

In addition to piezo artifacts, further complications may arise from unwanted movement between tip and sample such as vibrational noise or thermal expansion. Vibrational noise can range from low-frequency oscillations within a building structure up to high-frequency acoustic or electronic noise. Thermal expansion is caused by temperature changes of parts within the microscope. However, with careful site preparation, these factors can be controlled well and their disturbing influence kept to a minimum. In comparison with electron microscopes, atomic force microscopes are compact in design. The mechanical parts are often not larger than a fist, which makes vibrational damping easy and keeps thermal expansion to a minimum.

Relevant to adhesion measurements are contact techniques, which are used to study the force exchange between tip and sample. Next to the forces acting vertically between tip and sample, which will be discussed in detail later in this chapter, there are further dynamic frictional forces that may have an impact on the imaging process and thus should be mentioned here. During scanning, frictional forces act antiparallel to the movement between tip and sample. In addition to surface roughness, the reactivity and chemical nature of the surface influence the frictional force and thus may be useful for the study of biologically relevant samples. The frictional force results in a distortion of the cantilever parallel to its long axis. This distortion causes a "left-right" shift of the laser spot in the photodiode in addition to the "top-bottom" motion of the spot used to contact topography imaging. Thus, by acquisition of the photocurrent difference between the left and right quadrants of the photodiode, a lateral force image is obtained at the same time as the topography is mapped in contact mode.

C. Noncontact AFM

As well as measurements in which tip and surface are in immediate contact and interatomic repulsive forces are dominant, there are further possibilities for acquiring images in the regime of the much weaker van der Waals forces. These forces become dominant with increasing tip-to-sample spacing above 1 nm. Unfortunately, these noncontact regions above the sample surface are technically not accessible with contact AFM as discussed so far. Because of the extremely small spring constants, the attractive forces would pull a tip attached to a contact cantilever onto the sample until these forces were again balanced by repulsive forces.

Thus, to image with the help of far-ranging attractive forces, different tip-cantilever systems and imaging techniques have to be considered. For contact-free imaging, cantilevers with spring constants above 10 nN/nm have to be used to avoid contact caused by attractive forces. However, such a stiff cantilever responds very weakly to any change in force between tip and sample, and therefore more sensitive methods of acquiring the influence of forces on the tip have to be found. Instead of simply measuring the bending of the cantilever, the tip-cantilever system is brought into oscillation at its resonance frequency and the influence of the long-range forces on the resonance behavior is studied for positional feedback.

Thus, for noncontact operation the technical setup used for contact imaging has to be modified. The tip-cantilever system is now attached to a further piezoelectric ceramic, called a bimorph. A bimorph can be electronically excited into an oscillation of its thickness and is thus able to generate a mechanical oscillation of the cantilever. The driving frequency is typically chosen to be ν_{res}, the resonance frequency of the tip-cantilever system, at which the amplitude of oscillation reaches a maximum. The value of the resonance frequency itself varies with cantilever design and environment, typically ranging from 50 to 500 kHz. In the design of a typical instrument, the cantilever oscillation leads to an oscillation in the photocurrent difference between the top and bottom of the photodiode. This oscillation is then processed by a so-called lock-in amplifier, which generates an output proportional to the amplitude of the signal. At the same time it is possible to derive information about the phase shift between the driving oscillation at the bimorph and the mechanical oscillation of the cantilever.

Close to a surface, the resonance condition changes as the tip oscillates within the field of attractive forces. As the field decays within the oscillation amplitude, the tip is influenced by different forces during an oscillation cycle. The difference in force, more precisely expressed in terms of a force gradient $\partial F/\partial d$, enhances oscillation, resulting in a shift of the resonance frequency toward lower values. The system now behaves as if the spring constant k of the tip-cantilever system has been altered to

$$k' = k - \frac{\partial F}{\partial d}$$

This change in resonance properties is then monitored either as a decrease in oscillation amplitude or as a change in phase signal if the tip-cantilever system is still excited at the former resonance frequency v_{res}. Thus, the approach can be stopped at a given damping value of the amplitude or change in phase signal, which then serves as a set point for positional feedback. During scanning motion a protrusion in topography further reduces the amplitude, whereas a depression causes an increase. The system is able to respond to these topography changes if the amplitude serves as a signal for the feedback loop.

Next to van der Waals forces, the contamination layer, thin film of adsorbed water and hydrocarbon molecules, has a further impact on the resonance behavior of the tip. Depending on the chosen amplitude of the tip, its influence can be controlled and even utilized for imaging.

At a relatively high amplitude, the tip typically oscillates in and out of the contamination layer, and typically it cannot be avoided that the tip touches the surface during the oscillation cycle. In this mode the tip is in periodic contact with the surface, resulting in exchanged forces in the nanonewton regime. Thus, the exchanged force has the same order of magnitude as in contact techniques. The advantage of this mode is that topography and phase information can be acquired simultaneously. Being more sensitive to small changes in force, this phase information often reveals information about the fine structure of a surface or about the contrast in adhesion forces. The disadvantage is that the periodic contact of the tip and surface has the potential of destroying either tip or surface.

The development of more sensitive electronics led to the development of techniques in which smaller oscillation amplitudes can be used in the imaging process (9). In NearContact mode, oscillation amplitudes of typically 2 nm are used while the feedback loop reacts to the more sensitive change in phase signal. With this concept the tip can be held within the contamination layer while the tip touching the sample surface can be avoided. In this mode, there is no damage to tip or sample, and because of the small amplitude, extremely high resolution can be obtained. Acting forces are at least one order of magnitude smaller than those relevant to periodic contact imaging.

Finally, if small amplitudes are applied to the cantilever and the feedback is tuned to a high sensitivity to react to small changes in force during the approach to the surface, the system can be operated in a noncontact mode, in which the surface of the contamination layer is imaged (10). This mode has the advantage of extremely low force exchange, e.g., in the piconewton regime, but due to the relatively large distance between tip and surface the resolution is somewhat limited. This, however, allows imaging of most delicate samples, which would be destroyed or otherwise influenced, especially under periodic imaging conditions.

Examples of images obtained with these modes of noncontact operation are given later throughout this chapter.

III. FORCE MEASUREMENTS AND FORCE MAPPING

A. Relevant Forces in Atomic Force Microscopy

A thorough understanding of the various forces acting between tip and surface is of vital importance to the operation of an atomic force microscope. An in-depth theoretical understanding of all forces acting between such a tip and a probed surface is impossible, as a very large number of tip and sample atoms are involved within the interaction volume. Nevertheless, the natures and relative contributions of individual forces are well understood. Two contributing forces present in any system can be discussed by studying two neutral, nonpolar atoms in the gas phase. Separated by distances greater than several tens of nanometers, these atoms do not exchange any forces. As the distance between the atoms decreases, the atoms experience an attractive force, which is due to an electric dipole interaction between them. This attractive force between the two atoms is called the van der Waals force. The strength of the force is proportional to about $1/d^7$ for distances smaller than 10 nm.

As soon as the electron clouds of the two atoms interact directly, the resulting repulsive forces become stronger than the weak attractive van der Waals forces. The overlap of the electron shells results in incomplete shielding of the charge of the two atomic nuclei. This leads to an exchange of repulsive Coulomb forces. In addition, according to the Pauli exclusion principle, equal electron states can overlap only if the quantum mechanical state of one of the electrons changes, i.e., is brought to a higher energy level that causes an additional repulsive force. Thus, as the interatomic distance decreases to values below 1 nm, within the range of atomic radii, the atoms exchange strong interatomic repulsive forces. These forces easily reach a level of several nanonewtons and above. Mathematically, the dependence between force and interatomic distance can be derived from the Lennard-Jones potential:

$$V(d) = -3E_{\mathrm{Eq}} \left[\left(\frac{\sigma}{d}\right)^{12} - \left(\frac{\sigma}{d}\right)^{6} \right]$$

with E_{Eq} being the lowest potential energy at the equilibrium distance $d_{\mathrm{Eq}} = 2^{1/6}\sigma$. Of course, the Lennard-Jones potential is a rough approach to a tip and sample system. A complete description involves many tip and sample atoms within the interaction volume. A more detailed theory of van der Waals forces (11) yields

$$F_{vdW}(d) = -\frac{H}{6}\frac{R}{d^2}$$

for the van der Waals forces acting between a sphere of radius R, approximating a spherical tip, and a flat sample surface. The Hamaker constant, H, is itself a function of the refractive indices of sample, sphere, and immersion medium as well as of the absorption energies of these media. The decrease of force with distance is thus proportional to $1/d^2$, that is, slower than the distance dependence calculated from the Lennard-Jones potential. Typically, van der Waals forces are in the piconewton ($1\ pN = 10^{-12}\ N$) regime; thus they are three orders of magnitude lower than the repulsive interatomic forces used for feedback in contact imaging. Depending on the nature of the sample and tip, additional forces may be exchanged between tip and sample. On the one hand, far-ranging electrostatic or magnetic forces can be monitored a few hundred nanometers above a surface. On the other hand, chemical binding forces between tip and surface have to be taken into account whenever applicable—examples of such forces are discussed in Sec. V.

In an ambient air environment, surfaces are commonly covered by a thin contamination layer, which consists mainly of condensed water and hydrocarbon molecules. The absolute thickness of such a layer may reach values as high as 20 nm, with the absolute value depending on surface topography, chemical nature, and air humidity. If the tip is brought into the immediate vicinity of the surface, their contamination layers overlap and attractive capillary forces form while the system tries to minimize its total surface area. The strength of the capillary force can be calculated from thermodynamic equilibrium considerations, yielding

$$F_{Capillary}(d) = \frac{\pi RT\rho}{M}\ln\left(\frac{p}{p_s}\right)r(t-d)$$

where $2r$ is the diameter of the water bridge between tip and sample, which is about equal to the tip radius; R is the universal gas constant; T is the temperature; ρ and M are the mass density and the molar mass of the wetting liquid; p/p_s is the relative vapor pressure, which in ambient air is equal to the relative humidity; t is the thickness of the contamination layer; and d is the distance between tip and sample (from Ref. 12). Thus, the capillary force is proportional to the tip radius and can be quite large for a blunt tip. As a rule, assuming typical parameters, every nm of tip radius adds a nanonewton in force, so a tip with a radius of 50 nm yields a capillary force of about 50 nN. Therefore these capillary forces dominate van der Waals forces and falsify high-resolution adhesion force measurements significantly. The typical approach to solving this problem is either to operate the system in a dry gas atmosphere or to immerse the tip and sample completely in a liquid.

B. Contact AFM: Force-Distance Curves

Forces between a probing tip and a surface are accessed by force-distance curves. Such a curve displays the bending of the tip end of the cantilever versus the relative position between the free tip-cantilever system and a silicon sample. The force $F_{\text{Cantilever}}$ on the cantilever itself can then be calculated from Hooke's law:

$$F_{\text{Cantilever}} = k_{\text{Spring}}z$$

where k_{Spring} is the spring constant of the cantilever (e.g., 0.032 nN/nm) and z the bending of the cantilever.

At a large tip-to-sample distance there is no force exchange and thus no influence on the cantilever (horizontal line labeled (a) in Fig. 3). As soon as the contamination layers of tip and surface overlap, or, alternatively, as soon as van der Waals forces destabilize the tip-cantilever system, the tip is rapidly pulled onto the sample surface (dip labeled (b) in Fig. 3). Thus, in liquids under ultrahigh-vacuum conditions or in a dry protective gas atmosphere, i.e., conditions in which the strong capillary forces do not appear, information about van der Waals forces can be acquired. Once tip and sample are in contact, they exchange repulsive interatomic forces. The tip follows an upward motion of the sample, which leads to an increasing deflection (labeled (c) in Fig. 3). The upward motion of the tip is reversible upon retraction of the sample (upper part of line labeled (d) in Fig. 3). It is worthwhile to mention that the curve will deviate from its original upward path if there is inelastic deformation of the sample. When the former "jump to contact" position is surpassed, the tip still remains in contact with the surface as it is now held by capillary forces (lower part of line labeled (d) in Fig. 3) or adhesion forces, which play the major role in the results discussed later in this chapter. The cantilever is bent downward until the re-

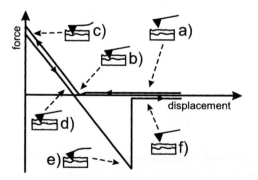

Figure 3   Schematic of a force-distance curve.

sulting force $F_{Cantilever}$ overcomes the adhesive forces. Then the cantilever snaps the tip away (line (e) in Fig. 3) from the surface until it reaches its original position (line (f) in Fig. 3).

IV. BIOLOGICAL APPLICATIONS

Microscopes are important and widespread tools in biological research in various fields, because they allow access to microscopic structures that are not directly accessible with the bare eye, such as cells or cell elements. Consequently, it is natural to apply scanning probe microscopy to biological specimens, even though this new class of microscopes was initially designed especially for applications in material science. Many SPM techniques have been used and still are in use for characterizing biological specimens, such as scanning near-field optical microscopy (SNOM) (6) or scanning tunneling microscopy (STM) (13) (a comprehensive introduction to SPM applications in biology can be found in, e.g. Ref 14). However, scanning force microscopy (SFM) is a very popular technique among scientists because it has several advantages in comparison with other SPM techniques. For example, no conductive specimen is required, and the investigations can be carried out in physiologic environments to name a few of them. Therefore, SFM is an extremely suitable and versatile technique for biological applications. Hence, the main part of this chapter focuses on SFM applications.

In principle, one can identify two major fields of application for SFM in biology: investigations of biocompatible materials and imaging of surface morphologies of specimens. The former is strongly linked with materials science, because in biology the same specimen properties are of interest as in materials science, such as roughness, friction, and adhesion properties (6). Because SFM was designed to measure those properties, it is also a valuable tool for determining the corresponding properties of biological specimens. But one should keep in mind that, in contrast to materials science, the special biological environment has to be taken into account when characterizing those surfaces with SFMs.

One interesting application of SFM is in the examination of molecular interactions that are manifest in forces (15). For this purpose the microscope tip is coated with organic monolayers, and the force interaction with a specially passivated surface can be measured (16). This measuring mode can be used to investigate antigen-antibody reactions (17,18). By performing these measurements in a two-dimensional manner, a mapping of the distribution of the binding partners on a surface is possible (19,20). A detailed discussion of this technique will be given in the next section.

Even if only the morphology of a specimen is of interest, the utilization of SFM is advantageous. In contrast to conventional light microscopy, which cannot provide any height information, SFM allows acquisition of the real three-dimensional geometry of a specimen with nanometer accuracy (Fig. 4). It is even superior to fairly new optical microscopes, such as confocal laser scan microscopes, which provide height information but lack sufficient spatial resolution. Microscopic techniques almost comparable with respect to resolution are scanning electron microscopy (SEM) and transmission electron microscopy (TEM). However, they do not provide any height information; they require extensive sample preparation, which probably induces artifacts; and they operate under vacuum conditions, which prevents investigation of living specimens. SFM, on the other hand, is a nondestructive technique; i.e., investigations can be performed with minimum interaction with specimens, allowing even examinations of living cells. The benefit of SPM in this field is that it enables the collection of nanoscopic surface structures of mesoscopic specimens, e.g., cells. Because the structure of a biological element is determined by its function, SPM can give deeper insight into biological functionality. The following micrographs demonstrate the efficiency of SFM investigations of the biology of selected specimens.

Figure 5 shows the topography of a human carcinogenic caco-2 cell measured with SFM in the so-called noncontact mode in a liquid. This mode prevents direct contact between specimen and probe, allowing an extremely

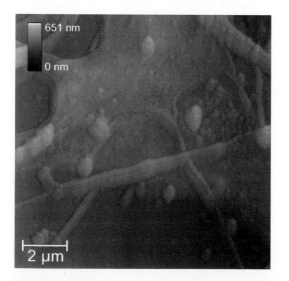

Figure 4 SFM image of a chicken cell.

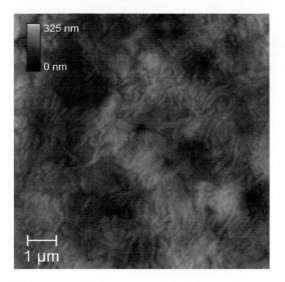

325 nm

0 nm

1 µm

Figure 5   SFM image of a caco-2 cell (non–contact mode measurement). (Courtesy of E. Haltner and C.-M. Lehr.)

careful examination of the cell surface. Besides the coarse overall topography of the cell, fine structures on its membrane are visible (microvilli). This example shows that SFM measures the whole topography of a surface in a physiological environment without any information loss and that it is possible with appropriate image processing procedures to emphasize and visualize fine structures that are normally covered by a coarse overall specimen topography. Figures 6 and 7 depict topographies of caco cell monolayers. Because of the fairly large scan range, the dynamic range of the topography, and the utilization of the so-called contact mode, the microvilli are not visible. However, these images indicate that a topography range from micrometers to nanometers can be handled by SFM. Because of its features, e.g., zoom capabilities, the instrument's operation is comparable to the operation of conventional microscopes; that is, easy switching from a macroscopic to a nanoscopic view is feasible. Figures 8 and 9 demonstrate the resolution capabilities of SFM. Figure 8 depicts chromosomes at different magnification levels, and Fig. 9 shows pores of a cell core. Both images delineate typical high-resolution measurements with SFM, and they show that even though the achievable spatial resolution cannot be better than the finite diameter of the probe apex, state-of-the-art probes are "sharp" enough to image even the smallest structures without artifacts.

SPM is not only a supplement to microscopic techniques already established in biology, it widens the field of possible microscopic applications

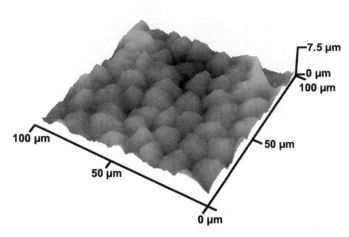

Figure 6   SFM image of a caco-2 cell monolayer (contact mode measurement). (Courtesy of E. Haltner and C.-M. Lehr.)

as well. With its high resolution and magnification capabilities, it provides the opportunity for imaging of single molecules. With SPM, it is possible to locate specific molecules within a living cell (21) or measure the shape of organic molecules. In this respect, SPM closes the gap between microscopic and molecular biology, as it allows the biologist to see single molecules and complex biological structures in the micrometer range.

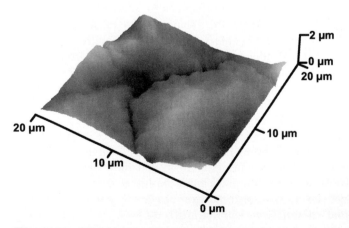

Figure 7   SFM image of a caco-2 cell monolayer (contact mode measurement). (Courtesy of E. Haltner and C.-M. Lehr.)

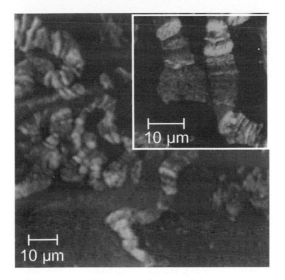

Figure 8 SFM image of chromosomes.

Figure 9 SFM image of nuclear pores of *Xenopus laevis*.

V. BIOADHESION MEASUREMENTS WITH SFM

The ability of SFM to measure extremely weak forces makes it most suitable for investigations of binding forces between different materials. Many phenomena in biology are based on binding forces between complex organic molecules, and SFM has been used extensively for the characterization of those forces. However, this application is not restricted to molecular force interaction. The force interaction between individual particles can be measured too (6). In all experimental setups dealing with molecular interaction, a probe is chemically modified by functional groups that interact with binding partners on a surface (functionalization of probes). Depending on the chemical modification of the probe, the force interaction is specific to the chemical structure of the surface. In this respect SFM has become a chemical (material) sensitive technique. Besides the experimental efforts to measure the forces, first attempts have been made to provide a theoretical understanding of the force interaction (22,23).

In the case of bioadhesion, the chemical modification of the probe is accomplished by coating it with a biomaterial. The choice of material is restricted only by the requirement that the binding of the material to the probe is much stronger than the binding forces present in the interaction; otherwise, the biomaterial will come off the probe and stick to the surface. But this is mainly an issue of appropriate probe preparation and is under control for various materials.

With the measurement of bioadhesion it is possible to characterize the binding forces between individual cells in a multicellular organism (24) as well as the binding characteristics of complementary strands of DNA (25) or nucleotide bases (26). In some cases SFM cannot measure the binding force between two adjacent molecules, as many probe and surface molecules are generally involved in the measurement. But by varying the experimental conditions one finds a quantization of binding forces depending on the number of molecules involved (27). This approach allows determination of the binding force between individual molecules.

This compilation reveals some of possible applications of SFM bioadhesion measurements. But this is by far not the complete spectrum of possible applications. One can expect new applications to emerge in parallel with the improvement of control in probe modification, making SFM a material-sensitive technique.

A. Biotin-Streptavidin

A prominent example of bioadhesion investigations is the characterization of binding forces between biotin and streptavidin molecules. This combi-

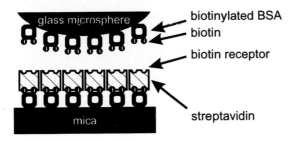

Figure 10 Experimental setup for biotin-streptavidin force interaction measurements.

nation has several advantageous features: it is robust and does not alter its binding behavior after transfer to probes or solid surfaces, and it has noncovalent binding similar to the important antibody–antigen binding, with comparable binding strength. A comprehensive description of biotin–streptavidin interactions can be found in, e.g., Ref. 28.

Figure 10 depicts the principal experimental setup used for binding force measurements. On a flat surface, e.g., mica, a thin layer of bovine serum albumin (BSA) is adsorbed. BSA is necessary because it tends to adsorb nonspecifically and irreversibly to glass or mica. Biotin is then attached to BSA by covalent binding initiated by appropriate conditions. The required biotin receptor (streptavidin) has four binding sites for biotin, which allows the streptavidin to stick to the BSA-biotin complex as well as to behave as an active receptor for the biotin layer on the respective glass or probe surface. In order to perform a reference experiment the same configuration is used, but the biotin receptors are deactivated—blocked—by additional biotin (Fig. 11). This setup is used for reference purposes, because in this configuration no binding forces between biotin and streptavidin

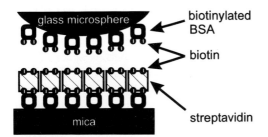

Figure 11 Experimental setup for biotin-streptavidin force interaction measurements with receptor blocking.

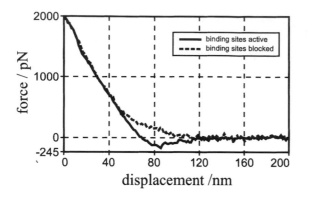

Figure 12 Measured force-distance curves of biotin-streptavidin binding.

should be present. Details of the sample preparation can be found in, e.g., Refs. 29 and 30.

A typical force-distance curve of the biotin-streptavidin system is shown in Fig. 12. The solid line represents the force-distance curve with the biotin receptor active, the dashed line the force-distance curve with the receptor blocked. It can be seen that the (solid) force curve crosses the zero force line, and then at a relative displacement of the tip of approximately 80 nm an adhesion force of roughly 200 pN is present. From this adhesion force information about the bond energy can be derived, which gives information about the binding mechanism (30). Numerical calculations simulating this adhesion process have been performed and provide deeper insight into the dynamics of the adhesion and rupture process (31).

B. Interaction Between Single-Chain Fv Antibody Fragments and Corresponding Antigens

Genetically engineered single-chain Fv antibody fragments (scFvs) (32) are ideal model proteins for studying antigen-antibody interactions by force spectroscopy experiments. They can be generated against all conceivable antigenic targets, and mutants with various binding properties can be engineered. Furthermore, an scFv is the smallest part of an antibody molecule that still contains the intact antigen binding site. It is crucial for the measurements of binding forces that the attachment of antibody and antigen to their respective surfaces is so strong that the antibody-antigen binding is correctly probed and no detachment of any partner occurs. To achieve stable and directed immobilization of the scFv on a flat gold surface, the molecule was designed with a cysteine at the carboxyl terminus, i.e., at the part opposite to the binding site. Immobilization of the scFv fragment via the thiol group of the cystein ensures free accessibility of the binding site for antigens. To avoid denaturation of the proteins in contact with the surface, the gold was treated with mercaptoethanesulfonate, yielding a negatively charged surface (33). In these model experiment the scFv molecules used were directed against the antigen fluorescein (34). For the force spectroscopy measurements, the antigen was covalently immobilized to the silanized silicon nitride AFM tip via a poly(ethylene glycol) linker about 40 nm in length.

Surfaces with immobilized scFvs were first scanned with the antigen-functionalized tip (Fig. 13A) in contact mode, with very low forces ($F <$ 500 pN) in order to avoid detachment or destruction of the antibody fragments. Well-separated proteins were then chosen for series of force-distance

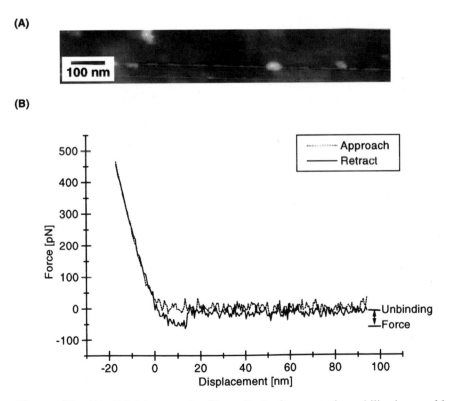

Figure 13 (A) AFM image of scFv antibody fragments immobilized on gold, scanned with an antigen-functionalized tip in contact mode. (B) Typical force-distance curve of an antigen-functionalized tip and a single, well-separated scFv molecule. (Courtesy of R. Ros.)

measurements (Fig. 13B). The unbinding force, i.e., the maximum force at the moment of detachment, was taken as a measure of the binding force between the scFv fragment and the antigen.

The histogram in Fig. 14A representing the probability distribution of unbinding forces shows a single peak with a mean value of about 50 pN and a number of events where no binding forces can be observed, so-called zero events. When free antigen was added in order to block the binding sites of the scFv fragments, the number of zero events drastically increased and the peak at higher force values disappeared (Fig. 14B). These blocking experiments prove that the forces determined do indeed result from interaction between the ligand and the receptor and not from unspecific adhesion.

The power of this novel measurement technique is that it can distinguish closely related molecules (35). Comparison of the binding forces of the scFv fragment just described and a mutant that has a single amino acid exchanged within the binding pocket (34) showed 20% lower binding forces for the mutant. In order to detect such small differences, errors related to cantilever calibration must be avoided. Therefore, measurements of the wild-type and the mutant molecules were carried out with the same tip.

With the current model system it is possible to measure binding forces of a sufficient number of mutants and to correlate the values with their

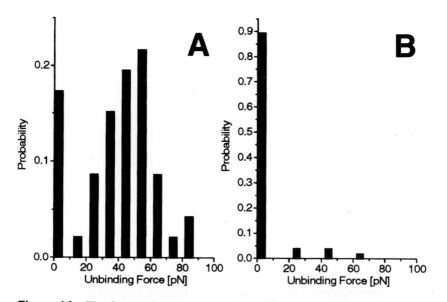

Figure 14   The force values determined from 50 force-distance curves as one single molecule in histograms. Distributions for scFv fragments (A) before and (B) after blocking with free antigen. (Courtesy of R. Ros.)

thermodynamic parameters in order to obtain deeper insight into the molecular recognition processes.

VI. OUTLOOK

Scanning probe microscopes are valuable supplements to conventional microscopes as used in biology, not only with respect to bioadhesion. In terms of possible applications, we are only at the beginning of exciting developments in microscopy, especially in biology. It is very likely that one important branch of biology, medicine and pharmacy, will benefit most in the near future from SPM. Many solutions to problems in biomedical sciences are based on nanoscopic engineering, and scanning probe microscopes are tools for nanoscopic modifications and analyses. An example will support this statement. An important problem in medicine and pharmaceutical sciences is the development of biocompatible materials, and one solution seems to be the use of self-organized monolayers between the material surface and the bioactive element, i.e., the body. SPM can be used to characterize and modify those monolayers very easily. Another application is in the development of nanoscopic drug carrier systems. Such nanoparticles can be analyzed in an efficient way with SPM (36). Finally, it does not seem to be too far fetched to imagine the SPM will some day be a standard diagnostic instrument in medicine and an important tool for the development of novel drug delivery systems.

REFERENCES

1. RNS Sodhi. Application of surface analytical and modification techniques to biomaterial research. J Electron Spectrosc Relat Phenom 81:269–284, 1996.
2. J Jahanmir, BG Haggar, JB Hayes. The scanning probe microscope. Scanning Microsc 6:625–660, 1992.
3. DA Bonnel. Microscope design and operation. In: Scanning Tunneling Microscopy and Spectroscopy. New York: VHC, 1993, pp 7–30.
4. G Binning, H Rohrer. Scanning tunneling microscope. Helv Phys Acta 55:726–735, 1982.
5. G Binning, CF Quate, C Gerber. Atomic force microscope. Phys Rev Lett 56:930–933, 1986.
6. MA Paesler, PJ Moyer. Near-Field Optics: Theory, Instrumentation and Applications. New York: John Wiley & Sons, 1996.
7. R Wiesendanger. Scanning Probe Microscopy and Spectroscopy. Cambridge: Cambridge University Press, 1994.

8. E zur Mühlen, H Niehus. In: RH Müller, W Mehnert. Particle and Surface Characterisation Methods. Medpharm Scientific, 1997, p 99.

9. H Ho, P West. Optimising AC mode AFM imaging. Scanning 18:339, 1996.

10. Y Martin, CC Williams, HK Wickramasinghe. Atomic force microscope— Microscopy mapping and profiling on a sub 100-Å scale. J Appl Phys 61, 1987, pp 4723–4729.

11. U Hartmann. In: R Wiesendanger, H-J Güntherodt, eds. Scanning Tunneling Microscopy III. Berlin: Springer, 1993, pp 293–359.

12. Nanotribology. MRS Bull 18(5), 1993.

13. R Wiesendanger, H-J Güntherodt. The scanning tunneling microscope in biology. In: Scanning Tunneling Microscopy II. Berlin: Springer Verlag, 1992, pp 51–98.

14. SM Lindsay. Biological applications of the scanning probe microscope. In: Scanning Tunneling Microscopy and Spectroscopy. New York: VHC, 1993, pp 335–408.

15. A Noy, V Vezenov, CM Lieber. Chemical force microscopy. Annu Rev Mater Sci 27:381–421, 1997.

16. GU Lee, LA Chrisey, CE O'Ferral, DE Pilloff, NH Turner, RJ Colton. Chemically-specific probes for the atomic force microscope. Isr J Chem 36:81–87, 1996.

17. U Dammer, M Hegner, D Anselmetti, P Wagner, M Dreier, W Huber, H-J Güntherodt. Specific antigen/antibody interactions measured by force microscopy. Biophys J 70:2437–2441, 1996.

18. S Allen, X Chen, J Davies, MC Davies, AC Dawkes, JC Edwards, CJ Roberts, J Sefton, SJB Tendler, PM Williams. Detection of antigen–antibody binding events with the atomic force microscope. Biochemistry 36:7457–7463, 1997.

19. M Ludwig, W Dettmann, HE Gaub. Atomic force microscope imaging contrast based on molecular recognition. Biophys J 72:445–448, 1997.

20. P Hinterdorfer, W Baumgartner, HJ Gruber, K Schilcher, H Schindler. Detection and localisation of individual antibody-antigen recognition events by atomic force microscopy. Proc Natl Acad Sci U S A 93:3477–3481, 1996.

21. PG Haydon, S Marchese-Ragona, TA Basarsky, M Szulczewski, M McCloskey. Near-field confocal optical spectroscopy (NCOS): Subdiffraction optical resolution for biological systems. J Microsc 182:208–216, 1996.

22. WA Ducker, TJ Senden, RM Pashley. Direct measurement of colloidal forces using an atomic force microscope. Nature 353:239–241, 1991.

23. JL Hutter, J Bechhoefer. Measurement and manipulation of van der Waals forces in atomic force microscopy. J Vac Sci Technol B 12:2251–2253, 1994.

24. U Dammer, O Popescu, P Wagner, D Anselmetti, H-J Güntherodt, GN Misevic. Binding strength between cell adhesion proteoglycans measured by atomic force microscopy. Science 267:1173–1175, 1995.

25. GU Lee, LA Chrisey, RJ Colton. Direct measurement of the forces between complementary strands of DNA. Science 266:771–773, 1994.

26. T Boland, BD Ratner. Direct measurement of hydrogen bonding in DNA nucleotide bases by atomic force microscopy. Proc Natl Acad Sci U S A 92: 5297–5301, 1995.

27. E-F Florin, VT Moy, HE Gaub. Adhesion forces between individual ligand-receptor pairs. Science 264:415–417, 1994.

28. AL Weisenhorn, F-J Schmitt, W Knoll, PK Hansma. Streptavidin binding observed with an atomic force microscope. Ultramicroscopy 42–44:1125–1132, 1992.

29. S Allen, J Davies, AC Dawkes, MC Davies, JC Edwards, MC Parker, CJ Roberts, J Sefton, SJB Tendler, PM Williams. In situ observation of streptavidin-biotin binding on an immunoassay well surface using an atomic force microscope. FEBS Lett 390:161–164, 1996.

30. GU Lee, DA Kidwell, RJ Colton. Sensing discrete streptavidin-biotin interactions with atomic force microscopy. Langmuir 10:354–357, 1994.

31. H Grubmüller, B Heymann, P Tavan. Ligand binding: Molecular mechanics calculation of the streptavidin-biotin rupture force. Science 271:997–999, 1996.

32. A Plückthun. Recombinant antibodies. Immunol Rev 130:151–188, 1994.

33. LX Tiefenauer, S Kossek, C Padeste, P Thiebaud. Towards amerometric immunosensor devices. Biosensors Bioelectron 12:213–223, 1997.

34. G Pedrazzi, F Schwesinger, A Honegger, C Krebber, A Plückthun. Affinity and folding properties both influence the selection of antibodies with the selectivity infective phage (SIP) methodology. FEBS Lett 415.289–293, 1997.

35. R Ros, F Schwesinger, D Anselmetti, M Kubon, R Schaefer, A Plückthun, L Tiefenauer. Antigen binding forces of individually addressed single-chain Fv antibody molecules. Proc Natl Acad Sci U S A 95:7402–7405, 1998.

36. A zur Mühlen, E zur Mühlen, H Niehus, W Mehnert. Atomic force microscopy studies of solid lipid nanoparticles. Pharm Res 13:1411–1416, 1996.

10

Direct Measurement of Molecular-Level Forces and Adhesion in Biological Systems

James Schneider* and Matthew Tirrell
University of Minnesota, Minneapolis, Minnesota

I. INTRODUCTION

A major challenge facing the biomaterials community is that of correlating molecular structure with biological response. Given this knowledge, the rational design of biocompatible materials, novel drug delivery systems, and artificial tissues can be achieved. The complexity of biological systems makes isolating molecular contributions to macroscopic phenomena difficult and inconclusive. Recent advances have introduced and refined a number of techniques that allow scientists to probe interactions at the molecular level, and these techniques have been applied to many systems of biological significance.

Direct force measurement (DFM) techniques are capable of obtaining distance-resolved force data between well-defined surfaces at molecular separations. This force-distance profile can be compared with theoretical predictions and previous results to gain insight into not merely the structure and composition of the surface but also the influence the surface has on molecules or surfaces in its proximity. With proper surface design, surface behavior can be meaningfully correlated with the molecular behavior underlying it.

**Current affiliation*: Carnegie Mellon University, Pittsburgh, Pennsylvania.

This chapter describes available DFM techniques, delineating their individual strengths and limitations. Representative and important experimental results will be reviewed. Because we restrict our discussion to DFM techniques, we neglect important techniques that make measurements of adhesion or surface energy only, such as shear flow assays of cell adhesion (1), peel tests of polymer adhesion (2), and wetting studies (3).

II. EXPERIMENTAL ASPECTS OF DIRECT FORCE MEASUREMENT

A. Intersurface Forces

To understand how DFMs are carried out and what information can be extracted from them, it is important to understand a few things about how particles and surfaces interact at ranges less than 1 μm. These "intersurface" forces have been studied in great detail theoretically and experimentally using the techniques described here. The physics underlying some of these forces is a subject of debate; the reader is referred to excellent summaries (4–7) and encouraged to keep abreast of new developments.

Van der Waals forces (8) are exerted between all surfaces in nature, and they are almost always attractive. They are brought about by the net attraction between molecular dipoles, be they permanent or "induced." Induced dipoles arise because of the dynamic nature of the electron cloud. As two surfaces are brought near each other, fluctuations in electron density induce dipoles in each cloud, leading to their mutual attraction. Van der Waals forces are extremely strong at very short range and cause almost all surfaces to adhere to each other if brought close enough together. The distance dependence (D) of the van der Waals interaction is a power law; the power law exponent is geometry dependent. For the interaction of two spheres, a sphere and a flat, or two crossed cylinders, we have

$$F_{vdw} = \frac{AR}{6D^2} \tag{1}$$

where A is the material-dependent Hamaker constant and R is the mean radius of curvature.

Electrostatic interactions between charged surfaces in water (or "double-layer forces") (9) include the osmotic penalty for compressing the counterion cloud near the charged surfaces. Because force measurements are typically made between surfaces with like charge, this is usually considered a repulsive force. The range of electrostatic interactions is modulated by the dielectric environment separating the surfaces and the presence of ions between the surfaces. Electrostatic repulsion between surfaces exponentially increases as the surfaces are brought closer together with a decay constant called the Debye screening length ($1/\kappa$). The Debye length is directly related to the concentration of ions (c) in the solution or gas between the surfaces:

$$F_{elect} \sim \exp(-\kappa D), \quad \kappa \sim \sqrt{c} \tag{2}$$

Steric forces, or "hard-core repulsions," occur when surfaces are brought into such intimate contact that the electron clouds of the individual surfaces begin to overlap; the Pauli exclusion principle prevents the surfaces from coming closer together. Steric repulsions are extremely steep. The additional steric force felt by molecules tightly adsorbed to the surface has been called a "solvation force" or, in the case of water or aqueous ions, "hydration force."

The DLVO theory contends that van der Waals, electrostatic, and steric forces acting simultaneously between two particles can be treated additively. Continuum DLVO descriptions successfully model the interactions of many colloidal particles—surprisingly, even down to molecular separations. An effective numerical method based on DLVO theory is provided by Chan et al. (10) to describe interparticle force versus distance profiles given the surface charge density, the Debye length, and the Hamaker constant. Fitting DFM data using this recipe is a useful way to obtain these quantities.

Intersurface forces can be greatly modified by the application of polymers (11). Not only do these molecules alter the van der Waals and electrostatic character of the surface, they "soften" the surface and introduce new interactions because of their mobility. As polymer-covered surfaces approach each other, eventually polymers from each surface begin to overlap, restricting their mobility. The resulting repulsive force has an entropic origin. Polymers are often applied to the surface of colloidal particles to prevent their uncontrolled aggregation due to van der Waals attraction. Accordingly, this force has been called "steric stabilization." Polymers can also cause a "bridging attraction" when single chains are adsorbed to two different particles at once in a poor solvent.

An attractive "hydrophobic force" (12) has been observed between nonpolar surfaces in water. The range of the attraction between lipid-covered surfaces is ionic strength dependent, suggesting an electrostatic origin (13,14). Others have ascribed the hydrophobic attraction to the interaction of shells of ordered water near each surface (4) or to cavition of the confined water (15). Measurements made on rigidly held, polymerized hydrophobic surfaces show a very short-range hydrophobic attraction that can be described by van der Waals interactions only (16). Even though the source of this attraction has been questioned, it is important for the experimenter to keep in mind because the long-range attraction phenomenon has been observed many times in many different systems.

In many biological systems, several of these intersurface forces may be operating at a certain position. Consider a hypothetical "cellular potential" (Fig. 1) existing between the membranes of living cells (1). Charged polysaccharides envelop the membrane in a "glycocalyx," which helps stabilize cell populations against uncontrolled aggregation. Cell adhesion receptors bound to the membrane can specifically bind to ligands in the extracellular matrix, leading to an attraction. As always, van der Waals attraction between the membranes will dominate at close range.

The presence of multiple interaction forces at the same position complicates greatly the ability of the experimenter to decouple the contributions of each individual force. Experimental conditions can be changed to attenuate or accentuate individual forces, as in the "salting out" of electrostatic forces. Because only a few functional forms can be clearly identified in a measured potential and very few direct force measurement techniques exist,

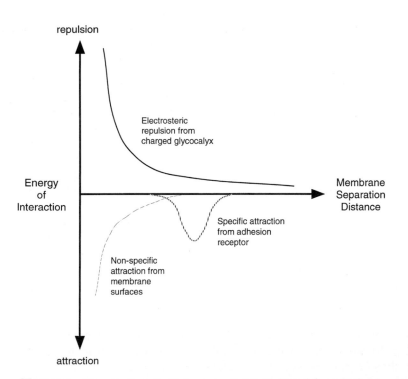

Figure 1 Hypothetical "cellular potential." (Adapted from Ref. 1.)

the experimenter must apply some molecular insight and common sense to interpret force data meaningfully.

B. Surfaces in Contact: Adhesion

The strength of adhesion between two particles is best described by the energy required to separate them. For an equilibrium separation process, we define the "work of adhesion" or "adhesion energy" (W):

$$W = \int_{\infty}^{D_0} F(D) \, dD \tag{3}$$

where D is the separation distance between the surfaces and D_0 is the separation distance at contact, which depends on the distance reference chosen. The surfaces of particles can be considered to be "in contact" when the further application of force acts only to deform the particle, bringing more surface into a region of equal separation distance called the "contact area." By this operational definition, the establishment of contact depends not only on surface forces but also on the bulk deformation properties of the supporting material.

Hertz (17) provided the first description of the deformation of elastic bodies in contact. The force required to create contact area for nonadhering, curved surfaces is given by

$$F_{\text{Hertz}} = \frac{Ka^3}{R} \tag{4}$$

where a is the radius of the circular contact area, R is the mean radius of curvature, and K is the elasticity of the particles, directly related to bulk moduli. Johnson, Kendall, and Roberts (JKR) (18) argued that surface forces act to create additional contact area between curved, elastic bodies. In this case, less force is required to establish the same contact area:

$$F_{\text{JKR}} = F_{\text{Hertz}} - 1.5W\pi R + \sqrt{3W\pi R F_{\text{JKR}} + (1.5W\pi R)^2} \tag{5}$$

Here W is the "adhesion energy," an energetic comparison of the interfacial energy in contact and the energies of the exposed surfaces when out of contact. Surface forces keep adhered bodies in contact even with no applied load; to separate the surfaces, a tensile "pull-off force" is required:

$$F_{\text{pull-off}} = -1.5W\pi R \tag{6}$$

Notice that the pull-off force in Eq. (6) does not depend on the bulk deformation properties embodied in K.

The importance of the JKR theory for direct force measurement that it relates the force required to separate surfaces to the adhesive interaction density on the surfaces. Given this quantity and the density of functional molecules on the surface, we can assess the contribution *per molecule*. It should be noted that assumptions in the JKR theory do not apply for highly deformable substrates (19), but materials typically used in direct force measurement techniques are sufficiently rigid that the JKR theory is valid.

Another opportunity provided by the JKR theory is that of evaluating W without completely separating the surfaces. Combining Eqs. (4) and (5), we have

$$a^3 = \frac{R}{K}(P + 3\pi WR + \sqrt{6\pi WRP + (3\pi WR)^2}) \tag{7}$$

with parameters defined as before. Contact area data collected during loading and unloading contact surfaces can be fit to Eq. (6) with K and W as adjustable parameters. In so doing, the adhesion energy required to *create* contact area can be compared with that required to *diminish* it. Differences in W on loading and unloading are referred to as ''adhesion hysteresis'' (Fig. 2).

Adhesion hysteresis can be traced to subtle molecular rearrangements that occur after contact has been established. Common examples of this include interpenetration of molecules on opposing surfaces (20) and rearrangements that expose unreacted functional groups on each surface for intersurface chemical bonding (21,22). Disruption or permanent deformation of the surface structure on loading, a nonequilibrium process not described

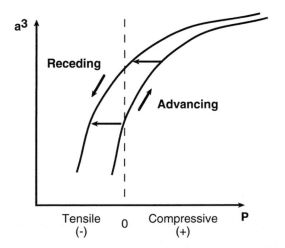

Figure 2 JKR adhesion analysis showing adhesion hysteresis. Arrows represent the data obtained for a loading-unloading cycle. (Adapted from Ref. 20.)

by JKR theory, is another example (23,24). "Capillary forces" (4) are brought about by the condensation of vapor in the gap between surfaces close enough together that the Laplace pressure falls below the vapor pressure. The condensed vapor imparts pressure forces on the surfaces, which resist separation. Capillary forces greatly complicate gas-phase force measurements. The reader is referred to detailed discussions concerning adhesion hysteresis (20,25,26).

C. Mechanical Force Measurement

Mechanical force measurement techniques obtain force information by the deflection of springs, whose spring constant is independently determined. The spring deflection can be monitored directly or inferred by tracking the position of both ends of the spring with respect to a known reference. Techniques using spring deflection as a means of measuring forces have inherent mechanical instabilities that prevent them from elucidating the entire interaction potential in many cases. Surfaces "jump" discontinuously at regions of the force versus distance profile at which the slope exceeds the spring constant.

Consider a typical DLVO interaction potential (Fig. 3). As the surfaces approach each other, "jump-to-contact" occurs as a repulsive barrier is surmounted, and surfaces abruptly snap together. As the surfaces in contact are unloaded in an attempt to separate them, eventually a second spring instability is observed and the surfaces jump apart. The "jump-out" typically moves the surfaces a great distance apart, such that the surfaces feel no interaction force whatsoever.

Jump events are beneficial in that the pull-off force measured on jump-out gives a useful measure of adhesion energy. Even though in principle we can always calculate W using Eq. (3) given the full interaction profile, in practice the attractive well may be so narrow that it is indistinguishable. Also, the nonequilibrium jump-out process may lend insight into realistic dissipative processes occurring on separation. In some systems, jump-to-contact and jump-out events may cause unintended surface rearrangement or damage. Mechanical force measurement is capable of testing surfaces with higher compressive forces and can measure stronger adhesion energies than other DFM techniques.

1. The Surface-Force Apparatus (SFA)

Among the first instruments developed for DFM is the surface-force apparatus (SFA), developed by Tabor and Winterton (27) in the late 1960s and later refined by Israelachvili (28). The SFA uses a multiple-beam interfer-

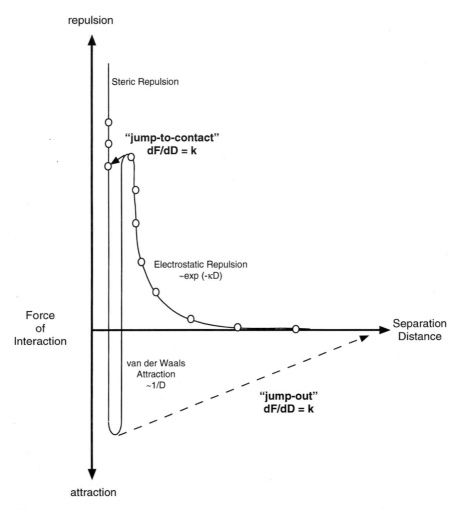

Figure 3 "Jump" phenomena in mechanical force measurement of a typical DLVO-type potential. Circles represent data obtainable on approach.

ometry technique (Fig. 4) to measure the separation distance between surfaces with angstrom resolution. The technique relies on the creation of an optical cavity between two semireflective surfaces. Only wavelengths of light that are integer multiples of the distance between the semireflective surfaces can be transmitted through the cavity. White light passing through the cavity will be filtered into appropriate discrete-wavelength bands that can be observed in a spectrometer. These bands are called "fringes of equal chromatic order (FECO)." Given the wavelength of two neighboring FECO

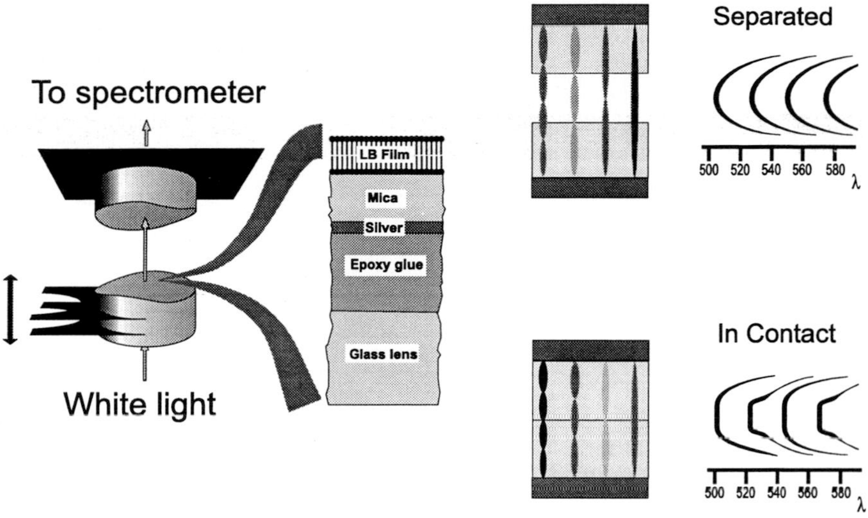

Figure 4 Multiple-beam interferometry in the SFA. The left panel shows the general composition of the interacting surfaces. The right panels show the optical cavity and the approximate shape of the FECO fringes when the surfaces are separated and in contact.

fringes, the distance between the semireflecting surfaces and the refractive index of the intervening material are easily calculated.

In order to take advantage of the angstrom level resolution afforded by multiple-beam interferometry, it is necessary to build the optical cavity using a molecularly smooth material that is optically transparent. SFA experiments typically use the aluminosilicate mineral mica for this purpose, although sapphire (29), silica (30), and thin polymer films (31) have been successfully implemented as well. Mica has a weak plane of interactions in its crystal structure, allowing it to be cleaved into molecularly smooth sheets. A thin, vapor-deposited layer of silver is applied to one side of freshly cleaved mica sheets to serve as the semireflective surface. The silvered mica sheets are glued, silver side down, onto cylindrically curved glass lenses for the experiment. Two of these lenses are opposed and brought together with the axes of curvature mutually perpendicular to form the optical cavity for the SFA experiment.

The "crossed-cylinders" orientation provides a number of experimental benefits. Crossed cylinders interact and contact over a small region (about 100 μm^2), and many new regions can be studied by shifting the relative position of the lenses. The use of crossed cylinders also circumvents the problem of aligning parallel surfaces separated by a few angstroms. The interaction force between crossed cylinders (F) is directly related to the interaction energy per unit area (E/A), a quantity amenable to theoretical prediction, through the Derjaguin approximation (32):

$$\frac{F}{2\pi R}\bigg|_{\text{cross.cyl.}} = \frac{E}{A}\bigg|_{\text{flat plates}} \tag{8}$$

where R is the mean radius of curvature of the crossed cylinders. Taking advantage of the Derjaguin approximation, SFA force data are normalized by R for presentation and comparison. Equation (8) also applies for sphere-on-flat interactions, with R the sphere radius.

When the separation distance between the two lenses is much smaller than the mean radius of curvature, the shape of the optical cavity is geometrically equivalent to that produced by a sphere and flat plate. The FECO fringes observed in the SFA experiment are shaped like the letter "C" and reflect the local curvature of this sphere. When the lenses are brought into contact, the FECO fringes appear blunted, as would a basketball pressed against the floor. By taking into account the magnification factor of the SFA optics, the shape of the FECO fringes gives the local mean radius of curvature and the area of contact, along with the separation distance data.

Many variations exist, and a representative schematic of the SFA is given in Fig. 5. The SFA itself is a stainless steel box housing a spring mechanism. One lens is mounted to the frame of the box; the other is mounted to a leaf spring. Windows are provided to allow white light to pass through the lenses. A differential spring mechanism provides for the nanoscopic control of the position of the leaf spring. Other designs employ piezoelectric crystals to approach and separate the lenses with nanofine control.

During a typical experiment, lenses are progressively moved closer together using a micrometer to compress the spring mechanism. At each step in the compression, the separation distance of the lenses is monitored by the FECO. When the lenses are far apart, no force is felt between the lenses, and the displacement of the springs is proportional to the separation distance of the lenses. Closer in, the supporting leaf spring will deflect in response to repulsive or attractive forces. This deflection is measurable as a deviation from the previously linear dependence of spring displacement and separation distance. Inherent in all SFA data, then, is the assumption that there are no interaction forces at the beginning of the measurement, where a calibrating "baseline" is established.

Several modifications of this basic design are available. A "shearing SFA" (33) has the lower lens supported by piezoelectric bimorph strips that deflect in response to applied voltage. Coupled with a function generator, the lower lens can be oscillated laterally for rheological measurements. The upper lens is supported by a conducting strip sandwiched between two oth-

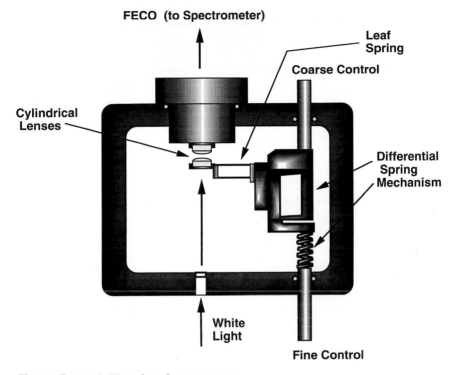

Figure 5 Mark II surface force apparatus.

ers. The response of the upper lens is evaluated by tracking the thickness-dependent capacitance of the gap between the strips (34). Alternatively, the response of the upper lens can be evaluated using piezoelectrics (33,35). The "bimorph SFA" (36) has the leaf spring made of a piezoelectric bimorph material; forces are accurately measured by the voltage built up in the bimorph as it deflects.

Freshly cleaved mica develops a negative charge when placed in water as surface potassium ions are solubilized. Measurements made between bare mica surfaces in water have been instrumental in understanding DLVO-type interactions in colloidal systems (37). Mica, by virtue of its surface chemistry and smoothness, also serves as an excellent substrate for Langmuir-Blodgett (LB) deposition (38,39), a powerful technique that transfers organized amphiphilic monolayers to solid surfaces. Because LB layers have a highly ordered and homogeneous composition, SFA measurements between them can be believably interpolated to individual molecular contributions. Functional molecules can be covalently attached to amphiphiles take advantage of the LB technique as a means of pinning them to mica in an organized way for SFA measurements. Functionality can be measurably "diluted" by mixing nonfunctional amphiphiles in the monolayer.

Potentially, the SFA experiment does give a very complete picture of the interaction of two well-defined surfaces. The force versus distance profiles can be compared with theoretical predictions of intersurface forces to gain insight into individual molecular contributions. Contact area is directly measured during adhesion measurements, giving a realistic measurement of molecular binding affinities. The position of the FECO fringes in contact reveals the amount of adsorbed material, given its refractive index. SFA experiments are quite tedious, as the rigidity and molecular smoothness of mica demand that contamination of any kind be eliminated. Because SFA data are averaged over a fairly large contact area, the spatial resolution is only on the micrometer level. The slow response time of the leaf spring and driving motors gives the SFA poor temporal resolution.

2. The Atomic Force Microscope (AFM)

The AFM was developed as an nanoscopic imaging tool in the early 1980s (40). AFM instruments are able to image a surface with nanometer resolution by rastering a nanofine tip over the surface of interest and monitoring the deflection of the tip as it passes over molecular features. A typical AFM setup is shown in Fig. 6.

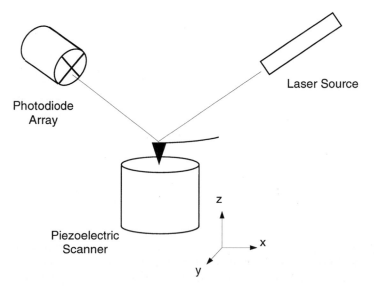

Figure 6   Simplified AFM schematic.

The sample is mounted on a piezoelectric-crystal scanner, which allows three-dimensional translation of the surface. The response of the scanner to applied voltage is a precalibrated function of the crystal size, shape, and composition. A pyramidal crystal, usually of silicon nitride, is mounted on a thin cantilever. The cantilever deflects as the crystal tip traverses surface features. Laser light is reflected off the top surface of the cantilever and onto a light-sensitive photodiode array (PDA). The PDA generates a beam position–dependent voltage in four separate quadrants when illuminated by the laser light. Both the linear and angular deflections of the cantilever can be precisely measured by monitoring the voltage in each of the quadrants. Before each experiment, the PDA must be calibrated by forcing the tip against a noncompliant surface and recording the change in PDA voltage with scanner position as the cantilever deflects.

Because images are obtained by a spring mechanism, the AFM tip undergoes jump-to-contact and jump-out phenomena as does the SFA. Care must be taken to obtain reliable images of soft surfaces, because the tip may damage the surface during scanning. Feedback loops control the vertical position of the scanner, keeping a constant cantilever deflection "set point" during the measurement. A "contact mode" AFM image can be obtained by bringing the tip in contact with sample and then translating the sample under the tip with the scanner. The three-dimensional coordinates of the image are specified by the x and y position of the scanner and the z-piezo voltage required to maintain the PDA setpoint. The image can reflect normal or frictional forces, depending on the PDA quadrants evaluated. The use of feedback loops allows noncontact imaging, e.g., at a constant deflection out of contact, or oscillating the tip while the sample translates ("tapping mode"). The vertical coordinate of the image can be tip amplitude or phase in the latter case.

Early in the development of the instrument, it was appreciated that AFM could be a powerful technique for measuring force profiles between tip and sample. To do so, reliable cantilever spring constants needed to be obtained to convert deflection to force; now many methods exist (41). AFM force profiles have distinct advantages over those obtained using the SFA. The nanofine tip has a countable number of molecules participating in the surface interaction, and individual molecular contributions can be directly measured by statistical analysis of multiple contacts. Attaching carbon nanotubes with a diameter as small as 5 nm to the AFM tip greatly improves imaging resolution (42) and potentially provides an exact account of a small number of molecules interacting with the surface. The high temporal resolution of the AFM can be exploited to observe dynamical effects (43). AFM force curves can be obtained on inhomogeneous surfaces at the same time an image of the surface is obtained, allowing direct comparisons between functional domains. A recent innovation, the "interfacial force microscope (IFM)," uses a feedback loop to stiffen the cantilever near contact, preventing the jump to contact (44). This method avoids damage to the surface and gives the entire interaction profile.

The AFM tip certainly does not provide an ideal surface for force measurements. The end of the pyramidal AFM tip is rounded to varying degrees, and identifying the exact shape of the tip is challenging. Furthermore, theoretical modeling of DLVO forces between irregularly shaped tips and flat surfaces has not been achieved. Researchers have addressed this by gluing small beads to the AFM tip for force measurements. Accurate DLVO-type profiles have been obtained using this methodology (45,46), but it should be noted that these spheres cannot truly be considered "molecularly smooth." The fine tip utilized obviates AFM measurements between *two* LB layers. Instead, organized self-assembled monolayer (SAM) films can be created by gold-thiol chemistry (39). As with LB films, functionality can be measurably diluted by placing nonfunctional thiols in the solution, creating a mixed SAM. The gold-thiol interaction is extremely strong; as a result, gold-thiol SAMs are generally more robust than LB-deposited layers. The reflective gold surface complicates—but does not prevent (47)—their use in the SFA.

AFM also does not provide a direct measurement of tip-surface separation distance or contact area as the SFA does. The operator is left to surmise that contact has been achieved when the deflection of the spring equals the vertical translation of the scanner. This principle will fail if the surfaces are at all deformable. A solution is to work with an inhomogeneous surface, with "hard" domains serving as a distance reference. JKR-type adhesion measurements can be made only by calculating an expected contact area given the probe shape, the applied load, and the displacement of the tip after contact.

AFM provides an experimentally simpler alternative to the SFA and at the same time supplies high-quality surface images and better temporal resolution. The AFM and SFA are perhaps best viewed as complementary mechanical force measurement techniques.

D. Force Measurement Using Suspended Microspheres

A much higher degree of force sensitivity can be achieved by making measurements using microspheres suspended in liquid, small enough that they remain levitated by Brownian motion. These microspheres are not rigidly held in place but thermally fluctuate about a mean position determined by a delicate balance of gravitational and colloidal forces. Thermal motions cause the sphere to sample regions of this potential well that holds it near the mean position. The well itself acts as a highly sensitive "spring" for these force measurements.

The range of measurement can be greatly improved by using "optical forces" to push the microsphere farther from the bottom of the potential well. Optical forces arise whenever light interacts with matter but in almost all situations are insignificantly small. As light is scattered by a small particle, some of its momentum is transferred to the object in the form of mechanical energy, pushing it directly away from the source. The magnitude of the scattering force is easily controlled by varying the light intensity. These "scattering forces" can be used to push a sphere in a preferred direction.

Optical forces can also be delivered to objects in a direction orthogonal to the propagation of light. Ashkin (48) took advantage of these "gradient forces" to hold microspheres with a focused laser beam in an "optical trap." The vertically oriented laser beam is focused on the sphere so that the light intensity varies significantly across the particle. Momentum is transferred unequally at different positions on the sphere (Fig. 7), leading to a horizontal component of the optical force. For spheres with refractive index higher than that of the medium, the gradient optical force pushes the sphere to brighter regions of the focus. At the brightest region, the center of focus, symmetry ensures that the momentum delivered at any position is balanced by that at another geometrically opposite to it. The balance of horizontal momentum helps maintain the sphere in the center of the trap in the presence

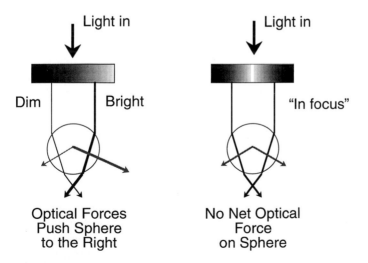

Figure 7 Ray optics description of the optical trap. (Adapted from Ref. 58.)

of small external forces. It should be noted that the scattering force also operates in the optical trap, and the laser must be focused tightly to prevent the sphere from being ejected from the trap.

The position of the trap is readily varied by tilting or translating the laser optics. Laser and white light can be simultaneously focused on the sample, allowing direct visualization of the manipulated objects. The optical trap has proved very useful as a micromanipulation tool, especially in biological systems (49), but care must be taken to choose wavelengths and intensities that do not harm biological samples. Single molecules can even be manipulated by anchoring them to trapped beads. Trapping is useful in force measurements not only to manipulate microspheres but also to stiffen the potential well that holds the sphere in place.

Microsphere techniques make a highly sensitive measurement of forces in a realistic, dynamic environment. Because they avoid the jump instabilities of mechanical measurements, individual bonding potentials can be measured, in principle. However, the use of microspheres limits surface chemistries available for modification, and measurements must be made in the aqueous phase.

1. Total Internal Reflectance Microscopy (TIRM)

The first DFM technique using microspheres was developed by Prieve and coworkers in the mid-1980s (50). Total internal reflectance microscopy measurements are made in liquid between a microsphere and a flat surface (Fig. 8). When the sphere is placed in the liquid, it settles and approaches the flat surface. Short-range repulsive interactions prevent the sphere from making intimate contact with the flat surface, and an equilibrium separation distance is established of a few hundred nanometers. Brownian forces cause the sphere to fluctuate about this mean separation.

The flat surface is made of a transparent, optically dense material such as glass. The surface is back-illuminated by laser light at an angle of incidence below the critical angle, the "total internal reflectance" condition. When this condition is satisfied, a low-intensity "evanescent wave" escapes from the glass surface. The intensity of the evanescent wave decreases exponentially from the surface; the decay constant ("penetration depth") is on the order of 100 nm. The moving sphere scatters this evanescent wave; detecting the intensity of the scattered evanescent wave at a prescribed angle gives the instantaneous separation distance with nanometer resolution (51).

For a typical TIRM experiment, more than 10,000 scattering intensity data are collected over a sampling period of 10 ms and condensed into histograms. In the limit of infinite observations, the histograms will fill a Boltzmann distribution of probabilities (p):

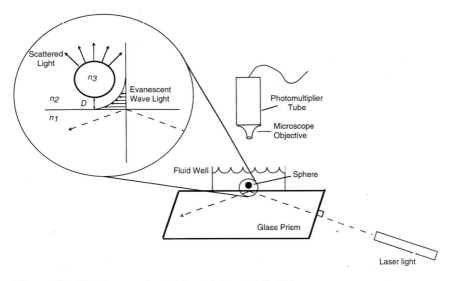

Figure 8 TIRM apparatus. (Adapted from Ref. 54.)

$$p(h) = A \, \exp \left(\frac{-\phi(h)}{kT} \right) \qquad (9)$$

where A is a normalization constant, kT is the thermal energy, and ϕ is the intersurface potential energy. The fitted potential will have two regions, an attractive gravitational restoring potential above the equilibrium separation and a repulsive colloidal restoring potential below it, usually electrostatic or steric in nature. The colloidal potential contains the important DFM information. TIRM has been successfully employed to measure DLVO-type interactions (52) and hydrodynamic interactions between the sphere and wall (53).

Based on this methodology, TIRM force data can be collected only at separation distances that are sampled by Brownian motion. The equilibrium separation distance can be decreased substantially by applying scattering optical forces on the particle using high-power lasers (54). However, their use can alter expected forces due to local heating of the liquid phase. Adhesion measurements are difficult to make using TIRM; the sphere must first settle into contact before being ejected by radiative power.

2. Optical Tweezers

A useful way to measure molecular-level adhesion is to establish contact between an optically trapped sphere and an anchored particle or macromolecule and eject the sphere from the trap under adhesive conditions (55) using hydrodynamic forces. Conversely, a moving particle can be stalled by optical trapping to assess its locomotive force (56,57). To analyze this "optical tweezers" data, a calibration curve must first be constructed plotting escape force versus laser intensity under nonadhesive conditions.

It is now possible to monitor accurately the displacement of the sphere from the center of the trap, allowing a more sensitive force measurement that does not require complete dislodging of the sphere from the trap (58). As with a mechanical spring, the restoring force acting to return the sphere to the center of the trap increases roughly linearly with displacement from the center of the trap. The proportionality constant, or trap "stiffness," can be assessed by tracking the displacement of the sphere due to Brownian motion or hydrodynamic drag. Again, a measurement similar to the calibration can be made between a sphere and an interacting particle or macromolecule that is anchored to a stationary surface. Given the trap stiffness, a force versus displacement curve can be constructed, yielding a true DFM.

The "optical trapping interferometer" (Fig. 9) accomplishes this by creating two overlapping laser beams, with mutually orthogonal linear polarization states (59). The combined beams function as an optical trap with characteristics similar to those of the single-beam variety. Sphere displacement from the center of the trap will obscure one beam more than another, and the recombined beams will be elliptically polarized as a result. The degree of ellipticity is directly related to the displacement from the trap center. Calibration of the interferometer is achieved by immobilizing a sphere on a translating piezoelectric stage and plotting ellipticity versus stage translation. Force measurements can be made between a trapped sphere and a particle or macromolecule anchored to the stage, which is typically translating in a sinusoidal motion. Hydrodynamic drag forces the sphere to sample different regions of the potential well.

A second configuration has a PDA of the kind used in AFM to detect the position of the trapping beam (60,61). The PDA response is fed back to an acousto-optic modulator, which shifts the beam optics to maintain the beam position on the photodiode, holding the particle in place. The feedback signal required to achieve this is a sensitive displacement measurement.

Multiple optical traps can be employed to stretch molecules prior to force measurement (60) or to hold two spheres in relative proximity for interparticle force measurements (62,63). A single laser beam chopped between two or more traps can be as effective as continuous illumination.

Optical tweezers have the same molecular-level force sensitivity as TIRM, without the requirement of a flat surface for evanescent wave propagation. Displacement measurements in an optical trap are separation distance measurements only if the center of the trap can be accurately located

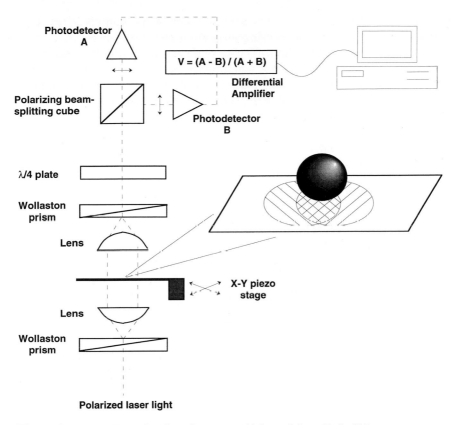

Figure 9 Optical trapping interferometer. (Adapted from Ref. 58.)

with respect to the second interacting molecule or surface; at present, this can be done only by optical microscopy. Although this limits its usefulness in interfacial force measurements, optical tweezers make indispensably sensitive adhesion measurements between single molecules.

E. Forces Between Lipid Bilayers in Water

A few techniques are available that take advantage of the self-assembly of amphiphilic molecules in solution (4) to measure forces or adhesion between lipid bilayers in water. The "thin-film balance" (36) measures interactions between lipid bilayers drawn across a gap with a tiny amount of intervening water. The pressure inside the gap is externally controlled and varied with respect to the pressure in the measurement cell. The pressure drop sets the

applied force. The separation distance is measured by multiple-beam interferometry, with the bilayers serving as semireflective surfaces.

The "osmotic stress device" (64) makes thickness measurements between ordered arrays of lipids or macromolecules in solution by x-ray scattering or ellipsometry. The measurement solution is in contact with a polymer solution across a membrane that excludes the polymer. The added
polymer establishes a calculable osmotic pressure acting to bring the bilayers
closer together. In addition, the osmotic pressure of the measurement solution can be increased by mechanical compression. Force versus distance
profiles obtained by osmotic stress measurements on fluid DLPC (65) bilayers show a short-range entropic repulsion ("undulation forces") as the freely
moving membranes are progressively confined. These forces are not observable when the bilayers are rigidly anchored to mica, as in the SFA.

"Micropipette aspiration (MPA)" (66) is an elaborate technique for
measuring adhesion between lipid vesicles or living cells (67). Two vesicles,
held by suction onto micropipettes, are carefully brought close together (Fig.
10). The vesicles are allowed to adhere by slowly relieving the suction on
one vesicle. During this process the area of contact between the vesicles is
monitored by video-enhanced microscopy. A detailed model balancing membrane tension and external pressure can be fit to these data, providing the
adhesion energy between the cells or vesicles.

Although intersurface forces cannot be measured by this technique, an
extension of MPA does make a DFM, effectively using a held vesicle as a
force transducer (68). Antibodies that recognize the vesicle are covalently
linked to a micrometer-sized bead, "gluing" it to the vesicle surface (Fig.
11). The bead is brought very near a test surface. The vesicle position is
controlled by a piezoelectric crystal, and bead-surface separation distance is
accurately measured by reflectance interference microscopy. The interaction
force is obtained by model calculations, given the length of the vesicle and
the pressure holding it in place that sets its membrane tension.

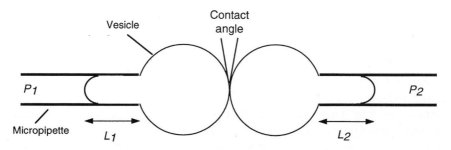

Figure 10   Micropipette aspiration (MPA) technique.

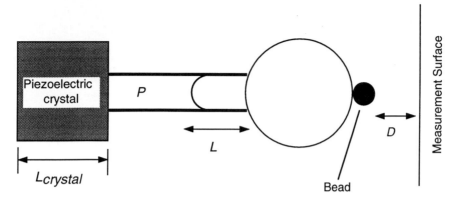

Figure 11 Direct force measurement with micropipette-probe combination.

This micropipette-probe technique has the ultrafine force resolution of TIRM but does not require the measurement surface to generate an evanescent wave. This opens the exciting possibility of making realistic force measurements between a molecular probe and a soft, irregular surface, even one hosting live cells. Its development is still in its infancy, however.

F. Summary of DFM Techniques

Table 1 summarizes the resolution, strengths, and weaknesses of the DFM techniques that have been described here. Broadly, the mechanical methods (SFA and AFM) are the most versatile and have the largest measurement range. The microsphere methods (TIRM and optical tweezers) are best suited for single-particle or single-molecule studies. The bilayer methods (osmotic stress and MPA) are appropriate when realistic membrane properties are important. Although the force resolution varies greatly among the methods, the resolution *per molecule* is about the same throughout. The more sensitive methods have the advantage that less (or no) averaging is required to obtain the molecular information.

III. BIOPOLYMERS AND PROTEINS

The first direct measurements performed in biological systems were made between protein layers adsorbed on mica substrates in the SFA. One goal was to understand the forces responsible for protein-induced steric stabilization by denatured proteins. Another was to probe sensitively the surface structure of adsorbed near-native proteins under different conditions, essen-

Table 1 Features of DFM Techniques

Technique	Force resolution (N)	Distance resolution (nm)	Comments
SFA	10^{-9}	0.1	Directly measures contact area, adsorbed amount, separation distance
			Poor lateral and temporal resolution, experimentally challenging
AFM	10^{-11}	1	Gives surface image, good lateral and temporal resolution
			Contact area and separation distance must be inferred
TIRM	10^{-14}	1	Highly sensitive, nonmechanical measurement
			Not all separation distances accessible, experimental systems limited
Optical tweezers	10^{-12}	0.1	Benefits and limitations similar to TIRM
			No direct measurement of separation distance
			Ideally suited for single-molecule or single-molecule measurements
Osmotic stress	0.1 (dyne/cm^2)	0.1	Measurements made in a realistic membrane environment
			Molecules must be assembled into ordered arrays
MPA	10^{-14}	5	Measurements made in a realistic membrane environment
			Highly sensitive, makes "soft" contact
			Experimentally challenging

tially to *image* the surface. AFM-derived images have largely replaced the SFA in this regard. Because many proteins denature on adsorption, and many denatured proteins retain some degree of order in their adsorbed structures, these two bodies of work are related and will be discussed simultaneously. We will summarize the important observations in this area and give a flavor for the capabilities of DFM methods; the reader is referred to more comprehensive reviews (36,69–71).

A. Polypeptides and Denatured Proteins

1. Polylysine

Measurements between adsorbed layers of synthetic polylysine in the SFA gave important insight into the behavior of charged polymers and denatured proteins at interfaces. The first work was performed by Luckham and Klein (72,73), who made measurements between polylysine (molecular weight = 90,000) layers adsorbed in 0.1 M KNO_3. Lysine monomers bear a positively charged ε-amino moiety at moderate pH. They initially observed a large steric interaction beginning at a separation distance about 15 times greater than the radius of gyration of the polymer under these conditions. After the surfaces were brought into contact, previous expansion or compression curves gave a much shorter range interaction that was easily reproduced. They concluded that this "first-compression hysteresis" was a result of the crushing of the polylysine layer at high loads, which brings more positively charged monomers into contact with negatively charged mica and establishes an irreversible surface ionic bond. This observation makes an important point with respect to these mechanical measurements: that the act of measuring the interaction forces itself can affect the properties of the layer.

Later work by Afshar-rad et al. (74,75) made measurements between polylysine layers adsorbed under low-salt conditions, where intrachain charge repulsion stiffens the polymer considerably. Most all interactions were well described by DLVO theory, and no first-compression hysteresis was observed. They explained that in this low-salt case polylysine adsorbed in a very flat conformation, providing little steric stabilization and effectively occupying all surface binding sites.

2. Gelatin

For years, gelatin (denatured collagen) has been used to stabilize silver halide emulsions in the photography industry, mostly due to its ready supply. Gelatin is a fairly complicated material, featuring acidic and basic functionalities. The isoelectric point (IEP) of gelatin is about pH 5, depending on the method of preparation. Hence, at moderate pH levels, the charge of gelatin can be net positive or net negative.

Kawanishi and coworkers (76) used the SFA to make measurements between gelatin layers adsorbed onto mica as a function of pH. At pH 3.5 (net positive charge), a first-compression hysteresis was observed, with subsequent compressions and expansions showing a steric interaction, much as with the polylysine. At pH 8.5 (net negative charge), no first-compression hysteresis was observed, and the range of steric interactions was decreased. The researchers concluded that the mica was only sparsely populated by gelatin, anchored at the few positively charged groups remaining on the gelatin. Later work by Kamiyama and Israelachvili (77) compared these profiles with theoretical descriptions and showed they exhibit an electrostatic repulsion at long range combined with a steric repulsion at short range.

The thickest adsorbed layers were achieved near the IEP. Interaction forces between them showed a significant electrostatic repulsion, even though the gelatin was charge neutral. They argued that the layer surface displayed a net negative charge, because a net positive charge must exist to balance the negative charge at the mica surface. On separation, a small adhesive well was observed between these layers. The well was located at a separation distance that was less than the onset of steric interactions during compression. This adhesion was ascribed to gelatin that bonded to the opposite surface on compression, reinforcing the interface by bridging.

B.   Near-Native Proteins

1.   Ribonuclease

Lee and Belfort (78) performed a detailed study of the adsorption behavior and activity of adsorbed ribonuclease A (RNase). Using the SFA, they measured forces between layers of adsorbed RNase during various steps in the adsorption process. Force profiles showed an exponential steric and/or electrostatic repulsion at long range and a steep steric repulsion at short range. The "knee" at the connection of the two was a measure of the layer thickness. The initial layer thickness corresponded to the shortest crystallographic dimension of ribonuclease and the final thickness to the longest dimension. They concluded that the proteins maintained their folded conformation on adsorption and went from a flat-on adsorbed state to an end-on state as the adsorption progressed. Early in the adsorption, the layers showed a sizable adhesion energy, mostly likely due to bridging effects.

2.   Cytochrome *c*

Cytochrome *c* was shown (79) to form multilayers on mica near its IEP (pH 10). SFA force profiles between mica surfaces in the presence of cytochrome *c* show a series of successive jump-ins superposed on a long-range repulsion described well by DLVO theory. The peak-to-peak distance between these "oscillations" is about the crystallographic diameter of the roughly spherical protein (33 Å). A steep steric repulsion at a separation distance of 33 Å prevented the mica surfaces from making contact. The researchers concluded that weak protein-protein interactions were responsible for the multilayer formation and strong mica-protein interactions firmly held a monolayer of cytochrome c in place. Using data obtained from the FECO fringes, they calculated a layer refractive index in accord with one expected for a monolayer of these hydrated proteins.

3. Insulin

Insulin is a flattened spheroid consisting of six wedge-shaped monomers that make a zinc- and concentration-dependent association. The adsorption of insulin is complicated by the presence of multiple aggregates at equilibrium. Nylander et al. (80) made SFA measurements between insulin layers adsorbed on mica. They found an electrostatic repulsion with a short-range steric repulsion. The layer thickness progressively increased from that of a monolayer of monomer to that of a monolayer of the assembled hexamer, even though the concentration of assembled hexamer was six orders of magnitude smaller in the bulk. The researchers pointed to a molecular weight–dependent displacement of monomers at the surface with time, reminiscent of the Vroman effect in blood-contacting biomaterials.

Claesson et al. (81) observed significant multilayer formation in SFA measurements between insulin layers adsorbed on a hydrophobic Langmuir-Blodgett monolayer. This was manifest as force oscillations with a peak-to-peak distance of one monomer or so. Later (69), it was speculated that the multilayer formation was enabled by the exposure of hydrophobic patches on the monomer to the solution, brought about by preferential adsorption on the hydrophobic LB surface.

4. Albumin

Albumin, the most abundant protein in blood, consists roughly of three globular subunits and has a low degree of conformational stability. Blomberg and coworkers (82–84) have made a series of SFA measurements between layers of human serum albumin (HSA) adsorbed at different concentrations. At low HSA levels, they observed an electrostatic repulsion at long range, which they ascribed to the negative charge of mica that remains unshielded by the sparse population of HSA. An adhesion due to bridging is observed on separation. At intermediate HSA levels, the interaction profile is shifted outward because of the improved layer thickness. At high loads, the HSA layers can be forced to intermingle and bridge the mica surfaces, leading to adhesion. At high HSA levels, the layer thickness increases further, and no intermingling or adhesion can be effected. These results were explained by pressure-induced conformational changes that occur between sparsely populated layers. Pincet et al. (85) extended this work by considering the effect of the denaturant urea on the force profiles. They found that the denatured HSA behaved very much like a synthetic polymer.

C. Polysaccharides

Polysaccharides are highly versatile biomolecules, conferring structural support, lubrication, steric stabilization, and cell binding properties on biological tissues, depending on the circumstances. Only recently have direct force measurement techniques been applied unravel structure-property relationships in these systems.

Claesson et al. (86) have studied the interactions of adsorbed layers of chitosan, a cationic polyelectrolyte, in the SFA. Chitosan is a useful derivative of chitin, the second most abundant natural polysaccharide, and has been applied industrially as a flocculating and metal ion binding agent. The extent of charging of chitosan is pH dependent because of the deprotonation of glucosamine segments at high pH. Chitosan was found to adsorb in a relatively flat configuration over a wide range of pH, and interaction forces were well described by DLVO theory.

Previous SFA work in our group (87) focused on interactions between layers of hyaluronic acid, a highly charged polysaccharide responsible for lubrication and structural stability in living tissues. Hyaluronic acid could not be made to adsorb directly onto mica, so an initial layer of cationic polymer was adsorbed to reverse the charge of mica. Much as in the case of polylysine, we observed a pronounced first-compression hysteresis in measurements on both the cationic polymer and the cationic polymer–hyaluronic acid complex. Interestingly, hyaluronic acid failed to adsorb onto cationic polymer that was previously compressed, suggesting that most positive charges were complexed with mica after compression.

Dammer and coworkers (88) have used the AFM to measure interactions between adsorbed cell adhesion proteoglycans (APs), composed of a protein core with adhesive polysaccharides radiating from it. Their particular sample was taken from the marine sponge *Microciona prolifera*, in which APs had been shown to recruit cell adhesion via AP-AP interactions in a calcium-dependent way. Using gold-thiol chemistry the APs were covalently linked to the AFM tip and a smooth surface. On approach, no interactions were detected, but on separation a strong adhesive force was observed as the bound polysaccharides stretched prior to separation. Multiple jump-outs were observed with added calcium, indicating the APs were bound at multiple sites.

D. Lipidated Biomolecules

Relying on adsorption to pin down biomolecules for direct force measurement has several drawbacks. It is difficult to control or even measure the density and conformation of adsorbed proteins. Nonspecific interactions that hold them in place may be broken or new ones created on compression or separation. A clever solution is to link the biomolecule covalently to an amphiphile, or "lipidate" it (89,90). The amphiphile can be used to anchor the protein in amphiphilic structures such as vesicles or LB layers, relying on the packing of the hydrophobic matrix to hold the protein. A study by Abe and coworkers (91) illustrates the power of this approach. Oligomers of poly(L-glutamic acid) were covalently linked to dialkyl amphiphiles and spread on a Langmuir trough. Fourier transform infrared (FTIR) spectra showed that layers deposited at high molecular areas had a predominantly α-helix structure, and those at low areas were mostly β-sheets. Each structure was deposited on mica for SFA measurements; in both cases a long-range electrostatic repulsion was observed followed by a short-range steric repulsion. Elastic compressibility moduli were calculated in the steric region; the value for the β-sheet structure was many times greater.

IV. LIPID BILAYERS

Force and adhesion measurements between lipid bilayers have greatly improved our understanding of the physicochemical behavior of cell membranes. Because the cell surface is such a complex structure, these measurements must be considered limiting yet important. Efforts are ongoing (92) to build synthetic models of the cell surface that more closely approximate realistic ones.

A. DLVO and Hydration Forces

Some of the first direct force measurements were made between lipid bilayers in the thin-film balance (93). Early on, it was evident that a short-range, non-DLVO "hydration repulsion" existed between the bilayers. Since then, numerous studies have been undertaken to investigate the basis of the hydration repulsion between lipid bilayers. One complication is that repulsive steric interactions resulting from restricted bilayer mobility on confinement cannot be distinguished from hydration forces arising from tightly adsorbed ions. An elaboration of this point, along with a complete summary of direct measurements between lipid bilayers and theoretical models, is provided by Israelachvili and Wennerström (94).

Horn (95) was the first to make SFA measurements between adsorbed bilayers of biologically derived lipids. Measurements made between bilayers of egg lecithin and the synthetic phospholipid DLPC showed electrostatic repulsion at long range and an extra hydration repulsion at 30 Å from bilayer contact. This is a slightly shorter range than that measured between free bilayers using the osmotic stress method (65). It is believed that the suppression of membrane fluctuations by the mica substrates in the SFA was responsible for the discrepancy.

This work was extended by Marra and Israelachvili (96), who made measurements between LB bilayers of the phospholipids DPPC, DLPC, and DPPE in various salts. Again, long-range electrostatic and short-range hydration repulsions were observed. Using the pull-off method, a small adhesion energy was measured between the bilayers ($0.2-1.0$ mJ/m^2) due to van der Waals interactions. Small variations in adhesion energy with electrolyte concentration were ascribed to changes in the magnitude of the hydration repulsion. In parallel work, Marra (97) made measurements between monogalactosyldiglyceride (MGDG) and digalactosyldiglyceride (DGDG). "Hydration thicknesses" were calculated by comparing the mica-mica separation distance in contact with reasonable estimates of the anhydrous bilayer thickness based on space-filling models. It was found that the hydration thickness was exactly twice as large for the disaccharide headgroup of DGDG as for the monosaccharide MGDG. The adhesion energy of DGDG was almost exactly halved, as well. Measurements have also been made between LB bilayers of gangliosides (98), with similar results.

B. Membrane Fusion

The mechanisms of membrane fusion are of obvious importance in cell biology. Using the SFA, fusion processes between mica-supported bilayers have been directly observed. Helm and coworkers (99) performed a detailed study in which phospholipid bilayers were brought into contact and further loaded. At sufficiently high loads, the FECO fringes typically developed a break in the center of contact, corresponding to a region with a separation distance one bilayer smaller. Observing the FECO made it possible to track the fusion process in real time. It was discovered that fusion could be induced only by stressing that bilayers such that hydrophobic portions of the amphiphiles were exposed to the bilayer-bilayer interface. This can be achieved by "depleting" the bilayers or incubating them in lipid-free solution for a long enough time to allow lipids in the LB layer to desorb. Bilayer stressing can also be brought about by introducing defects in the bilayer or by causing the bilayer to phase separate by the addition of calcium (100). Partial bilayers, LB deposited at surface pressures below that required for close packing, spontaneously fuse on contact. Large adhesion energies are generally required to separated fused membranes, and the magnitude of the adhesion energy can be correlated with the extent of membrane fusion that is allowed to occur prior to separation.

C. Adhesion Between Vesicles

As described in the first section, the MPA method holds the promise of making molecular-level adhesion measurements in a very realistic membrane environment. MPA has been applied predominantly to problems of membrane mechanics, in lipid vesicle systems and with real cells. Adhesion energies of 0.01–0.015 mJ/m^2 were recorded for fluid vesicles of the phospholipids DMPC and SOPC (101,102). Vesicles of the less hydrated phospholipid POPE had a much higher adhesion energy (0.12–0.15 mJ/m^2) (103). Adhesion energies as high as 0.22 mJ/m^2 were observed between vesicles of the galactolipid DGDG (102). These values appear to be slightly lower than those observed in the SFA experiment. This may be due to disturbance of the bilayer on jump-to-contact in the SFA.

V. MOLECULAR RECOGNITION FORCES

The specific recognition and binding of paired molecules make life possible, allowing protein assembly, DNA replication and conservation, and hormonal signaling. DFM techniques can be uniquely applied to determine the physicochemical mechanisms of these important, fundamental interactions. The ultrafine feature of the AFM probe has allowed researchers to probe directly single-molecule interactions and bonding forces. Truly intermolecular van der Waals, electrostatic, hydrogen bonding, and hydrophobic forces have now been measured (104,105), and this approach has been applied to more complex biological systems.

A more directly practical forum for these experiments is in the area of cell adhesion. It is now widely appreciated that cells are capable of making strong adhesions with substrates via "focal adhesions" (1) whose creation is dependent on molecular recognition processes between cell-surface adhesion receptors and the surrounding matrix. The bonding potential and dissociation rates as a function of applied load are critical parameters in theoretical predictions of cell adhesion, spreading, and migration (1). With the exception of the MPA assay, however, meaningful DFM experiments cannot be performed on live cells, and a major challenge to experimenters has been to build realistic mimics by self-assembly or LB deposition.

A.   Receptor-Ligand Binding (Streptavidin-Biotin)

The streptavidin-biotin system has been the receptor-ligand system most widely studied by direct force measurement. Biotin-conjugated lipids are readily prepared (106) and can be inserted into vesicles or applied to surfaces by LB deposition. Because streptavidin has four binding sites for biotin, it can link two biotin-functionalized surfaces. Streptavidin-biotin binding is unusually strong; for this reason, some dispute its usefulness as a model system. However, work with this system has been instrumental in understanding mechanisms of receptor-ligand bond formation as well as the dynamics and cohesive strength of supporting layers.

Building on binding studies in vesicles and Langmuir monolayers, the first direct measurements of receptor-ligand interactions were provided by Helm et al. (107), who made SFA force measurements between mixed LB layers of 95% DLPC and 5% DPPE-conjugated biotin in the fluid state. These layers showed the electrostatic repulsion characteristic of lipid bilayers, with no adhesion. A single surface was then removed from the SFA and incubated in a solution of streptavidin, which adsorbed onto the biotinylated surface. Force measurements made between these surfaces showed a very weak repulsion before the jump-to-contact. A strong adhesion energy (7.4 mJ/m^2) was observed on pull-off and ascribed to the linking of the biotin surfaces by multifunctional streptavidin. On second approach, the surfaces came into a weak contact greater than the initial contact at a separation distance about twice the thickness of a streptavidin-biotin complex. They reasoned that the binding was so strong that bound complexes were uprooted from the bilayers on separation. This observation prevented assignment of the molecular binding strength.

Leckband et al. (108) investigated more closely the interaction of streptavidin and biotin prior to contact. As a control, measurements were made between a biotin layer and a layer of streptavidin that had been incubated with soluble biotin, blocking its binding ability. These layers showed a much higher repulsion than in the unblocked case. They reasoned that a long-range "steering" attraction between the receptor and ligand lowered the repulsion. Later, the lowering of the repulsion was explained convincingly by the alteration of surface electrostatics by the bound, soluble biotin instead (109). Measurements were also made between streptavidin and biotin layers in the gel phase, showing a markedly reduced adhesion energy (1.1 mJ/m^2) as the receptors and ligands were trapped in place, unable to reorient themselves to create more bonds. In a very illuminating series of experiments (110), Leckband et al. made similar measurements on biotin analogues with weaker affinities for streptavidin. In some cases, the bound complex was uprooted on separation even when the binding energy was lower than the energy required to displace lipids from the layers. Only when the binding force (roughly calculated as energy per bond length) was equivalent between lipid uprooting and receptor-ligand dissociation was uprooting suppressed.

Single-molecule interactions between streptavidin and biotin have been measured with the AFM (111–113). Biotinylated BSA adsorbed onto AFM tips prior to incubation with streptavidin. The tip was rastered over a surface of biotinylated agarose beads, making approaches and retractions to obtain force curves. The pull-off forces measured differed by integer multiples; and this quantization was deemed to be due to individual molecular interactions with a binding force of 257 ± 25 pN per pair. Again, more weakly binding analogues were studied, and a strong correlation was demonstrated between enthalpies of binding (from calorimetry) and pull-off force per pair, suggesting that dissipative processes were not at work during pull-off. Multiple contacts at the same point gave the same magnitude of pull-off force, obviating the uprooting complications that plague the SFA studies.

Vesicles containing biotinylated DPPE were the subject of an MPA study (114). One vesicle, previously incubated with fluorescently labeled streptavidin, was brought into contact with another uncoated vesicle. Avidin was observed to accumulate in the contact area during the MPA assay. Results of the assay gave the concentration of avidin cross-bridges in the contact zone. The membranes failed in tension before they could be peeled apart, paralleling the SFA observations.

B. DNA Base-Pairing

Encouraged by the streptavidin-biotin work, similar studies using the SFA and AFM have been embarked upon to investigate DNA base-pairing interactions. The first attempt was that of Berndt et al. (115), who made SFA force measurements between LB layers of dialkyl amphiphiles bearing the nucleobase adenine and the uracil-like rotate. A long-range attraction and strong adhesion were observed between complementary and noncomplementary pairs of surfaces. These amphiphiles formed micelles under some conditions, clouding the interpretation of the data. An improvement was provided by Kurihara et al. (116) using alkylammonium molecules end functionalized with adenine and thymine nucleobases. LB layers of these amphiphiles did not form micelles and gave similar results. Adenine-adenine layers showed a long-range repulsion at neutral pH, whereas thymine-thymine and adenine-thymine layers showed a long-range attraction. The thymine-thymine system showed the largest adhesion energy (30 mJ/m^2, compared with 14 mJ/m^2 for adenine-thymine). Pincet and coworkers (117) completed a similar SFA study on interactions between LB layers of unsaturated amphiphiles bearing full nucleotides (with the adjacent sugar group).

The degree of unsaturation of the amphiphile gave the layer fluidity, allowing complementary pairs to reorient near contact. Here, long-range attractions and strong adhesions were observed in all cases, with the complementary interaction giving the highest adhesion energy (9.1 mJ/m^2). The reasons for the discrepancies are not immediately clear, nor is it clear whether the source of the attraction is related to the hydrophobic force discussed earlier. It should be noted that in none of these studies was it conclusively proved that uprooting of bound complexes did not occur.

Single-molecule interactions between nucleotides have been measured by AFM. Boland and Ratner (118) adsorbed nucleotides directly onto gold-coated tips and rastered them in tapping mode over a surface of nucleotides prepared in the same way. Again, long-range attractions and strong adhesions were observed in all cases. Complementary interactions exhibited a secondary minimum as the tip approached the surface. A force of 54 pN was assigned to a single adenine-thymine bond. Lee et al. (119) linked single strands of DNA with AFM tips and surfaces using gold-thiol chemistry. Complementary strands of repeating, four-base sequences showed a series of jump-out events at a discrete distances corresponding to integer multiples of the four-base contour length. This observation highlights the single-molecule potential of these experiments. Optical tweezers have been used to measure stretch-induced conformational changes in DNA as well (120).

C. Hydrogen-Bonded Networks of Amino Acid Amphiphiles

Using the SFA, surface force and JKR-type adhesion measurements were made on LB layers of amphiphiles conjugated to the amino acid glycine (121). On approach, the glycine headgroups interpenetrated to a small but measurable degree, allowing carbonyl and amide linkages on opposing surfaces to make complementary hydrogen bonds, reinforcing the interface. As with the streptavidin-biotin system, the bilayers were torn apart on separation, suggesting that the glycine-glycine interaction of β-sheets of polyglycine is effectively as strong. The layers were charged at high pH, preventing interpenetration and the resulting strong adhesion. JKR measurements showed significant adhesion hysteresis, indicating that the surfaces were brought into contact by van der Waals forces, and postcontact, stress-dependent rearrangements were responsible for the formation of interlayer hydrogen bonds.

D. Antigen-Antibody Recognition

Surprisingly few attempts have been made to measure antigen-antibody interactions using SFA or AFM (112) methods. Liebert and Prieve (122) used the TIRM method to track interactions between protein A and antibodies against it from different animal species. At all distances, the colloidal part of the potential energy profile shows a lower potential than expected from DLVO theory. The authors considered this to be a long-range "steering" attraction but also offered electrostatic explanations. The decay length of the attraction correlated well with the dissociation constants for the binding of protein A to antibodies obtained from rabbit, horse, and goat sources. Leckband (123) made SFA force and adhesion measurements between antigen/antibody conjugates, and observed long-range attractions as well.

E. Muscle Contractile Forces

Single-molecule interaction forces have been observed between myosin and an actin filament using optical tweezers with PDA displacement detection (60). Two free microspheres were anchored to either end of a single actin filament to maneuver it in proximity with myosin adsorbed onto surface-held microspheres. A relatively compliant trap was used for the initial experiments, allowing the myosin to ratchet the filament at observable 11-nm intervals in the presence of ATP. The trap was then stiffened, and PDA feedback was used to maintain the position of both beads. With ATP, 3- to 4-pN force transients were measured as the myosin attempted to move the filament. Svoboda and Block (59) used the enzyme kinesin as a motor rather than myosin. With the optical trapping interferometer, they trapped a kinesin-coated sphere and maneuvered it in contact with a surface-held actin filament. After calibration, the position of the sphere in the trap was monitored by interferometry while the stage was sinusoidally oscillated. Instantaneous force and velocity data from this work showed that the velocity of kinesin along the filament decreased markedly with applied load. This load-dependent change in enzyme kinetics is critical to the understanding of muscular tissue function.

VI. CONCLUSIONS

Direct force measurement techniques have a unique potential to reveal molecular mechanisms underlying biological processes at interfaces. However, they are not magical; the onus is on the experimenter to create and characterize well-defined surfaces for measurements and to apply molecular insight in the interpretation of results. In the future, we expect that these techniques will be applied to increasingly complex systems, guiding the rational design of interactive biomaterials and leading to deeper understanding of the mechanics of the cell.

REFERENCES

1. DA Lauffenburger, JJ Linderman. Receptors: Models for Binding, Trafficking, and Signalling and Their Relationship to Cell Function. New York: Oxford University Press, 1992.
2. S Wu. Polymer Interface and Adhesion. New York: Marcel Dekker, 1982.
3. AW Adamson. Physical Chemistry of Surfaces. New York: John Wiley & Sons, 1990.
4. JN Israelachvili. Intermolecular and Surface Forces. New York: Academic Press, 1992.
5. RJ Colton, WR Barger, DR Baselt, SG Corcoran, DD Koleske, GU Lee. In: WJ van Ooij, HR Anderson, eds. Mittal Festschrift. Amsterdam: VSP, 1998, pp 1–27.
6. DA Hammer, M Tirrell. Annu Rev Mater Sci 26:651–691, 1996.
7. VA Parsegian, EA Evans. Curr Opin Colloid Interface Sci 1:53–60, 1996.
8. J Mahanty, BW Ninham. Dispersion Forces. New York: Academic Press, 1976.
9. G Cevc. Biochim Biophys Acta 1031:311–382, 1990.
10. DYC Chan, RM Pashley, LR White. J Colloid Interface Sci 77:283–285, 1980.
11. SS Patel, M Tirrell. Annu Rev Phys Chem 40:597–635, 1989.
12. HK Christenson. In: ME Schrader, G Loeb, eds. Modern Approaches to Wettability: Theory and Applications. New York: Plenum, 1992, pp 29–51.
13. HK Christenson, PM Claesson, J Berg, PC Herder. J Phys Chem 93:1472–1478, 1989.
14. Y Tsao, DF Evans, H Wennerström. Science 262:547–550, 1993.
15. HK Christenson, PM Claesson. Science 239:390–392, 1988.
16. J Wood, R Sharma. Langmuir 11:4797–4802, 1995.
17. H Hertz. J Reine Angew Math 92:156–171, 1881.
18. KL Johnson, K Kendall, AD Roberts. Proc R Soc Lond A 324:301–313, 1971.
19. D Maugis. J Colloid Interface Sci 150:243–269, 1992.
20. YL Chen, CA Helm, JN Israelachvili. J Phys Chem 95:10736–10747, 1991.
21. P Silberzan, S Perutz, EJ Kramer, MK Chaudhury. Langmuir 10:2466–2470, 1994.
22. MK Chaudhury, T Weaver, CY Hui, EJ Kramer. J Appl Phys 80:30–37, 1996.
23. A Falsafi, P Deprez, FS Bates, M Tirrell. J Rheol 41:1–16, 1997.
24. M Tirrell. Langmuir 12:4548–4551, 1996.
25. K Kendall. Science 263:1720–1725, 1994.
26. MK Chaudhury. Curr Opin Colloid Interface Sci 2:65–69, 1997.
27. D Tabor, RHS Winterton. Proc R Soc Lond A 312:435–450, 1969.
28. JN Israelachvili. J Colloid Interface Sci 44:259–272, 1972.
29. RG Horn, DR Clarke, MT Clarkson. J Mater Res 3:413–416, 1988.
30. RG Horn, DT Smith. Science 256:362–364, 1992.
31. V Mangipudi, M Tirrell, AV Pocius. Langmuir 11:19–23, 1995.
32. B Derjaguin. Kolloid Z 69:155–164, 1934.

33. J Klein. Coll Surf A 86:63–76, 1994.
34. SM Kilbey, FS Bates, M Tirrell, H Yoshizawa, R Hill, J Israelachvili. Macromolecules 28:5626–5631, 1995.
35. J Peachey, J Van Alsten, S Granick. Rev Sci Instrum 62:463–473, 1991.
36. PM Claesson, T Ederth, V Bergeron, MW Rutland. Adv Colloid Interface Sci 67:119–183, 1996.
37. J Israelachvili, GE Adams. J Chem Soc Faraday Trans I 74:975–1001, 1978.
38. JA Zasadzinski, R Viswanathan, L Madsen, J Garnaes, DK Schwartz. Science 263:1726–1733, 1994.
39. A Ulman. An Introduction to Ultrathin Organic Films. New York: Academic Press, 1991.
40. G Binning, CF Quate, C Gerber. Phys Rev Lett 56:930–933, 1986.
41. JP Cleveland, S Manne, D Bocek, PK Hansma. Rev Sci Instrum 64:403–405, 1993.
42. H Dai, JH Hafner, AG Rinzler, DT Colbert, RE Smalley. Nature 384:147–150, 1996.
43. E Meyer, L Howald, R Overney, D Brodbeck, R Lüthi, H Haefke, J Frommer, H-J Güntherodt. Ultramicroscopy 42–44:274–280, 1992.
44. JE Houston, TA Michalske. Nature 356:266–267, 1992.
45. WA Ducker, TJ Senden, RM Pashley. Langmuir 8:1831–1836, 1992.
46. A Milling, P Mulvaney, I Larson. J Colloid Interface Sci 180:460–465, 1996.
47. JM Levins, TK Vanderlick. J Colloid Interface Sci 185:449–458, 1997.
48. A Ashkin. Phys Rev Lett 24:156–159, 1970.
49. A Ashkin. Proc Natl Acad Sci U S A 94:4853–4860, 1997.
50. DC Prieve, F Luo, F Lanni. Faraday Discuss Chem Soc 83:297–307, 1987.
51. DC Prieve, JY Walz. Appl Opt 32:1629–1641, 1993.
52. SG Flicker, SG Bike. Langmuir 9:257–262, 1993.
53. SG Bike, L Lazarro, DC Prieve. J Colloid Interface Sci 175:411–421, 1995.
54. JY Walz, DC Prieve. Langmuir 8:3073–3082, 1992.
55. A Ashkin, K Schütze, JM Dziedzic, U Euteneuer, M Schliwa. Nature 348:346–348, 1990.
56. SM Block, LSB Goldstein, BJ Schnapp. Nature 348:348–352, 1990.
57. JM Colon, P Sarosi, PG McGovern, A Ashkin, JM Dziedzic, J Skurnick, G Weiss, EM Bonder. Fertil Steril 57:695–698, 1992.
58. K Svoboda, SM Block. Annu Rev Biophys Biomol Struct 23:247–285, 1994.
59. Svoboda K, SM Block. Cell 77:773–784, 1994.
60. JT Finer, RM Simmons, JA Spudich. Nature 368:113–119, 1994.
61. RM Simmons, JT Finer, S Chu, JA Spudich. Biophys J 70:1813–1822, 1996.
62. JC Crocker, DG Grier. Phys Rev Lett 73:352–355, 1994.
63. AE Larsen, DG Grier. Nature 385:230–233, 1997.
64. VA Parsegian, RP Rand, NL Fuller, DC Rau. Methods Enzymol 127:400–416, 1986.
65. RP Rand, VA Parsegain. Biochim Biophys Acta 988:351–376, 1989.
66. E Evans. Adv Colloid Interface Sci 39:103–128, 1992.
67. E Evans, D Berk, A Leung. Biophys J 59:838–848, 1991.
68. E Evans. Biophys J 68:2580–2587, 1995.

69. PM Claesson, E Blomberg, JC Fröberg, T Nylander, T Arnebrant. Adv Colloid Interface Sci 57:161–227, 1995.
70. M Tirrell. Curr Opin Colloid Interface Sci 2:70–75, 1997.
71. PF Luckham, PG Hartley. Adv Colloid Interface Sci 49:341–386, 1994.
72. PF Luckham, J Klein. J Chem Soc Faraday Trans I 80:865–878, 1984.
73. J Klein, PF Luckham. Coll Surf 10:65–76, 1984.
74. T Afshar-Rad, AI Bailey, PF Luckham, W Macnaughtan, D Chapman. Coll Surf 25:263–277, 1987.
75. T Afshar-Rad, AI Bailey, PF Luckham, W Macnaughtan, D Chapman. Coll Surf 31:125–146, 1988.
76. N Kawanishi, HK Christenson, BW Ninham. J Phys Chem 94:4611–4617, 1990.
77. Y Kamiyama, J Israeachvili. Macromolecules 25:5081–5088, 1992,
78. C-S Lee, G Belfort. Proc Natl Acad Sci U S A 86:8392–8396, 1989.
79. P Kékicheff, WA Ducker, BW Ninham, MP Pileni. Langmuir 6:1704–1708, 1990.
80. T Nylander, P Kékicheff, B Ninham. J Colloid Interface Sci 164:136–150, 1994.
81. PM Claesson, T Arnebrandt, B Bergenståhl, T Nylander. J Colloid Interface Sci 130:457–466, 1989.
82. E Blomberg, PM Claesson, HK Christenson. J Colloid Interface Sci 138:291–293, 1990.
83. E Blomberg, PM Claesson, CG Gölander. J Disp Sci Tech 12:179–200, 1991.
84. E Blomberg, PM Claesson, RD Tilton. J Colloid Interface Sci 166:427–436, 1994.
85. F Pincet, E Perez, G Belfort. Macromolecules 27:3424–3425, 1994.
86. PM Claesson, BW Ninham. Langmuir 8:1406–1412, 1992.
87. S Dhoot, ED Goddard, DS Murphy, M Tirrell. Coll Surf 66:91–96, 1992.
88. U Dammer, O Popescu, P Wagner, D Anselmetti, H Güntherodt, GN Misevic. Science 267:1173–1175, 1995.
89. K Kurihara, T Kunitake, N Higashi, M Niwa. Langmuir 8:2087–2089, 1992.
90. P Berndt, G Fields, M Tirrell. J Am Chem Soc 117:9515–9522, 1995.
91. T Abe, K Kurihara, N Higashi, M Niwa. J Phys Chem 99:1820–1823, 1995.
92. E Sackmann. Science 271:43–48, 1996.
93. DM LeNeveu, RP Rand, VA Parsegian, D Gingell. Biophys J 18:209–230, 1977.
94. JN Israelachvili, H Wennerström. J Phys Chem 96:520–531, 1992.
95. RG Horn. Biochim Biophys Acta 778:224–228, 1984.
96. J Marra, J Israelachvili. Biochemistry 24:4608–4618, 1985.
97. J Marra. J Colloid Interface Sci 107:446–458, 1985.
98. P Luckham, J Wood, R Swart. J Colloid Interface Sci 156:173–183, 1993.
99. CA Helm, JN Israelachvili, PM McGuiggan. Science 246:919–922, 1989.
100. DE Leckband, CA Helm, J Israelachvili. Biochemistry 32:1127–1140, 1993.
101. E Evans, M Metcalfe. Biophys J 46:423–426, 1984.
102. E Evans, D Needham. J Phys Chem 91:4219–4228, 1987.
103. E Evans, D Needham. Faraday Discuss Chem Soc 81:267–280, 1986.

104. CD Frisbie, LF Rozsnyai, A Noy, MS Wrighton, CM Lieber. Science 265: 2071–2074, 1994.
105. RC Thomas, JE Houston, RM Crooks, T Kim, TA Michalske. J Am Chem Soc 117:3830–3834, 1995.
106. EA Bayer, B Rivnay, E Skuletsky. Biochim Biophys Acta 550:464–473, 1979.
107. CA Helm, W Knoll, JN Israelachvili. Proc Natl Acad Sci U S A 88:8169–8173, 1991.
108. DE Leckband, JN Israelachvili, F-J Schmitt, W Knoll. Science 255:1419–1421, 1992.
109. DE Leckband, F-J Schmitt, JN Israelachvili, W Knoll. Biochemistry 33:4611–4624, 1994.
110. D Leckband, W Müller, F-J Schmitt, H Ringsdorf. Biophys J 69:1162–1169, 1995.
111. VT Moy, EL Florin, HE Gaub. Science 266:257–259, 1994.
112. VT Moy, EL Florin, HE Gaub. Coll Surf A 93:343–348, 1994.
113. GU Lee, DA Kidwell, RJ Colton. Langmuir 10:354–357, 1994.
114. DA Noppl-Simson, D Needham. Biophys J 70:1391–1401, 1996.
115. P Berndt, K Kurihara, T Kunitake. Langmuir 11:3083–3091, 1995.
116. Kurihara K, T Abe, N Nakashima. Langmuir 12:4053–4056, 1996.
117. F Pincet, E Perez, G Bryant, L Lebeau, C Mioskowski. Phys Rev Lett 73: 2780–2783, 1994.
118. T Boland, BD Ratner. Proc Natl Acad Sci U S A 92:5297–5301, 1995.
119. GU Lee, LA Chrisey, RJ Colton. Science 266:771–773, 1994.
120. SB Smith, YJ Cui, C Bustamante. Science 271:795–799, 1996.
121. J Schneider, P Berndt, K Haverstick, S Kumar, S Chiruvolu, M Tirrell. J Am Chem Soc 120:3508–3509, 1998.
122. RB Liebert, DC Prieve. Biophys J 69:66–73, 1995.
123. DE Leckband, T Kuhl, HK Wang, J Herron, W Müller, H Ringsdorf. Bio chemistry 34:11467–11478, 1995.

11

A Centrifugation Method for Measurement of Two-Dimensional Binding Characteristics of Receptor-Ligand Interaction

Cheng Zhu and James W. Piper
Georgia Institute of Technology, Atlanta, Georgia

Robert A. Swerlick
Emory University School of Medicine, Atlanta, Georgia

I. INTRODUCTION

Among the various techniques of measuring cell adhesion, centrifugation is a simple and yet quantitative method. Since the original work of McClay et al. (1981), this method has been widely used. However, only recently has a mathematical model for the assay been developed and tested experimentally (Piper et al., 1998). This chapter describes how to use the centrifugation assay to determine the binding characteristics of receptor-ligand interactions via quantitative measurement and analysis of the cell adhesion that is mediated by such interactions. The presentation follows closely that of Piper et al. (1998) with some extension and modification.

The binding characteristic of interest is the two-dimensional (2-D) binding affinity or equilibrium association constant. Because the temporal resolution of the centrifugation assay is of the order of a minute, it is best suited for measurement of equilibrium properties of adhesion molecules that bind with fast kinetics, e.g., selectins interacting with their carbohydrate ligands (e.g., Lasky, 1992). In principle, the mathematical model described herein should be extendable to the analysis of the transient situation provided that time-dependent data can be obtained from the centrifugation assay. However, in Piper et al. (1998), only the steady-state assumptions and predictions of the model have been validated.

The 2-D binding affinity is so called because both of the interacting receptors and ligands are anchored to an apposing surface (Bell, 1978; Lauffenburger and Linderman, 1993). Hence, their motions are restricted to two dimensions (Fig. 1A). This is in contrast to the binding of a cytokine or growth factor molecule in solution to a cell surface receptor. In such a case the soluble ligand can move in three dimensions. The binding affinity for this latter case is referred to as 3-D affinity (Fig. 1B). The measurements of 2-D binding characteristics have become possible only recently and have been made mostly using the flow chamber technique (Kaplanski et al., 1993; Pierres et al., 1995; Alon et al., 1995, 1997; Chen et al., 1997). The method described here provides a simple way to make such measurements that is complementary to other existing methods.

In addition to the difference in dimension, in the case of cell adhesion, the receptor-ligand bonds are usually subjected to externally applied forces. These forces arise because the bonds enable an object a thousand times larger, the cell, to be bound to another cell or the substrate. By comparison, a 3-D bond is subject to no force other than that of thermal agitation, i.e., the Brownian force. The presence of externally applied forces is likely to

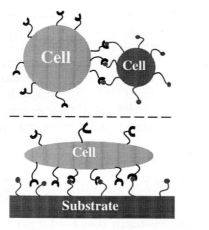
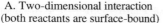
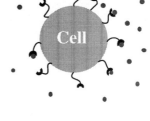

A. Two-dimensional interaction
(both reactants are surface-bound)

B. Three-dimensional interaction
(one reactant is in solution)

Figure 1 Illustration of two-dimensional versus three-dimensional receptor-ligand interactions.

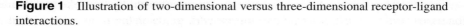

influence the value of binding affinity, just as the equilibrium association constant of a chemical reaction is usually affected by temperature and/or pressure. Bond breaking by the applied force can also be viewed as accelerated dissociation of the bond by the force. Thus, what should be determined is the *force dependence* of the equilibrium association *coefficient*, not just an equilibrium association constant. Such a dependence can be described mathematically by a constitutive equation and a set of intrinsic parameters. The centrifugation method can be used to determine the most appropriate form of the constitutive equation for a class of adhesion molecules and to evaluate the parameters that best describe the binding properties of a particular pair of receptor and ligand.

II. EXPERIMENT

A. The Centrifugation Assay

An obvious advantage of the centrifugation assay is its simplicity. The only piece of equipment required is a tabletop centrifuge, which is usually available in most biological laboratories. Common protocols for preparation of cells and multiwell plates as well as for execution of the experiment can be found in the literature (e.g., McClay et al., 1981; Chu et al., 1994; Piper et al., 1998). Briefly, prior to the experiment, target cells need to be labeled with either a radioisotope or a fluorescent dye to allow automated quantification. Cells are washed to remove any radioactivity or fluorescence not associated with the cells. They are added to the wells of multiwell plates at a given concentration (typically 20,000–40,000 cells per well). The total radioactivity (or fluorescence intensity) added per well is recorded. The plates are placed in a centrifuge and spun at low speed for a short time (e.g., 8 g, 30 seconds) to bring all cells into contact with the bottom surface, on which either specific molecules are coated or a confluent monolayer of cells is grown. After being incubated for a predetermined time (e.g., 30 minutes) to allow adhesion to occur, the plates are inverted and spun in the centrifuge to impose a defined force on the cells for a given duration (e.g., 1 minute). After removal of the cells that detached from the bottom surface, the radioactivity (or fluorescence intensity) of the remaining adherent cells in each well is collected and measured in a radiometric counter (or a fluorescent plate reader). The fraction of adhesion is calculated from $P = (C_a - C_b)/(C_t - C_b)$, where C_a, C_b, and C_t are, respectively, the adherent, background, and total counts (or intensity) read by the radiometric counter (or the fluorescent plate reader).

Caution must be exercised so that the measured radioactivity (or fluorescence intensity) faithfully reflects the viable cells. It is useful to generate a calibration curve to relate the radioactivity (or fluorescence intensity) to the number of cells in a sample. Prior to addition of cells to the multiwell plates, a trypan blue exclusion test should always be conducted to ensure that the cell viability is not lower than 90%. In addition, visual inspection of the well surface under a microscope should be performed on a regular basis to ensure that cells are in singlets (as opposed to aggregates) and are uniformly deposited (as opposed to concentrating at edges) before centrifucation and that, after centrifugation, adherent cells remain intact (as opposed to being lysed into cell debris by excessive forces; see Limitations of the Method).

B.   The Mechanical Aspects

Figure 2 shows a schematic of how the centrifugal force is applied to detach the cells. It should be noted that, for quantitative work, it is necessary to fill the wells with medium and seal them with plate sealers to form an air-free system. Otherwise, the force calculation will be complicated by the presence of surface tension on the water-air interface.

C.   Measurements of Cell Size and Density

To quantify the centrifugal force exerted on the cell requires knowledge of the density and size of the cell (Fig. 2). The latter can easily be measured directly under an optical microscope (Fig. 3). The former can be obtained by centrifuging the cells through a continuous density gradient along with calibration beads (Fig. 4).

D.   Control and Quantification of Receptor
Surface Density

To determine the numerical value of the binding characteristics, it is desirable to be able to quantify and vary the surface densities of the receptors and ligands. Also, the model developed in Piper et al. (1998) is for analysis of adhesion mediated by a single pair of receptor and ligand species, although it is possible to extend the model to include more complex situations in which multiple adhesion pathways coexist. These requirements can easily be satisfied when purified molecules are available, as several methods exist to coat proteins on plastic plates or beads (Fig. 5) and to quantify the surface density (Fig. 6). This is not to say the method is not applicable when cells are used. However, in such a case it is important that the nonspecific adhesion is much lower than the specific ones, respectively defined as binding

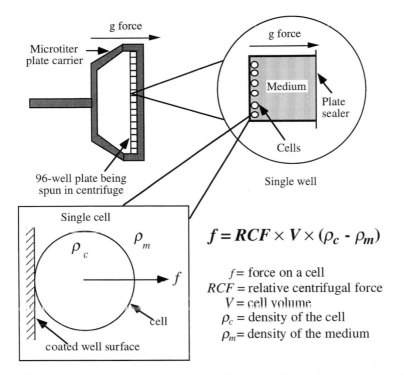

Figure 2 Schematic (not to scale) of how the dislodging force is applied to an adherent cell in the centrifugation experiment. The direction of the force is perpendicular to the contact area, and the magnitude of the force is easily controlled and quantified by the indicated equation (the unit for f is pN when the units of RCF, V, and ρ are m/s², μm², and g/cm³, respectively).

independent of and dependent on the particular receptor-ligand pair of interest (Piper, 1997).

III. THEORY

A. The General Kinetic Framework

In biology, quantitative descriptions of receptor-ligand interaction use primarily chemical terms. For a single-step reversible reaction of v_r receptors binding to v_l ligands to form v_b bonds, as given by the chemical equation

$$v_r M_r + v_l M_l \underset{k_r}{\overset{k_f}{\rightleftharpoons}} v_b M_b \tag{1}$$

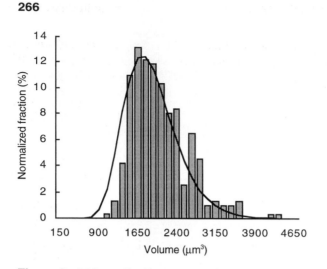

Figure 3   Volume distributions of Colo-205 cells. The volumes of 575 freshly tryp-sinized Colo-205 cells were measured microscopically. The measured normalized fractions (bars) are well described by a lognormal distribution (curve).

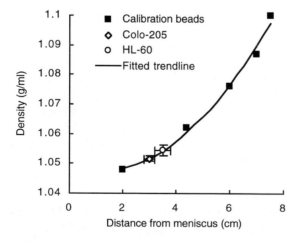

Figure 4   Measurement of cell density. A continuous gradient was used to deter-mine the density of Colo-205 (diamond) and HL-60 (circle) cells. Beads of known densities were added to the gradient along with the cells. The loaded gradient was centrifuged and the cell densities were determined by comparing the depth of the cell layers with those of the calibration beads.

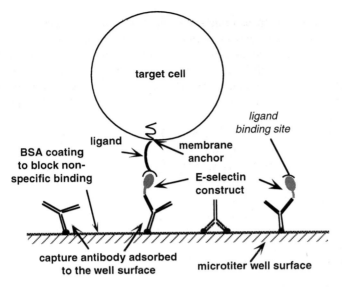

Figure 5   Schematic (not to scale) of the procedures for functionalization of the microtiter plate surface (Li et al., 1994). Physisorption of proteins to plastic usually results in randomly oriented binding sites, not all of which are functioning, as depicted by the capture antibody. By comparison, the E-selectin constructs captured by the antibodies have the proper orientation with the ligand binding sites pointing outward.

The deterministic form of the kinetic equation can be written as

$$\frac{dm_b}{dt} = k_f \left(m_r - \frac{v_r}{v_b} m_b \right)^{v_r} \left(m_l - \frac{v_l}{v_b} m_b \right)^{v_l} - k_r m_b^{v_b} \tag{2}$$

where M_r, M_l, and M_b designate the respective molecular species, the corresponding lowercase symbols denote their respective volumetric concentrations (in the 3-D case) or surface densities (in the 2-D case) and k_f and k_r are, respectively, the forward and reverse rate constants. In the 3-D case, k_f and k_r have the respective dimensions of $(\text{volume})^{v_r+v_l-1}/\text{time}$ and $(\text{volume})^{v_b-1}/\text{time}$, whereas in the 2-D case, their dimensions are $(\text{area})^{v_r+v_l-1}/\text{time}$ and $(\text{area})^{v_b-1}/\text{time}$, respectively. Note that if $v_b = 1$ (and/or $v_r + v_l = 1$), then k_r (and/or k_f) has the same dimension, per unit time in both the 2-D and 3-D cases.

   At steady state, $dm_b/dt = 0$, the ratio of the concentrations (or densities) of the product to those of the reactants gives rise to what is called the equilibrium association constant or binding affinity of the reaction:

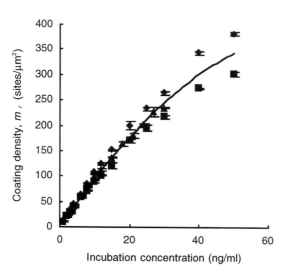

Figure 6 Control and quantification of surface density of adhesion molecules coated on the microtiter plate surface. The concentrations of E-selectin construct used to coat the wells were well correlated with its densities coated on the surface as determined by radioimmunoassays. Each data point represents the mean ± standard deviation of four wells. The continuous curve was a quadratic fit to the data. The data represent the combined results of three separate coating experiments as indicated by different symbols (squares, triangles, diamonds).

$$K_{a} \equiv \frac{k_{f}}{k_{r}} = \frac{m_{b}^{v_{b}}}{\left(m_{r} - \dfrac{v_{r}}{v_{b}} m_{b}\right)^{v_{r}} \left(m_{l} - \dfrac{v_{l}}{v_{b}} m_{b}\right)^{v_{l}}} \tag{3}$$

Note that m_{r} and m_{l} denote total concentrations (or densities) in Eqs. (2) and (3). By contrast, m_{b} is the instantaneous value in Eq. (2) but steady-state value in Eq. (3).

Traditional bulk chemistry approaches (Hulme and Birdsall, 1992), such as Scatchard analysis (1949), for measuring binding affinity are based on the definition of affinity as the concentration ratio of the bound to the free receptors and ligands [Eq. (3)], and it thus requires the ability to measure these concentrations separately. In adhesion assays, by contrast, only the fractions of adherent and detached cells are measured. Whereas there are no bonds associated with a detached cell, the number of bonds of an adherent cell can be any number allowable by the receptors and ligands in the contact area available for binding, which is very difficult, although not impossible (e.g., see Dustin et al., 1996), to measure. Because cell adhesion is mediated by the formation of receptor-ligand bonds, it is conceivable to derive the molecular binding characteristics directly from the cellular adhesion data. Accomplishing this requires a model to relate the kinetic rate and equilibrium constants to the measured fractions of adherent and detached cells and to infer the density of bonds from these fractions.

B. The Master Equations

In formulating such a model, it is important to realize that when a population of cells is assayed it is usually a fraction, not all or none of the cells, that are adherent. In addition, the fraction of adherent cells decreases with decreasing receptor and/or ligand density and increasing dislodging force. To account for this lack of all-or-none, or indeterministic, phenomenon requires a hypothesis for its underlying random events. Two such hypotheses were tested in Piper et al. (1998). The first hypothesis postulates that the fractionalization seen in adhesion assays is a manifestation of the stochastic nature inherent in receptor-ligand binding, which becomes significant when the number of bonds per cell is small. The second hypothesis postulates that the fractionalization is due to the heterogeneity of a cell population, which is commonly seen in cultured cells with asynchronous cycles. The data of Piper et al. (1998) were found to be consistent with the assumptions and predictions of the model based on the first but not the second hypothesis.

Because receptor-ligand binding is assumed to be stochastic, the number of bonds formed, n, in the contact area, A_c, fluctuates significantly when it is small. A single deterministic value (scalar) for the (averaged) surface density of bonds, $m_b = \langle n \rangle / A_c$, is no longer adequate for a complete description of the system. Instead, one considers a probability vector $\{p_0, p_1, \ldots, p_n, \ldots, p_{A_c m_{min}}\}$ to describe the state of the system. In other words, the adhesion could be mediated by any number of bonds ranging from 0 to $A_c m_{min}$ [$m_{min} = \min(m_r, m_l)$]. However, the likelihood of having n bonds is defined by p_n. The probabilistic counterpart of Eq. (2) is the master equations (McQuarrie, 1963) that govern the rate of change of these $A_c m_{min} + 1$ probability components.

$$
\begin{aligned}
\frac{dp_n}{dt} &= (n + 1)^{v_b} \frac{k_r^{(n+1)}}{A_c^{v_b-1}} p_{n+1} - \left[\left(A_c m_r - \frac{v_r}{v_b} n \right)^{v_r} \left(A_c m_l - \frac{v_l}{v_b} n \right)^{v_l} \right. \\
&\quad \left. \cdot \frac{k_f^{(n+1)}}{A_c^{v_r+v_l-1}} + n^{v_b} \frac{k_r^{(n)}}{A_c^{v_b-1}} \right] p_n + \left[A_c m_r - \frac{v_r}{v_b} (n - 1) \right]^{v_r} \\
&\quad \cdot \left[A_c m_l - \frac{v_l}{v_b} (n - 1) \right]^{v_l} \frac{k_f^{(n)}}{A_c^{v_r+v_l-1}} p_{n-1}
\end{aligned}
\tag{4}
$$

A special case of Eq. (4), i.e., one with unity stoichiometric coefficients, $v_r = v_l = v_b = 1$, has been studied in Piper et al. (1998).

In Eq. (4), $k_f^{(n)}$ and $k_r^{(n)}$ are allowed to be functions of the number of bonds instead of constants, as indicated by the superscript. They are therefore referred to as kinetic rate coefficients instead of kinetic rate constants as in Eq. (2). There are at least two reasons for the dependence of $k_f^{(n)}$ and $k_r^{(n)}$ on n. First, as discussed, bond formation and breakage are influenced by externally applied force. However, the more bonds, the less force per bond for a given force exerted on a cell. Second, because of possible cell motions, the formation of the first bond ($k_f^{(1)}$) and breakage of the last bond ($k_r^{(1)}$) are likely to be different from those for other numbers of bonds (Kaplanski et al., 1993; Piper et al., 1998).

The probability of simultaneously forming or breaking more than one bond at a time is assumed infinitesimal compared with that of forming or breaking one bond at a time. Therefore, terms associated with $p_{n\pm2}$, $p_{n\pm3}$, ... have been neglected in Eq. (4), leading to a tridiagonal system. Also, the process of bond formation and breakage is assumed to be Markovian. Thus, only the information for the current time, t, of the probabilities on the right-hand side of Eq. (4) is required, not their past histories. Other assumptions underlying Eq. (4) are that all free receptors and ligands within the contact area have equal opportunity to form a bond and that each bond in an adherent cell has an identical probability of dissociating.

C. Steady-State Solutions

Although in its general form, Eq. (4) can be used to describe transient adhesion (Piper, 1997), only the steady state is relevant to the centrifugation experiment. The system of finite-difference equations that resulted from setting the left-hand side of Eq. (4) to zero can be solved by means of mathematical induction (Piper, 1997), which yields

$$p_n = p_0 \left[\frac{\left(\frac{v_b}{v_r} A_c m_r \right)}{n} \right]^{v_r} \left[\frac{\left(\frac{v_b}{v_l} A_c m_l \right)}{n} \right]^{v_l} \prod_{i=1}^{n} \left(\frac{v_r}{v_b} \right)^{v_r} \left(\frac{v_l}{v_b} \right)^{v_l}$$
$$\cdot \left(\frac{i}{A_c} \right)^{v_r + v_l - v_b} K_a^{(i)} \tag{5a}$$

where $K_a^{(i)} \equiv k_f^{(i)}/k_r^{(i)}$. The probability of a cell having no bonds (detached) can be obtained via normalization, $\sum_0^{A_c m_{min}} p_n = 1$, which leads to

$$p_0 = \left\{ 1 + \sum_{n=1}^{A_c m_{min}} \left[\left(\frac{v_b}{v_r} A_c m_r \right) \frac{}{n} \right]^{v_r} \left[\left(\frac{v_b}{v_1} A_c m_1 \right) \frac{}{n} \right]^{v_1} \prod_{i=1}^{n} \left(\frac{v_r}{v_b} \right)^{v_r} \right.$$
$$\left. \cdot \left(\frac{v_1}{v_b} \right)^{v_1} \left(\frac{i}{A_c} \right)^{v_r + v_1 - v_b} K_a^{(i)} \right\}^{-1}$$

(5b)

As expected, the steady-state solution, Eq. (5), no longer depends on the forward and reverse rate coefficients separately but depends on their ratio, the binding affinity, as a whole. It should be noted that the general mathematical structure of this solution depends only on the definition of the binding affinity (as the ratio of forward and reverse rate coefficients), not on its specific functional form. It is therefore very convenient to use Eq. (5) to determine the formulation and parameters most appropriate for a particular set of data (see the following).

Piper et al. (1998) discussed two limiting cases, which correspond to simplified kinetic mechanisms. When one of the molecular species excessively outnumbers the other, the density of the former can be approximated as constant because the reaction is limited by the availability of the latter. Hence, the kinetic mechanism can be reduced to one of a v_{min}th order reversible reaction between free and bound states of the limiting species. The master equations for this simplified binding mechanism can be obtained from Eq. (4) by replacing $[A_c m_{max} - (v_{max}/v_b)n]$ and $[A_c m_{max} - (v_{max}/v_b)(n-1)]$ by $A_c m_{max}$, where $m_{max} = \max(m_r, m_1)$. If $m_{max} = m_r$, then $v_{max} = v_r$ and $v_{min} = v_1$. But if $m_{max} = m_1$, then $v_{max} = v_1$ and $v_{min} = v_r$. The $v_r = v_1 = v_b = 1$ case of such simplified master equations was used by Cozens-Roberts et al. (1990). The steady-state solution is reduced to

$$p_n = p_0 \left[\left(\frac{v_b}{v_{min}} A_c m_{min} \right) \frac{}{n} \right]^{v_{min}} \prod_{i=1}^{n} \left(\frac{v_{min}}{v_b} \right)^{v_{min}} \left(\frac{i}{A_c} \right)^{v_{min} - v_b} (m_{max})^{v_{max}} K_a^{(i)}$$

(6a)

where

$$p_0 = \left\{ 1 + \sum_{n=1}^{A_c m_{min}} \left[\left(\frac{v_b}{v_{min}} A_b m_{min} \right) \frac{}{n} \right]^{v_{min}} \prod_{i=1}^{n} \left(\frac{v_{min}}{v_b} \right)^{v_{min}} \right.$$
$$\left. \cdot \left(\frac{i}{A_c} \right)^{v_{min} - v_b} (m_{max})^{v_{max}} K_a^{(i)} \right\}^{-1}$$

(6b)

As expected, the affinity for the simplified mechanism is $(m_{max})^{v_{max}} K_a$.

When the number of bonds having nonvanishing probabilities is much smaller than the numbers of both receptors and ligands in the contact area available for binding, the formation of a small number of bonds will not significantly deplete the free receptors and ligands. The system of master equations can then be approximated by one that neglects, respectively, n and $(n - 1)$ in $[A_c m_i - (v_i/v_b)n]$ and $[A_c m_i - (v_i/v_b)(n - 1)]$ (subscript i = r or l) in Eq. (4). Such simplified master equations have been discussed in Long et al. (1999). The $v_r = v_l = v_b = 1$ case (with constant kinetic rates) was employed by Kaplanski et al. (1993). The steady-state solution to the reduced equations is

$$p_n = p_0 \prod_{i=1}^{n} \left(\frac{A_c}{i}\right)^{v_b} (m_r)^{v_r}(m_l)^{v_l} K_a^{(i)} \tag{7a}$$

where

$$p_0 = \left[1 + \sum_{n=1}^{A_c m_{min}} \prod_{i=1}^{n} \left(\frac{A_c}{i}\right)^{v_b} (m_r)^{v_r}(m_l)^{v_l} K_a^{(i)}\right]^{-1} \tag{7b}$$

As expected, the per molecular density binding affinity K_a appears in Eq. (7) together with the densities of the receptors and ligands as a grouped quantity, i.e., the per cell binding avidity $(m_r)^{v_r}(m_l)^{v_l} K_a$.

When $v_r = v_l = v_b = 1$, the steady-state solutions given by Eqs. (5)–(7) reduce to the respective solutions obtained by Piper et al. (1998), as required. If $K_a^{(n)} = K_a^0$ is a constant independent of the bond number n (the case in which the externally applied force is zero; see later), the results of Eqs. (6) and (7) are related to the well-known binomial and Poisson distributions (Piper et al., 1998). In the case of $v_r = v_l = v_b = 1$, Eq. (6) reduces to a binomial distribution with parameters $[1 + (m_{max}K_a^0)^{-1}]^{-1}$ and $A_c m_{min}$. The physical meaning of the former parameter is the probability of a receptor binding a ligand. In the case of $v_b = 1$, Eq. (7) reduces to a Poisson distribution with the average number of bonds being $\langle n \rangle = (A_c)^{v_b}(m_r)^{v_r}(m_l)^{v_l} K_a^0$. These relationships are not surprising, as the conditions underlying the binomial and Poisson distributions are equivalent to the assumptions on which the master equations and their corresponding simplifications are based. Both distributions have been suggested by others to describe the formation of a small number of bonds (Bell, 1981; Capo et al., 1982; Evans, 1995). However, these previous works assumed the parameters involved as given a priori. In contrast, our closed-form solutions reveal how these parameters are related to the binding affinity, the densities of receptors and ligands, the stoichiometric coefficients, and the contact area. In addition, our derivation of these distributions from the master equations enabled their generalization to cases in which the kinetic mechanism is more general (i.e., when the stoichiometric coefficients are not unity), the bonds are stressed (i.e., when $K_a^{(n)}$ depends on n), and the simplifying assumptions are removed (i.e., when $\langle n \rangle$, $A_c m_r$, and $A_c m_l$ are comparable).

D. Reduction to Deterministic Limits

The relationship between the probabilistic master equations, Eq. (4), and the deterministic kinetic equation, Eq. (2), can be revealed in the following derivation. Multiplying Eq. (4) by n and summing the resulting equation from 0 to $A_c m_{\min}$, we obtain

$$
\frac{d\langle n \rangle}{dt} = \left\langle \frac{k_f^{(n+1)}}{A_c^{v_r + v_1 - 1}} \left(A_c m_r - \frac{v_r}{v_b} n \right)^{v_r} \left(A_c m_l - \frac{v_1}{v_b} n \right)^{v_1} \right\rangle
$$

$$
- \left\langle \frac{k_r^{(n)}}{A_c^{v_b - 1}} n^{v_b} \right\rangle \tag{8}
$$

where $\langle \cdot \rangle$ denotes averaging. This can be viewed as the generalization of Eq. (2). If $k_f^{(n)} = k_f^0$ and $k_r^{(n)} = k_r^0$ are independent of n, the preceding equation can be rewritten as

$$
\frac{d}{dt} \left(\frac{\langle n \rangle}{A_c} \right) = k_f^0 \left[\left(m_r - \frac{v_r}{v_b} \frac{\langle n \rangle}{A_c} \right)^{v_r} \left(m_l = \frac{v_1}{v_b} \frac{\langle n \rangle}{A_c} \right)^{v_1} + \frac{\sigma_n^{(v_r + v_1)}}{A_c^{v_r + v_1}} \right]
$$

$$
- k_r^0 \left[\left(\frac{\langle n \rangle}{A_c} \right)^{v_b} + \frac{\sigma_n^{(v_b)}}{A_c^{v_b}} \right] \tag{9a}
$$

where

$$
\sigma_n^{(v_r + v_1)} \equiv \sum_{i=0}^{v_r} \sum_{j=0}^{v_1} \binom{v_r}{i} \binom{v_1}{j} (A_c m_r)^{v_r - i} (A_c m_l)^{v_1 - j}
$$

$$
\cdot \left(-\frac{v_r}{v_b} \right)^i \left(\frac{v_1}{v_b} \right)^j (\langle n^{i+j} \rangle - \langle n \rangle^{i+j}) \tag{9b}
$$

and

$$
\sigma_n^{(v_b)} \equiv \langle n^{v_b} \rangle - \langle n \rangle^{v_b} \tag{9c}
$$

are measures of fluctuations in the bond number. It can readily be shown the $\sigma_n^{(v_b)} = 0$ when $v_b = 1$, $\sigma_n^{(v_r + v_1)} = 0$ when $v_r + v_1 = 1$, $\sigma_n^{(v_b)} = \sigma_n^2$ when $v_b = 2$, and $\sigma_n^{(v_r + v_1)} = \sigma_n^2$ when $v_r = v_1 = 1$, where σ_n^2 is the variance of n. It can be seen that, for large systems in which fluctuations are small, Eq. (9a) reduces to Eq. (2) with $m_b = \langle n \rangle / A_c$.

The analytical solution for the probability distribution of having bonds, Eq. (5), also enables one to calculate various statistical aspects, including the mean, $\langle n \rangle$, and variance, σ_n^2, of the bond number. It follows from direct calculations of $\langle [A_c m_r - (v_r/v_b)n]^{v_r} [A_c m_l - (v_l/v_b)n]^{v_l} \rangle$ that (Piper, 1997)

$$\frac{\langle n^{v_b} \rangle}{\langle n^{v_b}/K_a^{(n)} \rangle} = \frac{[\langle n \rangle^{v_b} + \sigma_n^{(v_b)}]/A_c^{v_b}}{\left(m_r - \dfrac{v_r}{v_b} \dfrac{\langle n \rangle}{A_c} \right)^{v_r} \left(m_l - \dfrac{v_l}{v_b} \dfrac{\langle n \rangle}{A_c} \right)^{v_l} + \dfrac{\sigma_n^{(v_r+v_l)}}{A_c^{v_r+v_l}}} \tag{10}$$

This is an interesting result because it reveals how the binding affinity, defined as the ratio of forward to reverse rate coefficients based on the detailed balance between formation and breakage of a small number of bonds for each n value, is related back to its traditional deterministic definition for large systems. In general, statistical fluctuations (as measured by $\sigma^{(v_r+v_l)}$ and $\sigma^{(v_b)}$) become smaller as the system becomes larger. Thus, the right-hand side of Eq. (10) approaches, as $\langle n \rangle \to \infty$, the deterministic definition of binding affinity, as given by the far right-hand side of Eq. (3). If $K_a^{(n)} = K_a^0$ is a constant, the left-hand side of Eq. (10) is equal to K_a^0. As expected, this result can also be directly obtained by setting the left-hand side of Eq. (9a) to zero. Thus, the deterministic equation [Eq. (3)] is recovered from Eq. (10).

On the other hand, if $K_a^{(n)}$ depends on n, the left-hand side of Eq. (10) can be viewed as a weighted statistical average of the (stressed) binding affinity over all subpopulations of cells that are adhered via different numbers of bonds and hence have different binding affinities. This result, by contrast, cannot be directly obtained by setting the left-hand side of Eq. (8) to zero.

IV. RESULTS

The experimental data (Piper et al., 1998; Piper, 1997) presented here were obtained for adhesion of Colo-205 human colon carcinoma cells expressing carbohydrate ligands (Daneker et al., 1996) to surfaces coated with E-selectin construct (Erbe et al., 1992; Li et al., 1994).

A. Determining the Valence of the Receptors

It follows from Eq. (7b) that, when m_r or m_l is small,

$$P_a = 1 - p_0 \approx (A_c)^{v_b} (m_r)^{v_r} (m_l)^{v_l} K_a^{(1)} \tag{11}$$

This prediction suggests an experimental design for determination of the stoichiometric coefficients by measuring the dependence on the densities of the receptors and ligands of adhesion when it is low. The results of one such experiment are shown in Fig. 7, in which the adherent fraction is plotted against the E-selectin density (m_r) at various applied relative centrifugal forces (RCFs). As expected, the detached fraction increased with increasing centrifugal force. For each given force level and over a wide range, the adherent fraction increased nearly linearly with the E-selectin density when it was low (Fig. 7A), which suggests that the E-selectin binding is monovalent (i.e., $v_r = 1$). Binding became saturated when E-selectin density reached ~ 100 sites/μm^2 (Fig. 7B), which indicates a ligand density on the target cells of the order of 100 sites/μm^2. If the density of the ligand was also able to be varied, its valence could have been determined using a similar experiment as well.

B. Direct Measurement of $K_a^{(1)}(f)$

Because the data in Fig. 7A support the approximation given by Eq. (11) with $v_r = 1$, the ratio of P_a to m_r can be used to determine the force dependence of the binding affinity for the single bond subpopulation, $(A_c)^{v_h}(m_l)^{v_l}K_a^{(1)}$. The average P_a/m_r for each set of constant RCF data obtained from two separate experiments is plotted in Fig. 8. Note that no specific functional form for $K_a^{(1)}$ needs to be assumed here.

C. Constitutive Equation for Binding Affinity

Based on the data shown in Fig. 8, a constitutive equation can be formulated to describe the coupling between mechanics (separation force) and chemistry (binding affinity). Upon assuming the dislodging force acting on a cell to be equally shared by all (n) bonds, the following functional form is proposed for $K_a^{(n)}$.

$$K_a^{(n)} = K_a\left(\frac{f}{n}\right) = K_a^0 \left\{1 + c\left[\frac{a(f/n)}{k_B T}\right]^d\right\}^{-1} \exp\left\{-\left[\frac{a(f/n)}{k_B T}\right]^b\right\} \quad (12)$$

where K_a^0 (in $\mu m^{2(v_r+v_l-v_b)}$) is the affinity in the absence of force (zero-load affinity), k_B the Boltzmann constant, and T the absolute temperature. The a (in Å) can be viewed as the range of the energy well that defines the bound state; a is referred to as the interaction range of a bond or simply bond range. The ratio $k_B T/a$ provides a reference scale for the bond force. The $b-d$ are dimensionless parameters.

Several variations of the relationships between affinity, kinetic rates, and force have been proposed. Bell (1978) proposed that the reverse rate of a bond could be increased by the presence of a separation force. This formulation was based on an analogy to experimental results on statistical failure in solids. Other authors have adapted Bell's (1978) exponential model

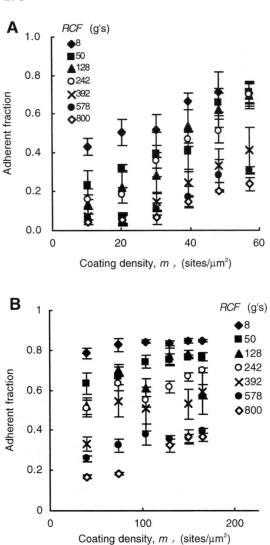

Figure 7 Dependence of Colo-205 cell adhesion on the E-selectin coating density (m_r) and relative centrifugal force (RCF). (A) Adhesion is nearly proportional to m_r at low E-selectin densities; (B) adhesion is saturable at high densities of E-selectin.

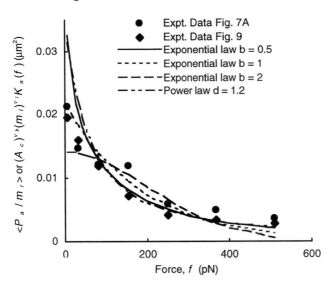

Figure 8 Relation between binding affinity and force on a cell. The average ratio of adhesive fraction, P_a, to site density, m_r, for the points of the constant RCF curves shown in Figs. 7A and 9 are plotted against force on a cell that is produced at that RCF (points). Theoretical predictions (curves) made using various forms of the re-relationship between affinity and force [Eq. (12)] are also shown.

for $k_r^{(n)}$ but assumed $k_f^{(n)}$ to be a constant (Hammer and Lauffenburger, 1987; Cozens-Roberts et al., 1990). This is the $b = 1$ and $c = 0$ case in Eq. (12). Bell et al. (1984) suggested including the bond elastic energy (in its simplest linear spring form) in the Gibbs free energy change of the binding reaction in the absence of force. This corresponds to the case of $b = 2$ and $c = 0$ in Eq. (12), and $k_B T/2a^2$ is the spring constant. Dembo (1994; Dembo et al., 1988) suggested two exponential laws for $k_f^{(n)}$ and $k_r^{(n)}$ but required their ratio to satisfy the equation of Bell et al. (1984) for K_a. Such a formulation was also used by Hammer and Apte (1992), who introduced the notion of re-active compliance to describe the responsiveness of the rate of a chemical reaction to applied forces. Evans (1995; Evans et al., 1991) proposed a power law for $k_r^{(n)}$ [with $k_f^{(n)}$ assumed constant, this becomes the $b = 0$, $c = 1$ case in Eq. (12)*]. He described the bonds as brittle if the power $d \gg 1$ and ductile if $d \sim 1$. Evans and Ritchie (1997) placed the relationship be-tween reverse rate and bond force on a more sound foundation by deriving

---

*Evans' original form was $k_r(f/n) = k_r^0(af/nk_B T)^d$ and hence $k_r(0) = 0$. Our modified version includes a crossover to a nonzero reverse rate at zero force.

it from Kramers' (1940) theory of escape of thermally agitated particles from an energy well tilted by an applied force. Under greatly simplified conditions, the result obtained by Evans and Ritchie (1997) was a combined power and exponential model (i.e., $c = 1$ and nonzero b and d values, again with $k_f^{(n)}$ assumed constant). Although the ranges of the model parameters have been estimated, their values cannot be determined theoretically. This is due to the fact that, for the particular interacting receptor-ligand pair, no information is available on either the detailed energy profile that determines the transition state or the work mode that couples the external force to energy. Here Eq. (12) is viewed as a constitutive equation for the binding affinity whose parameters will be evaluated by comparison with the experimental data.

D. Determining the Most Appropriate Formulation and Parameters for $K_a^{(n)}$

Because of the simplicity and the easiness of the centrifugation assay, a large amount of data has been generated in our laboratory (Piper et al., 1998). This allowed us to conduct a comparative study to determine the abilities of different formulations (e.g., exponential versus power laws) to account for the data for the E-selectin–carbohydrate ligand system. The results are summarized in Tables 1 and 2. The goodness of fit of the curves

Table 1   Model Parameters Calculated Using the Exponential Law ($b \neq 0$, $c = 0$)

Data sets shown in Fig.	Number of data points N	Number of fitting parameters M	b	$A_c m_1^{v_1} K_a^0$ (μm^2)	a (Å)	Reduced chi square χ_v^2
7A	41	3	0.6	0.0320 ± 0.0015	0.280 ± 0.017	2.3
		2	**1**[a]	0.0236 ± 0.0009	0.157 ± 0.006	2.7
		2	**2**	0.0175 ± 0.0004	0.108 ± 0.006	5.4
7B	40	3	0.4[b]	0.0511 ± 0.0018	1.240 ± 0.056	13.1
		2	**1**	0.0226 ± 0.0002	0.211 ± 0.004	21.6
		2	**2**	0.0175 ± 0.0003	0.139 ± 0.001	39.4
9	42	3	0.5	0.0445 ± 0.0028	0.775 ± 0.052	1.5
		2	**1**	0.0218 ± 0.0010	0.225 ± 0.008	2.3
		2	**2**	0.0140 ± 0.0005	0.142 ± 0.003	5.8

[a]Boldface indicates that the values were held constant during the fitting.
[b]Further reduction of b resulted in a slight reduction of χ_v^2, but gave an unreasonable value for a.

Table 2 Model Parameters Calculated Using the Power Law ($b = 0$, $c = 1$, $d \neq 0$)

Data sets shown in Fig.	Number of data points N	Number of fitting parameters M	Predicted model parameters $\pm$ estimated standard deviation			Reduced chi square χ_v^2
			d	$A_c m_1^{v_1} K_a^0$ (μm^2)	a (Å)	
7A	41	3	1.2	0.0273 ± 0.0013	0.350 ± 0.024	2.4
7B	40	3	0.8	0.0644 ± 0.0047	3.40 ± 0.38	12.3
9	42	3	1.2	0.0332 ± 0.0023	0.720 ± 0.059	1.1

that resulted from using different models is quantified by the best-fit values of the reduced chi square statistic, χ_v^2, which reflects both the appropriateness of the model and the quality of the data (Bevington and Robinson, 1992).

We first examined the exponential law (the $b \neq 0$, $c = 0$ case). It is evident from direct visual inspection that the exponent $b = 1$ (Fig. 9A) clearly represented our data better than $b = 2$ (Fig. 9B). This conclusion holds true for all data sets (Table 1). The exponent b was also allowed to vary freely to arrive at a value that best fit the data, which consistently yielded $b \approx 0.5$ (Fig. 9C). Again, the χ_v^2 at $b \approx 0.5$ was smaller than the χ_v^2 at $b = 1$ for all sets of data summarized in Table 1.

The power law (the $b = 0$, $c = 1$ case) was next examined. That model fit the data (Fig. 9D) slightly better than the exponential model with equal number of free parameters ($M = 3$). This also holds true for all but one of the data sets (Tables 1 and 2). The average best fit d was about 1.1 (range from 0.8 to 1.2, Table 2), suggesting ductile bonds for the E-selectin–carbohydrate ligand interaction (Evans, 1995). The more general model (the $b \neq 0$, $c = 1$, $d \neq 0$ case) did not significantly reduce the χ_v^2 value (not shown). Using this more complex model for our data appears not to be warranted because the number of freely adjustable parameters was increased with no improvement in the goodness of fit.

The various functional forms of the K_a versus f relationship examined are plotted in Fig. 8 along with the $\langle P_a/m_r \rangle$ versus f data. It is evident that good agreement was found in such a comparison for the cases $b = 0.5$ and $c = 0$, $c = 1$ and $d = 1.2$, and $b = 1$ and $c = 0$, and the discrepancy is significant for the model with $b = 2$ and $c = 0$, which are consistent with the χ_v^2 results shown in Tables 1 and 2. It is worth mentioning that the theoretical curves, with parameters obtained from fitting of data of Fig. 9, fit both sets of data (the other set was computed from the data in Fig. 7A).

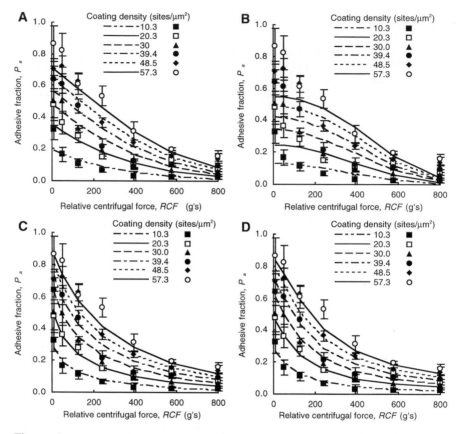

Figure 9 Comparison of the abilities of various special forms of Eq. (12) to account for the data. Different forms of Eq. (12) were used in connection with Eq. (5b) to predict (curves) the same set of experimental data (points) using the best-fit parameter values (see Tables 1 and 2) to compare the appropriateness of their application. The exponential law ($c = 0$) with $b = 1$ (A) is a better model than that with $b = 2$ (B). If b is allowed to vary freely, then the best fit value is $b \approx 0.5$ (C). The power law ($b = 0$, $c = 1$, and $d = 1.2$) also fits the data well (D).

E. Validation of Model Prediction

In addition to fitting the entire collection of data with a single set of parameters, each subset of P_a versus RCF data (for a given m_r) shown in Fig. 9 was used to evaluate the binding characteristics. The parameters so predicted [$A_c(m_l)^{v_l}K_a^0$ and a, for given values of $b = 0.5$ and $c = 0$] are plotted against the E-selectin site density, m_r, in Fig. 10. As can be seen, no dependence of

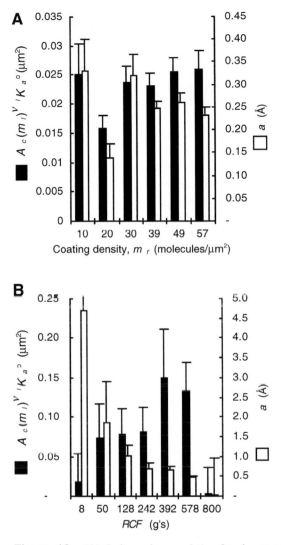

Figure 10 (A) Independence of the fitted parameters on E-selectin site density. Each binding curve (one m_r but various RCF) of the data in Fig. 9 was used to evaluate the best fit parameters of $A_c(m_l)^{v_l}K_a^0$ (solid bars, left ordinate) and a (open bars, right ordinate), and the results were plotted against the E-selectin site density, m_r. (B) When the data were grouped according to RCF (one RCF, various m_r), the calculations of $A_c(m_l)^{v_l}K_a^0$ and a were not as consistent between groups.

the parameter values on m_r was found, supporting the validity of the model and indicating that these are indeed intrinsic parameters. In other words, the two model parameters evaluated by best fitting any one set of adhesion versus centrifugation force data in the family (for a given m_r) enabled us to *predict* accurately, without any fudge factor, other sets of P_a versus RCF data in the family (for other m_r values), as shown in Fig. 11.

When individual P_a versus m_r data (for a given RCF) were fitted to evaluate the binding characteristics, however, variations were seen in the predicted values of a at small RCF (e.g., 8g) values (Fig. 10B). Such a result pointed out a limitation of the method. Because the parameter a determines how the applied force influences adhesion, data generated from experimental situations in which the separation force is low and does not play a significant role are not suited for its evaluation.

F. Confirming the Order of Dissociation

In the previous two sections, the order of dissociation, v_b, was assumed to have a value of 1. The reason for this is that the three stoichiometric coefficients should be constrained by a relation $1 \leq v_b \leq \min(v_r, v_l)$, and we have determined from Fig. 7A that $v_r = 1$. That the order of the reverse reaction should be unity can be further confirmed by comparing the ability of the model prediction based on different values of v_b to fit the data. The results of such a comparison based on the data shown in Fig. 9 are plotted in Fig. 12. It is evident that v_b is indeed equal to unity, as the minimum χ_v^2 value is smallest when $v_b = 1$.

G. Relating the Strength to the Probability of Adhesion

The probability and the strength of cell adhesion have been regarded as two separate physical quantities (Bongrand et al., 1982). The coupling of mechanics and chemistry via the force dependence of binding affinity, Eq. (12), allows the two to be related. For the small systems under consideration, the deterministic notion of adhesion strength (defined as the force required to detach an adherent cell) is no longer applicable and needs to be extended. The reason is that, in the present probabilistic framework, the detachment of a given cell is a random event and can happen at any force, even at zero force! It is this stochastic nature of the individual cell detachment that is assumed to give rise to the fractionalization of adhesion seen in a population of cells. However, the probability of adhesion, defined as the probability of a cell having at least one bond, is predictable and decreases with the applied force, $P_a = 1 - p_0$. The probability of an initially adherent cell (i.e., adherent

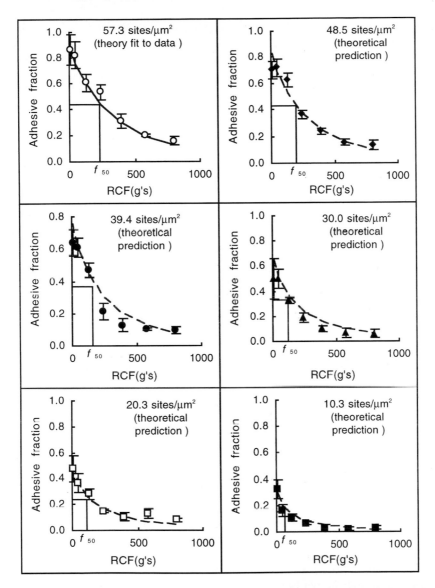

Figure 11 Prediction of adhesion behavior. Equations (5b) and (12) ($b = 0.5$, $c = 0$) was used to calculate the best fit adhesion parameters [$a = 0.86$ Å, $A_c(m_1)^{v_1}K_a^0 = 0.055$ μm^2] based on the data shown in the upper left panel ($m_r = 57.3$ sites/μm^2). The model, with those calculated parameters, was then used to predict the adhesion behavior of the system when the different coating densities were used, and very good agreement between the theory (curves) and experiment (points) was found. Also shown in each panel is how one of the statistical measures of adhesion strength, f_{50}, defined by Eq. (14b), can be determined by interpolation of the data.

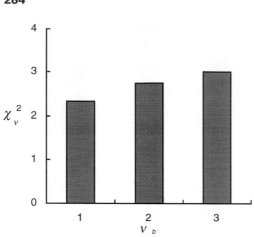

Figure 12 Comparison of the goodness of fit to data of models assuming various orders of dissociation. The error (as indicated by the reduced chi square statistic) between the theoretical predictions of Eq. (5b) with various values of v_b and the experimental data is shown.

when $f = 0$) to remain adherent after it is subjected to a force is $P_a(f)$ renormalized by $P_a(0)$ to discount the initially nonadherent cells. The cumulative probability for an adherent cell to be detached by a force not exceeding f is therefore given by

$$P_d(f) = 1 - P_a(f)/P_a(0) \tag{13}$$

The probability density for an adherent cell to be detached by an applied force can be obtained by differentiating Eq. (13), $p_d(f) = dP_d(f)/df$. This, in general, is a broad distribution (see Fig. 13). There is no one force at which a cell will shift abruptly from adherent to nonadherent. Nevertheless, various statistical definitions for the strength of cell adhesion can be obtained. Three possible definitions are f_m, the force at which detachment occurs most frequently, f_{50}, the force at which 50% of the adherent cells have been detached, and $\langle f \rangle$, the average force required to detach an adherent cell. These are given respectively by

$$\frac{dp_d(f_m)}{df_m} = 0 \qquad P_d(f_{50}) = 50\% \qquad \langle f \rangle = \int_0^\infty f p_d(f)\, df \tag{14a–14c}$$

which correspond, respectively, to the mode, median, and mean of the probability distribution $P_d(f)$. These are shown graphically in Fig. 13.

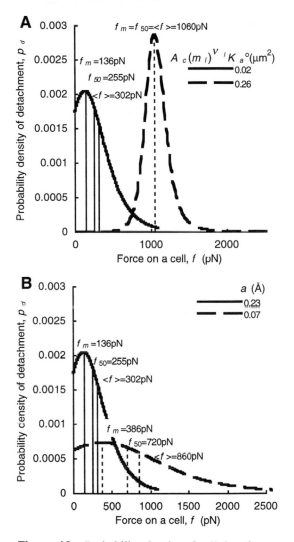

Figure 13 Probability density of cell detachment as a function of dislodging force (parameters: $b = 1$, $c = 0$, $m_r = 60$ sites/μm^2). Three statistical definitions of the critical detachment force are shown: most probable force f_m, 50th percentile force f_{50}, and average force $\langle f \rangle$ of cell detachment. (A) The density curve shifts rightward and narrows with an increase in $A_c(m_l)^{v_l}K_a^0$. (B) The density curve is right shifted and broadened by a decrease in a. Also shown are the corresponding changes in the statistical definitions of adhesion strength.

H. Dependence of Adhesion Strength on Molecular Parameters

The formulation for the binding affinity includes two parameters (for the exponential law with $b = 1$), i.e., the no-load binding affinity, K_a^0, and the range of the bond, a. The dependence of the strength of cell adhesion on the former has been suggested to be of a logarithmic form (Dembo et al., 1988; Zhu, 1991; Kuo and Lauffenburger, 1993). Such a weak dependence was derived from thermodynamic models of cell adhesion (Dembo et al., 1988; Zhu, 1991), which required a large number of continuously distributed molecules for this prediction to be valid. Piper et al. (1998) examined whether such a logarithmic relationship would still hold true for adhesion processes that are mediated by a small number of discretely distributed molecules, as is in the present case (see the following). The analysis and results are presented here.

Using model parameters calculated from experiments, the probability densities of detachment for target cells were computed and are shown in Fig. 13A and B (solid curves). Adhesion strength can be defined as either f_m, f_{50}, or $\langle f \rangle$, via Eqs. (14a)–(14c), respectively. Also demonstrated in these figures is how the probability density of detachment is affected by the no-load binding affinity, K_a^0, and by the bond range, a. With all other parameters held constant, an increase in K_a^0 causes a rightward shift of the p_d versus f curve, indicating the ability of cells to remain adherent at greater dislodging forces (Fig. 13A). The curve also changes shape as it is shifted to the right. It becomes bell shaped and less broadly distributed, which diminishes the differences of the three definitions of adhesion strength. This shift and change of shape resulted in a dependence of the adhesion strength on K_a^0 that is stronger than logarithmic. Such a relationship is plotted in Fig. 14A.

The effect of the bond interaction range on the single-bond strength was envisioned by Bell (1978), who argued that different values of a could alter the order of bond strengths as suggested by the order of interaction energies. In the absence of forces, the high energy barrier a cell surface receptor has to overcome in order to escape from the membrane linkage ensures a stable anchor of the receptor for a seemingly infinitely long time, whereas the low binding energy of the receptor for a ligand results in spontaneous dissociation in an observable time. However, the force required to extract the receptor from the cell membrane is estimated to be of the same order of magnitude as that required to break a receptor-ligand bond, because the distance over which the force acts (i.e., the a value) is much larger in the former case than in the latter case (Bell, 1978). The same reasoning can be applied to delineate the dependence of cell adhesion strength on the range of a bond. As shown in Fig. 13B, a smaller a value causes a rightward shift

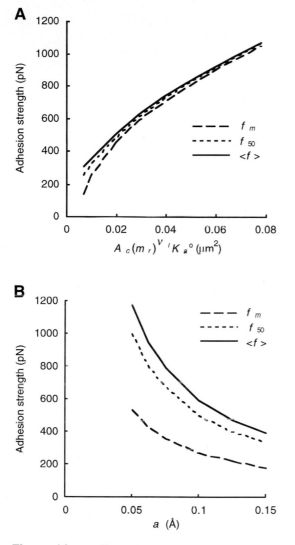

Figure 14 Predicted changes in the three statistical definitions of cell adhesion strength (f_m, f_{50}, and $\langle f \rangle$) with change in either intrinsic parameters, no-load affinity $A_c(m_1)^{v_1}K_a^0$ (A), or bond range a (B), when all other parameters remain constant (b = 1, c = 0, m_r = 60 sites/μm^2).

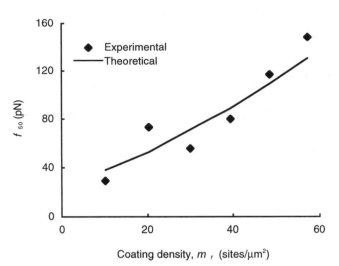

Figure 15 Comparison of predicted (curve) with measured (points) cell adhesion strength (f_{50}) as a function of the density, m_r, of the E-selectin construct coated on the 96-well plate when all other parameters remain constant [$a = 0.775$ Å, $b = 0.5$, $c = 0$, $A_c(m_l)^{v_l}K_a^0 = 0.0445$ μm²]. The adhesion strength data were obtained by interpolation of the measured P_a versus f curves at various m_r levels, as shown in Fig. 11.

of the p_d versus f curve. It also broadens the distribution. Figure 14B illustrates that a serves as a measure of the ease with which the binding energy landscape can be tilted (and hence the energy barrier can be abolished) by the externally applied forces. For the same interaction energies (same K_a^0), an increase in the bond range decreases the adhesion strength.

Similarly, the dependence of adhesion strength on the receptor density can be examined. Compared in Fig. 15 are the f_{50} versus m_r relationship measured from the data shown in Fig. 11 and that predicted by the present theory. It is evident that good agreement exists between the theory and the experiment.

V.   DISCUSSION

A.   Effect of Cellular Heterogeneity

Cultured cells with asynchronous cycles usually exhibit heterogeneous properties in a population. Although Piper et al. (1998) have shown that a model based on cellular heterogeneity alone is not adequate for the data presented in the Results section, the possible effect of cellular heterogeneity on the predictions of the stochastic kinetic model presented in the Theory section was not discussed.

From the equation shown in Fig. 2, the cellular property that is variable in a population that affects the centrifugal force applied to a cell can be identified as its volume (Fig. 3), as the cell density has a very narrow distribution (Fig. 4) and the spinning speed is well controlled. The probability distribution obtained from the solution of Eq. (5), or Eq. (6) or (7), is a function of the applied force and, thereby, of the cell volume. It can therefore be viewed as the conditioned probability for a given volume, $p_{n|V}$, and the unconditioned probability can be obtained using the total probability formula

$$p_n = \int_0^\infty p_{n|V} \psi(V) \, dV \tag{15}$$

where $\psi(V)$ denotes the lognormal distribution of volume in a population:

$$\psi(V) = \frac{1}{\sqrt{2\pi \ln(1 + \rho_V^2)}V} \exp\left\{ -\frac{\ln^2[\sqrt{1 + \rho_V^2}(V/\langle V \rangle)]}{2 \ln(1 + \rho_V^2)} \right\} \tag{16}$$

where $\rho_V^2 = \sigma_V^2/\langle V \rangle^2$ (≈ 0.082 for Colo-205 cells) is the coefficient of variance of the volume distribution. In the comparisons with the data described so far, the following approximation was used for the predictions:

$$p_n = p_{n|\langle V \rangle} \tag{17}$$

where $\langle V \rangle = \int_0^\infty V\psi(V) \, dV$. That is, the average of the functional was approximated by the functional of the average. This approximation was tested by comparing the p_0 given by Eqs. (15) and (17), as shown in Fig. 16A. It is evident that Eq. (17) is a good approximation of Eq. (15), which justifies neglecting the variation of cell volume in a population.

Another property that is variable in a cell population is the level of expression of the ligands, m_1. It affects the probability distribution of bonds and, in turn, the magnitude of the centrifugal force that an adherent cell can sustain. Using similar lines of reasoning, the left-hand side of Eq. (15) can be viewed as the conditional probability for a given m_1, $p_{n|m_1}$. The unconditional probability can be calculated, again, using the total probability function

$$p_n = \int_0^\infty p_{n|m_1} \phi(m_1) \, dm_1 \tag{18}$$

where $\phi(m_1)$, the lognormal distribution of ligand expression, can be expressed by the same functional form as that given in the right-hand side of Eq. (16), except that V, $\langle V \rangle$, and ρ_V^2 are respectively replaced by m_1, $\langle m_1 \rangle$,

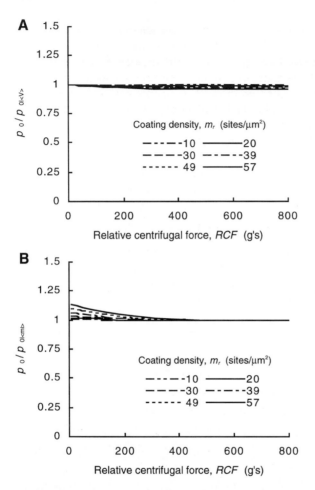

Figure 16   Effect of (A) volume variation and (B) ligand density variation on the predicted probability of detachment. The ratio of p_0 predicted from the total probability formula to that predicted from the conditional probability calculated using average values of V or m_l is plotted against RCF for various m_r levels.

and $\rho_{m_l}^2$. The latter two values are assumed to be $\langle m_l \rangle = 100$ sites/μm^2 and $\rho_{m_l}^2 = 0.25$, respectively (Piper et al., 1998). To test whether the probability conditioned at the average ligand expression level, $p_{n|\langle m_l \rangle}$, could approximate the total probability given by Eq. (18), the ratio of the latter to the former was computed and is shown in Fig. 16B as a function of the relative centrifugal force, RCF, and the receptor density, m_r. It can be seen that only at low force levels do small differences between the two become detectable.

These results validate the assumption of neglecting the variation of ligand expression in a cell population.

It should be noted that, because the two variables, V and m_1, should be independent, the order in which the present calculations were conducted and the fact that they were examined separately should have no bearing on the conclusion (i.e., neither variation significantly affected the results obtained when these variations were ignored).

It should be emphasized that the conclusions regarding the relative contributions to adhesion fractionalization from molecular stochasticity and cellular heterogeneity are based on data presented herein and in Piper et al. (1998) and thus have been validated only for the experimental system used in those studies. It is possible that the relative contributions from cellular heterogeneity may vary from system to system, as opposed to our system, which is dominated by the contribution from molecular stochasticity. An example in which cellular heterogeneity is dominant can be found in Saterbak et al. (1993). Also, it is likely that there are factors other than cell size and ligand expression level that may be variable in a cell population. An example of heterogeneous adhesive ability demonstrated in granulocytes will be described in a later section.

B. Justifying the Probabilistic Formulation

The generalization from a deterministic kinetic model to its corresponding probabilistic formulation necessitates solving a system of coupled ordinary differential equations, Eq. (1), instead of just one such equation, Eq. (2), which represents a major increase in the mathematical complexity of the problem. Because both the deterministic and probabilistic models provide the same information as the system size becomes large, it is of interest to examine the number of bonds mediating cell adhesion in our centrifugation assay to see if the probabilistic formulation is warranted. Using the fitted parameters $[A_c(m_1)^{v_1}K_a^0$ and $a]$ from Table 1 (for $b = 1$ and $c = 0$), the sub-populations of cells having various numbers of bonds were calculated using Eq. (5) and are plotted in Fig. 17A. It can be seen that, even when the adherent fraction was as high as 73%, the majority of cells were bound by only a few (<5) bonds (to the surface coated with an E-selection density of 60 sites/μm^2 and subject to no force). Also predicted was how the average number of bonds, $\langle n \rangle$, per adherent cell and its fluctuation (represented by the standard deviation, σ_n) would vary with changes in applied force (Fig. 17B). Again, the average number of bonds was small (<2 bonds per adherent cell), even with no applied separation force. By contrast, the standard deviation was large (comparable to $\langle n \rangle$). Such a surprisingly small bond num-

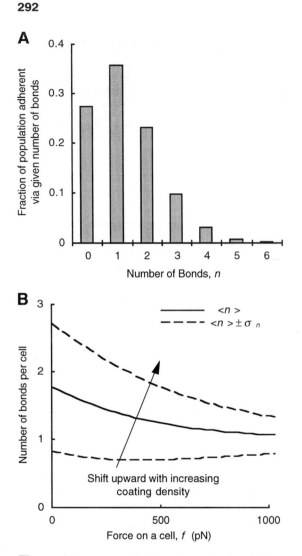

Figure 17 (A) Predicted probability distribution, p_n, of the number of bonds, n, formed between a Colo-205 cell and the E-selectin−coated surface [parameters $b =$ 1, $c = 0$, $A_c(m_l)^n K_a^0 = 0.0218$ μm^2, $a = 0.225$ Å, $m_r = 60$ sites/μm^2, and $f = 0$]. (B) The average number of bonds per adherent cell $\langle n \rangle$ (solid curve) $\pm$ standard deviation σ_n (dashed curves) as functions of the dislodging force (parameters: same as in A except that f was allowed to vary freely).

ber and significant fluctuation evidently point to the inadequacy of the deterministic model and argue for the use of the probabilistic model.

C. Limitations of the Method

The range of relative centrifugal force (RCF) that can be generated by a tabletop centrifuge is typically $1-1000$ g. For a typical cell of diameter 10 μm and density 1.05 g/cm^3, this translates to $2.5-250$ pN (see Fig. 2). This places a practical limitation on the kind of cell adhesion behavior that this method is suited to analyze, namely that of "weak adhesion." For a typical "bond strength" of several tens of piconewtons (Evans et al., 1991), the number of bonds that can be broken by centrifugation must be relatively small (see Fig. 17). If the adhesion process involves metabolically dependent activation and spreading, the centrifugation method may not be applicable, as a large number (~ 1000) of bonds usually participate in such adhesions, which prevents the adherent cells from being detached by a tabletop centrifuge even at its high-speed end. In such a case, the measured detached fraction reflects, instead of cells whose adhesive bonds have been broken, cells that are not yet activated to form such bonds. This measurement bears no relation to the binding characteristics of the adhesion molecules of interest. Even when a high-speed centrifuge is used to generate sufficiently large dislodging forces, the bonds are likely to be stressed nonuniformly, as detachment must occur by peeling at the edge of the contact zone. Also, severe deformation and even tearing of the cells may occur under such large forces. All of these violate the simplifying assumptions underlying the present mathematical model and prevent it from being applicable.

Also, the method does not work when the cell population exhibits heterogeneity in adhesive ability, possibly associated with the cells' activation state. This is exemplified by experiments using freshly isolated human granulocytes (Fig. 18). In addition to there being a significant fraction of granulocytes that did not bind to E-selectin even at very high surface density, adhesion mechanisms other than the E-selectin–carbohydrate ligand pathway may have influenced the adhesion of a subpopulation of granulocytes in our experimental system. This alternative adhesion pathway manifested itself as significant granulocyte adhesions to 96-well plates coated with or without the E-selectin construct (capture antibody and BSA alone or BSA only). Direct microscopic observation indicated that this adherent fraction corresponded to granulocytes that had spread on the substrate surface, which could not be detached by even the highest RCF (~ 800 g) employed in the present study. Simon et al. (1995) have demonstrated that several proteins including BSA, when derivatized on a plastic surface, became ligands for β_2-integrins on granulocytes. This is consistent with our findings.

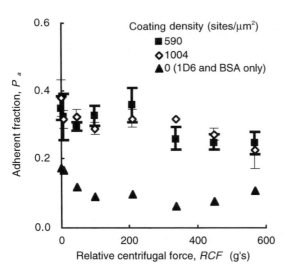

Figure 18   Adhesion of granulocytes to plastic wells coated with E-selectin construct and subjected to various RCFs. Maximum binding of granulocytes remained low (<40%) even at very high coating densities (1000 sites/μm^2), and the adherent population did not detach with increasing RCF. The adhesion to wells coated with 1D6 and BSA was only greater ($\sim$10–20%) than that seen in experiments with Colo-205 and HL-60 cells ($\sim$1–10%).

D.   Other Issues

An implicit assumption of the method is that cell detachment occurs by dissociation of receptor-ligand bonds. Without this assumption, it would not be possible to relate the fraction of detached cells directly to the affinity of receptor-ligand binding, as in Eq. (5b). It has been shown that cell detachment could also occur by disruption of the cytoskeleton and/or membrane anchor of the receptors or ligands (Evans et al., 1991; Chesla et al., 1997). To ensure applicability of the method to the E-selectin–carbohydrate ligand system, Piper (1997) used a micropipette assay to show that it was very unlikely to uproot the carbohydrate ligands from the Colo-205 cell membrane. If molecular extraction is significant in one's experimental system, the present method can no longer be applied. To circumvent this problem, one would need to use a low centrifugal force. This would still allow measurement of K_a^0 but would reduce the measurability of how K_a depends on force. Alternatively, a new model have to be developed to relate the fraction of detached cells to the strength of cytoskeleton and/or membrane anchor.

A common procedure for functionalizing the substrate surface with an adhesion molecule of interest is to use a linker molecule, such as the capture antibody shown in Fig. 5. This introduces another link point in series with the receptor-ligand bond in question, just like the cytoskeleton and/or membrane linkage, which also is in series with the receptor-ligand linkage (Fig. 5). By contrast, the mathematical model developed in Piper et al. (1998) and herein assumes a single receptor-ligand linkage. We were able to argue convincingly that the values shown in Tables 1 and 2 are indeed parameters for E-selectin–carbohydrate ligand interaction. The argument was based on micropipette measurements of kinetics using identical cellular and molecular systems (Piper, 1997) and on comparison of results with other independent kinetic measurements of selectin and antibody bindings (see Piper et al., 1998). Selectin–carbohydrate ligand interaction follows much faster kinetics and has a much lower afffinity than those of a typical antibody-antigen interaction. This is a case in which the receptor-ligand bond of interest is much weaker than the other serial linkages. Hence, the rupture of the receptor-ligand bond dominates the cell detachment. Hence, the detachment can be treated approximately by the present single-linkage theory. If multiple serial linkages have similar binding characteristics, a more general multiple linkage theory is required for data interpretation. Initial discussion of such a theory can be found in Saterbak and Lauffenburger (1996) (also Long et al., 1999).

VI. CONCLUSION

This chapter describes a recently developed method for measuring the two-dimensional binding parameters, a and K_a^{r0}, of a receptor-ligand pair mediating adhesion. The method should be easy to implement, as the experiment is a commonly used centrifugation assay and the theory resulted in simple closed-form solutions. These solutions are an extension of those of Piper et al. (1998) from the special case of unity stoichiometric coefficients to the general case of their arbitrary values. The simplicity of the assay afforded well-controlled experimental conditions and a large amount of high-quality data. This allowed decisive selections of not only the kinetic mechanism but also various constitutive equations relating bond dissociation to force. The analytical solution provides important predictions about a variety of physical and chemical parameters relevant to cell adhesion.

ACKNOWLEDGMENTS

We thank Drs. K. S. Huang and B. Wolitzky of Hoffmann LaRoche and Dr. D. Burns of Galaxo for their generous donation of the E-selectin construct and 1D6 capture antibody. This work was supported by National Institutes of Health (NIH) training grant GM08433, NIH grant R29AI38282, National Science Foundation (NSF) grant BCS9350370, and a grant from the Whitaker Foundation.

REFERENCES

Alon R, Chen S, Puri KD, Finger EB, Springer TA. 1997. The kinetics of L-selectin telhers and the mechanics of selectin-mediated rolling. J Cell Biol 138:1169–1180.

Alon R, Hammer DA, Springer TA. 1995. Lifetime of the P-selectin–carbohydrate bond and its response to tensile force in hydrodynamice flow. Nature 374: 539–542.

Bell GI. 1978. Models for the specific adhesion of cells to cells. Science 200:618–627.

Bell GI. 1981. Estimate of the sticking probability for cells in uniform shear flow with adhesion caused by specific bonds. Cell Biophys 3:289–304.

Bell GI, Dembo M, Bongrand P. 1984. Cell adhesion: Competition between nonspecific repulsion and specific bonding. Biophys J 45:1051–1064.

Bevington PR, Robinson DK. 1992. Data Reduction and Error Analysis for the Physical Sciences. New York: McGraw-Hill.

Bongrand P, Capo C, Depieds R. 1982. Physics of cell adhesion. Prog Surf Sci 12: 217–286.

Capo C, Garrouste F, Benoliel AM, Bongrand P, Ryter A, Bell GI. 1982. Concanavalin A–mediated thymocyte agglutination: A model for a quantitative study of cell adhesion. J Cell Sci 56:21–48.

Chen S, Alon R, Fuhlbrigge RC, Springer TA. 1997. Rolling and transient tethering of leukocytes on antibodies reveal specializations of selectins. Proc Natl Acad Sci U S A 94:3172–3177.

Chesla S, Marshal BT, Zhu C. 1997. Measuring the probability of receptor extraction from the cell membrane. Adv Bioeng Proc ASME 36:129–130.

Chu L, Tempelman LA, Miller C, Hammer DA. 1994. Centrifugation assay of IgE-mediated cell adhesion to antigen-coated gels. AIChE J 40:692–703.

Cozens-Roberts C, Lauffenburger DA, Quinn JA. 1990. Receptor-mediated cell attachment and detachment kinetics: Probabilistic model and analysis. Biophys J 58:841–856.

Daneker GW, Lund SA, Caughman SW, Stanley CA, Wood WC. 1996. Antimetastatic prostacyclins inhibit the adhesion of colon carcinoma to endothelial cells by blocking E-selectin expression. Clin Exp Metastasis 14:230–238.

Dembo M. 1994. On peeling an adherent cell from a surface. In: Lectures on Mathematics in the Life Sciences, Some Mathematical Problems in Biology. Providence, RI: American Mathematical Society, Vol 24, pp 51–77.

Dembo M, Tourney DC, Saxman K, Hammer D. 1988. The reaction-limited kinetics of membrane-to-surface adhesion and detachment. Proc R Soc Lond 234:55–83.

Dustin ML, Ferguson LM, Chan P-Y, Springer TA, Golan DE. 1996. Visualization of CD2 interaction with LFA-3 and determination of the two-dimensional dissociation constant for adhesion receptors in a contact area. J Cell Biol 132:465–474.

Erbe DV, Wolitzky BA, Presta LG, Norton CR, Ramos RJ, Burns DK, Rumberger JM, Rao BN, Foxall C, Brandley BK, Lasky LA. 1992. Identification of an E-selectin region critical for carbohydrate recognition and cell adhesion. J Cell Biol 119:215–227.

Evans E, Berk D, Leung A. 1991. Detachment of agglutinin-bonded red cells. I. Forces to rupture molecular-point attachments. Biophys J 59:838–848.

Evans EA. 1995. Physical actions in biological adhesion. In: Lipowsky R, Sackmann E, eds. Structure and Dynamics of Membranes. handbook of Physics of Biological Systems. Vol 1. Amsterdam: Elsevier Science, pp 723–754.

Evans EA, Ritchie K. 1997. Dynamic strength of molecular adhesion bonds. Biophys J 72:1541–1555.

Hammer DA, Apte SM. 1992. Simulation of cell rolling and adhesion on surfaces in shear flow: General results and analysis of selectin-mediated neutrophil adhesion. Biophys J 63:35–57.

Hammer DA, Lauffenburger DA. 1987. A dynamical model for receptor-mediated cell adhesion to surfaces. Biophys J 52:475–487.

Hulme EC, Birdsall NJM. 1992. Strategy and tactics in receptor-binding studies. In: Hulme EC, ed. Receptor-Ligand Interactions: A Practical Approach. New York: Oxford University Press, pp 63–176.

Kaplanski G, Farnarier C, Tissot O, Pierres A, Benoliel A-M, Alessi M-C, Kaplanski S, Bongrand P. 1993. Granulocyte-endothelium initial adhesion: Analysis of transient binding events mediated by E-selectin in a laminar shear flow. Biophys J 64:1922–1933.

Kramers HA. 1940. Brownian motion in a field of force and the diffusion model of chemical reactions. Physica (Utrecht) 7:284–304.

Kuo SC, Lauffenburger DA. 1993. Relationship between receptor/ligand binding affinity and adhesion strength. Biophys J 65:2191–2200.

Lasky LA. 1992. Selectins: Interpreters of cell-specific carbohydrate information during inflammation. Science 258:964–969.

Lauffenburger DA, Linderman JJ. 1993. Receptors: Models for Binding, Trafficking, and Signaling. New York: Oxford University Press.

Li SH, Burns DK, Rumberger JM, Presky DH, Wilkinson VL, Anostario M Jr, Wolitzky BA, Norton CR, Familletti PC, Kim KJ, Goldstein AL, Cox DC, Huang K-S. 1994. Consensus repeat domains of E-selectin enhance ligand binding. J Biol Chem 269:4431–4437.

Long M, Goldsmith HL, Tees DFJ, Zhu C. 1999. Probabilistic modeling of shear-induced formation and breakage of doublets cross-linked by receptor-ligand bonds. Biophys J 76:1112–1128.

McClay DR, Wessel GM, Marchase RB. 1981. Intercellular recognition: Quantitation of initial binding events. Proc Natl Acad Sci U S A 78:4975–4979.

McQuarrie DA. 1963. Kinetics of small systems. I. J Chem Phys 38:433–436.

Pierres A, Benoliel A-M, Bongrand P. 1995. Measuring the lifetime of bonds made between surface-linked molecules. J Biol Chem 270:26586–26593.

Piper JW. 1997. Force dependence of cell bound E-selectin/carbohydrate ligand binding characteristics. PhD thesis, Woodruff School of Mechanical Engineering, Georgia Institute of Technology. Atlanta.

Piper JW, Swerlick RA, Zhu C. 1998. Determining force dependence of two-dimensional receptor-ligand binding affinity by centrifugation. Biophys J 74:492–513.

Saterbak A, Kuo SC, Lauffenburger DA. 1993. Heterogeneity and probabilistic binding contributions to receptor-mediated cell detachment kinetics. Biophys J 65:243–252.

Saterbak A, Lauffenburger DA. 1996. Adhesion mediated by bonds in series. Biotechnol Prog 12:682–699.

Simon SI, Burns AR, Taylor AD, Gopalan PK, Lyman EB, Sklar LA, Smith CW. 1995. L-selectin (CD62L) cross-linking signals neutrophil adhesive functions via the Mac-1 (CD11b/CD18) beta2-integrin. J Immunol 155:1502–1514.

Zhu C. 1991. A thermodynamic and biomechanical theory of cell adhesion: General formalism. J Theor Biol 150:27–50.

12

Multifunctional Polymers for the Peroral Delivery of Peptide Drugs

Henrik L. Luessen*
LTS Lohmann Therapie-Systeme GmbH, Andernach, Germany

J. Coos Verhoef, A. (Bert) G. de Boer, and H. E. Junginger
Leiden/Amsterdam Center for Drug Research, Leiden University,
Leiden, The Netherlands

Bas J. de Leeuw
Rotterdam School of Management, Rotterdam, The Netherlands

Gerrit Borchard and Claus-Michael Lehr**
Saarland University, Saarbrücken, Germany

I. INTRODUCTION

A. Peroral Peptide Drug Delivery

Recent years have seen enormous advances in the field of peptide and protein engineering by means of biotechnology and recombinant DNA techniques, and today's possibilities include the production of significant quantities of a wide variety of biologically active peptides and proteins that are therapeutically applicable. In most cases such compounds are indicated for chronic therapy, and they will need to be administered by an appropriate delivery system. At present, all possible alternative routes of peptide and protein administration—avoiding the parenteral route—are being investi-

---

Current affiliation: OctoPlus, B.V., Leiden, The Netherlands.
**Current affiliation*: Leiden/Amsterdam Center for Drug Research, Leiden University, Leiden, The Netherlands.

gated with great efforts in both industry and academia. They include the nasal, transdermal, pulmonal, rectal, buccal, vaginal, ocular, and peroral routes (Epstein and Longenecker, 1988; Verhoef et al., 1990; Lee et al., 1991a,b; Lee, 1991), and for each route special drug delivery systems have been or are being designed.

The peroral route of peptide and protein administration still offers the greatest ease of application. However, particular difficulties are met in designing effective delivery systems for gastrointestinal (GI) application, i.e., to achieve predictable and reproducible absorption in therapeutic doses without wasting a major fraction of the drug. Successful peroral delivery of peptide drugs, therefore, can be achieved only by taking care of the particular physiological conditions of the gastrointestinal tract. Whereas the stomach is a very unfavorable environment for peptides and proteins, because of its low pH and high proteolytic activity, the intestine seems to have some possibilities of being used as an absorption site for peptide drugs. The main absorption barriers present in the intestine are shown schematically in Fig. 1 and can be divided into three different parts.

First, the metabolic barrier plays an important role by inactivating the peptide drug before it reaches its site of absorption. It consists of luminal

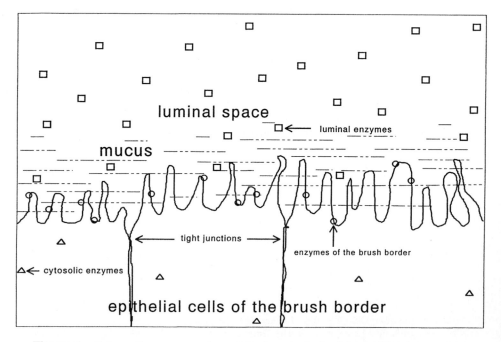

Figure 1   Schematic representation of the intestinal physiological barriers to peptide drug absorption.

proteases (such as trypsin, α-chymotrypsin, elastase, carboxypeptidases), brush border peptidases that are incorporated in or attached to the membrane of epithelial cells, and cytosolic enzymes.

The second barrier against peptide drug absorption is presented by the paracellular epithelial integrity of the intestinal mucosa. Hydrophilic macromolecules, such as peptides, will most likely choose the paracellular route rather than permeation through the lipophilic cell membranes. Passive diffusion of substances in between the cells is controlled by the integrity of intercellular junctions, such as tight junctions.

Third, the mucus, covering the epithelial cell surface, forms an efficient barrier hampering the diffusion of peptide drugs. The continuous secretion of glycoproteins by goblet cells into the intestinal lumen creates a highly viscous gel whose viscosity increases strongly toward the cell surface (Strous and Dekker, 1992).

B. Mucoadhesive Polymers in Peroral Peptide Delivery

In the past decade, mucoadhesive polymers received considerable attention for controlled (peptide) drug delivery (Gu et al., 1988; Gurny et al., 1984; Smart et al., 1984; Lenaerts and Gurny, 1989; Gurny and Junginger, 1990; Junginger, 1990). The main reason for this interest is the role mucoadhesive polymers might play in the following desirable features of a controlled drug delivery system:

> Prolonged residence time at the site of drug absorption, e.g., by controlling GI transit
> Increased contact with the absorbing mucosa, resulting in a steep concentration gradient favoring drug absorption
> Localization in specified regions to improve and enhance the bioavailability of the drug (e.g., targeting to the colon).

Several classes of polymers were found to display pronounced adhesive properties in contact with mucosal surfaces. Among them were poly(acrylates), chitosans, poly(glucan) derivatives (Esposito et al., 1994; Maggi et al., 1994), and hyaluroran derivatives (Pritchard et al., 1996; Sanzgiri et al., 1994). In principle, these polymers are hydrophilic and spread easily over mucosal tissues. For long-term mucoadhesion they should also display good swelling properties but should not dissolve in aqueous environments such as intestinal lumen.

Because of these physical properties, many polymers are already commonly used as pharmaceutical excipients for different purposes, such as matrix-forming agents for sustained drug release in solid dosage forms (Hosny, 1993; Kristl et al., 1993) and viscosity-adjusting or/and gel-forming compounds in semisolids (Knapczyk, 1992). The underlying mechanisms for mucus-polymer association has been investigated by a number of approaches, including interfacial energy thermodynamics (Lehr et al., 1993) and rheological studies (Caramella et al., 1993; Mortazavi and Smart, 1994). The contribution of interpenetration of poly(acrylate)-mucin chains to mucoadhesion was studied by Jabbari et al. (1993).

    The investigations discussed in this chapter focus mainly on two classes of mucoadhesive polymers.

1. Poly(acrylic acid) Derivatives

Among the poly(acrylates), especially the cross-linked derivatives polycarbophil (Noveon AA1, weakly cross-linked with divinylglycol) and carbomer (Carbopol, cross-linked with allyl sucrose, with 931P and 971P relatively weakly and 934P and 974P relatively strongly cross-linked) display pronounced binding interactions with intestinal mucosal tissues (Gu et al., 1988). Unfortunately, exact data on the structure and degree of cross-linking are not available from the manufacturer (BF Goodrich, Cleveland, OH). For further characterization of carbomer 934, the reader is also referred to Craig et al. (1994) and Péres-Marcos et al. (1993). The chemical structure of cross-linked poly(acrylates) is shown in Fig. 2.

2. Chitosans

Chitosan is obtained by deacetylation of the natural product chitin present in the cell wall of most of the higher organized fungi and in the outer skeleton of insects and crustaceous species. Its chemical structure is presented in Fig. 3, and it can be described as a β-(1,4) 2-deoxy-2-amino D-glucan polymer. Depending on the degree of deacetylation, it shows basic

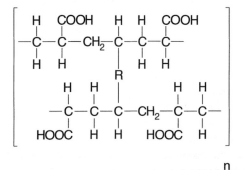

Figure 2   Structure of cross-linked poly(acrylic acid) derivatives. The cross-linking agent (R) for polycarbophil is divinylglycol and for carbomer is allylsucrose.

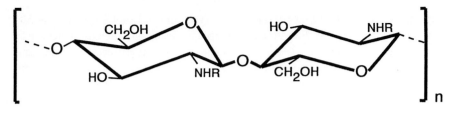

Figure 3 Structure of chitosan (R = H) and chitin (R = acetyl).

properties that lead in slightly acidic media to a polycationic and swellable structure of the polymer. The mucoadhesiveness of chitosan has been evaluated by Lehr et al. (1992b).

For peroral application of non–specifically binding mucoadhesive drug delivery systems, however, it turned out that a residence time of approximately 24 hours, to permit once-daily dosing, could not be achieved in humans (Koshla and Davis, 1987; Harris et al., 1990a), despite the first promising results in animal studies (Longer et al., 1985; Harris et al., 1990b). In a pilot study in rats, poly(hydroxymethyl methacrylate) (pHEMA) particles coated with the weakly cross-linked poly(acrylic acid) derivative polycarbophil showed no difference in transit time between the pylorus and the ileocecal junction in comparison with uncoated pHEMA particles (Lehr, 1991; Junginger et al., 1990). There may be several reasons for this effect:

> Drug delivery systems coated with a mucoadhesive layer undergo rapid deactivation of their mucoadhesive properties through absorption of soluble mucins, mucus degradation products, or other constituents present in the intestinal tract before reaching the mucosal lining.
> The delivery systems designed until now do not possess the ability to renew their mucoadhesive surface for renewed adhesion to the intestinal mucosa.
> The turnover rate of mucus is far too high to achieve long-term adhesion at the mucosal surface of the intestinal tract (Lehr et al., 1990).

C. Additional Possibilities with Mucoadhesive Polymers in Intestinal Peptide Drug Absorption

In general, poor absorption of peptides across mucosal surfaces is caused by the high polarity and high molecular weight of this class of compounds and their susceptibility to proteolytic degradation by luminal, brush border, and cytosolic enzymes. Intestinal peptide absorption is furthermore reduced by the hostile environment of the gastrointestinal tract, i.e., the strong pH extremes and the abundant presence of very potent luminal enzyme systems. On the other hand, peptide absorption in the duodenal intestinal part may have some advantageous features because of the minor brush border and cytosolic enzyme activity in comparison with the jejunal and ileal parts and the large surface area of the upper intestinal part for rapid peptide absorption in comparison with the colon.

Examples of strategies for the peroral absorption of polypeptides, which were proposed earlier, are listed in Table 1 and include:

The use of absorption enhancers, such as surfactants, bile salts, or calcium chelators

The use of enzyme inhibitors, such as aprotinin, bestatin, or soybean trypsin inhibitor

Encapsulation of the drug in particular or vesicular carriers, such as microspheres, nanoparticles, artificial chylomicrons, or liposomes.

In this context, a novel approach may consist of the use of mucoadhesive polymers. As will be shown in this chapter, these polymers are multifunctional macromolecules that are able to increase the permeability of epithelial tissues and simultaneously to inhibit proteolytic enzymes. By virtue of their mucoadhesive properties, these polymers make close contact with the mucosa, thereby exerting such effects at locally high concentrations within a confined area. Because of their large molecular weight, the polymers themselves are not absorbed and are not expected to have any undesired systemic side effects.

II. INFLUENCE OF MUCOADHESIVE POLYMERS ON THE INTESTINAL METABOLIC BARRIER

The intestinal proteolytic enzymes can be classified according to several criteria. Based on their origin and location of action, they can be classified as either luminal, brush border membrane, or intracellular enzymes. Among these enzymes, the luminal and brush border enzymes play the most important role in the digestion of peptide drugs. Cytosolic enzymes such as lysozymes (Hasilik, 1992; Bohley and Seglen, 1992; Seglen and Bohely, 1992) are more relevant in processes of internalization of the peptide drug into the epithelial cell such as endo- and transcytosis. Furthermore, because of the hydrophilic macromolecular structure of peptides, resulting in a preference for paracellular transport, the exposure to intracellular enzymes is very limited. A more traditional classification, based on the proteolytic action of the respective enzymes, defines them as being either endopeptidases or exopeptidases. Endopeptidases hydrolyze the bond interior to the terminal bonds of the peptide chain, and exopeptidases hydrolyze the bond linking the NII$_2$-terminal or the COOH-terminal amino acid to the peptide chain. Table 2 summarizes the important proteolytic enzymes in the gastrointestinal tract according to this classification.

Luminal endopeptidases such as trypsin and α-chymotrypsin often initiate the degradation of perorally administered peptides. The resulting fragments are then further digested by a variety of exopeptidases, such as carboxypeptidases, aminopeptidases, and several di- and oligopeptidases, which are mainly embedded in the brush border membrane of the intestinal epithelium but are also present in the lumen of the gut (Bai, 1994; Lee et al. 1991a).

It has been shown that ligation of the pancreatic duct or the use of protease inhibitors such as aprotinin and soybean trypsin inhibitor can lead to increased peptide drug absorption (Lee at al., 1991a). A major drawback of these inhibitors, however, is their high toxicity, especially in chronic drug therapy. In addition, the non-site-specific intestinal application of such compounds will markedly change the metabolic pattern in the gastrointestinal tract due to reduced digestion of food proteins. Furthermore, their activity is mainly limited to luminal enzymes with a preference for endopeptidases. Proteases embedded in the mucus layer or located in the apical membrane of the epithelial cells are not easily affected, because direct interaction between enzyme and inhibitor is difficult to achieve. This holds particularly for high-molecular-weight structures such as soybean trypsin inhibitor, aprotinin, and Bowman-Birk inhibitor, for which diffusion is hampered by the mucus layer (MacAdam, 1993; Matthes et al., 1992; Strous and Dekker, 1992).

A. Effects of Mucoadhesive Polymers on the Proteolytic Activity of Intestinal Enzymes

A locally acting drug delivery system that is able to change the physiological surroundings of the intestine in a small restricted area will be advantageous for chronic peptide drug delivery over the inhibitors. One approach to designing such a dosage form involves the use of mucoadhesive excipients. Their sticking capability is rather limited under physiological conditions because of the high turnover rate of the mucus layer (Lehr et al., 1990), but they may be applicable for short-term delay of transit time and for intensifying contact between the dosage form and the site of peptide drug absorption. In previous studies, improved intestinal absorption of the peptide drug 9-desglycinamide, 8-arginine vasopressin (DGAVP) was observed in rats in vitro as well as in vivo using the weakly cross-linked poly(acrylate) derivative polycarbophil dispersed in physiological saline (Lehr et al., 1992a). A similar effect was shown with another class of mucoadhesive polymers, the chitosans, in a vertically perfused intestinal loop model of the rat (Rentel et

Table 1 Strategies for Improving Peroral Delivery of Peptide Drugs

Strategy	Peptide	Method	Reference
Absorption enhancers	dDAVP[a]	Sodium taurodihydrofusidate	Lundin et al., 1990
	Insulin	Crown ethers	Touitou, 1992
	Insulin	β-Cyclodextrin derivatives	Shao et al., 1994
	Insulin	Salicylate-microcrystalline cellulose dispersions with hydroxypropyl cellulose	Mesiha and Sidhom, 1995
	Human calcitonin	Monoolein:sodium taurocholate (1:1 mixture)	Hattewell et al., 1994
Enzyme inhibitors	Insulin	Bestatin, aprotinin, soybean trypsin inhibitor, streptozotocin	Morishita et al., 1992a, b, 1993
	Vasopressin and analogues	Aprotinin	Saffran et al., 1988
	Insulin	Soybean trypsin inhibitor, aprotinin, bacitracin, camostat mesilate, sodium glycocholate	Yamamoto et al., 1994
Carrier systems	Insulin	Microspheres	Morishita et al., 1992a, b
	Insulin, calcitonin	Nanocapsules	Lowe and Temple, 1994
	Insulin	Nanoparticles	Michel et al., 1991
	HIV protease inhibitor (CGP57813)	pH-sensitive nanoparticles	Oppenheim et al., 1982 Leroux et al., 1996
	Cyclosporin	Lipid microemulsions	Constantinides, 1995
		Microemulsions	Sarciaux et al., 1995
	Insulin	Emulsion (water-in-oil-in-water)	Engel et al., 1968 Shichiri et al., 1974, 1975
	Calcitonin, dDAVP	Mucoadhesive submicron emulsions	Ilan et al., 1996 Baluom et al., 1997

Insulin	Liposomes	Patel et al., 1978, 1982 Ilan et al., 1996
DGAVP	Niosomes	Yoshida et al., 1992
Leucine enkephalin	Sugar coupling with cellobiose and gentiobiose	Mizuma et al., 1996
Insulin	Biomembranes of erythrocytes	Al-Achi et al., 1992
Insulin	Polyacrylic polymer–coated soft gelatin capsules for targeted delivery to the colon	Touitou and Rubinstein, 1986
Insulin	Azopolymer-coated hard gelatin capsules for targeted delivery to the colon	Saffran et al., 1990
HG-CSF	Modification by conjungation to PEG	Jensen-Pippo et al., 1996

[a]Abbreviations: dDAVP, 1-desamino, 8-arginine vasopressin; HIV, human immunodeficiency virus; DGAVP, 9-desglycinamide, 8-arginine vasopressin; HG-CSF, human granulocyte colony-stimulating factor; PEG, poly(ethylene glycol).
Adapted from Luessen et al. (1994).

Table 2 Important Enzymes of the Gastrointestinal Tract

Enzyme	Type	Cleavage preferences[a]
Endopeptidases		
Trypsin	Serine	Basic amino acids
α-Chymotrypsin	Serine	Aromatic amino acids
Elastase	Serine	Uncharged, nonaromatic amino acids
Pepsin	Aspartic	Hydrophobic amino acids
Exopeptidases		
Carboxypeptidase A	Metallo	Hydrophobic (except Arg, Lys, and Pro)
Carboxypeptidase B	Metallo	Lys, Arg
Carboxypeptidase Y	Serine	Nonspecific
Carboxypeptidase P	Serine	Nonspecific
Leucine aminopeptidase	Metallo	Leu (not Lys, not Arg)
Pyroglutamyl aminopeptidase	Cysteine	S-oxoproline (or pyroglutamic acid)

[a]Cleavage usually occurs at the carboxylic terminus of the mentioned amino acids.

al., 1993). The improved DGAVP absorption could not be explained by mucoadhesion alone. In addition, an influence on the physiological absorption barriers, such as inhibition of proteolytic enzyme activities and enhanced paracellular permeability, was proposed.

The aim of this section of the chapter is to discuss the mucoadhesive poly(acrylates) polycarbophil and carbomer with regard to their ability to reduce the proteolytic activity of the endopeptidases trypsin and α-chymotrypsin as well as the exopeptidases carboxypeptidase A and B, microsomal and cytosolic leucine aminopeptidase, and pyroglutamyl aminopeptidase.

1. Endopeptidases

Among the luminal endopeptidases, trypsin, α-chymotrypsin, and elastase were investigated to determine whether they might be inhibited by poly(acrylic acid) derivatives. All these enzymes belong to the group of Ca^{2+}-containing serine proteases. Trypsin shows a strong affinity for basic amino acids such as lysine and arginine. α-Chymotrypsin, however, cleaves preferably at aromatic amino acids such as L-tyrosine and L-phenylalanine (Table 2). Except for elastase, all experiments were performed at pH 6.7; the procedures are described in more detail by Luessen et al. (1996b).

a. Trypsin

The effects of different mucoadhesive polymers on trypsin activity are displayed in Table 3. Among all the polymers tested, polycarbophil and carbomer uniquely showed a strong concentration-dependent inhibitory effect

Table 3 Viscosity and Trypsin Inhibition Factor of a Number of Mucoadhesive Polymers (mean, $n = 3$)

Mucoadhesive polymer	Concentration (% w/v)	η (mPas)	IF[a]
Mes/KOH buffer pH 6.7 (control)		0.70	1.0
Polycarbophil	0.25	4.66	1.1
	0.35	18.11	7.2
Carbomer	0.10	1.62	3.7
	0.25	7.62	10.4
Methylcellulose	0.50	12.39	0.8
	1.0	29.68	1.1
	2.0	312.63	3.9
Chitosan-glutamate	0.5	2.87	0.8
Chitosan-lactate			
(medium molecular weight)	0.5	9.47	0.7
(high molecular weight)	0.5	5.07	0.8
Gantrez 119 AN	1.0	5.13	0.7
EDTA	7.5	1.36	1.0

[a]IF, inhibition factor. IF was calculated from the formula $AUC_{control}/AUC_{polymer}$ determined from the metabolite time curve (AUC = area under the curve).
From Luessen et al. (1996b).

on the hydrolytic activity of trypsin with N-α-benzoyl-L-arginine ethyl ester (BAEE) used as a model substrate specific to trypsin.

The corresponding degradation profile of the metabolite-time curve revealed a nonlinear decrease of trypsin activity in the first 20 minutes (see Fig. 5). Carbomer had a more pronounced inhibitory effect than polycarbophil. The minimal polymer concentrations resulting in complete inhibition of trypsin activity after a period of 20 minutes were 0.35% and 0.15% (w/v) for polycarbophil and carbomer, respectively.

b. α-Chymotrypsin

α-Chymotrypsin could also be inhibited by the poly(acrylates) polycarbophil and carbomer (Luessen et al., 1996b). Inhibition was found to be dependent on the polymer concentration. Just as for trypsin inhibition, carbomer was more efficient in inhibiting α-chymotrypsin than polycarbophil. Whereas a concentration of 0.25% (w/v) polycarbophil showed only weak inhibition, a concentration of 0.25% (w/v) carbomer was able to block markedly the hydrolytic activity of α-chymotrypsin toward N-acetyl-L-tyrosine ethylester (ATEE).

c. Elastase

As reported by Bernkop-Schnürch et al. (1997b), elastase activity could be reduced by Carbopol 940 and a poly(acrylic acid–divinyl glycol) derivative. Although other experimental conditions were used (pH 7.8, incubation time of 5 minutes), the results indicate an effect comparable to that in the studies with trypsin and α-chymotrypsin.

2. Exopeptidases

a. Carboxypeptidase A

Polycarbophil and carbomer are able to inhibit carboxypeptidase A activity and both poly(acrylates) showed a quite similar concentration dependence (Luessen et al., 1996b). A polymer concentration of 0.005% (w/v) was sufficient to reduce carboxypeptidase A activity to about 15% of the control value.

b. Carboxypeptidase B

A concentration-dependent inhibition of carboxypeptidase B activity was observed with both polycarbophil and carbomer (Luessen et al., 1996a). Carbomer was found to be more potent than polycarbophil (Table 4).

c. Leucine Aminopeptidases

Microsomal leucine aminopeptidase M activity was not inhibited by the poly(acrylic acid) derivatives polycarbophil and carbomer (Luessen et al., 1996b). In contrast to the enzymes of microsomal origin, the cytosolic leucine aminopeptidase showed a strong reduction of its proteolytic activity in the presence of either polycarbophil or carbomer (Luessen et al., 1996b). At a concentration of 0.5% (w/v) the proteolytic activity was reduced to 13% for the carbomer and to 22% for the polycarbophil preparation as compared with the control value.

d. Pyroglutamyl Aminopeptidase

The cysteine protease pyroglutamyl aminopeptidase could not be inhibited by polycarbophil and carbomer (Luessen et al., 1996b).

3. Rheological Properties of Mucoadhesive Polymers

Another aspect of enzyme activity studies in viscous systems is the possibility of immobilization of both enzyme and substrate in the gel matrix of the different polymers investigated. The viscosities of the inhibiting polycarbophil and carbomer preparations, however, were quite low in comparison with those of the other polymers studied, which were not able to inhibit trypsin activity [e.g., 1% (w/v) methylcellulose; Table 3]. Only under the

Table 4 Inhibition Factors of Peptide Degradation by Brush-Border Membrane Vesicles (BBMV) in the Presence of Polycarbophil and Carbomer

Peptide drug or substrate	Polymer	Polymer concentration (% w/v)	IF[a]
Buserelin	Control	—	1.0
	Polycarbophil	0.25	1.1
		0.5	1.1
	Carbomer	0.25	1.0
		0.5[b]	1.0
DGAVP	Control	—	1.0
	Polycarbophil	0.25	1.8
		0.5	1.6
	Carbomer	0.25	1.5
		0.5	1.7
		0.5[c]	2.0
		0.5[b]	2.4
Metkephamid	Control	—	1.0
	Polycarbophil	0.25	1.3
		0.5	1.7
	Carbomer	0.25	1.2
		0.5	2.1
PGNA[d]	Control	—	1.0
	Polycarbophil	0.1	0.8
		0.25	0.7
		0.5	1.1
	Carbomer	0.1	0.7
		0.2	0.7
		0.5	0.7
Hipp-Arg[e]	Control	—	1.0
	Polycarbophil	0.25	1.7
		0.5	2.2
	Carbomer	0.25	2.0
		0.5	4.4

[a]Inhibition factors (IF), calculated from the formula $AUC_{polymer}/AUC_{control}$ from the substrate-time curve (for buserelin, DGAVP and metkephamid). For areas under the percentage of metabolite-time curve (in the case of pyroglutamyl aminopeptidase) the IF was calculated by the formula $AUC_{control}/AUC_{polymer}$.
[b]30 min preincubation.
[c]15 min preincubation.
[d]PGNA (L-pyroglutamic acid p-nitroanilide) was used as substrate for pyroglutamyl aminopeptidase.
[e]Hipp-Arg (hippuryl-L-arginine) was used as substrate for carboxypeptidase B.
From Luessen et al., (1996a).

influence of 2% (w/v) methylcellulose could less enzyme activity be found. At this concentration, however, the viscosity was so high (about 300 mPa*s) that reduced trypsin activity was most probably due to hampered diffusion of both the enzyme and the substrate.

4. Ca^{2+} and Zn^{2+} Binding by the Poly(acrylates)

Because bivalent cations are incorporated in or associated with the structure of a major part of proteolytic enzymes, a possible inhibition of proteases by the strong binding properties of poly(acrylic acid) derivatives toward bivalent cations was proposed. The ability of polycarbophil and carbomer to bind Ca^{2+} and Zn^{2+} ions was investigated at different pH values. The results are displayed in Fig. 4.

The zinc and calcium binding capacities of polycarbophil and carbomer were strongly pH dependent (Luessen et al., 1996b). The ability to bind Zn^{2+} and Ca^{2+} ions increased with higher pH values. An explanation for this phenomenon may be that at higher pH values the carboxylic groups of poly(acrylates) dissociate, forming a polyanionic polymer that tends to salt out in the presence of cations. For Zn^{2+} a molar binding ratio (Zn^{2+}:

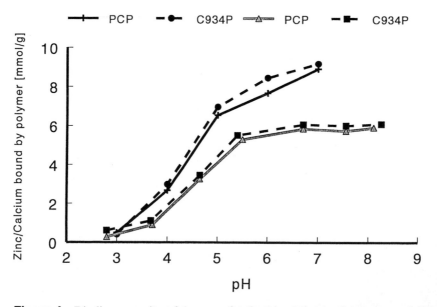

Figure 4 Binding capacity of 1 gram of polycarbophil or carbomer toward either calcium or zinc ions at different pH values. Calcium binding curves: (▲) polycarbophil; (■) carbomer. Zinc binding curves: (+) polycarbophil; (●) carbomer. (Mean ± SD, n = 3.) (From Luessen et al., 1996b.)

carboxylic group) of 1:1.8 to 1:1.4 and for Ca^{2+} of 1:2.3 (Ca^{2+}:carboxylic group) was found. This suggests that two carboxylic groups bind one calcium cation to form a neutralized charge complex. Consequently, this $Ca_n[acrylate]_{2n}$ complex reduces the hydrophilicity of the polymer, resulting in an insoluble precipitate.

Under the conditions of the degradation experiments at pH 6.7, carbomer showed a slightly higher Zn^{2+} and Ca^{2+} binding capacity than polycarbophil. However, a significantly higher binding affinity of poly(acrylate) toward Ca^{2+} could be found for carbomer 934P in comparison with polycarbophil (data not shown; see Luessen et al., 1995). The higher calcium binding affinity found for carbomer may be ascribed to a different way of cross-linking. Polycarbophil is cross-linked by divinylglycol and to a lower degree than carbomer 934P, which is cross-linked by allylsucrose. This suggests that the flexibility of the molecular structure of this particular carbomer to avoid regions of higher negative charge densities, caused by proton dissociation from carboxylic groups, is reduced compared with polycarbophil. Regions of higher negative charge densities within the poly(acrylate) structure, as expected for carbomer, may explain the higher binding affinity of this polymer for bivalent cations such as calcium.

5. Discussion of the Effect of Polycarbophil and Carbomer on Proteolytic Enzyme Activities

The results presented earlier showed that the poly(acrylate) derivatives polycarbophil and carbomer are potent inhibitors of the proteolytic enzymes trypsin, α-chymotrypsin, and carboxypeptidase A and B as well as cytosolic leucine aminopeptidase, whereas the enzyme activities of microsomal leucine aminopeptidase and pyroglutamyl aminopeptidase are not affected by these polymers. Furthermore, the activity of elastase was found to be reduced by carbomer 940 and a poly(acrylic acid–divinyl glycol) derivative (Bernkop-Schnürch et al., 1997b). The inhibitory properties of poly(acrylates) toward intestinal proteases were first reported by Hutton et al. (1990). They found a strong reduction of albumin degradation by a mixture of proteases in the presence of carbomer 934P. Inhibition was pH dependent; the inhibitory effect of the polymer was strong at pH 4.5 and 7.5 but minor at pH 11. At pH 7.5 V_{max} decreased and k_m increased, indicating that inhibition could not be ascribed to either classical competitive or noncompetitive interactions.

As already mentioned, many proteases have bivalent cations such as zinc and calcium as either essential cofactors for their activity (metallopeptidases such as carboxypeptidase A + B) or stabilizing factors (e.g., trypsin) within their structure. The endoproteases trypsin and α-chymotrypsin were chosen as representatives of the group of Ca^{2+}-containing serine proteases. The enzymes carboxypeptidase A and B and microsomal and cytosolic aminopeptidase belong to the group of Zn^{2+}-dependent exopeptidases. A classification of the family of zinc metalloproteases has been given by Hooper (1994). Depletion of zinc reduces or completely inhibits their activity (Himmelhoch, 1969; Salvesen and Nagase, 1989; DiGregorio et al., 1988). Pyroglutamyl aminopeptidase was chosen as a member of the group of cysteine proteases (Armentrout and Doolittle, 1969). Our results revealed a nonlinear metabolite versus time profile, thus a decrease of the degradation activity of trypsin with time. In the case of 0.35% polycarbophil and 0.15% and 0.25% carbomer, a time period of 20–30 minutes was required before complete loss of trypsin activity could be detected. This time dependence indicates that trypsin inhibition is due not to a rapid enzyme-inhibitor interaction but to a more complex pattern of different kinetic parameters. The efficacy of the poly(acrylates) is underlined by the observation that 7.5% (w/v) of the chelating agent EDTA was not sufficient to inhibit trypsin activity at pH 6.7 (Table 3).

The second Ca^{2+}-containing serine protease studied, α-chymotrypsin, shows high similarities to trypsin in its tertiary structure (Tsukada and Blow, 1985; Hedstrom et al., 1994). In comparison with the trypsin degradation studies, higher polymer concentrations were required to inhibit α-chymotrypsin in order to display a comparable reduction of enzyme activity. For example, 0.1% carbomer in the trypsin experiment showed an inhibitory profile comparable to that of 0.5% carbomer in the α-chymotrypsin study. As found for trypsin, α-chymotrypsin showed a nonlinear reduction of its activity under the influence of the poly(acrylates), which could be observed in the first 20 minutes after addition of the enzyme to the incubation medium. This indicates that some steps in the inhibition kinetics are slow, such as dissociation of Ca^{2+} from the enzyme, and assumes a more complex inhibitory mechanism, which is discussed in more detail in Sec. II.B of this chapter.

Carboxypeptidase A and B, requiring Zn^{2+} for their activity (Vallee et al., 1960), could also be inhibited by polycarbophil and carbomer at neutral pH values. In principle, the concentrations of the two polymers needed to achieve carboxypeptidase A inhibition were quite comparable, but the concentrations required for inhibition were much lower compared with trypsin and α-chymotrypsin. Carboxypeptidase A activity was reversible upon addition of zinc, indicating a strong effect of this bivalent cation on the inhibitory properties of the poly(acrylates) (data not shown; Luessen et al., 1996b). Binding of zinc to the poly(acrylates) by depletion from the secondary structure of the enzyme may explain the time-dependent inactivation of carboxypeptidase A + B. Although there is a high structural similarity between carboxypeptidase A and B (Guasch et al., 1992; Chan and Pfuetzner, 1993), the inhibitory effect of the poly(acrylates) on carboxypeptidase B was not as strongly pronounced as that on carboxypeptidase A. An explanation for this phenomenon may be that the binding affinity of carboxypeptidase B for its cofactor Zn^{2+} is higher than that of carboxypeptidase A. The lower polymer concentrations required for an inhibitory effect on carboxypeptidase A as compared with the degradation studies with trypsin and α-chymotrypsin may be ascribed to either a higher dissociation constant of the cation from the enzyme structure or a higher binding affinity of the poly(acrylates) for zinc in comparison with calcium.

Another zinc-dependent protease that could be inhibited by the poly(acrylates) was cytosolic leucine aminopeptidase, but the polymer concentrations required to achieve inhibition were higher than for carboxypeptidase A. However, carboxypeptidase B appeared to be less sensitive to carbomer-mediated inhibition than cytosolic leucine aminopeptidase. This may be due to a lower dissociation constant between Zn^{2+} and cytosolic leucine aminopeptidase than between Zn^{2+} and carboxypeptidase A. In contrast, microsomal leucine aminopeptidase, also a zinc metalloenzyme, could not be inhibited by polycarbophil and carbomer, suggesting that the binding ability of the polymers toward Zn^{2+} is not high enough to deplete cations from the enzyme structure. These two aminopeptidases represent two different Zn^{2+}-dependent aminopeptidases. Microsomal leucine aminopeptidase binds only one Zn^{2+} ion per subunit (Himmelhoch, 1969; DiGregorio et al., 1988), whereas cytosolic leucine aminopeptidase binds two bivalent cations (Van Wart and Lin, 1981; Taylor, 1993). In the commercially available form, in which $MgCl_2$ is added to the stock suspension, the metalloenzyme complex can be described as $[(LAP)Zn_xMg_y]$ (Allen et al., 1983). A tenfold lower k_{cat} was reported for the non–magnesium-containing enzyme (Van Wart and Lin, 1981). Thus, the reduced but not completely inhibited enzyme activity may also be due to binding of the more easily dissociated Mg^{2+} cation than Zn^{2+} to the poly(acrylate), resulting in a still measurable amount of enzyme activity.

Pyroglutamyl aminopeptidase belongs to the family of cysteine exoproteases (Armentrout and Doolittle, 1969). The enzyme does not contain any bivalent cations but exhibits increased activity in the presence of chelating agents such as EDTA (Szewczuk and Mulczyk, 1969). Accordingly, polycarbophil and carbomer increased rather than inhibited the activity of pyroglutamyl aminopeptidase. This enzyme is of pharmaceutical importance, because it is involved in the intestinal degradation of luteinizing hormone–releasing hormone (LHRH) and its analogs such as the peptide drug buserelin (Sandow, 1989). For further information see Sec. II.C of this chapter.

B. Poly(acrylates) as Inhibitors of Trypsin: Mechanism of Action

From Table 3 it is evident that among all the mucoadhesive polymers studied only the poly(acrylates) polycarbophil and carbomer are able to inhibit trypsin activity. This indicates that mucoadhesive properties are not primarily responsible for trypsin inhibition. Direct binding interactions between the enzyme structure, as a hydrophilic macromolecule, and the mucoadhesive polymer may occur, leading to partial inactivation of protease activity. It might also be speculated that the polyanionic poly(acrylates) interact with the enzyme, causing conformational changes in the structure that lead to a facilitated dissociation of the bivalent cation from the protease. However, pronounced binding between proteins and both polycarbophil and carbomer could be observed only at pH 4, not at pH 6.7, at which the enzyme inhibition studies were performed (Luessen et al., 1995).

1. Calcium-Dependent Inhibitory Effect of Poly(acrylates) on Trypsin Activity

A fixed concentration of 0.25% (w/v) carbomer (Fig. 5) was chosen to study the effect of calcium addition at different time intervals in the degradation experiment. Without addition of Ca^{2+} to the polymer preparations, slight metabolite formation was observed only during the first 10 to 20 minutes of the degradation experiment and did not increase during the following 4 hours. However, the remaining trypsin activity appeared to be dependent on the time of Ca^{2+} addition. After addition of Ca^{2+} to both polymer preparations just before trypsin incubation, trypsin activity was not inhibited by the polymers. In contrast, the trypsin activity partly recovered when Ca^{2+} was added after 10 minutes of incubation but not when Ca^{2+} was added after 240 minutes.

2. Simultaneous Determination of the Effects of Poly(acrylates) on Trypsin Activity and Protein and Calcium Binding to These Polymers

In this study three parameters were determined in one experiment: the influence of the poly(acrylates) on (a) Ca^{2+} depletion from the trypsin structure, (b) trypsin binding to the polymers, and (c) enzymatic activity of trypsin. Following incubation with both polymers, Ca^{2+} was depleted from trypsin to 10–15% of the initial content, whereas the free protein concen-

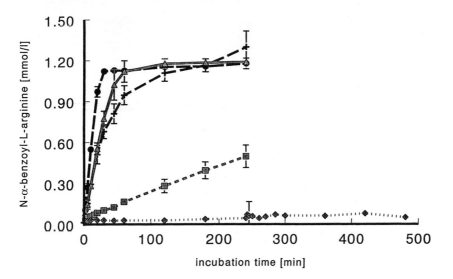

Figure 5 Formation of the metabolite N-α-benzoyl-L-arginine (BA) following incubation of the substrate N-α-benzoyl-L-arginine ethylester (BAEE) with trypsin with or without 0.25% (w/v) carbomer (C934P). A concentration of 14.3 mM Ca^{2+} was added at predetermined time intervals (mean $\pm$ SD, $n = 3$). ($+$) Control (no C934P); ($\bullet$) control with Ca^{2+} (no C934P); ($\blacktriangle$) Ca^{2+} added just before trypsin incubation (0.25% C934P); ($\blacksquare$) added after 10 minutes of incubation (0.25% C934P); ($\blacklozenge$) Ca^{2+} added after 240 minutes of incubation (0.25% C934P). (From Luessen et al., 1995.)

tration in the supernatants was still 60–70% of its original value. However, the remaining enzyme activity of trypsin was 7.5% compared with the non–polymer-treated control (Fig. 6).

3. Circular Dichroism Studies

Circular dichroism spectra revealed that the secondary structure of trypsin was changed under the influence of the poly(acrylates) (Fig. 7). The trypsin spectrum showed a maximum at 185 nm and two minima at 194 and 209 nm, respectively. However, the spectrum of trypsin that was allowed to undergo autodegradation for 30 minutes at 37°C revealed a pronounced minimum in the wavelength area between 195 and 200 nm. This minimum was even more pronounced for the polymer-treated samples, and carbomer showed a stronger decrease of the minimum than polycarbophil.

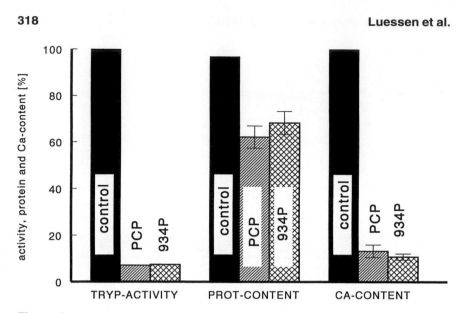

Figure 6   Effects of 0.35% (w/v) polycarbophil (PCP) and 0.25% (w/v) carbomer (934P) on trypsin activity, protein content, and Ca^{2+} depletion from trypsin following a 30-minute incubation at 20°C (mean $\pm$ SD, $n = 3$). Control incubations contained trypsin without poly(acrylates). (From Luessen et al., 1995.)

4.   Discussion of the Mechanism of Trypsin Inhibition by Poly(acrylates)

Trypsin is a serine protease with a binding site for the bivalent cation Ca^{2+} (Bartunik et al., 1989). Calcium plays an important role in maintaining the thermodynamic stability of this particular enzyme (Delaage and Lazdunski, 1967). It has been found that depletion of Ca^{2+} from the enzyme structure by chelating agents such as EDTA can affect the activity of trypsin (Salvesen and Nagase, 1989). Therefore, it is important to determine whether poly(acrylic acid) derivatives, which are able to bind large amounts of bivalent cations in the dissociated state, can also inhibit trypsin activity.

As shown in the present study, the inhibitory effect of polycarbophil and carbomer is dependent on the time interval of Ca^{2+} addition to the trypsin incubation medium. An explanation for this phenomenon may be either time-dependent association of trypsin with the poly(acrylate) structure or time-dependent denaturation of the enzyme under the influence of the mucoadhesive polymers. When Ca^{2+} was added at $t = 10$ minutes, due to salting out of the polymers, no strong inclusion or adsorption of trypsin and thus no inactivation of the enzyme occurred, which may explain the irreversible effect on trypsin inhibition by adding Ca^{2+} at $t = 240$ minutes. In

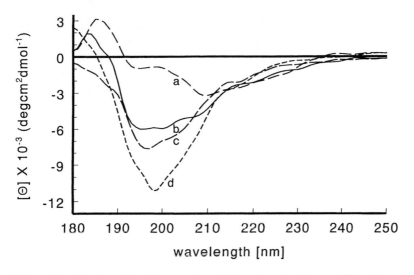

Figure 7 Circular dichroism spectra of (c) polycarbophil- and (d) carbomer-treated trypsin preparations. Two trypsin control samples were also included: (a) "nontreated trypsin," trypsin dissolved in buffer, stored at −20°C until use, and then thawed directly before CD measurements; (b) "treated trypsin," trypsin incubated for 30 minutes at 37°C. (From Luessen et al., 1995.)

contrast, these observations suggest that the poly(acrylates) irreversibly modify the enzyme structure itself with time.

The higher Ca^{2+} binding affinity of carbomer as described by Luessen et al. (1995) may cause more complete depletion of Ca^{2+} from the trypsin structure than with polycarbophil. It is suggested that the dissociation constant for the polymer and Ca^{2+} should be low in comparison with the dissociation constant for trypsin and Ca^{2+} in order to reach sufficient enzyme inhibition. Such a relationship might also explain the lower carbomer concentration required for the inhibition of trypsin activity. For optimization of the inhibitory effect of Ca^{2+}-binding polymers, affinity values may represent an important parameter in predicting their potency to affect the enzymatic activity of Ca^{2+}-dependent proteases.

The extent of trypsin inactivation by polycarbophil and carbomer seems to be comparable with the reduced Ca^{2+} content in the supernatant under the influence of the polymers. The Ca^{2+} content in the supernatant of the control samples can be attributed to trypsin, which contains Ca^{2+} to build up its tertiary structure (Bartunik et al., 1989). The reduced Ca^{2+} concentration in the supernatant after polymer treatment may be attributed to depletion of Ca^{2+} from the trypsin structure. The free protein content in the same supernatants was still 60–70% of that of the control samples, indicating that only a small amount of trypsin was adsorbed to the polymers, which can hardly explain the almost complete inhibition of enzyme activity. However, the remaining amount of protein showed strongly reduced trypsin activity, which fits with the observed Ca^{2+} depletion from the trypsin structure.

The circular dichroism spectra of polymer-treated trypsin showed a minimum at 194–209 nm, which was comparable to that of the spectrum of trypsin allowing to undergo autodegradation for 30 minutes. This suggests that the observed irreversible inhibition of trypsin activity by poly(acrylic acid) derivatives, which deprives the enzyme of Ca^{2+}, is in parallel with structural changes similar to those in autodegradation. Calcium has an essential role in the thermodynamic stability and the resistance to degradation and autodegradation of trypsin (Rypniewski et al., 1994; Kretsinger, 1976; Gabel and Kasche, 1973). Hence, the release of Ca^{2+} from trypsin also leads to changes of the tertiary structure of the enzyme (Bartunik et al., 1989). It may be suggested that exposure of arginine and lysine bonds, which are normally protected from the external environment by being embedded inside the trypsin structure, increases the chance for enhanced autodegradation. This implies that the mechanism of trypsin inhibition does not require a direct interaction between polymer and enzyme, which would be necessary in the sense of classical Michaelis-Menten kinetics. In the case of trypsin, the inactivation of the enzyme is not directly related to the "inhibitor" but depends on the degree of degradation of the enzyme itself as a consequence of Ca^{2+} deprivation. This makes it difficult to predict the time course of trypsin inactivation by compounds such as poly(acrylates), because many different parameters influence the inhibitory profile. These include the degree of Ca^{2+} binding affinity of the polymers, pH, ionic strength, activity pattern of trypsin, Ca^{2+}-free trypsin and its metabolites, type and amount of nutrients, and concentration and type of different cations in the intestinal lumen.

C. Enzymatic Degradation of Peptide Drugs by Intestinal Brush Border Membrane Vesicles: Effect of Poly(acrylates)

The membrane-bound enzymes lining the mucosal epithelium form a highly efficient obstacle to peptide drug absorption, which is difficult to influence with classical enzyme inhibitors of high molecular weight. Besides being sterically protected by being partly embedded in or attached to a cell membrane, the enzymes are protected from the luminal contents of the gut by the overlaying highly viscous mucus layer (Strous and Dekker, 1992). The mucus is a diffusion barrier not only for the peptide drug to be delivered but also for protease inhibitors, which require direct contact with the enzyme (Matthes et al., 1992).

A novel class of enzyme inhibitors that are able to reduce the proteolytic activity of enzymes of the intestinal brush border without requiring direct contact would be of great interest for improving intestinal peptide drug absorption. As discussed in Secs. II.A and II.B of this chapter, poly(acrylate) derivatives have inhibitory effects on proteases by depriving the enzyme structure of bivalent cations. With this mechanism a direct inhibitor-enzyme interaction is not necessary and inhibition may also be achieved through a "far distant effect" through the mucus layer.

The potential of the mucoadhesive poly(acrylates) polycarbophil and carbomer to protect different peptide drugs from degradation by membrane-bound peptidases is evaluated in this present section.

1. Intestinal Brush Border Membrane Vesicles

Brush border membrane vesicles (BBMVs) were derived from the luminal epithelium of rat small intestine as described by Biber et al (1981), using a Mg^{2+} precipitation method. Mainly unilamellar vesicles of different sizes were obtained, as shown in Fig. 8.

Dispersing the vesicles in poly(acrylate) dispersions up to 0.35% (w/v) did not affect their appearance macroscopically. Therefore, possible inhibitory effects observed in degradation studies with BBMV performed at similar conditions could not directly be ascribed to macroscopical changes of the vesicles by the polymers.

2. Peptide Degradation Studies

The results of the degradation studies with BBMV are displayed in Table 4. The corresponding inhibition factors (IFs) of the respective peptide drugs or model substrates in the different poly(acrylate) preparations are given as a relative measure of the change in proteolytic activities of membrane-bound enzymes.

Among all the peptide drugs studied, only DGAVP (9-desglycinamide, 8-L-arginine vasopressin) degradation could be partly inhibited by the poly(acrylic acid) derivatives polycarbophil and carbomer 934P in concentrations of 0.25 and 0.5% (w/v). The inhibitory effect of carbomer 934P on DGAVP degradation can be improved if the BBMV-attached peptidase is preincubated in the polymer environment for about 30 minutes (Table 4). In the case of metkephamid, only carbomer at a concentration of 0.5% (w/v) had an inhibitory effect, whereas buserelin degradation was not inhibited by both poly(acrylates) in concentrations up to 0.5% (w/v). DGAVP is a vasopressin fragment that was originally developed for the treatment of mem-

Figure 8   Electron micrograph of BBMV: BBMV suspension mixed with MES/ KOH buffer (pH 6.7) in a 1:1 ratio (control, magnification 1:11,200). (From Luessen et al., 1996a.)

ory disorders (Bruins et al., 1990). It is highly stabilized against metabolic degradation, and cleavage by carboxypeptidase B or prolyl-endopeptidase– like activities occurs preferentially at the carboxy-terminal L-arginine (Verhoef et al., 1986).

Metkephamid is an enkephalin analogue with the following amino acid sequence: Tyr-D-Ala-Gly-Phe-N-Me-Met-NH$_2$. This peptide shows high stability against proteolytic degradation and is degraded mainly by aminopeptidase activity (Langguth et al., 1994a). The high resistance to pancreatic endo- and exopeptidases makes metkephamid an interesting model peptide with which to study enzyme activities of the intestinal brush border (Langguth et al., 1994b). Of the poly(acrylate) preparations studied, only 0.5% (w/v) carbomer dispersion had an inhibitory effect on metkephamid-degrading BBMV enzyme activities. Assuming that removal of Zn^{2+} from the metkephamid-degrading aminopeptidase structure by carbomer serves as a mechanism for the observed inhibitory effect (Luessen et al., 1995, 1996b), it may be suggested that the higher binding affinity of carbomer for bivalent cations such as Zn^{2+} and Ca^{2+} is responsible for the differences in efficacy between carbomer and polycarbophil.

Insulin appeared to be relatively stable against degradation by BBMV (data not shown). This may be due to the conformational structure of this polypeptide drug, where the terminal ends of the A and B chains can be sterically protected from exopeptidase degradation. The observed stability of insulin may also be explained by the formation of insulin associates and agglomerates (Cleland et al., 1993). The fact that insulin is highly resistant to degradation by BBMV makes insulin an interesting peptide drug for the design of peroral delivery systems based on polycarbophil or carbomer, because the major insulin-degrading proteases, α-chymotrypsin and trypsin (Schilling and Mitra, 1991), are strongly inhibited by these poly(acrylates). This is also a possible explanation for the increased recovery of insulin in the presence of 0.25% carbomer. Another aspect is the possibility of binding the Zn^{2+} ions, which supports the association of insulin monomers (Cleland et al., 1993). Because of their Zn^{2+}-binding properties, poly(acrylates) such as polycarbophil and carbomer may stabilize insulin by preventing the formation of irreversible insulin agglomerates. In this respect, the recently developed insulin derivatives of Eli Lilly, which do not show such pronounced association and aggregation behavior, are of interest (Ter Braak et al., 1993).

Buserelin degradation is mainly initiated by α-chymotrypsin, neutral endopeptidases, or pyroglutamic acid aminopeptidase. In the intestinal brush border and in the systemic circulation, pyroglutamic aminopeptidase is the major enzyme responsible for the inactivation of this peptide drug (Sandow, 1989). The effect of poly(acrylates) is also discussed in a previous part of this chapter (Sec. II.A.5).

The observed inhibitory activities of the poly(acrylates) on metkephamid and DGAVP degradation by BBMV were not as pronounced as previously shown for trypsin, α-chymotrypsin, carboxypeptidase A and B, and cytosolic aminopeptidase. One reason for this may be the bivalent cation–containing peptidase itself. The binding affinity of the poly(acrylate) should be much higher than the binding affinity of the enzyme toward bivalent cations. Thus, if the dissociation constant of the enzyme-X^{2+} (X = bivalent cation, such as Zn^{2+} or Ca^{2+}) complex is very small, the ability of the X^{2+}-binding polymer to remove the bivalent cation from the enzyme structure is decreased. Another reason for the relatively weak inhibitory effect of the poly(acrylates) on the brush border–bound enzymes may be the environment of the peptidases. Being attached to or embedded in the lipid bilayer of the cell membrane, the X^{2+}-bearing part of the enzyme may be sterically protected, resulting in an apparent lower dissociation constant of the X^{2+}-enzyme complex. Both effects may occur simultaneously for the BBMV, which make it very difficult to inhibit membrane-bound proteolytic activities by X^{2+}-binding polymers such as polycarbophil and carbomer.

III. MODULATION OF THE MUCOSAL TRANSPORT BARRIER BY MUCOADHESIVE POLYMERS

The absorption barrier in the intestine is represented by epithelial cell membranes, interconnected by proteinaceous tight junctions. The structure of these tight junctions is influenced by a variety of factors, e.g., cyclic AMP (cAMP), protein kinase C, cytochalasin D, and hormones (Duffey et al., 1981; Madara et al., 1986; Coleman and Kan, 1990). There is also evidence that tight junctions are linked to the cytoskeleton, so that intracellular processes might lead to changes in the dynamic structure of the tight junctions (Schnittler et al., 1990). In addition, their intactness is linked to the presence of Ca^{2+} and Mg^{2+} ions (Pitelka et al., 1983; Gonzales-Mariscal et al., 1990), a reduction of the extracellular Ca^{2+} concentration resulting in an opening of tight junctions (Noach et al., 1993).

The linear polysaccharide chitosan [$(1\rightarrow4)$-2-amino-2-deoxy-β-D-glucan] with its basic primary amino groups displays polycationic properties after being dispersed in aqueous solutions. Studies have shown that chitosan also improves the absorption of peptides and proteins across nasal (Illum et al., 1994) and intestinal epithelia (Artursson et al., 1994; Lehr et al., 1992b; Lehr, 1994). The mechanism of this effect is probably a combination of ionic interactions between the negatively charged tight junctions, the positive charge of the polymer, and the mucoadhesive properties of the polycationic polymer.

As discussed previously in this chapter, the poly(acrylic acid) derivatives carbomer 934P and polycarbophil possess strong binding affinities toward bivalent cations. We postulated that poly(acrylates) might also influence the intactness of epithelial tight junctions by binding of extracellular Ca^{2+} to the polyanionic polymer (Luessen et al., 1994, 1995). Here, the two mucoadhesive polymers of different molecular structures and net ionic charges will be evaluated with respect to affecting the structure of epithelial tight junctions, allowing the paracellular route of peptide uptake. Well-characterized Caco-2 cell cultures (Hidalgo et al., 1989; Hilgers et al., 1990), derived from a human colon carcinoma cell line, served as an in vitro model for the intestinal epithelium. The alteration of transepithelial electrical resistance (TEER), a measure of the tightness of the cell layer, was determined. In addition, paracellular transport of the hydrophilic macromolecular compound fluorescein isothiocyanate–dextran 4400 (FD-4) was visualized using confocal laser scanning microscopy (CLSM). This noninvasive method allows optical sectioning and three-dimensional imaging of the transport pathways through the cell monolayer without disrupting its structure. Furthermore, the potential of chitosans and the poly(acrylic acid) derivative polycarbophil to improve the intestinal transport of the peptide drug DGAVP in vitro was investigated.

A. Effect of Poly(acrylates) and Chitosans on TEER

Results of the TEER measurements at pH 7 are depicted in Figs. 9 and 10. As our results show, both polymers were indeed able to decrease TEER of Caco-2 cell monolayers (a measure of the tightness of the cell layer) at the highest concentrations used. The TEER slightly recovered after 120 minutes on changing back to the culture medium.

Transport studies with both FD-4 and ^{14}C-labeled mannitol revealed (Borchard et al., 1996; data not shown) that the TEER of epithelial mono-

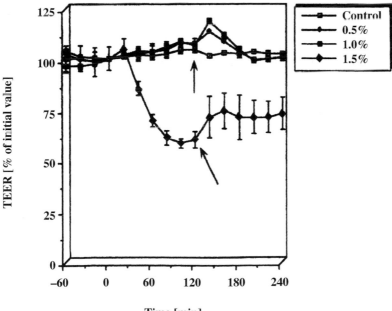

Figure 9 Alteration of TEER of Caco-2 cell monolayers treated with 0.5 (♦), 1.0 (■), and 1.5% (▲) chitosan-glutamate, compared with the control (+) and expressed as percentage of the initial value (↓: cells were washed and returned to culture medium). (From Borchard et al., 1996.)

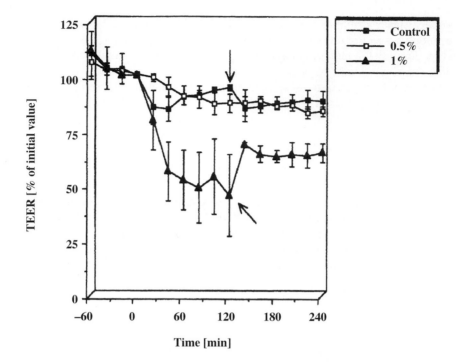

Figure 10 Alteration of TEER of Caco-2 cell monolayers treated with 0.5 (□) and 1.0% (▲) carbomer, compared with the control (■) and expressed as percentage of the initial value (↓: cells were washed and returned to culture medium). (From Borchard et al., 1996.)

layers has to decrease below a certain threshold value if paracellular transport is to be established. In this study, a decrease of TEER to a value not exceeding 50% of the initial value did not seem to be sufficient to permit paracellular transport of the marker substances used.

B. Visualization of Paracellular Transport of Hydrophilic Macromolecules

Visualization studies were performed by applying a fluorescent marker on the apical side of the Caco-2 cell monolayer and measuring fluorescence after incubation of the cells, as described by Borchard et al. (1996) and Luessen et al. (1994).

In the control preparation, fluorescence was detected only on top of the monolayer. The same result was obtained after incubation with chitosan-

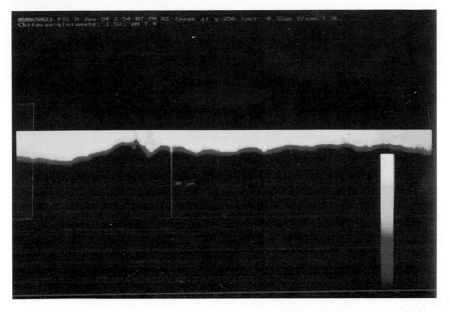

Figure 11 Optical section through a cell monolayer after treatment with 1.5% chitosan-glutamate. Pictures showing *XZ* cross sections as visualized by a confocal laser scanning microscope. (From Borchard et al., 1996.)

glutamate at 1.5% concentration (Fig. 11). After treatment with carbomer at 1.0%, however, fluorescence could also be detected in between the cells (Fig. 12). A concentration of 0.5% polycarbophil was not able to induce paracellular permeation of FD-4 (data not shown). This was well observed after basolateral application of the polymer (Fig. 13). These results indicate a side-dependent sensitivity of the cells to the decrease in extracellular Ca^{2+} concentration, which corroborates the observation by Noach et al. (1993) that Ca^{2+} complexation by EDTA at the basolateral side of Caco-2 cell monolayers was much more effective in opening the tight junctions than complexation of Ca^{2+} at the apical side.

Staining with trypan blue (data not shown) and confocal microscopy showed no nuclear uptake of the fluorescence label, indicating that the polymers do not damage the cells.

C. Improvement of Peptide Drug Transport by Mucoadhesive Polymers

As described by Borchard et al. (1996), paracellular drug transport enhancement was achieved with both chitosans and poly(acrylates). These studies

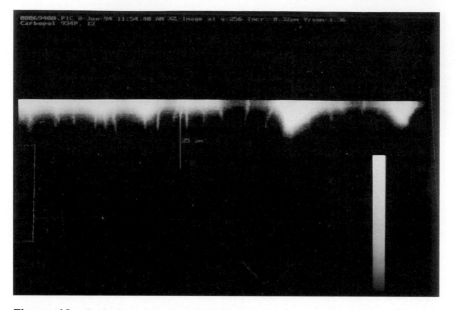

Figure 12 Optical section through a cell monolayer after treatment with 1.0% carbomer. Note that FD-4 can be detected in between the cells down to a depth of about 15 μm. (From Borchard et al., 1996.)

were performed at neutral pH values (between 6.5 and 7.4), at which the poly(acrylates) are mostly dissociated and no longer display pronounced mucoadhesive properties. At neutral pH, however, chitosans have poor swelling abilities, which restrict their use in the development of mucoadhesive dosage forms. At slightly acidic pH values the mucoadhesiveness of the poly(acrylate) polymers is much more pronounced (Lehr et al., 1993; Rillosi and Buckton, 1995). Therefore, experiments described in this section (see also Luessen et al., 1997) were performed at pH 5.6 to evaluate whether transport enhancement could also be achieved under conditions in which both polymers display sufficient swelling and mucoadhesive properties.

Apical application of polycarbophil slightly improved the transport of the peptide drug DGAVP through Caco-2 cell monolayers (Fig. 14). In the case of 1% (w/v) polycarbophil, substantial variation was observed, and the average transport was only 0.2% of the total dose after 4 hours. With chitosan glutamate at concentrations of 0.4 and 1% (w/v), DGAVP was transported at 1.2% of the total dose after 4 hours, which was substantially higher than both the control amount and the amount with 1% (w/v) polycarbophil.

The similarity in transport at the two different chitosan glutamate concentrations of 0.4 and 1% (w/v) indicates that at 0.4% (w/v) a maximum in

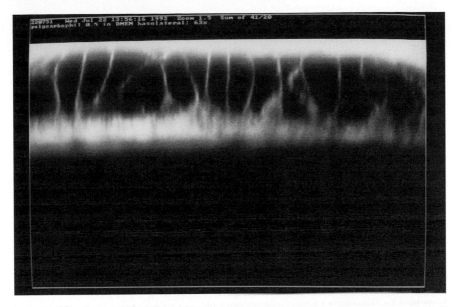

Figure 13 Intercellular permeation of FD-4 after apical application on a Caco-2 cell monolayer. A concentration of 0.5% polycarbophil was applied to the basolateral side. (From Luessen et al., 1994.)

the transport rate has already been reached. This observation is in accordance with the results obtained with chitosan glutamate at pH values ranging between 4.9 and 6.0, where a plateau level in P_{app} was reached at polymer concentrations of 0.25 and 0.5% (w/v) (Artursson et al., 1994). Such a plateau effect has also been found for nasal insulin absorption in sheeps, where at a pH value of 4.4, chitosan glutamate in concentrations exceeding 0.4% (w/v) did not result in a stronger reduction of glucose levels (Illum et al., 1994).

Polycarbophil, however, showed only a shallow increase in DGAVP transport, which might be explained by the different mechanism through which it affects the integrity of intercellular epithelial junctions, e.g., tight junctions. Whereas the opening of tight junctions by chitosan derivatives is attributed to the interaction of the positively charged amino groups (pK_a of chitosan approximately 5.6) with the negatively charged sialic groups of the membrane-bound glycoproteins (Artursson et al., 1994), the effect of polycarbophil is ascribed to its high Ca^{2+}-binding abilities (Luessen et al., 1994). At neutral pH, however, opening of tight junctions and permeation of FD-4 into the paracellular space could be observed only when polycarbophil was applied to the basolateral side of the Caco-2 cell monolayers. On the

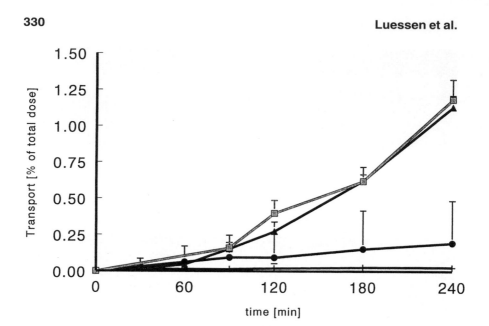

Figure 14   Transport of DGAVP across Caco-2 cell monolayers. Transport data
are expressed as percentage of the total dose of DGAVP applied to the apical side
of the cell monolayer. ($+$) Control; ($\bullet$) 1% (w/v) polycarbophil; ($\blacktriangle$) 0.4% (w/v)
chitosan glutamate; ($\blacksquare$) 1% (w/v) chitosan glutamate (mean $\pm$ SD, $n = 3$). (From
Luessen et al., 1997.)

other hand, the tightness of intercellular junctions might be affected by adhesive interaction between the poly(acrylate) and the surface of the epithelial
cell monolayer. This hypothesis might be further underlined by the fact that
at pH 5.6 the Ca^{2+} binding properties of poly(acrylates) are relatively low
compared with those at neutral pH, because the dissociation of carboxylic
groups of poly(acrylates) is not optimal (Luessen et al., 1996b). Opening of
tight junctions by extracellular withdrawal of Ca^{2+} is therefore not expected
to be the only underlying mechanism of action for the transport enhancement
of DGAVP observed at pH 5.6.

In summary, our studies show that the two mucoadhesive polymers
chosen were able to influence the permeability of epithelial cell monolayers
in vitro. Because of their different structures and ionic charges, the mechanisms underlying this alteration of permeability have to be of different char-
acters. A direct interaction of the cationic polymer molecule with the negatively charged cell membrane might be presumed for chitosan-glutamate,
as also discussed by Schipper et al. (1997), whereas depletion of extracel-

lular Ca^{2+} plays the major role in the studies with poly(acrylic acid) derivatives (Luessen et al., 1995).

IV. ENHANCEMENT OF INTESTINAL PEPTIDE DRUG ABSORPTION BY MULTIFUNCTIONAL POLYMERS IN VIVO

Considering the promising in vitro performance of the poly(acrylates) and chitosans in lowering the barriers to peroral peptide absorption, as presented in the preceding part of this chapter, both polymers were evaluated in in vivo absorption studies with the nonapeptide buserelin. Besides carbomer 934P (C934P) and chitosan-HCl, as the most promising representatives of each polymer class, the freeze-dried sodium salt of carbomer, FNaC934P, was also included in these studies. This fast-swelling modification of C934P showed fast dispersion in mixtures with suitable disintegrants or after microdispersion in microparticles of polyglycerol esters of fatty acids (Akiyama et al., 1996a, 1996b). FNaC934P was prepared in accordance with Akiyama's formulations. Dispersions of the LHRH superagonist buserelin (500 μg per rat) in the respective polymers were administered intraduodenally to rats. All polymer dispersions displayed a considerable increase in intestinal absorption in comparison with control solutions of buserelin. Application of the control solution resulted in low plasma concentrations and an absolute bioavailability of 0.1%. The maximum buserelin plasma concentration after administration of poly(acrylate) dispersions reached levels of 45.8 and 112.1 ng/mL for 0.5% (w/v) dispersions of FNaC934P and C934P, respectively (Fig. 15). The calculated bioavailabilities for the respective polymers increased 6- and 19-fold compared with the control value (Table 5).

Application of buserelin in a 1.5% (w/v) chitosan-HCl dispersion highly increased plasma levels to a maximum of 364 ng/mL, and the bioavailability increased 51-fold compared with the control solution (Fig. 16; Table 5).

In addition to calculation of pharmacokinetic parameters such as C_{max} and the area under the curve (AUC), the data were subjected to numerical deconvolution analysis. By means of this analysis it could be demonstrated that the t_{90} value (time to reach 90% of total absorption) for C934P was considerably higher for the C934P dispersion than for the chitosan-HCl dispersion (Table 5). No statistically significant differences in t_{90} were observed between FNaC934P and C934P.

Of the two poly(acrylates), the fast-swelling FNaC934P performed less well than native C934P. This difference in efficacy is ascribed to the presence

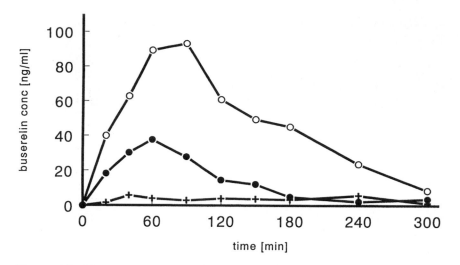

Figure 15 Mean serum concentrations of buserelin after intraduodenal application (500 µg/rat). (+) Control (MES/KOH buffer, pH 6.7); (●) 0.5% (w/v) FNaC934P; (○) 0.5% C934P.

Table 5 Pharmacokinetic Parameters After Intraduodenal Administration of Buserelin (500 µg/rat)[a]

Polymer	T_{max} (min)	T_{90} (min)	C_{max} (ng/mL)	F (%)	n
Control	60–90	ND	6.7 ± 1.7	0.1 ± 0.1	6
Poly(acrylates)					
FNaC934P (0.5%)	40–60	95 ± 25	45.8 ± 20.8[2]	0.6 ± 0.2[2]	5
C934P (0.5%)	40–90	165 ± 80	112.1 ± 53.4[2,3]	1.9 ± 1.3[2,3]	5
Chitosan					
Chitosan-HCl (1.5%)	40–90	76 ± 29[1]	364.0 ± 140.0[2,4]	5.1 ± 1.5[2,4]	6

[a]Data are presented as mean ± SD for the number of animals (n) indicated. T_{max}, time to reach serum peak concentration; T_{90}, time to reach 90% of total amount absorbed; C_{max}, serum peak concentration; F, absolute bioavailability; ND, not determined.
[1]Significantly different from C934P ($P < .05$).
[2]Significantly different from control ($P < .01$).
[3]Significantly different from FNaC934P ($P < .05$).
[4]Significantly different from C934P ($P < .005$).

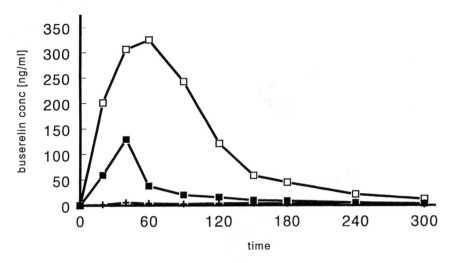

Figure 16 Mean serum concentrations of buserelin after intraduodenal application (500 μg/rat). (+) Control MES/KOH buffer, pH 6.7; (■) 0.5% (w/v) FNaC934P/ 1.5% (w/v) chitosan-HCl mixture (1:1); (□) 1.5% (w/v) chitosan-HCl.

of sodium in FNaC934P, as a counterion for the dissociated carboxylic group. It is assumed that the presence of this counterion decreases the binding affinity toward Ca^{2+} and Zn^{2+} and, consequently, decreases the inhibitory effect on proteolytic enzymes and the efficacy in opening intercellular junctions.

Chitosan-HCl has shown to be a more effective polymer for increasing the bioavailability of buserelin, although it was previously reported that chitosan improves mucosal peptide drug absorption only by opening intercellular junctions and that this class of polymers does not exert inhibitory effects on proteolytic enzymes (Luessen et al., 1996b). The higher effectiveness of chitosan in these in vivo studies, however, is in accordance with in vitro transport studies of De Leeuw et al. (1996), which demonstrated a higher efficacy of chitosan in improving buserelin transport through Caco-2 cell monolayers than of poly(acrylates). It is therefore suggested that in the case of buserelin, the transport-enhancing effect plays a predominant role in improving peroral peptide drug absorption.

A mixture of chitosan with the poly(acrylate) was also included in the present absorption studies (Fig. 16). However, no pronounced improvement in the buserelin bioavailability was found compared with the poly(acrylate) or chitosan formulation alone. An approach to combining the Ca^{2+}-chelating effect on enzyme inhibition and absorption enhancement with the absorption-promoting properties of chitosan was studied by binding EDTA covalently to chitosan (Bernkop-Schnürch et al., 1997a). It was found that the bioadhesive properties of the polymer conjugate were still maintained and that the activity of aminopeptidase N could be inhibited. The duration of absorption (expressed by the t_{90} parameter) of buserelin dispersed with C934P, on the other hand, was extended compared with the chitosan-HCl preparations. This indicates that C934P was able to prolong the intactness of buserelin in the intestine by enzyme inhibition. Relating this observation to previous in vitro demonstrations of the time dependence of enzyme inhibition by the poly(acrylates) in both enzyme inhibition studies (Sec. II of this chapter) and combined enzyme inhibition-transport studies (De Leeuw et al., 1996), it can be argued that a delayed release of the peptide drug 20 to 30 minutes later than poly(acrylate) release and subsequent swelling may substantially improve the absorption enhancement effect of the poly-(acrylates).

V. CONCLUSIONS

Although the optimal mucoadhesive polymer still does not exist among the available polymers, the classes of poly(acrylic acid) derivatives and chitosans show a variety of favorable properties as possible new classes of absorption enhancers:

> They are not likely to be absorbed in view of their very high molecular weight and therefore not expected to display systemic toxicity.
> They protect the peptide drug from proteolytic degradation [poly-(acrylates)].
> They improve peptide transport across the epithelial barrier [penetration enhancement by both poly(acrylates) and chitosans].
> They intensify the contact between dosage form and the site of absorption, thereby reducing the luminal diffusion pathway of the drug (bioadhesion).

VI. FUTURE PERSPECTIVES

Among the vasopressins, the octapeptide analogue desmopressin was introduced on the market as the first perorally available peptide drug. However, reported bioavailabilities of about 2% are still low and one would expect huge variations in plasma levels. Nevertheless, the variations seem to be tolerable, because manyfold higher plasma levels exceeding the minimal therapeutic plasma concentration did not show any toxic or other undesired side effects. Furthermore, at time of preparation of this chapter a peroral salmon calcitonin formulation (Macritonin) was reported to yield therapeutic blood levels in clinical phase III studies (Cortecs, 1997). These events show that peroral peptide drug delivery is no longer an impossible task, and the future goals of pharmaceutical scientists are now (a) to include other peptide drugs in peroral delivery systems and (b) to improve the predictability and reproducibility of drug absorption as well as the bioavailability of peptide drugs. The challenge to improve peroral peptide drug absorption has been successfully met by using highly accepted polymers such as cross-linked poly(acrylates) and chitosans as new classes of absorption enhancers. Knowing more about the underlying mechanisms of action of these polymers will lead to systematic modifications of their chemical structure in order to improve their enhancing effects on peptide drug absorption. One possibility may be to increase the negative charge density of the dissociated and hydrated poly(acrylates) to display a higher affinity for bivalent cations such as Ca^{2+} and Zn^{2+}. Another approach may be to increase the basic properties of chitosans and to characterize in more detail their mechanisms for affecting the tightness of intercellular junctions. For incorporation in solid dosage forms, these new polymers should display sufficient swelling rates to exert an optimal effect on the intestinal barriers against peptide drug absorption. In combination with their mucoadhesive properties that allow localization of the absorption-enhancing effects in the gut, these multifunctional properties of the polymers will provide a highly promising tool for mastering the peroral delivery of peptide drugs in the near future.

REFERENCES

Akiyama Y, Luessen HL, de Boer AG, Verhoef JC, Junginger HE. 1996a. Int J Pharm 136:155.

Akiyama Y, Luessen HL, de Boer AG, Verhoef JC, Junginger HE. 1996b. Int J Pharm 138:13.

Al-Achi A, Greenwood R, Walker B. 1992. Introduodenal administration of biocarrier insulin. Proceedings of 19th International Symposium on Controlled Release of Bioactive Materials, Vol 20, p 526.

Allen MP, Yamada AH, Carpenter FH. 1983. Biochemistry 22:3778.

Armentrout RW, Doolittle RF. 1969. Arch Biochem Biophys 132:80.

Artursson P, Lindmark T, Davis SS, Illum L. 1994. Pharm Res 11:1358.

Bai JPF. 1994. Pharm Res 11:897.

Baluom M, Friedman DI, Rubinstein A. 1997. Int J Pharm 154:235.

Bartunik HD, Summers LJ, Bartsch HH. 1989. J Mol Biol 210:813.

Bernkop-Schnürch A, Paikl C, Valenta C. 1997a. Pharm Res 14:917.

Bernkop-Schnürch A, Schwarz GH, Kratzel M. 1997b. J Controlled Release 47:113.

Biber J, Stiegler B, Haase W, Murer H. 1981. Biochim Biophys Acta 647:169.

Bohley P, Seglen PO. 1992. Experientia 48:151.

Borchard G, Luessen HL, De Boer AG, Verhoef JC, Lehr C-M, Junginger HE. 1996. J Controlled Release 39:131.

Bruins J, Kumar A, Schneider-Helmert D. 1990. Neuropsychobiology 23:82.

Caramella C, Rossi S, Bonferoni MC, La Manna A. 1993. Rheological characterization of some bioadhesive systems. Proceedings of International Symposium on Controlled Release of Bioactive Materials 23:240.

Chan WW-C, Pfuetzner RA. 1993. Eur J Biochem 218:529.

Cleland JL, Powell MF, Shire SJ. 1993. Crit Rev Ther Drug Carrier Syst 10:307.

Coleman R, Kan KS. 1990. Biochem J 266:622.

Constantinides PP. 1995. Pharm Res 12:1561.

Cortecs. 1997. Cortecs to submit Macritonin in EC. Scrip 2258:17.

Craig DQM, Tamburic S, Buckton G, Newton JM. 1994. J Controlled Release 30:213.

Delaage M, Lazdunski M. 1967. Biochem Biophys Res Commun 28:390.

de Leeuw BJ, Luessen HL, Kotzé AF, De Boer AG, Verhoef JC, Junginger HE. 1996. Protection against enzymatic degradation and transport enhancement of a peptide drug by mucoadhesive excipients. Proceedings of International Symposium on Controlled Release of Bioactive Materials, Vol 22, p 845.

DiGregorio M, Pickering DS, Chan WW-S. 1988. Biochemistry 27:3613.

Duffey ME, Hainau B, Ho S, Bentzel CJ. 1981. Nature 294:451.

Engel RH, Riggi SJ, Fahrenbach MJ. 1968. Nature 219:856.

Epstein DA, Longenecker JP. 1988. Crit Rev Ther Drug Carrier Syst 5:99.

Esposito P, Colombo I, Lovrecich M. 1994. Biomaterials 15:177.

Gabel D, Kasche V. 1973. Acta Chem Scand 27:1971.

Gonzalez-Mariscal L, Contreras RG, Bolivár JJ, Ponce A, Cháez de Ramirez B, Cereijido M. 1990. Am J Physiol 259:C978.

Gu J-M, Robinson JR, Leung S-HS. 1988. Crit Rev Ther Drug Carrier Syst 5:21.

Guasch A, Coll M, Avilés FX, Huber R. 1992. J Mol Biol 224:141.

Gurny R, Junginger HE, eds. 1990. Bioadhesion—Possibilities and Future Trends. Paperback APV 25. Stuttgart: Wissenschaftliche Verlagsgesellschaft.

Gurny R, Meyer JM, Peppas NA. 1984. Biomaterials 5:336.

Harris AS. 1993. J Drug Targeting 1:116.

Harris D, Fell JT, Sharma HL, Taylor DC. 1990a. J Controlled Release 12:45.

Harris D, Fell JT, Taylor DC, Lynch J, Sharma HL. 1990b. J Controlled Release 12:55.

Hasilik A. 1992. Experientia 48:130.

Hattewell J, Lynch S, Fox R, Williamson I, Skelton-Stroud P, Mackay M. 1994. Int J Pharm 101:115.

Hedstrom L, Perona JJ, Rutter WJ. 1994. Biochemistry 33:8757.

Hidalgo I, Raub T, Borchardt RT. 1989. Gastroenterology 96:736.

Hilgers AR, Conradi RA, Burton PS. 1990. Pharm Res 9:902.

Himmelhoch SR, 1969. Arch Biochem Biophys 134:597.

Hooper NM. 1994. FEBS Lett 354:1.

Hosny EA. 1993. Int J Pharm 98:235.

Hutton DA, Pearson JP, Allen A, Foster SNE. 1990. Clin Sci 78:265.

Ilan E, Amselem S, Weisspapir M, Schwarz J, Yogev A, Zawoznik E, Friedman D. 1996. Pharm Res 13:1083.

Illum L, Farraj LF, Davis SS. 1994. Pharm Res. 11:1186.

Jabbari E, Wiesniewski N, Peppas NA. 1993. J Controlled Release 26:99.

Jensen-Pippo KE, Whitcomb KL, DePrince RB, Ralph L, Habberfield AD. 1996. Pharm Res 13:102.

Junginger HE. 1990. Acta Pharm Technol 36:110.

Junginger HE, Lehr C-M, Bouwstra JA, Tukker JJ, Verhoef JC. 1990. In: Gurny R, Junginger HE, eds. Bioadhesion—Possibilities and Future Trends. Paperback APV 25. Stuttgart: Wissenschaftliche Verlagsgesellschaft, p 117.

Knapczyk J. 1992. Int J Pharm 93:233.

Koshla R, Davis SS. 1987. J Pharm Pharmacol 39:47.

Kretsinger RH, 1976. Annu Rev Biochem 45:239.

Kristl J, Šmid-Korbar J, Štruc E, Schara M, Rupprecht H. 1993. Int J Pharm 99:13.

Langguth P, Bohner V, Biber J, Merkle HP. 1994a. J Pharm Pharmacol 46:34.

Langguth P, Merkle HP, Amidon GL. 1994b. Pharm Res 11:528.

Lee VHL. 1991. Peptide and Protein Drug Delivery. New York: Marcel Dekker.

Lee VHL, Traver RD, Taub ME. 1991a. In: Lee VHL, ed. Peptide and Protein Drug Delivery. New York: Marcel Dekker, p 303.

Lee VHL, Yamamoto A, Kompella UB. 1991b. Crit Rev Ther Drug Carrier Syst 8: 91.

Lehr C-M. 1991. Bioadhesive drug delivery systems for oral application. Thesis, Leiden University.

Lehr C-M. 1994. Crit Rev Ther Drug Carrier Syst 11:119.

Lehr C-M, Boddé HE, Bouwstra JA, Junginger HE. 1993. Eur J Pharm Sci 1:19.

Lehr C-M, Bouwstra JA, Kok W, De Boer AG, Tukker JJ, Verhoef JC, Breimer DD, Junginger HE. 1992a. J Pharm Pharmacol 4:402.

Lehr C-M, Bouwstra JA, Schacht EH, Junginger HE. 1992b. Int J Pharm 78:43.

Lehr C-M, Poelma FGJ, Junginger HE, Tukker JJ. 1990. Int J Pharm 70:235.

Lenaerts VM, Gurny R. 1989. Bioadhesive Drug Delivery Systems. Boca Raton, FL: CRC Press.

Leroux J-C, Cozens RM, Roesel JL, Galli B, Doelker E, Gurny R. 1996. Pharm Res 13:485.

Longer MA, Ch'ng HS, Robinson JR. 1985. J Pharm Sci 74:406.

Lowe PJ, Temple CS. 1994. J Pharm Pharmacol 46:547.

Luessen HL, Bohner V, Pérard D, Langguth P, Verhoef JC, De Boer, AG, Merkle HP, Junginger HE. 1996a. Int J Pharm 141:39.

Luessen HL, de Leeuw BJ, Langemeijer MWE, de Boer AG, Verhoef JC, Junginger HE. 1996c. Pharm Res 13:1668.

Luessen HL, de Leeuw BJ, Pérard D, Lehr C-M, de Boer AG, Verhoef JC, Junginger HE. 1996b. Eur J Pharm Sci 4:117.

Luessen HL, Lehr C-M, Rentel C-O, Noach ABJ, de Boer AG, Verhoef JC, Junginger HE. 1994. J Controlled Release 29:329.

Luessen HL, Rentel C-O, Kotzé A, Lehr C-M, de Boer AG, Verhoef JC, Junginger HE. 1997. J Controlled Release 45:15.

Luessen HL, Verhoef JC, Borchard G, Lehr C-M, De Boer AG, Junginger HE. 1995. Pharm Res 12:1293.

Lundin S, Pantzar N, Hedinand L, Weström BR. 1990. Int J Pharm 59:263.

MacAdam A. 1993. Adv Drug Delivery Rev 11:201.

Madara JL, Barenberg B, Carlson S. 1986. J Cell Biol 102:2125.

Maggi L, Carena E, Torre ML, Giunchedi P, Conte U. 1994. STP Pharm Sci 4:343.

Matthes I, Nimmerfall F, Sucker H. 1992. Pharmazie 47:609.

Mesiha M, Sidhom M. 1995. Int J Pharm 114:137.

Michel C, Aprahamian L, Defontaine P, Couvreur P, Devissaguet JP. 1991. J Pharm Pharmacol 43:1.

Mizuma T, Ohta K, Koyanagi A, Awazu S. 1996. J Pharm Sci 85:854.

Morishita I, Morishita M, Takayama K, Machida Y, Nagai T. 1992b. Int J Pharm 78:9.

Morishita I, Morishita M, Takayama K, Machida Y, Nagai T. 1993. Int J Pharm 91: 29.

Morishita M, Morishita I, Takayama K, Machida Y, Nagai T. 1992a. Int J Pharm 78:1.

Mortazavi SA, Smart JD. 1994. J Pharm Pharmacol 46:86.

Noach ABJ, Kurosaki Y, Blom-Roosemalen MCM, de Boer AG, Breimer DD. 1993. Int J Pharm 90:229.

Oppenheim RC, Steward NF, Gordon L, Patel HM. 1982. Drug Dev Ind Pharm 8: 531.

Patel HM, Harding NGL, Logue F, Kesson C, MacCuish AC, MacKenzie JC, Ryman BE, Scobie I. (1978). Biochem Soc Trans 6:784.

Patel HM, Stevenson RW, Parsons JA, Ryman BE. 1982. Biochim Biophys Acta 716:188.

Péres-Marcos M, Martínez-Pacheco R, Gómez-Amoza JL, Souto C, Concheiro A, Rowe RC. 1993. Int J Pharm 100:207.

Pitelka DR, Taggart BN, Hamamoto ST. 1983. J Cell Biol 96:613.

Pritchard K, Lansley AB, Martin GP, Helliwell M, Marriott C, Benedetti LM. 1996. Int J Pharm 129:137.

Rentel C-O, Lehr C-M, Bouwstra JA, Luessen HL, Junginger HE. 1993. Enhanced peptide absorption by the mucoadhesive polymers polycarbophil and chitosan. Proc Int Symp Controlled Release Bioact Mater 23:446.

Rillosi M, Buckton G. 1995. Pharm Res 12:669.

Rubinstein A, Tirosh B, Baluom M, Nassar T, David A, Radai R, Gliko-Kabir I, Friedmann M. 1997. J Controlled Release 46:59.

Rypniewski WR, Perrakis A, Vorgias CE, Wilson KS. 1994. Protein Eng 7:57.

Saffran M, Breda C, Kumar GS, Neckers DC. 1988. J Pharm Sci 77:33.

Saffran M, Kumar GS, Neckers DC, Peña J, Jones RH, Field JB. 1990. Biochem Soc Trans 18:752.

Salvesen G, Nagase H. 1989. In: Benyon RJ, Bond JS, eds. Proteolytic Enzymes— A Practical Approach. Oxford: IRL Press, p 83.

Sandow J. 1989. In: Schindler AE, Schweppe KW, eds. Endometriose—Neue Therapiemöglichkeiten durch Buserelin. Berlin: Walter de Gryter, p 23.

Sanzgiri YD, Topp EM, Benedetti L, Stella VJ. 1994. Int J Pharm 107:91.

Sarciaux JM, Acar L, Sado PA. 1995. Int J Pharm 120:127.

Schilling RJ, Mitra A. 1991. Pharm Res 8:721.

Schipper NGM, Olsson S, Hoogstraate JA, DeBoer AG, Vårum KM, Artursson P. 1997. Pharm Res 14:923.

Schnittler HJ, Wilke A, Gress T, Suttorp N, Drenckhahn D. 1990. J Physiol 431: 379.

Seglen PO, Bohley P. 1992. Experientia 48:158.

Shao Z, Li Y, Chermak T, Mitra AK. 1994. Pharm Res 11:1174.

Shichiri M, Kawamori R, Yoshida M, Etani N, Hishi M, Izumi K, Shigeta Y, Hiroshi A. 1975. Diabetes 24:971.

Shichiri M, Shimizu Y, Yosida Y, Kawamore R, Fukuchi M, Shigeta Y, Abe H. 1974. Diabetologia 10:317.

Smart JD, Kellaway IW, Worthington HEC. 1984. J Pharm Pharmacol 36:295.

Strous GJ, Dekker J. 1992. Crit Rev Biochem Mol Biol 27:57.

Szewczuk A, Mulczyk M. 1969. Eur J Biochem 8:63.

Takeuchi H, Yamamoto H, Niwa T, Hino T, Kawashima Y 1996. Pharm Res 13: 896.

Taylor A. 1993. Trends Pharmacol Sci 18:167.

Ter Braak EWTM, Bianci R, Erkelens DW. 1993. Diabetes 42:207A.

Touitou E. 1992. J Controlled Release 29:139.

Touitou E, Rubinstein A. 1986. Int J Pharm 30:95.

Tsukada H, Blow DM. 1985. J Mol Biol 184:703.

Vallee BL, Rupley JA, Coombs TL, Neurath H. 1960. J Biol Chem 235:64.

van Wart HE, Lin SH. 1981. Biochemistry 20:5682.

Verhoef J, Van den Wildenberg HM, van Nispen JW. 1986. J Endocrinol 110:557.

Verhoef JC, Boddé HE, de Boer AG, Bouwstra JA, Junginger HE, Merkus FWHM, Breimer DD. 1990. Eur J Drug Metab Pharmacokinet 15:83.

Yamamoto A, Tanigushi T, Rikyuu K, Tsuji T, Fujita T, Murakami M, Muranishi S. 1994. Pharm Res 11:1496.

Yoshida H, Lehr C-M, Kok W, Junginger HE, Verhoef JC, Bouwstra JA. 1992. J Controlled Release 21:145.

13

Chitosan and Chitosan Derivatives as Absorption Enhancers for Peptide Drugs Across Mucosal Epithelia

A. F. Kotzé
Potchefstroom University for Christian Higher Education, Potchefstroom, Republic of South Africa

Henrik L. Luessen*
LTS Lohmann Therapie-Systeme GmbH, Andernach, Germany

M. Thanou, J. Coos Verhoef, A. (Bert) G. de Boer, and H. E. Junginger
Leiden/Amsterdam Center for Drug Research, Leiden University, Leiden, The Netherlands

Claus-Michael Lehr
Saarland University, Saarbrücken, Germany

I. INTRODUCTION

Enormous advances in drug design and biotechnology have made it possible to produce significant quantities of several new classes of drugs such as peptide and protein drugs. These substances could be seen as mere templates for more advanced peptidomimetic agents that will have a major impact on the treatment of medical disorders in modern humans, now and in the future. These new compounds are mostly indicated for chronic administration, and

---

Current affiliation: OctoPlus, B.V., Leiden, The Netherlands.

suitable delivery systems have to be developed to utilize the full benefit of these technologically advanced drugs. However, the progress in developing new drug entities is currently not matched by that in developing effective delivery systems for this new generation of drugs. This is currently a major challenge for the pharmaceutical scientist and is studied with great effort in both academia and industry. Furthermore, the emphasis in dosage form development has also shifted toward the design of more effective delivery systems for both new generation drugs and older drugs with regard to optimal therapeutic responses, controlled and site-specific release of the active ingredients, compliance of patients, and cost-effectiveness.

For most therapeutic agents, administration via a nonparenteral route is still the preferred choice. Although the oral route is regarded as the most convenient route of drug administration (Borchard et al., 1996), alternative routes such as the nasal, buccal, pulmonal, and rectal routes are becoming increasingly important in dosage form development (Lee, 1991). Successful peptide drug delivery has long been regarded as almost impossible (Borchard et al., 1996), but in recent years significant progress has been made in identifying agents that increase the absorption of hydrophilic drugs that are poorly absorbed when administered perorally (Artursson and Magnusson, 1990; Hurni et al., 1993; Hochman and Artursson, 1994; Boulenc et al., 1995; Kriwet and Kissel, 1996; Kotzé et al., 1996, 1997a–b, 1998; Luessen et al., 1996a; Tomita et al., 1996). However, particular difficulties are met in designing effective delivery systems by any of these routes, especially for peptide drugs. Apart from physical and chemical instabilities and susceptibility to enzymatic degradation, which negatively influence absorption kinetics, poor absorption at the site of administration is also a major limiting factor in the development of delivery systems for peptide drugs (Lee, 1991).

Permeation across any epithelial membrane is a fundamental step for absorption and systemic availability. The permeation process can be passive, active, or carrier-mediated and transcellular or paracellular. Among other factors, lipophilicity and size considerations are probably the most important factors for transmembranous transport (Lipka et al., 1996). Therapeutic compounds are absorbed by two routes, namely the transcellular and paracellular transport pathways. Most drugs are small-molecule pharmaceuticals and lipophilic in nature. They partition rapidly from the luminal fluid into the cell membranes (transcellular transport pathway) and are normally absorbed in sufficient amounts from the gastrointestinal tract to obtain the necessary therapeutic response (Madara and Trier, 1987). Peptide drugs are mostly large-molecule pharmaceuticals and are highly hydrophilic in nature. They do not partition to a large extent into the cell membranes and are consequently excluded from the transcellular transport pathway (Fix, 1987; Madara and Trier, 1987). Their size and polarity frequently exclude normal passive diffusion across cell membranes, and therefore administration of these classes of drugs mostly results in inadequate absorption and very poor, if any, systemic availability (Lee, 1991). The absorption of peptide drugs is for the most part limited to the alternative paracellular pathway (Hochman and Artursson, 1994), but the entry of these molecules through the paracellular route is nearly completely restricted by the tight junctions (Gumbiner, 1987). Furthermore, the paracellular route occupies only a very small surface area compared with the transcellular route (Madara, 1989).

One approach to overcoming the restriction of the paracellular transport pathway is the coadministration of absorption-enhancing agents (Fix, 1987) or agents that regulate the integrity of the tight junctions (Muranishi, 1990). Many compounds that enhance drug transport across the cellular monolayer have been studied, but most of them proved to have toxic effects and in general do not enhance drug transport by interaction with the tight junctions (Muranishi, 1990; Tomita et al., 1996). In the past decade, numerous investigations highlighted the potential of these agents in controlled drug delivery (for review see Swenson and Curatolo, 1992). Significant progress has been made in identifying agents that increase the absorption of drugs through the paracellular transport pathway (Borchard et al., 1996; Luessen et al., 1997). These absorption enhancers can be divided into two general classes: calcium chelators such as EDTA and surfactants such as bile salts and palmitoylcarnitine (Hochmann and Artursson, 1994). Calcium chelators do not act directly on the tight junctions but rather induce changes in the cells such as disruption of actin filaments (Citi, 1992). Surfactants are rather lytic in nature (Hochman and Artursson, 1994), and concern regarding the toxicity of these compounds has generally excluded them for pharmaceutical use. The ideal enhancer should be nontoxic and act in a reversible way on the tight junctions. Furthermore, it should be effective in the low pH of the stomach, the neutral pH of the small intestine, as well as the basic pH of the colon. Reliability in site-specific drug release is an additional factor that could greatly improve the potential use of such an absorption enhancer.

Chitosan, a mucoadhesive polymer, has attracted a great deal of attention as a potential absorption enhancer across mucosal epithelia. Chitosan is a linear polysaccharide derived by N-deacetylation of the natural polymer chitin, which is the second most abundant naturally occurring polymer in nature. Chitosan consists of linear 1−4 linked 2-acetamido-2-deoxy-β-D-glucopyranose and 2-amino-β-D-glucopyranose units (Vårum et al., 1991). Chitosan has already been approved as a food additive in Japan and is believed to be nontoxic (Hirano et al., 1998). Chitosan is a polycationic polymer at acidic pH values (Muzzarelli, 1973) with numerous applications in the food, agricultural, and cosmetic industries. Advantages of this polymer include high availability, low cost, high biocompatibility, biodegradability, and ease of chemical modification (Arai et al., 1968; Hirano et al., 1988; Skaugrud, 1989; Li et al., 1992). Although chitosan has been used for a long time in these industries, it was only in recent years that its potential application in the pharmaceutical field has been highlighted by several scientists. This chapter will focus on chitosan and chitosan derivatives and their potential use in novel dosage form design.

II. PHARMACEUTICAL APPLICATIONS OF CHITOSAN

Several studies have highlighted the potential use of chitosan in various dosage forms. Pharmaceutical applications include use as a tablet binder (Upadrashta et al., 1992) and a disintegrant (Ritthidej et al., 1994). Chitosan has gel-forming properties in the low pH range and is used as a drug carrier in hydrocolloids and gel formulations (Knapczyk, 1993; Kristl et al., 1993). Chitosan membranes, obtained by controlled cross-linking with glutaraldehyde, were used to regulate the release of propranolol hydrochloride in membrane permeation–controlled drug delivery systems (Thacharodi and Rao, 1995). Chitosan is also applied as a constituent in polymeric matrix systems, microspheres, and microcapsules for the sustained release of water-soluble drugs (Nigalaye et al., 1990; Chithambara et al., 1992; Polk et al., 1994; Berthold et al., 1996). From these studies it is clear that chitosan has a wide field of application in pharmaceutical technology. Perhaps more significant is the use of chitosan as a mucoadhesive excipient and as an absorption enhancer for the increased absorption of several hydrophilic drugs. This will be discussed in more detail in the next sections.

A. Mucoadhesive Properties of Chitosan

The mucoadhesive properties of chitosan were first evaluated by Lehr et al. (1992). Table 1 compares the mucoadhesive character of chitosan with that of several other mucoadhesive compounds. According to Peppas and Burim (1985), a number of polymer characteristics are necessary for mucoadhesion: (a) strong hydrogen-bonding groups (—OH, —COOH), (b) strong anionic charges, (c) high molecular weight, (d) sufficient chain flexibility, and (e) surface energy properties favoring spreading onto mucus. However, chitosan is a polycationic polymer and does not have any anionic charge. Instead, a positively charged hydrogel is formed in acidic environments that could develop additional molecular attraction forces by electro-

Table 1 Survey of Mucoadhesive Properties of Various Polymers

Polymers	Force of detachment[a] (mN/cm^2)
Cationic polymers	
Chitosan (Wella "low viscosity")	3.9 (1.2)
Chitosan (Wella "high viscosity")	6.7 (0.7)
Chitosan (Dr. Knapczyk)	5.7 (1.1)
Daichitosan H	8.0 (5.7)
Daichitosan VH	9.5 (2.5)
Sea Cure 240	4.1 (2.9)
Sea Cure 210+	9.5 (2.5)
Chitosan (Sigma)	6.6 (3.0)
Polycarbophil/Diachitosan VH blend	11.9 (2.5)
DEAE-dextran	0
Aminodextran	0
Nonionic polymers	
Scleroglucan	2.8 (2.8)
HE-starch	0.6 (0.8)
HPC	0
Anionic polymers	
CMC (low viscosity)	1.8 (1.1)
CMC (medium viscosity)	0.3 (0.3)
CMC (high viscosity)	1.3 (1.0)
Pectin	0
Xanthan gum	0
Polycarbophil	17.6 (3.6)

[a]Indicated is the mean (SD) force of detachment in two to six measurements.
Abbreviations: CMC, carboxymethylcellulose; DEAE, diethylaminoethyl; HE, hydroxyethyl; HPC, hydroxypropylcellulose.
From Lehr et al. (1992).

static interactions with negatively charged mucosal surfaces or the negatively charged sialic end groups of the mucus network. It was also shown that the mucoadhesive properties of chitosan are dependent on its molecular weight, with the best results obtained at the higher molecular weights (Lehr et al., 1992).

Miyazaki et al. (1988) reported prolonged absorption of indomethacin in rabbits after oral administration of granules prepared from a 1:2 mixture of drug and chitosan. They explained their result by prolonged gastric retention of floating granules but did not discuss possible bioadhesive properties of their formulation. The mucoadhesive properties of chitosan have led to its use as a coating material for multilamellar liposomes (Takeuchi et al., 1994) and application in formulations aimed at controlled drug delivery to specific sections of the gastrointestinal tract (Geary and Schlameus, 1993) and across other mucosal surfaces such as the nasal, buccal, and vaginal epithelia (Knapczyk, 1992; Illum et al., 1994).

B. Chitosan as an Absorption Enhancer for Hydrophilic Drugs

Because of its mucoadhesive character (Lehr et al., 1992) and favorable toxicological properties, chitosan has been studied as a potential absorption enhancer across intestinal epithelia. It has been shown that chitosan glutamate is able to reduce the transepithelial electrical resistance (TEER) in vitro of a cultured intestinal epithelial cell line (Caco-2) (Borchard et al., 1996). Chitosan glutamate was able to increase the transport of hydrophilic molecules such as [^{14}C]mannitol [molecular weight (MW) 182.2] and a fluorescein-dextran (MW 4400) significantly in Caco-2 cell monolayers (Artursson et al., 1994; Borchard et al., 1996; Schipper et al., 1996). Similarly, the transport of the peptide drug 9-desglycinamide, 8-arginine vasopressin (DGAVP, MW 1412) was increased markedly after coadministration with chitosan glutamate in Caco-2 cell monolayers (Luessen et al., 1997). Chitosan salts such as chitosan glutamate and chitosan hydrochloride have been used in vivo as absorption enhancers for peptide drugs. The nasal application of insulin with chitosan glutamate led to a significant reduction in blood glucose levels of rats and sheep (Illum et al., 1994), and the intraduodenal application of buserelin (MW 1299.5) and chitosan hydrochloride in a gel formulation increased the absolute bioavailability of buserelin from 0.1 ± 0.1 to 5.1 ± 1.5% (Luessen et al., 1996a). These increases in absorption could be attributed to the effect of chitosan on the integrity of the epithelial tight junctions. Tight junctions play a crucial part in maintaining the selective barrier function of cell membranes and in sealing cells together to form a continuous cell layer through which even small molecules cannot penetrate. However, tight junctions are permeable to water, electrolytes, and other charged or uncharged molecules up to a certain size (Madara, 1989; Wilson and Washington, 1989). Tight junctions are known to respond to changes in calcium concentrations, cyclic AMP (cAMP), osmolarity, pH, and the status of the cytoskeleton (Cereijido et al., 1993).

It has been proposed that chitosan salts open the tight junctions in a concentration- and pH-dependent way to allow paracellular transport of large hydrophilic compounds. The increase in the transport of these compounds could be attributed to an interaction of a positively charged amino group on the C-2 position of chitosan with negatively charged sites on the cell membranes and tight junctions of the mucosal epithelial cells to allow opening of the tight junctions. It has been shown that chitosan glutamate can induce changes in the F-actin distribution (Artursson et al., 1994). It is also known that pharmacological agents that interact with cytoskeletal F-actin simultaneously increase the paracellular permeability (Meza et al., 1982). This is in agreement with the hypothesis that F-actin is directly or indirectly associated with the proteins in the tight junctions such as ZO-1 (Madara, 1987). Schipper et al. (1997) have shown that chitosan induces a redistribution of cytoskeletal F-actin and the tight junction protein ZO-1. Confocal laser scanning microscopy has confirmed that chitosan is able to open the tight junctions to allow the paracellular transport of large hydrophilic compounds (Borchard et al., 1996; Schipper et al., 1997). Mucoadhesion may play an additional role in this process by increasing the residence time of the drugs on the cell surfaces.

III. THE NEED FOR CHITOSAN DERIVATIVES

In all the studies that have been mentioned, absorption enhancement was found only in acidic environments in which the pH was less or of the order of the PK_a value of chitosan (5.5 to 6.5). This suggests that the charge density of chitosan is an important factor for the enhancement of mucosal transport (Schipper et al., (1996). Figures 1 and 2 show the decrease in the TEER of Caco-2 cell monolayers during incubation with chitosan glutamate and chitosan hydrochloride at a pH of 6.20 and 7.40, respectively. In Table 2 apparent permeability coefficients (P_{app}) and transport enhancement ratios (R) for the radioactive marker [^{14}C]mannitol are given after incubation with these chitosan salts at a pH of 6.20 and 7.40.

Measurement of the TEER is believed to give a good indication of the tightness of the junctions between cells and has been used in previous studies to predict the paracellular transport of hydrophilic compounds (Boulenc et al., 1995; Borchard et al., 1996; Schipper et al., 1997). Incubation on the apical side of the monolayers with both polymers, at a pH of 6.20, led to an immediate and pronounced reduction in TEER values compared with the control group. The reduction in TEER 20 minutes after the start of the experiment, with 0.5% concentrations of the polymers, was in the following order: chitosan hydrochloride (80 ± 3% reduction) > chitosan glutamate (65 ± 2% reduction). Prolonged incubation, up to 2 hours, resulted in only a slight decrease in TEER values, compared with the initial reductions in TEER measured after 20 minutes, which were 89 ± 2% (chitosan hydrochloride) and 69 ± 2% (chitosan glutamate). Increasing the polymer con-

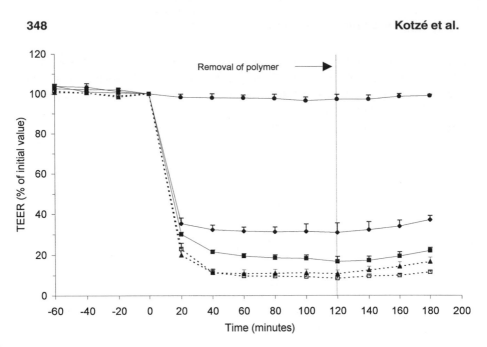

Figure 1 Effect of chitosan glutamate and chitosan hydrochloride on the TEER of Caco-2 cell monolayers at a pH of 6.20. Each point represents the mean ± SD of three experiments. Control (●); chitosan glutamate 0.5% (◆); chitosan glutamate 1.5% (■); chitosan hydrochloride 0.5% (▲); chitosan hydrochloride 1.5% (□); dotted line represents start of reversibility experiment.

centration to 1.5% did not result in any major decreases in TEER compared with the reduction measured with 0.5% concentrations of these polymers.

The effect on TEER values seems to be saturable, because no major differences were found between the different concentrations of the chitosan salts used in this experiment. This is in agreement with the findings of a previous study, in which plateau levels were reached between 0.25 and 0.5% concentrations of chitosan glutamate (Artursson et al., 1994). After removal of the polymer solutions, the cell monolayers started to recover slowly and a very slight increase in resistance toward the initial values was found. Because of the high viscosity and mucoadhesive properties of these polymer solutions, it was almost impossible to remove all the polymer molecules from the cell surfaces and therefore the reversibility observed was only gradual. At a pH of 7.40 both chitosan salts do not form clear solutions, and apical incubation with these dispersions did not result in any decreases in the TEER values compared to the control group, as evident from Fig. 2.

In agreement with the results of the TEER experiments, no increase in the transport of [^{14}C]mannitol was found at a pH of 7.40 in the presence

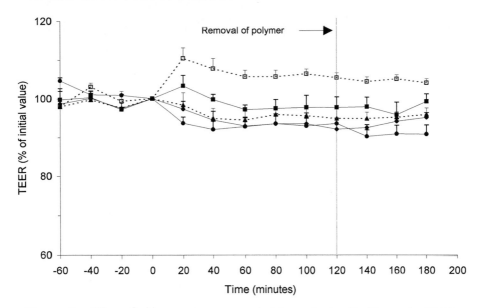

Figure 2 Effect of chitosan glutamate and chitosan hydrochloride on the TEER of Caco-2 cell monolayers at a pH of 7.40. Each point represents the mean ± SD of three experiments. Control (●); chitosan glutamate 0.5% (♦); chitosan glutamate 1.5% (■); chitosan hydrochloride 0.5% (▲); chitosan hydrochloride 1.5% (□); dotted line represents start of reversibility experiment.

of both chitosan salts. At a pH of 6.20, incubation with these polymers resulted in a marked accumulation of [14C]mannitol in the acceptor compartments. The cumulative amounts transported up to 4 hours after incubation with 0.5% concentrations of the polymers were 14.3 ± 1.0% (chitosan hydrochloride) and 8.9 ± 0.2% (chitosan glutamate) of the total dose applied. At 1.5% concentrations of the polymers no major increases in the transport of [14C]mannitol, compared with the transport at 0.5% concentrations, were found. The cumulative amounts transported were 17.4 ± 1.2% (chitosan hydrochloride) and 11.8 ± 1.1% (chitosan glutamate) of the total dose applied, and this represents 36-fold and 25-fold increases in the transport rate compared with the control groups, as indicated by the P_{app} values in Table 2.

These results clearly demonstrated that chitosan hydrochloride and chitosan glutamate are able to decrease the TEER of Caco-2 monolayers in an acidic environment to allow the paracellular transport of a hydrophilic marker, [14C]mannitol. It has been shown that a low pH in itself does not lead to changes in membrane permeability (Anderberg and Artursson, 1993).

Table 2 Effect of Chitosan Hydrochloride and Chitosan Glutamate on the Permeability of $[^{14}C]$Mannitol Across Caco-2 Cell Monolayers at pH 7.40 and 6.20

pH	Polymer concentration (% w/v)	Chitosan Hydrochloride		Chitosan Glutamate	
		$P_{app} \times 10^{-7a}$ (cm/s)	R^a	$P_{app} \times 10^{-7}$ (cm/s)	R
7.40	Control	0.45 + 0.04	1.0	0.45 + 0.04	1.0
	0.5	0.53 + 0.08	1.2	0.55 ± 0.04	1.2
	1.5	0.39 + 0.02	0.9	0.44 + 0.03	1.0
6.20	Control	0.72 ± 0.08	1.0	0.72 ± 0.08	1.0
	0.5	23.28 ± 1.00	32.3	14.17 ± 0.45	19.7
	1.5	26.16 ± 1.86	36.3	18.29 ± 1.53	25.4

[a] P_{app}, apparent permeability coefficient; R, transport enhancement ratio.

Therefore, the reduction in TEER could be attributed to an interaction of these polymers with the cell surfaces or the tight junctions. Chitosan is a weak base, and at a pH of 6.20 the amino group on the C-2 position is protonated and positively charged. The chitosan molecule exists in an un-coiled configuration because of positive charges on the repeating units of the polymer backbone. It has been proposed that chitosan acts on the neg-atively charged sites at the cell surfaces and tight junctions (Artursson et al., 1994; Borchard et al., 1996). Considering the size of a chitosan molecule and the relatively small surface area of the tight junctions compared with the surface area of the cell membranes (Madara, 1989), it probably acts mainly on the cell surface. It has been shown that chitosan is able to induce changes in the F-actin distribution (Artursson et al., 1994; Schipper et al., 1997), and the theory exists that F-actin is directly or indirectly associated with the proteins in the tight junctions (Madara, 1987). Chitosan most prob-ably allows the paracellular transport of hydrophilic compounds by an in-direct mechanism, whereby the integrity of the tight junctions is altered by changes in intracellular F-actin that are associated with proteins such as ZO-1 in the tight junctions.

At a neutral pH of 7.40, both chitosan salts are insoluble and precip-itate from solution. The molecules most likely exist in a coiled configuration without any protonated amino groups; therefore, chitosan proves to be in-effective as an absorption enhancer at this pH value. Chitosan is a weak base and requires a certain amount of acid to transform the glucosamine units into the positively charged water-soluble form. In neutral and basic environments the chitosan molecules lose their charge and precipitate from solution. Under these conditions chitosan will be ineffective as an absorption enhancer, thus limiting its use in neutral and basic environments.

Most macromolecular pharmaceuticals such as peptide drugs are in-dicated for chronic administration, and therefore possibilities for the poten-tial use of chitosan in the more basic environment of the large intestine and colon are limited. In this regard chitosan derivatives with different physi-cochemical properties, especially water solubility at neutral and basic pH values, will be of particular interest. Our hypothesis is that polymers such as unmodified chitosan with a primary amino group may not be the optimal ones but that polymers or derivatives with different substituents, different basicities, or different charged densities will have the same or even increased efficacy in opening the tight junctions. Chitosan derivatives with different physicochemical properties, especially water solubility at neutral and basic pH values, might prove to be useful for absorption enhancement in the more alkaline environment of the large intestine, i.e., colon and rectum. The pH in the nasal cavity is of the order of 7.40, and therefore possibilities also exist for the use of such chitosan derivatives as nasal absorption enhancers.

Derivatives with retained mucoadhesive properties will have the additional advantage of increased residence time at any mucosal surface. Potential reactive groups make chitosan a versatile polymer, and chitosan derivatives with increased basicity and aqueous solubility such as derivatives with secondary, tertiary, or quaternary amino groups have been synthesized (Domard et al., 1986; Le Dung et al., 1994). These derivatives have not yet been evaluated for pharmaceutical applications. In particular, derivatives with quaternary amino groups such as N-trimethyl chitosan chloride might be useful as absorption enhancers in neutral and basic environments because they are water soluble and positively charged over a wide pH range.

IV. POTENTIAL USE OF N-TRIMETHYL CHITOSAN CHLORIDE AS AN ABSORPTION ENHANCER

A. Synthesis and Properties of N-*Trimethyl Chitosan Chloride (TMC)*

Figure 3 shows the chemical structure of chitosan and TMC, and Table 3 gives some of the physical and chemical properties of chitosan, chitosan hydrochloride, chitosan glutamate, and quaternized chitosan. TMC was synthesized from sieved fractions (<500 μm) of chitosan (degree of acetylation about 25%) based on the method of Domard et al. (1986). Briefly, the experimental conditions are reductive methylation of chitosan for 60 minutes with iodomethane in a strong basic environment at 60°C. The counterion (I^-) was exchanged to Cl^- by dissolving the quaternized polymer in a small quantity of water, followed by the addition of HCl in methanol.

The initial chitosan used to synthesize TMC was soluble only in acidic solutions, but after quaternization it became perfectly soluble in water. TMC solutions of 10% w/v could be prepared in either acidic or basic media. In general, TMC is soluble at a degree of quaternization as low as 10%, as determined from ^1H nuclear magnetic resonance (NMR) spectra. Chitosan hydrochloride and chitosan glutamate are soluble only at acidic pH values and do not dissolve at neutral and basic pH values. Even in acidic medium, solutions of 1.5% w/v proved difficult to prepare because of the high viscosity of such solutions. A pronounced decrease in the intrinsic viscosity of TMC, compared with the starting material, was observed, and this correlated well with the degradative reaction conditions in alkaline medium. A slight increase in pK_a from 5.5 to 6.0 was also found. The increase in solubility and basicity could be attributed to the replacement of the primary amino group on the C-2 position of chitosan with quaternary amino groups (Domard et al., 1986; Le Dung et al., 1994).

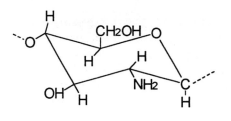

(A) Chitosan

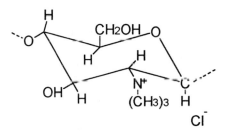

(B) N-trimethyl chitosan chloride

Figure 3 Chemical structure of (A) chitosan and (B) *N*-trimethyl chitosan chloride.

Polymers synthesized after a reaction time of 6 hours did not differ from polymers prepared after a reaction time of 60 minutes. Repeating the reaction several times under the same conditions, with the polymer recovered after an initial 60 minutes, gave polymers with degrees of quaternization of 40–60% (data not shown). However, at these degrees of quaternization ^{13}C and ^{1}H NMR spectra showed a high extent of methylation on the 3 and 6 hydroxyl groups of chitosan. Complete quaternization of chitosan will probably be difficult because of the presence of some acetyl groups (from chitin) and possible steric effects of the attached methyl groups on adjacent quaternary amino groups. Furthermore, ^{1}H NMR spectra showed that a high proportion of the amino groups were dimethylated and could still be protonated in acidic environments. However, even at very low degrees of quaternization this polymer was soluble at every pH and an increase in basicity was found. These results show that TMC is a derivative of chitosan with

Table 3 Physical and Chemical Properties of Chitosan Salts and *N*-Trimethyl Chitosan Chloride (TMC)

Polymer	Degree of Quaternization (%)	pK_a	Viscosity[a] (0.1% w/v) (mPa/s)	Solubility[b]			
				pH 4	pH 6	pH 7	pH 9
Chitosan	—	5.5 ± 0.0	19.57 ± 0.00	√	√	X	X
Chitosan hydrochloride	—	5.7 ± 0.1	4.46 ± 0.01	√	√	X	X
Chitosan glutamate	—	5.7 ± 0.1	5.49 ± 0.01	√	√	X	X
TMC	12.28	6.0 ± 0.1	1.27 ± 0.01	√	√	√	√

[a]Each value represents the mean ± SD of three experiments in 0.1% aqueous acetic acid.
[b]Solubility: √, soluble; X, insoluble for chitosan, chitosan hydrochloride, chitosan glutamate (1.5% w/v), and TMC (1–10% w/v). Measurements performed in water (pH adjusted with 0.1 M HCl or 0.1 M NaOH).

supcrior solubility and basicity, even at a low degree of quaternization, compared with other chitosan salts.

B. Comparison of the Effect of Different Chitosan Salts and *N*-Trimethyl Chitosan Chloride on the Permeability of Intestinal Epithelial (Caco-2) Cells

1.   Effect on the TEER of Intestinal Epithelial Cells

The effect of chitosan glutamate, chitosan hydrochloride, and TMC (degree of quaternization 12.28%) on the TEER of Caco-2 cell monolayers at a pH of 6.20 is summarized in Figs. 4 and 5. Incubation on the apical side of the monolayers with 0.25–1.5% w/v of the polymers resulted in a pronounced and immediate reduction in TEER values compared with the control group.

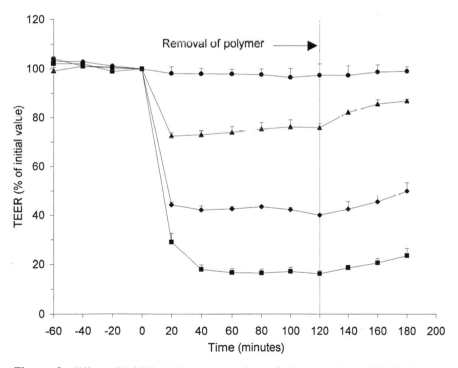

Figure 4   Effect of 0.25% (w/v) concentrations of chitosan salts and TMC (degree of quaternization 12.28%) on the TEER of Caco-2 cell monolayers at pH 6.20. Each point represents the mean ± SD of three experiments. Control (●); chitosan hydrochloride (■); chitosan glutamate (◆); TMC (▲); dotted line represents start of reversibility experiment.

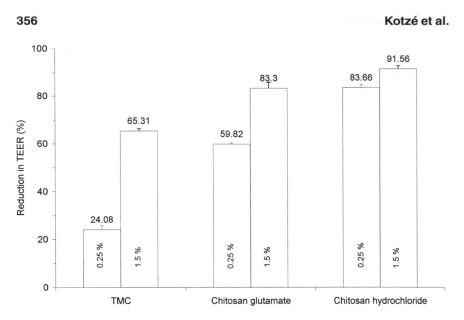

Figure 5 Reduction in TEER 2 hours after incubation with 0.25 and 1.5% (w/v) concentrations of chitosan hydrochloride, chitosan glutamate, and TMC (degree of quaternization 12.28%) at pH 6.20.

The decrease in TEER at 0.25% w/v concentrations 20 minutes after incubation was in the order chitosan hydrochloride (71 $\pm$ 4% reduction) > chitosan glutamate (56 $\pm$ 1% reduction) > TMC (28 $\pm$ 1% reduction). Prolonged incubation, up to 2 hours, resulted in only a slight decrease in TEER, compared with the initial reduction in TEER after 20 minutes, except with TMC, for which the reductions in TEER were 84 $\pm$ 1% (chitosan hydrochloride), 60 $\pm$ 1% (chitosan glutamate), and 24 $\pm$ 2% (TMC). Increased polymer concentrations led to further decreases in TEER values, especially with TMC. The decrease in TEER values was less pronounced with chitosan glutamate; the reductions in TEER at a concentration of 1.5% w/v 2 hours after incubation were 83 $\pm$ 2% (chitosan glutamate) and 65 $\pm$ 1% (TMC). The decrease in TEER with chitosan hydrochloride was even less pronounced: 92 $\pm$ 1% (1.5% w/v) compared with 84 $\pm$ 1% (0.25% w/v). Chitosan hydrochloride and chitosan glutamate solutions above 1.5% w/v could not be prepared at a pH of 6.20, because of low solubility and high viscosity. However, TMC easily dissolved at this pH in much higher concentrations and a further decrease in TEER values was found at concentrations above 1.5%. The reductions in TEER 2 hours after incubation started were 72 $\pm$ 1% (2.0% w/v) and 85 $\pm$ 1% (2.5% w/v), respectively.

These results clearly demonstrate that all these polymers are able to decrease the TEER of Caco-2 monolayers. Apparently, chitosan hydrochloride is more effective than chitosan glutamate, followed by TMC at similar weight concentrations. However, TMC is superior in solubility, and similar results could be obtained at higher concentrations of this polymer. Because the degree of deacetylation of each salt was the same (certificate of analysis), this difference in effect between these polymers could probably be explained in terms of the equivalent weight of each repeating unit in the polymer backbone of the respective polymers (theoretically 197.62 for chitosan hydrochloride, 308.30 for chitosan glutamate, and 239.80 for TMC), thus determining the amount of free chitosan base and therefore the density of the amino groups available for protonation at similar weight concentrations. About 50% of chitosan glutamate by weight is the glutamate salt, whereas for chitosan hydrochloride the salt part by weight is only a small fraction (5–10%). In addition, the attached methyl groups on the C-2 position of TMC probably causes steric effects and also partially hide the positive charge on the quaternary amino groups, thereby altering the time needed for interaction with the negatively charged cell membranes and tight junctions. Furthermore, the exact molecular weight of TMC, in terms of methylation, could not be determined because of the presence of some dimethylated and monomethylated amino groups on the C-2 position of chitosan. In this regard, a calculation for chitosan in terms of equivalent base of these polymers, to enable a clearer comparison between them, could not be made.

These results clearly demonstrated that the charge, charge density, and structural features of chitosan salts and TMC play an important role in their ability to act as absorption enhancers. TMC, as a partially quaternized derivative of chitosan, shows superior solubility compared with other chitosan salts. The higher solubility may compensate for its lower effectivity compared with chitosan salts at similar weight concentrations. Even at very low degrees of quaternization this polymer is soluble at every pH and an increase in basicity has been found. TMC with higher degrees of quaternization will probably result in similar effects on the TEER at lower concentrations as used in this study. This will be discussed in more detail in another section of this chapter.

The reversibility of the effect of these polymers on the TEER of the cell monolayers can also be seen from Fig. 4. With removal of the polymer solutions, repeated washing, and replacement of the apical medium again by medium without added polymers, monolayers started to recover slowly and a slight increase in resistance toward the initial values was found. At all concentrations tested, reversibility of the effects could be demonstrated after removal of the polymer solution. However, because of the high viscosity and adhesive character of these polymers (Lehr et al., 1992), probably not all the polymer material could be removed from the cell surface without damaging the cells; therefore the reversibility observed was only gradual. These results are in agreement with those of previous studies (Borchard et al., 1996). Staining with trypan blue, after completion of all the TEER experiments, did not result in any visible intracellular uptake of this marker. The absence of intracellular trypan blue after prolonged incubation with the different polymers implies that the Caco-2 cell monolayers remained undamaged and functionally intact. As the dye was excluded from the cells, we concluded that the viability of the monolayers was not affected by any of the polymers.

2. Transport of Model Compounds Across Caco-2 Cell Monolayers

From the permeation profiles of the radioactive markers, [^{14}C]mannitol (MW 182.2) and [^{14}C]polyethylene glycol 4000 ([^{14}C]-PEG-4000, MW 4000), apparent permeability coefficients (P_{app} values) and transport enhancement ratios (R) after incubation with chitosan glutamate, chitosan hydrochloride, and TMC (degree of quaternization 12.28%) at pH 6.20 were calculated. These results are presented in Table 4 for [^{14}C]mannitol and Table 5 for [^{14}C]-PEG-4000. Under the conditions described, very low baseline permeabilities were found for both compounds and only negligible amounts were transported in the control groups. Mannitol and PEG-4000 are metabolically inert and highly hydrophilic in nature, and mannitol has been used previously to follow changes in the epithelial integrity of mucosal cells (Artursson et al., 1994; Borchard et al., 1996; Schipper et al., 1996). Both compounds do not diffuse to a large extent into the cell membranes, but they are absorbed through the alternative aqueous paracellular pathway and are therefore ideal substances for detecting changes in permeability in studies of absorption enhancement.

Incubation of the monolayers with all the polymers resulted in a marked accumulation of [^{14}C]mannitol and [^{14}C]-PEG-4000 in the acceptor compartments. The increase in the transport of these compounds is also in good agreement with a decrease in the TEER, which was measured directly in the filters at several time points during the transport studies (data not shown). In all the control groups no change in the resistance was observed, whereas incubation with the different polymers resulted in reduction of the TEER. [^{14}C]Mannitol, with the lowest molecular weight, exhibited the highest permeability, and the highest cumulative amounts transported up to 4 hours at 0.25% w/v concentrations of the polymers were 15.2 ± 1.5% (chitosan hydrochloride), 11.2 ± 0.8% (chitosan glutamate), and 5.2 ± 0.1% (TMC) of the total dose applied, respectively (control: 0.5 ± 0.1%). The

Table 4 Effect of TMC (Degree of Quaternization 12.28%), Chitosan Glutamate, and Chitosan Hydrochloride on the Permeability of [^{14}C]Mannitol at pH 6.20

Marker	Chitosan concentration (% w/v)	TMC		Chitosan glutamate		Chitosan hydrochloride	
		$^{a}P_{app} \times 10^{-7}$ (cm/s)	R	$P_{app} \times 10^{-7}$ (cm/s)	R	$P_{app} \times 10^{-7}$ (cm/s)	R
[^{14}C]Mannitol	Control	0.72 ± 0.08	1	0.72 ± 0.08	1	0.72 ± 0.08	1
	0.25	8.11 ± 0.21[b]	11	18.25 ± 1.10[b]	25	24.65 ± 2.13[b]	34
	0.50	9.26 ± 0.35[b]	13	14.17 ± 0.45[b,c]	20	23.28 ± 1.00[b]	32
	1.00	14.00 ± 0.40[b]	19	20.82 ± 0.30[b]	29	25.56 ± 2.95[b]	36
	1.50	7.52 ± 0.86[b]	10	18.29 ± 1.53[b]	25	26.16 ± 1.86[b]	36
	2.00	12.35 ± 0.43[b]	17	ND[d]	ND	ND	ND
	2.50	15.21 ± 1.37[b]	21	ND	ND	ND	ND

[a]Each value represents the mean ± DS of three experiments.
[b]Significantly different from control ($P < .05$).
[c]Significantly different from all other treatments in group ($P < .05$).
[d]ND, not determined because of insolubility of chitosan salts.

Table 5   Effect of TMC (Degree of Quaternization 12.28%), Chitosan Glutamate, and Chitosan Hydrochloride on the Permeability of [^{14}C]-PEG-4000 at pH 6.20

Marker	Chitosan concentration (% w/v)	TMC $P_{app} \times 10^{-7a}$ (cm/s)	R	Chitosan glutamate $P_{app} \times 10^{-7}$ (cm/s)	R	Chitosan hydrochloride $P_{app} \times 10^{-7}$ (cm/s)	R
PEG-4000	Control	0.09 ± 0.02	1	0.09 ± 0.02	1	0.09 ± 0.02	1
	0.25	0.39 + 0.02[b]	4	0.91 ± 0.03[b]	10	2.44 ± 0.14[b]	27
	0.50	0.38 ± 0.03[b]	4	1.00 ± 0.05[b]	11	2.30 ± 0.19[b]	26
	1.00	0.57 ± 0.08[b]	6	1.28 ± 0.29[b]	14	3.54 ± 0.38[b,c]	39
	1.50	0.46 ± 0.06[b]	5	1.02 ± 0.19[b]	11	4.36 ± 0.29[b,c]	48
	2.00	0.67 ± 0.08[b]	7	ND[d]	ND	ND	ND
	2.50	0.80 ± 0.03[b]	9	ND	ND	ND	ND

[a]Each value represents the mean ± SD of three experiments.
[b]Significantly different from control ($P < .05$).
[c]Significantly different from all other treatments in group ($P < .05$).
[d]ND, not determined because of insolubility of chitosan salts.

permeability decreased with an increase in molecular weight and the cumulative amounts of [^{14}C]-PEG-4000 transported up to 4 hours, at 0.25% concentrations of the polymers, were 1.52 ± 0.10% (chitosan hydrochloride), 0.60 ± 0.04% (chitosan glutamate), and 0.26 ± 0.01% (TMC) of the total dose applied, respectively (control: 0.06 ± 0.01%). This suggests that the permeation of these compounds across intestinal epithelial cells is dependent on their molecular size and structural conformation, among other factors.

In agreement with the TEER results described in the previous section, higher P_{app} and R values were found for chitosan hydrochloride. The apparent better effect of chitosan hydrochloride could, as with the TEER results, be explained by the higher concentration of equivalent chitosan base in this chitosan salt, on a weight basis, compared with chitosan glutamate and TMC. An increase in polymer concentration did not result in major increases in P_{app} and R values compared with the initial increases at 0.25% w/v polymer concentrations, except for [^{14}C]-PEG-4000 and only with chitosan hydrochloride. In general, plateau levels were reached at concentrations between 0.25 and 0.50%. This is in agreement with a previous study (Artursson et al., 1994) in which the pH and concentration dependence of chitosan glutamate on the transport enhancement of [^{14}C]mannitol was investigated. Exposure of the apical side of the monolayers to 0.25% concentrations of these polymers gave 34-fold (chitosan hydrochloride), 25-fold (chitosan glutamate), and 11-fold (TMC) increases in the transport rate of [^{14}C]mannitol compared with the control group as indicated by the P_{app} values (Table 4). This changed to 36-fold (chitosan hydrochloride), 25-fold (chitosan glutamate), and 10-fold (TMC) increases in the presence of 1.5% concentrations of the respective polymers. Similar results were obtained for [^{14}C]-PEG-4000 (Table 5).

It should be explained that TMC was able to increase further the P_{app} and R values of both [^{14}C]mannitol and PEG-4000 at higher concentrations of this polymer. A 17-fold and a 21-fold increase in R were found for [^{14}C]mannitol at 2.0 and 2.5% w/v concentrations of TMC, respectively. The same tendency was also seen with [^{14}C]-PEG-4000. These results indicate that additional factors play a role in the absorption enhancement mechanism of TMC. This could most likely be explained in terms of the charge density, the equivalent weight of each repeating unit in the polymer backbone, and possible steric effects of the attached methyl groups and partial hiding of the positive charge on the quaternary amino groups.

In all the concentration curves a short lag time was observed, which is indicative of the size dependence of opening of tight junctions and the time each molecule needs to diffuse into the intercellular spaces before it can reach the acceptor compartments. After this short lag time, in the range of 0–40 minutes, the transport of the compounds from the donor to the acceptor sides was relatively steady, as evident from the slope of the individual concentration curves, indicating unhindered paracellular diffusion of these hydrophilic compounds through the opened tight junctions. No evidence of trypan blue inclusion in the intracellular spaces of the cells was found when cells were stained with this dye after completion of the transport studies, proving that the cells were still viable and not affected by incubation at any concentration of all the polymers used.

3. Visualization of the Transport Pathway

Confocal laser scanning microscopy was used to visualize the transport pathway of the fluorescent marker FD-4 (fluorescein isothiocyanate–labeled dextran, MW 4400) across Caco-2 cell monolayers at a pH of 6.20 after incubation with chitosan glutamate, chitosan hydrochloride, and TMC (degree of quaternization 12.28%). Optical cross sections of cell monolayers, after incubation with a control solution containing only FD-4, show no evidence of any intracellular or intercellular fluorescence (Fig. 6). Fluorescence was detected only on top of the monolayers. After 60 minutes of incubation with 0.5% concentrations of the different polymers, fluorescence was detected in the intercellular spaces, as evident from Fig. 6, which represents vertical scans (XZ images) through the monolayers. Similar results were obtained in horizontal (XY) images (Fig. 7). These confocal images clearly show that the tight junctions are open and that FD-4 is able to permeate into the paracellular spaces. No fluorescence could be found with the cells, which further proves that incubation with any polymer does not damage the cell membranes.

Similar results were obtained after incubation with a fluorescein isothiocyanate–labeled dextran with an average molecular weight of 19,600 (FD-20) (Fig. 8). Although no quantitative evaluation of the amount of fluorescence detected between FD-4 and FD-20 was made, it is clear that all the polymers were able to open the tight junctions to allow the transport of a large hydrophilic molecule such as FD-20. These results suggest that large hydrophilic molecules up to 20,000 daltons, such as peptide and protein drugs, can pass through the paracellular transport pathway. The mechanism of action of chitosan has been discussed in the previous section. In principle, the effect of chitosan will not be a direct interaction, because chitosan will most probably not enter the cells due to its high molecular weight and hydrophilicity. Therefore, it can be concluded that the effect of chitosan on the tight junctions must be indirect. TMC most likely has the same mechanism of action on the junctional complex as the chitosan salts.

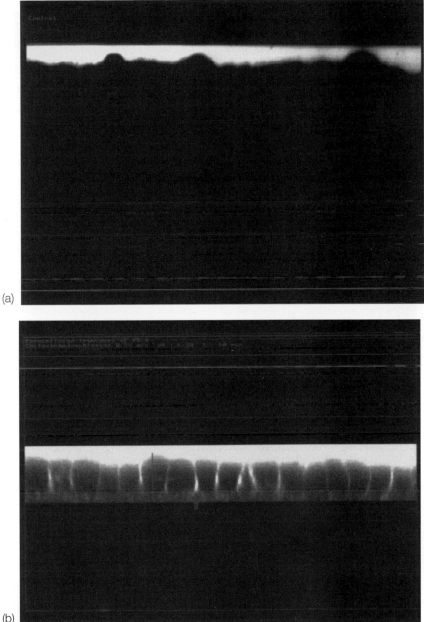

Figure 6 Optical vertical cross sections (*XZ* images) through Caco-2 cell monolayers. Top is apical and bottom is basolateral. Arrows indicate paracellular transport. (a) No polymer and FD-4 (1 mg/mL). (b) FD-4 (1 mg/mL) and chitosan hydrochloride (0.5%). (Figure continues.)

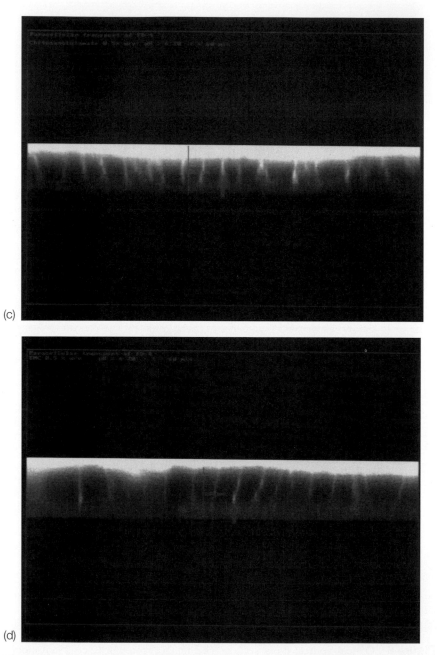

Figure 6 (Continued). (c) FD-4 (1 mg/mL) and chitosan glutamate (0.5%). (d) FD-4 (1 mg/mL) and TMC (0.5%, degree of quaternization 12.28%).

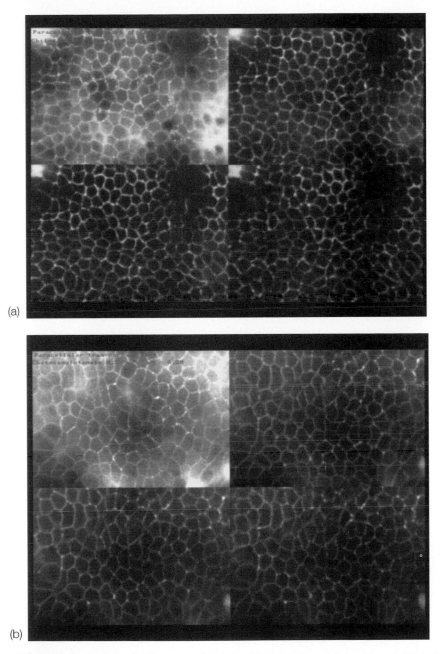

Figure 7 Optical horizontal cross sections (*XY* images, step size 3 μm) through Caco-2 cell monolayers. (a) FD-4 (1 mg/mL) and chitosan hydrochloride (0.5%). (b) FD-4 (1 mg/mL) and chitosan glutamate (0.5%). (Figure continues.)

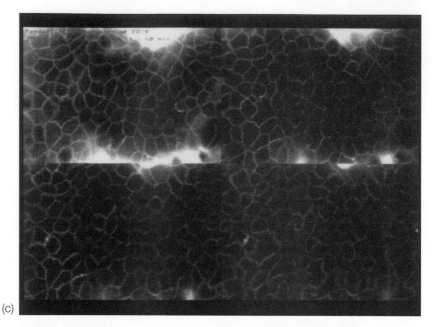

(c)

Figure 7 (Continued). (c) FD-4 (1 mg/mL) and TMC (0.5%, degree of quaternization 12.28%).

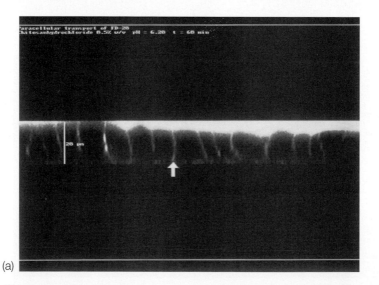

(a)

Figure 8 Optical vertical cross sections (*XZ* images) through Caco-2 cell monolayers. Top is apical and bottom is basolateral. Arrows indicate paracellular transport. (a) FD-20 (1 mg/mL) and chitosan hydrochloride (0.5%). (Figure continues.)

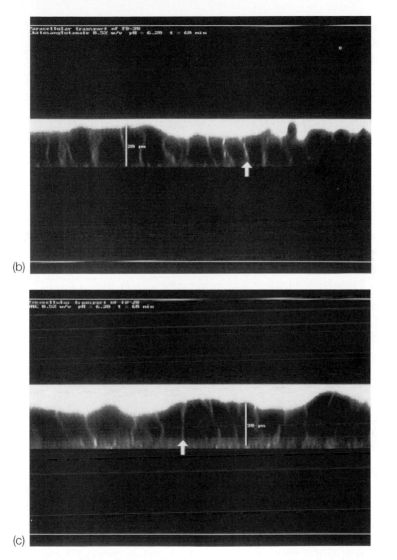

Figure 8 (Continued). (b) FD-20 (1 mg/mL) and chitosan glutamate (0.5%). (c) FD-20 (1 mg/mL) and TMC (0.5%, degree of quaternization 12.28%).

C. Transport of Peptide Drugs and Enzyme Inhibitory Effects of Chitosan and *N*-Trimethyl Chitosan Chloride

1. Effect of *N*-Trimethyl Chitosan Chloride on α-Chymotrypsin Activity

The profile of degradation of *N*-acetyl-L-tyrosine ethylester (ATEE) to *N*-acetyl-L-tyrosine (AT) by α-chymotrypsin at a pH of 6.70, in the presence and absence of TMC (degree of quaternization 12.28%), is shown in Fig. 9. Apparently, TMC is able to inhibit the conversion of ATEE to AT by α-chymotrypsin to a low extent. In the control group, no ATEE could be detected after 60 minutes of incubation with α-chymotrypsin. In the presence of TMC (1.5–2.5%) ATEE was completely converted to AT after 90 minutes of incubation with α-chymotrypsin. This minor enzyme inhibition by TMC is considered to be of no therapeutic importance compared with other well-known enzyme inhibitors such as carbomer and polycarbophil (Luessen et al., 1996b,c). Luessen et al. (1997) have already shown that chitosan glutamate is not able to prevent the degradation of *N*-α-benzoyl-L-arginine ethylester by trypsin. Therefore, it can be concluded that neither chitosan nor TMC has substantial enzyme inhibitory effects.

2. Peptide Transport by Chitosan Salts and *N*-Trimethyl Chitosan Chloride Across Caco-2 Cell Monolayers

a. Buserelin

The transport of buserelin (MW 1299.5) across Caco-2 cell monolayers at pH 6.20 in the presence of chitosan hydrochloride, chitosan glutamate, TMC (degree of quaternization 12.28%) and carbomer (Carbophol, C934P) is depicted in Fig. 10. The ability of chitosan hydrochloride and C934P to transport buserelin, in the presence of α-chymotrypsin, is also presented in this figure. Figure 11 shows the decrease in the TEER of the Caco-2 cell monolayers measured in the filters during the transport study with buserelin. Incubation on the apical side of the monolayers with the different polymers leads to a pronounced reduction in TEER values compared with the control group. The reduction in TEER 1 hour after the start of the experiment with 1.5% concentrations of the polymers was in the following order: chitosan hydrochloride (71 ± 4% reduction) > chitosan glutamate (64 ± 6% reduction) > TMC (55 ± 4% reduction) > C934P (39 ± 4% reduction). Prolonged incubation, up to 4 hours, resulted in only a slight decrease in TEER values compared with the initial reduction measured after 1 hour. At a concentration

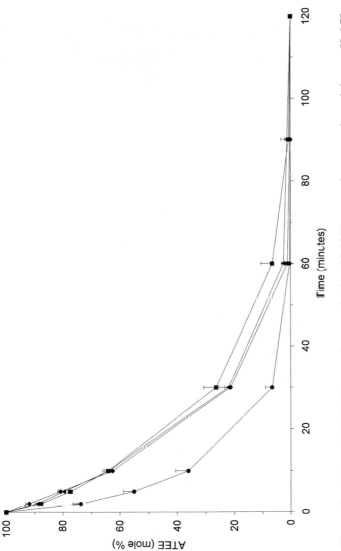
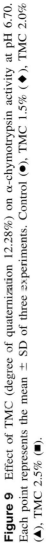

Figure 9 Effect of TMC (degree of quaternization 12.28%) on α-chymotrypsin activity at pH 6.70. Each point represents the mean ± SD of three experiments. Control (●), TMC 1.5% (◆), TMC 2.0% (▲), TMC 2.5% (■).

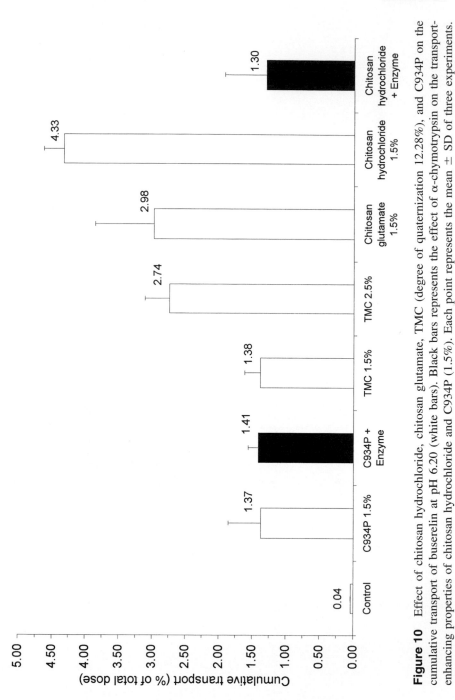

Figure 10 Effect of chitosan hydrochloride, chitosan glutamate, TMC (degree of quaternization 12.28%), and C934P on the cumulative transport of buserelin at pH 6.20 (white bars). Black bars represents the effect of α-chymotrypsin on the transport-enhancing properties of chitosan hydrochloride and C934P (1.5%). Each point represents the mean ± SD of three experiments.

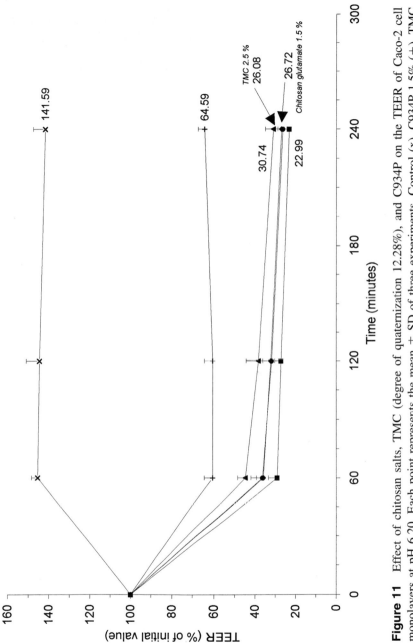

Figure 11 Effect of chitosan salts, TMC (degree of quaternization 12.28%), and C934P on the TEER of Caco-2 cell monolayers at pH 6.20. Each point represents the mean ± SD of three experiments. Control (x), C934P 1.5% (+), TMC 1.5% (▲), TMC 2.5% (♦), chitosan glutamate 1.5% (●), chitosan hydrochloride 1.5% (■).

of 2.5%, TMC was able to cause a further decrease in the TEER compared with the reduction measured with the 1.5% concentration.

All the polymers were able to improve the transport of buserelin significantly, as evident from Fig. 10. In the control group almost no buserelin was transported up to 4 hours of incubation (0.04% of the total dose applied). In agreement with the reduction measured in the TEER with 1.5% concentrations of the chitosan salts and TMC, the increase in the transport of buserelin up to 4 hours was in the following order: chitosan hydrochloride (4.3 ± 0.3% of the total dose applied) > chitosan glutamate (3.0 ± 0.9%) > TMC (1.4 ± 0.2%). However, at a 2.5% concentration of TMC the transport of buserelin was improved further (2.7 ± 0.3% of the total dose applied), compared with the transport found at 1.5% concentrations. This increase in the transport of buserelin was comparable to the transport found with 1.5% chitosan glutamate.

In the presence of C934P, 1.4 ± 0.2% of the total buserelin dose applied was transported across the cell monolayers. This is comparable to the transport found with TMC at a similar concentration (1.5%). In the presence of α-chymotrypsin, C934P was able to transport the same amount of buserelin (1.4 ± 0.2% of the total dose applied), compared with the group in which no enzyme was present. In contrast to C934P, the transport of buserelin was decreased markedly after 4 hours in the presence of α-chymotrypsin at a similar concentration (1.5%) of chitosan hydrochloride; the transport was decreased from 4.3 ± 0.3% to 1.3 ± 0.6% of the total dose of buserelin applied. Carbomer is a poly(acrylate) derivative and a polyanionic polymer. It has been reported that carbomer has a high affinity for binding of Ca^{2+} ions, especially above a pH of 4 (Luessen et al., 1996b). The improvement in the transport of buserelin at pH 6.20 could be explained by complexation of Ca^{2+} ions by the carboxylic groups of the polymer, thereby altering the integrity of the tight junctions, which is sensitive to changes in calcium concentration (Cereijido et al., 1993). In a similar way, carbomer was able to inhibit the activity of α-chymotrypsin. This enzyme is an endoprotease and a representative of the group of serine proteases, containing Ca^{2+} as an essential cofactor in its structure (Luessen et al., 1996b). In contrast to carbomer, chitosan hydrochloride and the novel derivative TMC were not able to inhibit α-chymotrypsin activity. Although both classes of polymers were able to improve the transport of buserelin, the chitosan salts and TMC proved to be more effective absorption enhancers. These results clearly suggest that they act by different mechanisms to improve the paracellular transport of peptide drugs. Whereas carbomer acts mainly as a protease inhibitor and calcium complexing agent, the chitosans improve peptide transport mainly by increasing the paracellular permeability.

b. Insulin

Figure 12 shows the transport of insulin (porcine insulin, sodium salt, MW 5777.6) at pH 4.40 across Caco-2 cell monolayers. No transport in the control group could be detected up to 4 hours. Incubation for 4 hours with both chitosan glutamate and chitosan hydrochloride (1.5%) and TMC (1.5%, degree of quaternization 12.28%) resulted in insulin transport in the following order: chitosan hydrochloride (1.2 ± 0.4% of the total dose applied) > chitosan glutamate (0.6 ± 0.2%) > TMC (0.3 ± 0.1%). At a concentration of 2.5% w/v, TMC was also able to increase the transport of insulin to 0.8 ± 0.1% of the total dose applied, comparable to the transport found in the presence of chitosan glutamate.

c. 9-Desglycinamide, 8-L-Arginine Vasopressin (DGAVP)

Figure 13 shows that both chitosan glutamate and TMC (degree of quaternization 12.28%) were able to improve the permeation of DGAVP (MW 1412) at a pH of 5.60 across Caco-2 cell monolayers. In the control group, very low permeability (0.19 ± 0.29% of the total DGAVP dose applied) was found. Chitosan glutamate (0.4 and 1.0%) improved the transport of DGAVP to 1.13 ± 0.08% and 1.19 ± 0.14% of the total dose applied, respectively. Similar results were obtained with TMC. TMC (1.5 and 2.5%) increases the transport to 0.96 ± 0.28 and 1.09 ± 0.08% of the total DGAVP dose applied, respectively.

D. Effect of the Degree of Quaternization of *N*-Trimethyl Chitosan Chloride on the Permeability of Intestinal Epithelial Cells (Caco-2)

In previous sections it has been shown that the partially quaternized (12.28%) derivative of chitosan, *N*-trimethyl chitosan chloride (TMC), was able to decrease the TEER of Caco-2 cell monolayers and to increase the transport of several hydrophilic compounds such as [^{14}C]mannitol and [^{14}C]polyethylene glycol and the peptide drugs buserelin, DGAVP, and porcine insulin significantly in Caco-2 cell monolayers at acidic pH values (4.40–6.20). It was proposed that TMC most likely has the same mechanism of action on the junctional complex as other chitosan salts. Although TMC was not as effective as other chitosan salts, such as chitosan glutamate and chitosan hydrochloride, its lower efficiency was explained by its charge density, determined by the degree of quaternization, and by partial hiding of the positive charge on the amino group by the attached methyl groups. However, the much better solubility of TMC may compensate for its lesser efficacy. TMC proved to be very soluble over a wide pH range (pH 1–9) up to 10% w/v concentrations, even at degrees of quaternization as low as 10%.

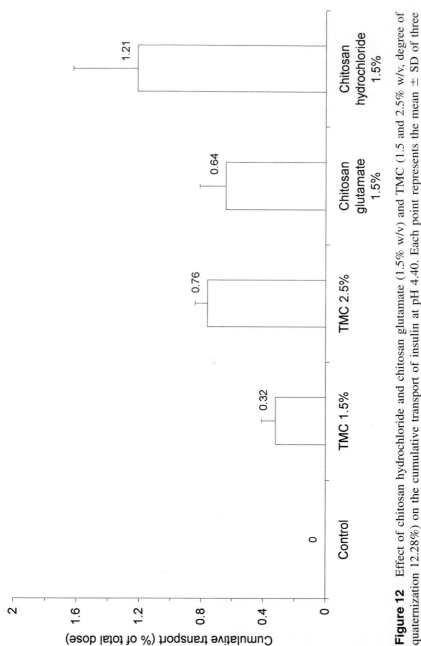

Figure 12 Effect of chitosan hydrochloride and chitosan glutamate (1.5% w/v) and TMC (1.5 and 2.5% w/v, degree of quaternization 12.28%) on the cumulative transport of insulin at pH 4.40. Each point represents the mean ± SD of three experiments.

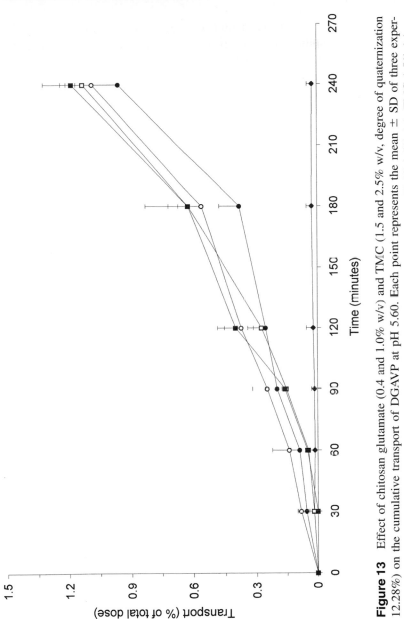

Figure 13 Effect of chitosan glutamate (0.4 and 1.0% w/v) and TMC (1.5 and 2.5% w/v, degree of quaternization 12.28%) on the cumulative transport of DGAVP at pH 5.60. Each point represents the mean ± SD of three experiments. Control (◆), chitosan glutamate 1.0% (■), chitosan glutamate 0.4% (□), TMC 2.5% (○), TMC 1.5% (●).

In this section the effects of highly quaternized TMC on the permeability of intestinal epithelial cells will be discussed. Our hypothesis is that TMC with higher degrees of quaternization may be more effective as an absorption enhancer and might be able to increase the paracellular transport of hydrophilic compounds at neutral and basic pH values.

1. Effect of Highly Quaternized *N*-Trimethyl Chitosan Chloride on the TEER of Caco-2 Cell Monolayers

Figures 14 and 15 show the effect of 0.5% w/v concentrations of TMC-H (degree of quaternization 61.2%), TMC-L (degree of quaternization 12.3%), and chitosan hydrochloride on the TEER of the Caco-2 cell monolayers at pH values of 6.20 and 7.40, respectively. In agreement with results presented previously for a pH of 6.20, incubation with all these polymers resulted in a pronounced reduction in TEER values compared with the control group.

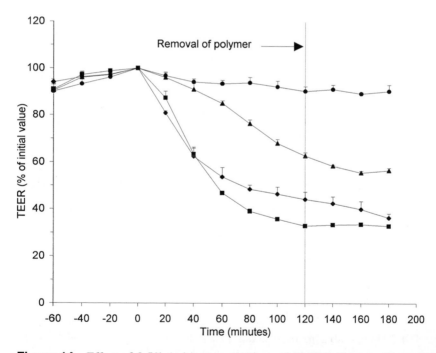

Figure 14 Effect of 0.5% (w/v) concentrations of TMC-L (degree of quaternization 12.3%), TMC-H (degree of quaternization 61.2%), and chitosan hydrochloride on the TEER of Caco-2 cell monolayers at pH 6.20. Each point represents the mean ± SD of three experiments. Control (●), TMC-L (▲), TMC-H (◆), chitosan hydrochloride (■); dotted line represents start of reversibility experiment.

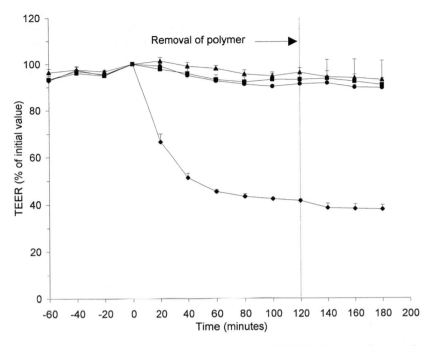

Figure 15 Effect of 0.5% (w/v) concentration of TMC-L (degree of quaternization 12.3%), TMC-H (degree of quaternization 61.2%), and chitosan hydrochloride on the TEER of Caco-2 cell monolayers at pH 7.40. Each point represents the mean ± SD of three experiments. Control (●), TMC-L (▲), TMC-H (♦), chitosan hydrochloride (■); dotted line represents start of reversibility experiment.

The reduction in TEER was in the order chitosan hydrochloride (67.1 ± 0.6% reduction, 2 hours after incubation started) > TMC-H (55.9 ± 3.1% reduction) > TMC-L (37.3 ± 1.4% reduction). However, at a pH of 7.40, only TMC-H was able to decrease the TEER of the cell monolayers, as evident from Fig. 15. Neither TMC-L nor chitosan hydrochloride, even at concentrations of 1.5% w/v, was able to cause any significant decreases in TEER values compared with the control group. Incubation of the monolayers with TMC-H resulted in a pronounced decrease in the TEER of the cell monolayers (34–63% reduction). Even at a concentration as low as 0.05% w/v, TMC-H was able to reduce the TEER by 34.8 ± 4.1%. TEER measurements suggest that a plateau level was reached at a 0.5% w/v concentration of this polymer, because the reduction in TEER measured at different concentrations was as follows: 43.1% reduction at a 0.10% concentration, 45.6% reduction at a 0.25% concentration, 58.6% reduction at a 0.50% concentration, 60.1% reduction at a 1.00% concentration, and 63.1% reduction at a 1.50% concentration. Staining with trypan blue after completion of the TEER experiment at pH 7.40 with TMC-H did not result in any visible intracellular uptake of the marker, indicating that the viability of the monolayers was not affected by incubation with this highly quaternized polymer.

2. Effect of Highly Quaternized N-Trimethyl Chitosan Chloride on [^{14}C]Mannitol Transport in Caco-2 Cell Monolayers

In Fig. 16 the cumulative amounts of [^{14}C]mannitol transported 4 hours after incubation started with 0.5% w/v concentrations of chitosan hydrochloride, TMC-L (degree of quaternization 12.3%), and TMC-H (degree of quaternization 61.2%) are given at pH 6.20 and 7.40. From the permeation profiles of [^{14}C]mannitol at different polymer concentrations, apparent permeability coefficients (P_{app}) values and transport enhancement ratios (R) were calculated and these are given in Table 6. In agreement with results discussed previously, incubation at pH 6.20 resulted in a marked accumulation of the marker molecule in the acceptor compartments. In agreement with the TEER results, only TMC-H was able to increase the transport of [^{14}C]mannitol at a pH of 7.40. Even at a concentration as low as 0.05% w/v, the permeability of [^{14}C]mannitol was increased 31-fold. In further agreement with the TEER results, the data suggest that a plateau level was reached at a 0.5% w/v concentration of TMC-H.

The results of these experiments show that both chitosan hydrochloride and TMC, with different degrees of quaternization, are potent absorption enhancers at a pH of 6.20. These polymers are able to decrease the TEER of the Caco-2 cell monolayers markedly. The apparent difference in effect between chitosan hydrochloride and TMC at a pH of 6.20 has already been explained. NMR spectra confirm that there is still a high amount of dimethylated amino groups present in TMC-L that can be protonated at a pH of 6.20 for interaction with the cell membranes or tight junctions. At a pH of 7.40 the charge density of TMC-L is still too low for interaction with the anionic components of the glycoproteins at the surface of the cells or with the fixed negative charges within the aqueous tight junctions. In addition, the attached methyl groups may partially hide the positive charge on the quaternized amino groups from significant interaction with the cell membranes or tight junctions. This may be the reason why TMC-L does not show any effect at a pH of 7.40. TMC-H, on the other hand, has a much higher proportion of quaternary amino groups that seems to be sufficient for interaction with the anionic components of the glycoproteins on the cell surfaces or the negative sites within the tight junctions. These results show that the insolubility of chitosan hydrochloride prevents this polymer from

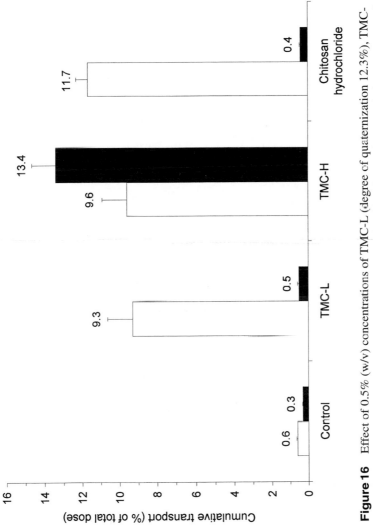

Figure 16 Effect of 0.5% (w/v) concentrations of TMC-L (degree of quaternization 12.3%), TMC-H (degree of quaternization 61.2%), and chitosan hydrochloride on the cumulative transport of [^{14}C]mannitol in Caco-2 cell monolayers at pH 6.20 (white bars) and pH 7.40 (black bars). Each point represents the mean ± SD of three experiments.

Table 6 Effect of TMC-H (Degree of Quaternization 61.2%), TMC-L (Degree of Quaternization 12.3%), and Chitosan Hydrochloride on the Permeability of [^{14}C]Mannitol

pH	Polymer concentration (% w/v)	TMC-H $P_{app} \times 10^{-7a}$ (cm/s)	R	TMC-L $P_{app} \times 10^{-7}$ (cm/s)	R	Chitosan hydrochloride $P_{app} \times 10^{-7}$ (cm/s)	R
6.20	Control	0.72 ± 0.09	1	0.72 ± 0.09	1	0.72 ± 0.09	1
	0.10	12.81 ± 2.56	18	ND		ND	
	0.25	13.32 ± 0.58	19	ND		ND	
	0.50	14.28 ± 2.14	20	13.15 ± 1.54	18	16.38 ± 0.68	23
	1.00	16.14 ± 0.72	22	10.73 ± 1.87	15	20.82 ± 1.07	29
	1.50	14.44 ± 1.80	20	11.32 ± 0.80	16	31.03 ± 3.16	43
7.40	Control	0.47 ± 0.04	1	0.47 ± 0.04	1	0.47 ± 0.04	1
	0.05	14.69 ± 2.90	31	ND^b		ND	
	0.10	14.06 ± 2.85	30	ND		ND	
	0.25	15.10 ± 2.20	32	ND		ND	
	0.50	20.14 ± 1.44	43	0.63 ± 0.08	1	0.56 ± 0.09	1
	1.00	22.57 ± 1.89	48	0.85 ± 0.08	2	0.51 ± 0.02	1
	1.50	15.13 ± 1.76	32	1.07 ± 0.89	2	0.67 ± 0.04	1

[a] P_{app}, apparent permeability coefficient. Each value represents the mean ± SD of three experiments; R, transport enhancement ratio.
[b] ND, not determined.

being effective as an absorption enhancer at neutral and basic pH values. The degree of quaternization of TMC is demonstrated to play an important role in its ability to open the tight junctions of intestinal epithelial cells. Highly quaternized TMC proves to be a very potent absorption enhancer, especially at neutral and basic pH values.

E. Nasal Absorption Enhancement of Insulin by Chitosan and *N*-Trimethyl Chitosan Chloride

The use of chitosan in nasal drug delivery to enhance the absorption of the peptide drug insulin was first described by Illum et al. (1994). The use of chitosan as a bioadhesive compound for the nasal delivery of several drugs, together with a number of toxicity studies, is discussed in detail in Sec. IV of this book. In the present section the potential use of *N*-trimethyl chitosan chloride as an absorption enhancer for nasal delivery of peptide drugs will be discussed briefly.

Figure 17 shows the reduction in blood glucose levels in rats 30 minutes after the administration of semisynthetic human insulin (4 IU/kg body weight) at a pH of 4.40 with chitosan hydrochloride, TMC-L (degree of quaternization 12.3%), and TMC-H (degree of quaternization 61.2%). In agreement with the results described in the previous section, all these polymers were able to reduce the blood glucose levels. Major increases in plasma insulin levels were also found after coadministration with these polymers. Apparently, chitosan hydrochloride is more effective than both TMC-L and TMC-H. The reason for this has already been explained in terms of the charge density of these polymers. No major differences in effect between TMC-L and TMC-H could be demonstrated. Histological evaluations of the nasal epithelia after 60 minutes of exposure with these polymers did not show any significant changes in the epithelial cells or submucosa for both chitosan hydrochloride and TMC-L (0.5%). However, examination of the nasal cavity at the site of application showed a very slight increase in mucus production. With TMC-H (0.5%) there was also no change in the epithelial cells, but there was some evidence of slight congestion of the submucosa and a mild to moderate increase in mucus content in the nasal cavity.

At a pH of 7.40 only TMC-H was able to decrease the blood glucose levels of the rats significantly, compared with the control situation. Neither chitosan hydrochloride nor TMC-L was able to produce any hypoglycemic response. A 0.5% concentration of TMC-H led to a reduction in blood glucose levels, 30 minutes after coadministration with insulin, of about 34%. Figure 18 shows the increase in the plasma insulin levels. Histological examination of the nasal epithelia did not reveal any significant changes in the epithelial cells, although there was some evidence of slight congestion of

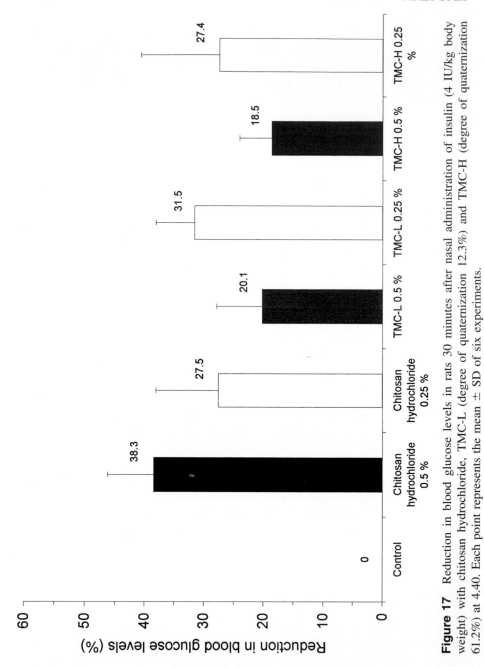

Figure 17 Reduction in blood glucose levels in rats 30 minutes after nasal administration of insulin (4 IU/kg body weight) with chitosan hydrochloride, TMC-L (degree of quaternization 12.3%) and TMC-H (degree of quaternization 61.2%) at 4.40. Each point represents the mean ± SD of six experiments.

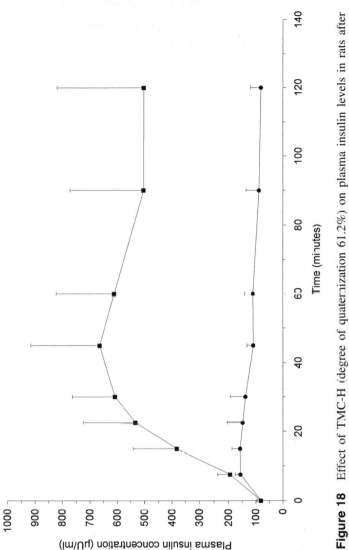

Figure 18 Effect of TMC-H (degree of quaterization 61.2%) on plasma insulin levels in rats after nasal administration of insulin (4 IU/kg body weight) at pH 7.40. Each point represents the mean ± SD of six experiments. Control (●), TMC-H 0.5% w/v (■).

the submucosa. Examination of the content of the nasal cavity exposed to this polymer showed a mild increase in mucus content. No difference in the histological effects of this polymer at both pH values could be found. The results of these experiments clearly demonstrate that TMC-H is a potent absorption enhancer in the neutral environment of the nasal cavity. Whether a degree of quaternization of TMC of about 60%, as used in these experiments, is the optimal degree of quaternization is still uncertain. Further investigations may indicate a direct relationship between the degree of quaternization of TMC, its effects as an absorption enhancer in neutral and basic environments, and the toxicity profile of this polymer.

V. CONCLUSION AND FUTURE PROSPECTS

The present chapter has given an overview of the pharmaceutical application of chitosan and the novel chitosan derivative *N*-trimethyl chitosan chloride as potential absorption enhancers across mucosal surfaces. Chitosan is able to open the tight junctions of epithelial cells to allow the paracellular transport of hydrophilic compounds such as peptide drugs. Furthermore, the mucoadhesive character of chitosan makes it a promising substance in the development of bioadhesive dosage forms and drug delivery systems. However, the use of chitosan as an absorption enhancer in the more neutral and basic environment of the large intestine and colon is limited because of its insolubility in these environments.

It has been shown that the partially quaternized derivative *N*-trimethyl chitosan chloride has excellent water solubility over a wide pH range, and this polymer was able to reduce the TEER of intestinal epithelial cells and to increase the transport of hydrophilic compounds at neutral and basic pH values. Promising results with the nasal administration of insulin in combination with this derivative highlighted the future potential of this polymer in novel dosage form design.

Functional groups make chitosan a versatile polymer, and possibilities also exist for the development of other derivatives with different properties such as the ability to inhibit proteolytic enzymes. The potential use of chitosan and chitosan derivatives can be an important contribution toward the development of selective and effective delivery systems for peptide and protein drugs and other large hydrophilic molecules.

REFERENCES

Anderberg EK, Artursson P. 1993. J Pharm Sci 82:392–398.
Arai K, Kinumaki T, Fujita T. 1968. Bull Tokai Reg Fish Lab 43:89–94.

Artursson P, Magnusson C. 1990. J Pharm Sci 79:595–600.

Artursson P, Lindmark T, Davis SS, Illum L. 1994. Pharm Res 11:1358–1361.

Berthold A, Cremer K, Kreuter J. 1996. J Controlled Release 39:17–25.

Borchard G, Luessen HL, De Boer AG, Verhoef JC, Lehr C-M, Junginger HE. 1996. J Controlled Release 39:131–138.

Boulenc X, Roques C, Joyeux H, Berger Y, Fabre G. 1995. Int J Pharm 123:13–24.

Cereijido M, Ruiz O, Gonzalez-Mariscal L, Contreras RG, Balda MS, Garcia-Villegas MR. 1993. In: Audas KL, Raub TJ, eds. Biological Barriers to Protein Delivery. New York: Plenum, pp 3–21.

Chithambara Thanoo B, Sunny MC, Jayahrishnan A. 1992. J Pharm Pharmacol 44:283–286.

Citi S. 1992. J Cell Biol 117:169–178.

Domard A, Rinaudo M, Terrassin C. 1986. Int J Biol Macromol 8:105–107.

Fix JA. 1987. J Controlled Release 6:151–156.

Geary RS, Schlameus HW. 1993. J Controlled Release 23:65–74.

Gumbiner B. 1987. Am J Physiol. 253:749–758.

Hirano S, Seino H, Akiyama Y, Nonaka I. 1988. Polym Eng Sci 59:897–901.

Hochman J, Artursson P. 1994. J Controlled Release 29:253–267.

Hurni MA, Noach ABJ, Blom-Roosemalen MCM, De Boer AG, Nagelkerke JF, Breimer DD. 1993. Pharmacol Exp Ther 267:942–950.

Illum L, Farraj NF, Davis SS. 1994. Pharm Res 11:1186–1189.

Knapczyk J. 1992. Int J Pharm 88:9–14.

Knapczyk J. 1993. Int J Pharm 93:233–237.

Kotzé AF, de Leeuw BJ, Luessen HL, De Boer AG, Verhoef JC, Junginger HE. 1997b. Int J Pharm 159:243–253.

Kotzé AF, Luessen HL, De Leeuw BJ, De Boer AG, Verhoef JC, Junginger HE. 1997a. Pharm Res 14:1197–1202.

Kotzé AF, Luessen HL, De Leeuw BJ, De Boer AG, Verhoef JC, Junginger HE. 1998. J Controlled Release 51:35–46.

Kotzé AF, Van Wyk CJ, Verhoef JC, De Boer AG, Junginger HE. 1996. Proc Int Symp Controlled Release Bioact Mater 23:425–426.

Kristl J, Smid-Korbar J, Struc E, Schara M, Rupprecht H. 1993. Int J Pharm 99:13–19.

Kriwet B, Kissel T. 1996. Eur J Pharm Biopharm 76:233–240.

Le Dung P, Milas M, Rinaudo M, Desbrières J. 1994. Carbohydr Pol 24:209–214.

Lee VHL. 1991. In: Lee VHL, ed. Peptide and Protein Drug Delivery. New York: Marcel Dekker, pp 1–56.

Lehr C-M, Bouwstra JA, Schacht EH, Junginger HE. 1992. Int J Pharm 78:43–48.

Li Q, Dunn ET, Grandmaison EW, Goosen MFA. 1992. J Bioact Compat Polym 7:370–397.

Lipka E, Crison J, Amidon GL. 1996. J Controlled Release 39:121–129.

Luessen HL, Bohner V, Pérard D, Langguth P, Verhoef JC, De Boer AG, Merkle HP, Junginger HE. 1996c. Int J. Pharm 141:39–52.

Luessen HL, De Leeuw BJ, Langemeÿer MWE, De Boer AG, Verhoef JC, Junginger HE. 1996a. Pharm Res 13:1666–1670.

Luessen HL, De Leeuw BJ, Pérard D, Lehr C-M, De Boer AG, Verhoef JC, Junginger HE. 1996b. Eur J Pharm Sci 4:117–128.

Luessen HL, Rental C-O, Kotzé AF, Lehr C-M, De Boer AG, Verhoef JC, Junginger HE. 1997. J Controlled Release 45:15–23.

Madara JL. 1987. Am J Physiol 253:C171–C175.

Madara JL. 1989. J Clin Invest 83:1089–1094.

Madara JL, Trier JS. 1987. In: Johnson LR, ed. Physiology of the Gastrointestinal Tract. 2nd ed. New York: Raven Press, pp 1251–1266.

Meza I, Sabanero M, Stefani E, Cereijido M. 1982. J Cell Biochem 18:407–421.

Miyazaki S, Yamaguchi H, Yokouchi C, Takada M, Hou WM. 1988. Chem Pharm Bull 36:4033–4038.

Muranishi S. 1990. Crit Rev Ther Drug Carrier Syst 7:1–27.

Muzzarelli RAA. 1973. In: Muzzarelli RAA, ed. Natural Chelating Polymers. Oxford: Pergamon Press, pp 144–176.

Nigalaye AG, Adusumilli P, Bolton S. 1990. Drug Dev Ind Pharm 16:449–467.

Peppas NA, Burim PA. 1985. J Controlled Release 2:257–275.

Polk A, Amsden B, De Yao K, Peng T, Goosen MF. 1994. J Pharm Sci 83:178–185.

Ritthidej GC, Chomto P, Pummamgura S, Menasveta P. 1994. Drug Dev Ind Pharm 20:2109–2134.

Schipper NGM, Olsson S, Hoogstraate JA, De Boer AG, Vårum KM, Artursson P. 1997. Pharm Res 14:923–929.

Schipper NGM, Vårum KM, Artursson P. 1996. Pharm Res 13:1686–1692.

Skaugrud O. 1989. Manuf Chem 60:31–35.

Swenson ES, Curatolo WJ. 1992. Adv Drug Delivery Rev 8:39–92.

Takeuchi H, Yamamoto H, Niwa T, Hino T, Kawashima Y. 1994. Chem Pharm Bull 42:1954–1956.

Thacharodi D, Rao KP. 1995. Biomaterials 16:145–148.

Tomita M, Hayashi M, Awazu S. 1996. J Pharm Sci 85:608–611.

Upadrashta SM, Katikaneni PR, Nuessle NO. 1992. Drug Dev Ind Pharm 18:1701–1708.

Vårum KM, Anthonsen MW, Grasdalen H, Smidsrød O. 1991. Carbohydr Res 211:17–23.

Wilson CG, Washington N. 1989. In: Wilson CG, Washington N, eds. Physiological Pharmaceutics: Biological Barriers to Drug Absorption. Chichester, UK: Ellis Horwood, pp 1–186.

14
Plant Lectins for Oral Drug Delivery to Different Parts of the Gastrointestinal Tract

Arpad Pusztai and Susan Bardocz
The Rowett Research Institute, Bucksburn, Aberdeen, Scotland

Stanley W. B. Ewen
University Medical School, Aberdeen, Scotland

I. INTRODUCTION

The specific, site-directed and efficacious delivery of drugs, particularly peptides, proteins, or other macromolecules, to different compartments of the gastrointestinal tract through the oral route is a major challenge in pharmaceutical science (1). The main reason for this is that without specific and direct interaction between the drug and the epithelial surface and/or the availability of specific carrier systems, absorption of the drug into the systemic circulation or lymphatics is mainly by passive diffusion. This is usually slow and can be rather inefficient, particularly with higher molecular weight and lipid-insoluble pharmaceuticals. Another impediment is that after oral administration the drug is quickly dispersed and diluted by peristalsis in the small intestinal lumen, where it remains for a relatively short time (2–3 hours) before being cleared into the large intestine. A variable but potentially large proportion of the drug therefore cannot make the best use of the high surface area and absorptive potential of the small bowel and is potentially wasted unless its destination was the colon in the first place.

The situation is even more serious with macromolecular drugs (2). Although it is generally accepted that macromolecules can cross the intestinal epithelium and are transported into the systemic circulation, the extent of their absorption is slight and unsuitable for effective drug delivery. Moreover, as most macromolecules are degraded with relative ease by pancreatic, luminal, and brush border hydrolases, their effective luminal concentration can be substantially reduced during passage through the digestive tract, further limiting the extent of their systemic absorption. Although incorporation of sensitive macromolecules into microparticles, liposomes, or other forms of carrier systems can afford considerable protection against degradation, this encapsulation may further hinder the access of the drug to the mucosal surface (3,4). However, it is possible to increase the bioadhesiveness of microparticles. There are a number of bioadhesive systems based on hydrophyllic, water-insoluble, high-molecular-weight, self-dissolving polymers that adhere nonspecifically to the gut wall and particularly its protective mucus layer (reviewed in Ref. 4; see also other chapters of this book). Although most of these systems are relatively ineffective in practice, some polymers offer considerable promise of high bioadhesiveness (5,6).

One of the most attractive strategies for overcoming the limitations inherent in oral drug administration is to use chemical agents that can effectively target the pharmaceutical to adhere to the mucosal surfaces of the different functional compartments of the gut. The expectation is that this will delay the transit of the drug through the gut, deliver it to specific cells and sites within the alimentary tract, and, optimally, increase its rate of endo- or transcytosis by mediation by the targeting agent.

By directly linking these targeting molecules to the drug or to self-dissolving delivery systems containing the drug, not only should the residence time of the drug be increased in the targeted compartment of the gut but also, by bringing the drug close to the epithelial surface, its transepithelial transport rate should be considerably enhanced. However, a one-to-one association of the drug with a targeting agent appears to be of low efficiency for practical drug delivery because there is some limitation on the amount of the pharmaceutical compound that can be delivered into the systemic circulation. Moreover, implicit in this technique is that relatively large amounts of the targeting agent are also systemically absorbed and this may not be desirable. However, with macromolecular drugs, such as hormones, growth factors, vaccines, and other expensive highly bioactive agents, this may not be a major constraint, particularly if their linkage to the targeting agent will protect the sensitive macromolecule from digestion during its passage through the alimentary canal. The use of microparticles appears to overcome some of the problems associated with the direct coupling of the drug to the targeting agent. However, this mode suffers from the apparent disadvantage that, due to steric hindrance, even the smallest microparticles may have difficulties in gaining close enough access to the mucosal surface to allow effective transepithelial transport to occur (3).

II. LECTIN-MEDIATED ADHESION TO GUT MUCOSAL SURFACES

As the luminal surface of the gut is one of the most extensively glycosylated tissues of the body, considerable advantages may be gained by using drug delivery systems whose bioadhesiveness is increased by coupling sugar-binding plant lectins or various bacterial and viral adhesion factors to the delivery vehicle (1,2,7). Unfortunately, bacterial lectins such as the type 1 fimbriae from *Escherichia coli* are relatively difficult to obtain in sufficient quantities. Despite similar difficulties, viral attachment proteins appear to be particularly attractive because of their potential involvement in the invasiveness of these organisms. Animal endogenous lectins also appear to have potential for targeting, but the outcomes of the few practical applications have been disappointing. However, over 200 plant lectins of various sugar specificities have now been described and some of these are available in relatively large quantities. Furthermore, the genes of most of these have been cloned and are available for producing lectins on a large scale by recombinant technology. Most important, some lectins have high, selective, and predictable physiological reactivity with tissues of the different functional compartments of the mammalian digestive tract and are therefore well suited for use as gut-targeting agents.

Orally administered lectins are highly resistant to proteolytic breakdown and bind in a key-lock fashion to the luminal surface of the gut with affinities in excess of K_d values of 10^4–10^6 when suitable carbohydrate structures are present on the glycocalix and/or epithelial cell membranes (7). Accordingly, plant lectins appear to be promising targeting agents for the oral delivery of chemically conjugated and/or encapsulated drugs to sites in the gastrointestinal tract (8) or the nasal mucosa (9). However, surface glycosylation is not necessarily the same in the gut of all mammals, and it is possible that different lectins need to be used to achieve the same targeting objectives in different animal species. As surface glycosylation varies in different compartments of the gut, lectins with suitable specificities are also expected to show selectivity in delivering compounds to different functional parts of the alimentary tract (8). Indeed, one of the main objectives of this chapter is to examine critically the validity of this concept in the rat gastrointestinal tract.

III. TRANSEPITHELIAL TRANSPORT OF LECTINS AND LECTIN-CONJUGATED DRUGS

As opposed to the rather low-level sampling by the gut immune system and the somewhat restricted nonspecific pinocytosis, plant lectins such as wheat germ agglutinin are transported through the absorptive enterocytes of the small intestinal epithelium in relatively large quantities (10,11) by receptor-mediated endocytosis, followed by transcytosis (3,4). There has been a great deal of interest in this process, particularly as it can be conveniently studied with filter-grown Caco-2 cells in vitro (1). The prospects of using bacterial and viral factors involved in colonization and invasion are particularly bright. With the intensive interest in these targeting agents, progress in the construction of practical delivery systems is likely to be fast. It is hoped that it may be possible not only to deliver macromolecular drugs to the gastrointestinal epithelium but also to transport them into the systemic circulation so efficiently that they may cost-effectively replace the parenteral route of administration. Although it is not yet certain which particular type of lectins will be most suitable for oral drug delivery, no other factors can offer the same advantages or the prospect of selective delivery of drugs to different parts of the gut because of differences in the glycosylation of their luminal surfaces (8).

IV. SURFACE GLYCOSYLATION IN DIFFERENT FUNCTIONAL COMPARTMENTS OF THE RAT DIGESTIVE TRACT

In studies aimed at exploring whether lectins are suitable targeting agents for selective drug delivery to different parts of the gut, rats and mice have been good in vivo models. However, before expensive and labor-intensive animal experimentation, the establishment of a low-resolution glycosyl map of the epithelium of the alimentary tract, particularly the stomach, small intestine, and caecum or colon (12,13), has been shown to be an essential preliminary step. Using suitably labeled [e.g., with digoxigenin, biotin, fluorescein isothiocyanate (FITC), and other tags] lectin derivatives, it is possible to identify in vitro which particular lectin is likely to have high in vivo reactivity and selectivity with the epithelium of the desired part of the gut. However, it is essential to confirm whether this high in vitro reactivity and binding of a particular lectin in the initial histological examination also occurs in vivo. This is best achieved by immunohistology using antilectin antibodies (coupled with a suitable staining methodology) on tissue sections taken from rats gavaged with appropriate lectins and killed 1–2 hours later. It is also advisable to follow up this first qualitative test with quantitative measurements of both free and tissue-bound lectin. The amount of free lectin in the lumen and, more important, the amount of lectin actually bound to the epithelium and extractable from it with solutions containing appropriate haptenic sugars can best be estimated by enzyme-linked immunosorbent assay (ELISA) using antilectin antibodies. In this way, not only is the binding activity of a particular lectin validated but also its likely efficacy in an in vivo targeting application can be gauged.

A. Stomach

Although the absorptive potential for nutrients of the stomach epithelium is low, there are a number of instances in which targeting of drugs to the stomach and increasing their residence time therein could offer distinct advantages. These include facilitating the binding of drugs or antibodies effective against *Helicobacter pylori* to stomach surface tissues and delivering antitumor agents and drugs to actual stomach tumor tissue. It should also be possible to improve the protection of the stomach epithelium against the erosive action of acid by coating the surface with lectin-conjugated antiacid protective agents and drugs. Such lectin applications have the advantage that, because of negligible absorption of macromolecules through the stomach epithelium, the systemic absorption of the targeting lectin is likely to be slight, even when the lectin increases the residence time of the drug in the stomach.

Reactivity of the stomach epithelium is particularly strong with mannose (Man)-specific lectins such as *Galanthus nivalis* agglutinin (GNA) both in vitro and in vivo. The reaction is not confined to the surface and neck region, as GNA also reacts with cells farther down in the gastric pits (Fig. 1A). Reactivity with lectins specific for N-acetylglucosamine (GlcNAc) such as wheat (*Triticum aestivum*) germ agglutinin (WGA) is also strong (Fig. 1B), whereas the binding of lectins specific for galactose/N-acetylgalactosamine (Gal/GalNAc), such as soybean (*Glycine max*) agglutinin (SBA), to the stomach epithelium and to some extent the parietal cells is relatively weak (not illustrated). Lectins specific for complex glycosyl side chains such as phytohemagglutinin (PHA; *Phaseolus vulgaris* isoagglutinins), bind extensively to the surface cells of the stomach of rats gavaged with this lectin (Fig. 1C). This in vivo binding to the stomach is long lasting as it was detectable by immunohistochemistry 24 hours after rats were given a single dose of PHA (not illustrated). A particularly distinctive feature of PHA binding is the strong and apparently selective staining of the parietal cells located halfway down the gastric pits (Fig. 1C). This is particularly long lasting and was detectable in large amounts even 48 hours after the rats received a single oral dose of PHA, whereas it virtually disappeared from surface cells of the gastric epithelium by this time (Fig. 1D). The time-dependent selectivity and reactivity of PHA with parietal cells offer a potential means for selective targeting and delivery of drugs to these cells, provided that their coupling to PHA does not affect this reactivity.

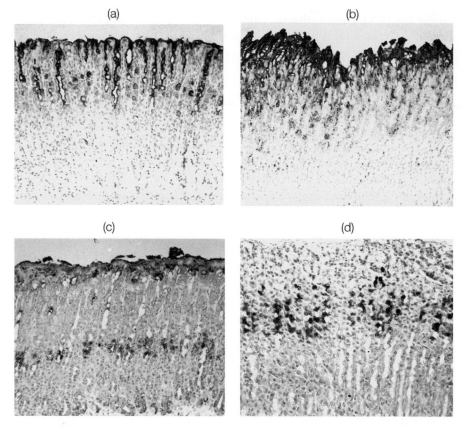

Figure 1 In vivo binding of orally administered lectins to the stomach epithelium of the rat. Groups of rats (five rats per group) were given different lectins as part of their daily diet (42 mg/rat). After 10 days, they were killed and dissected and sections were taken from their stomach, fixed in 4% buffered (pH 7) paraformaldehyde, embedded in paraffin wax, sectioned, and then, after reaction with the appropriate polyclonal monospecific antibody, processed by antibody peroxidase-antiperoxidase (PAP) staining. The following lectins were used: (a) *Galanthus nivalis* agglutinin, GNA; (b) wheat (*Triticum aestivum*) germ agglutinin, WGA; (c) *Phaseolus vulgaris* agglutinin, PHA. In (d), the rats were first exposed to PHA and then switched to control diet without PHA for 3 days. However, the sections were still reacted with anti-PHA antibody before processing with the PAP method. (Magnification ×10 for (a), (b), and (c) and ×20 for (d).)

B. Small Intestine

The small intestine is not only the most important compartment of the gastrointestinal tract from a nutritional standpoint but also the main target for oral drug delivery. The high surface area of the villous epithelium ensures relatively high rates of systemic absorption of drugs, even in the absence of active transport carriers to facilitate their transepithelial transport into the circulation. In addition to the high absorptive capacity of the numerous villus enterocytes, antigen sampling by M cells of Peyer's patches of the lymphoid tissue of the gut presents an opportunity for the absorption of macromolecular drugs. Although systemic absorption of compounds by this route differs from general transepithelial transport, it is particularly suitable for the uptake of antigens for vaccination and possibly for the systemic absorption of macromolecules with high biological activity. Even though only limited amounts of materials can be absorbed through M cells, this is still a physiologically important route. Accordingly, lectin reactivity needs to be investigated with cells from both the small intestinal villous epithelium and the gut lymphoid regions. Selectivity in lectin reactivity is of particular importance.

Reactivity of the small intestinal epithelium with GNA and other lectins specific for mannose is slight, particularly in the jejunum (14) (Fig. 2A). The M cell binding of these lectins is slight to moderate (not illustrated). In contrast, lectins with Gal/GalNAc specificity, such as the lectin ML-1 from mistletoes (*Viscum album*), bind strongly and persistently to membranes of epithelial cells and are endocytosed extensively by villus enterocytes and even goblet cells (15). Although ML-1 also binds appreciably to M cells, its subsequent transepithelial transport and uptake by antigen-presenting cells are slight (15).

Fucose-specific lectins have assumed great importance in drug delivery studies, particularly after many reports that they selectively bind to and are transcytosed by M cells, at least in the mouse follicle-associated epithelium overlying gut-associated lymphoid tissues (16). However, glycosylation of M cells in different species shows major differences, and no lectin has yet been identified that selectively binds to the membranes of M cells of small intestinal Peyer's patches in rats, rabbits, and humans (16,17). In the present work using a rat model these findings were fully confirmed. Moreover, the reactivity of fucose-specific lectins such as AAA (*Aleuria aurentia*) agglutinin with luminal membranes of villus enterocytes of healthy adult rats is slight to virtually nil, depending on the previous dietary history of the animals (12). In contrast, these lectins react extensively and avidly with small intestinal goblet cells and goblet cell mucins (Fig. 2B). This high mucin reactivity in the rat small intestine probably represents a major barrier to the attachment of fucose-specific lectins to epithelial cells, including M cells, that would be even more serious for lectin-conjugated microparticles.

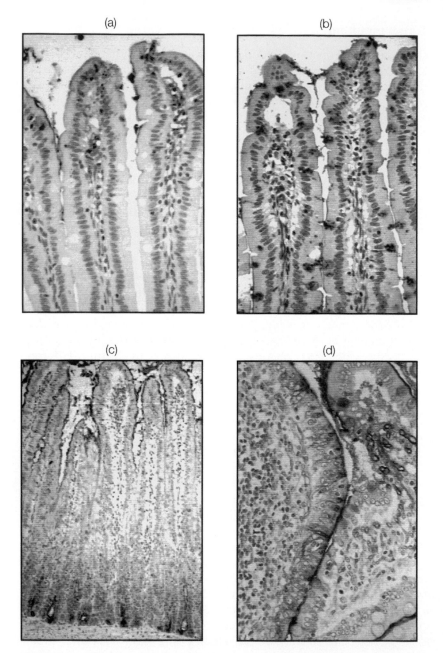

Figure 2 Binding of lectins of different carbohydrate specificities to the small intestinal epithelium. (a, c, and d) Groups of rats (five rats per group) were given

(caption continues on opposite page)

The in vitro reactivity of lectins specific for sialylated (NeuAc) terminal glycosyl groups with the small intestinal epithelial surface has been shown by histology to be slight and somewhat variable (12). Thus, MAA, a lectin isolated from the bark of *Maackia amurensis* and specific for NeuAc α2–3Gal/GalNAc, reacts with luminal membranes with moderate avidity. However, it binds to goblet cells and mucins almost as strongly as AAA, indicating that rat mucins are both terminally sialylated and fucosylated (12). Accordingly, as the in vivo binding of this lectin to the epithelium following oral administration is also at best only moderate (not illustrated), its potential for use in drug delivery applications is limited. SNA-I, a type 2 RIP lectin from the bark of elderberry (*Sambucus nigra*) that is specific for NeuAC α2,6Gal/GalNAc, binds to only a negligible extent to the membranes of epithelial cells in the rat small intestine and to goblet cells and their constituent mucins. Although cytoplasmic glycoconjugates in small intestinal enterocytes react strongly with SNA-I (12), this may have little relevance to drug delivery applications.

Lectins with specificity for GlcNAc such as WGA are avidly bound to the epithelial surface throughout the entire small intestine (10) (Fig. 2C). Similarly strong surface binding has been observed with LEA, a lectin from tomatoes (*Lycopersicon esculentum*) that is widely used in drug delivery research (1) (not illustrated). Moreover, both WGA and LEA show definite M cell binding. They are also transported across the epithelium of Peyer's patches and taken up by antigen-presenting cells in subepithelial tissues (Fig. 2D).

The membrane glycoconjugates of small intestinal villus enterocytes and goblet cells of healthy adult rats almost exclusively contain complex glycosyl side chains. Consequently, it is not surprising that lectins specific for complex glycans such as PHA or RPA, a lectin isolated from the bark

---

different lectins as part of their daily diet (42 mg/rat). After 10 days, they were killed and dissected and the sections taken from their small intestine (5 cm from pylorus) were first reacted with the appropriate antilectin antibody and then processed by PAP staining. In (a) the binding of *Galanthus nivalis* agglutinin, GNA, was slight to jejunal villi, but the strong staining in (c) indicated that wheat (*Triticum aestivum*) agglutinin, WGA, was bound avidly to the jejunum. Tomato (*Lycopersicon esculentum*) lectin gave moderately strong staining of Peyer's patch M cell region, but its endocytosis was slight (d). In sections taken from rats fed with control diet, fixed, and embedded in paraffin wax (b), goblet cells of jejunal villi stained strongly with digoxigenin labelled *Aleuria aurantia* agglutinin (a fucose-specific lectin) after exposure to antidigoxigenin antibody and processing by the PAP method. (Magnification ×25 in (a), (b), and (d) and ×10 in (c).)

of false acacia (*Robinia pseudoacacia*) tree, are powerful reagents for the rat small intestine (7,12). They exhibit strong and reversible surface binding to both villous and crypt epithelia of the small bowel (Fig. 3A–C) and are extensively endo- and transcytosed by the epithelial cells. PHA is also a powerful reagent for M cells of rat small bowel Peyer's patches. It is bound avidly and extensively transepithelially transported through M cells into subepithelial tissue, where it is taken up by antigen-presenting cells (Fig. 3D), leading to the development of a powerful systemic immune response (7). Similar findings were made with RPA (A. Pusztai, unpublished results).

Clearly, the high gut reactivity of lectins that are specific for a complex carbohydrate moiety makes them ideal candidates in targeted oral drug delivery, particularly as neither PHA nor RPA reacts with goblet cell mucins, which can create a barrier to the successful delivery of the lectin and the drug to the epithelial surface. These lectins are extensively endo- and transcytosed by both enterocytes and M cells, indicating that they offer two possible routes of entry into the systemic blood circulation. Although PHA is reputed to have harmful effects when used at very high concentrations (above 0.2 g/kg body weight), it should be completely safe at doses that are likely to be encountered in drug delivery applications.

C. Large Intestine

In rats the large intestine is subdivided into two functionally different compartments, the cecum and the colon. For completeness, the glycosylation of both cecal and colonic epithelia and their lectin reactivity will be discussed here even though this subdivision has little meaning in the human context.

In contrast to the poor reactivity of mannose-specific lectins with surface cells of the small intestine, their in vivo binding to the large intestinal epithelium is strong and relatively long lasting. Thus, orally administered GNA binds equally well to both cecal and colonic epithelia (Fig. 4A and B). GlcNAc-specific lectins behave similarly (not illustrated). The reaction of lectins of Gal/GalNAc specificity is diffuse, probably as a result of the

---

Figure 3 In vivo binding of *Phaseolus vulgaris* agglutinin, PHA, to the rat small intestine. Rats were given PHA (42 mg/rat) as part of their daily diet. In panels (a) (upper part of a jejunal villus), (c) (jejunal crypt), and (d) (Peyer's patch M cells) the sections taken from the small intestine of rats and processed through anti-PHA antibody treatment followed by the PAP method showed extensive endo- or transcy-

(caption continues on opposite page)

(a) (b)

(c) (d)

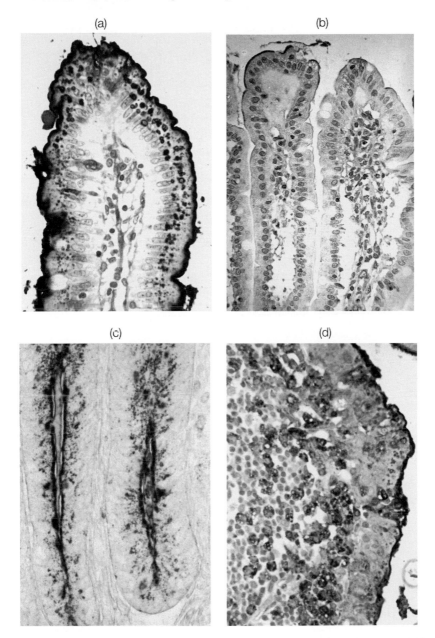

tosis. (b) After receiving PHA for 10 days the rats were switched to control diet for 48 hours but the jejunal sections were processed as above; the lack of staining indicates that binding of PHA was fully reversible. (Magnification ×40 in (a) and (d) and ×25 in (b) and (c).)

(a)   (b)

(c)   (d)

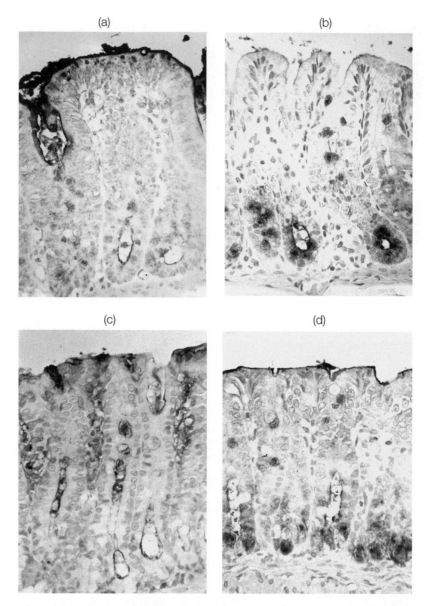

Figure 4 In vivo binding of different lectins to the rat large intestine. Groups of rats (five rats per group) were given different lectins as part of their daily diet, killed on the 10th day, and dissected, and sections of their large intestines were fixed, embedded in paraffin wax, sectioned, reacted with the appropriate antilectin antibody, and stained by the PAP method. (a) and (d) show strong binding of *Phaseolus*
(caption continues on opposite page)

presence of large amounts of mucus in the lumen (not illustrated). In contrast, PHA (complex specificity) reacts strongly with the large intestinal epithelium in rats gavaged with this lectin (Fig. 4C). The binding is long lasting as the presence of the lectin is discernible, particularly in the colonic crypts even 3 days after a single oral dose of PHA (Fig. 4D).

In conclusion, both in vitro and in vivo lectin binding studies indicate that there are sufficiently clear differences in surface glycosylation of the different parts of the digestive tract to enable appropriate lectins to be used for targeted and selective binding. Thus in the rat, mannose-specific lectins can be used to target the stomach and large intestine, without appreciable binding to the small intestine with the possible exception of a slight reactivity with M cells. Conversely, Gal/GalNAc-specific lectins are excellent and moderately selective reagents for the small intestine but not for the stomach or the large intestine. Most of these lectins have only slight reactivity with Peyer's patch M cells. GlcNAc-specific lectins show no selectivity; they react with all parts of the gastrointestinal tract. Fucose- and sialic acid–specific lectins react mainly with mucins and goblet cells wherever they occur. Unfortunately, in our rat model they have no selective binding to M cells. Lectins with complex specificity react strongly with the surface membranes in the whole digestive tract and also with Peyer's patch M cells. However, this poor selectivity is more than compensated by PHA's lack of mucin binding, the strength of its reactivity with all epithelial cell types, and its high rate of endo- and transcytosis by both enterocytes and M cells.

V. LECTINS AS ORAL DRUGS

Lectins are proteins or glycoproteins with high and specific biological activities. It is therefore not surprising that they can be considered not only as

---

vulgaris agglutinin, PHA, to the cecum and colon, respectively, with appreciable endocytosis of the lectin in (a). The surface binding of PHA was fully reversible (b). However, appreciable binding of PHA to the cecal crypts developed 48 hours after the rats were switched to control diet (b). Surface binding of the *Galanthus nivalis* agglutinin, GNA, to the colon was patchy but with clear penetration of the lectin into the colonic crypts (c). GNA binding to the cecum was similar (not illustrated). (Magnification ×25.)

targeting agents but also as drugs in their own right. They survive gastro-intestinal passage in functionally and immunologically intact form (18) and interact extensively and specifically with glycan moieties of epithelial cell membrane components, including receptors of growth factors, hormones, and other biologically active factors. Lectins therefore have the potential to be used as oral drugs (7,8). Thus, orally administered PHA (*P. vulgaris* agglutinin) and SBA (*G. max* agglutinin) are both potent stimulants of cholecystokinin (CCK) release from duodenal enteroendocrine cells in rats, leading to increased secretion of digestive enzymes from the acinar pancreas (19,20).

Although the involvement of CCK in the regulation of pancreatic metabolism in humans is not clear-cut, lectins may, like trypsin inhibitors, stimulate pancreatic enzyme secretion through a different pathway, with possible applications in acute pancreatitis in humans. Some lectins are excellent hormone mimics (7). Like insulin, PHA stimulates the oxidation of glucose and the synthesis of triglycerides by isolated fat cells in vitro (7,21). Moreover, by reducing insulin but not glucose levels in the blood circulation, PHA is a potent stimulant of the catabolism of body fat in both normal (22) and genetically obese (Zucker) rats (23) at or below nontoxic oral doses. It is therefore possible that PHA may in future be used in the clinical control of hyperglycemia and obesity. Furthermore, atrophy of the small intestine in parenterally fed rats (M. Jordinson et al., unpublished) and the lethal damaging effects of irradiation and/or chemotherapy agents such as 5-fluorouracil (24) can be successfully reversed by the oral administration of PHA or other lectins with powerful growth factor activity for the gut (7). Such use of these lectins as drugs in clinical trials is imminent. PHA reacts selectively with parietal cells in the rat stomach (13) and suppresses the stomach's production of acid (25), offering a possible therapeutic means for the reduction of excess stomach acidity, perhaps even in pathogenicity. Lectins that have the same carbohydrate specificity of binding as some bacterial adhesins are potential and selective blockers of infection by bacteria expressing the same adhesins. Thus, the oral administration of GNA, the agglutinin from snowdrop (*G. nivalis*) bulbs, can significantly reduce the colonization of the rat intestines by type 1 fimbriated bacteria such as *E. coli* (14). Furthermore, the use of abrin, a potent A-B–type toxin, as an oral immunomodulating drug in cancer and human immunodeficiency virus (HIV) therapy has been patented (26). Moreover, mistletoe lectin, ML-I, a type 2 ribosome-inactivating protein (RIP) (27), and PHA (28) are already being used parenterally in clinical tumor therapy in some countries and their oral use (15,29) in tumor suppression is also being explored.

VI. LECTINS DIRECTLY COUPLED TO DRUGS

It is possible to cross-link drugs chemically to lectins. However, only a few examples of this type of drugs, particularly *stable drug-lectin conjugates* in which the drug is unaffected by the lectin, have been described, even though their oral use may offer some advantages. Many chemical linkages can be used to form such conjugates, including peptide bonds, thioethers, chlorambucyl-piperazines, and homo- and heterobifunctional cross-links, depending on the chemical nature of the drug and the availability of reactive groups on it. For a full description of these the reader is referred to detailed textbooks (e.g., Ref. 30). A few examples of these conjugates have been described, including those in which concanavalin A (Con A) is conjugated with non-protein/peptide drugs such as daunomycin (31), methotrexate (32), or other similar drugs. Most of these are intended mainly for parenteral and not oral use. Unfortunately, on oral administration of these stable conjugates both the targeting lectin and the drug should be absorbed together into the blood circulation. Therefore they can suffer from the disadvantage that relatively large amounts of lectins can enter into the circulation with possibly undesirable side effects.

It appears to be more advantageous to use *labile conjugates* from which the drug is released in vivo. In these the lectin is used mainly as a targeting agent that is not necessarily taken up into the circulation, at least not in significant amounts. These conjugates are usually based on cleavable chemical bonds such as intrinsic and/or generated disulfide bonds, acid-labile cross-links, or photocleavable cross-links. These labile conjugates, particularly those containing polypeptide or protein drugs bound to lectins, are based on the A-B–type natural toxins, where a toxic subunit A is covalently linked, usually by a disulfide bridge(s), to a B lectin subunit. These are of great potential importance and include type 2 RIP toxins, bacterial toxins, and other similar toxic compounds that are nature's own drug delivery combinations. However, most of these lectin-toxin drug conjugates have been intended for parenteral use in clinical tumor therapy. Unfortunately, lectins react with all types of cells and soluble glycoconjugates in the blood circulation and therefore do not have sufficient selectivity to target only tumor cells. Thus, these drugs have now been largely superseded for parenteral use by the development of cytotoxic immunotoxins (Ehrlich magic bullets; reviewed in Refs. 7, 33, and 34) in which a target cell–specific (usually tumor cells) antibody is linked instead of the lectin subunit to the toxic A subunit. Despite this, magic bullets based on targeting lectins may still have important applications in colon tumor therapy because lectins are far more stable in the gastrointestinal tract than antibodies and have sufficient selectivity to deliver the lectin-toxin conjugate and kill the tumor cell.

VII. LECTINS COUPLED TO DELIVERY VEHICLES CONTAINING DRUGS

The composition and properties of various microparticles, liposomes, and other potential drug delivery vehicles that can be used for lectin coupling are described in detail in other chapters in this book. Therefore, only a few examples of drug delivery systems will be described here in which the efficacy of the targeting and absorption of the drug has been improved by the use of lectin coupling.

Chemical conjugation to lectins has been shown to enhance the extent of uptake of polystyrene particles by both isolated and fixed pig erythrocytes and monolayers of human Caco-2 cell cultures in vitro (35), and many other studies have also demonstrated that lectins are useful agents for increasing the bioadhesiveness of microparticles. Thus, tomato lectin (TL) is one of the most studied examples of plant lectins in drug delivery studies, with the benefit that it is apparently nontoxic for mammals and binds strongly to intestinal villi (36) and to everted rat intestinal rings (37). Unfortunately, orally administered microspheres coupled to TL were not retarded appreciably by the rat gut in vivo (1), possibly because of the high reactivity of this lectin with intestinal mucus (1,38), whose rate of clearance in the small bowel is relatively fast (3–4 hours). However, in a reinvestigation of the effect of TL on the uptake of polystyrene microparticles (500 nm) by the small intestine of rats dosed daily for 5 days, the absorption of particles covalently attached to TL was over 10 times greater than that of plain microspheres. Moreover, the locus of uptake shifted from lymphoid to normal absorptive epithelial tissue in the small intestine (39). It was also shown that when female Wistar rats were gavaged daily for 5 days with 500-nm fluorescent microparticles chemically linked to TL, 23% of the dose was systematically absorbed through small intestinal nonlymphoid tissues. This was in contrast to less than 0.5% uptake of the same particles coapplied with N-acetylchitotetraose (40).

In similar studies using PHA to increase the bioadhesiveness of microparticles, PHA-coupled 2-μm latex microspheres not only were quantitatively bound by rat brush border membrane preparations in vitro but also reacted in vivo with the surface of the small intestine, resulting in retardation during their small intestinal passage in vivo. Thus, about 10% of the initial dose of PHA-coupled particles was still present in the small bowel 5 hours after oral administration, whereas uncoated particles or those conjugated with control proteins were undetectable (1). Retardation of lectin-latex conjugates was dependent on the sugar specificity of the lectin (1). Control microspheres in the size range 0.05–2 μm or those conjugated to lectins specific for mannose/glucose (Con A) or GlcNAc (WGA or TL) reached the cecum 2 hours after oral administration. In contrast, microparticles chemically coupled to lectins specific for Gal/GalNAc (SBA) or to PHA were retarded, with 20–30% of the initial dose distributed throughout the small intestine as far back as the duodenum (1). However, in contrast to TL-linked microparticles (39,40), none of these lectin-conjugated fluorescent microspheres were systemically absorbed in measurable quantities.

Bacterial lectins were also shown to facilitate microparticle adherence to the intestinal surface. In fact, mannose-sensitive type 1 fimbriae were among the first bacterial adhesins to be shown to increase the association of polystyrene microspheres with the surface of the rat intestine (41). However, no transepithelial transport of these particles apparently occurred.

The reported discrepancies in the extent of systemic absorption of lectin-conjugated particles may be due to the different sizes of the microparticles used in different laboratories. Thus, FITC-labeled or dinitrophenylated LTB (heat-labile *E. coli* B subunit) and Con A, given orally or injected into intestinal loops, were not only bound to the surface of intestinal cells but also readily transported across these cells (42,43). Similarly, soluble FITC-labeled CTB (Cholera toxin B subunit) with a diameter of 6.4 nm was extensively bound to apical membranes of all epithelial cells, whereas CTB coupled to 14-nm colloidal gold particles (diameter of 28.8 nm) adhered to M cells but not to enterocytes in vivo. When CTB was coupled to fluorescent nanoparticles with a final diameter of 1.13 μm, the conjugate did not attach to either enterocytes or M cells (3). Interestingly, lectin-bearing polymerized liposomes with an average diameter of 100 nm also appeared to have high affinity for small intestinal Peyer's patch M cells. Thus, liposomes modified with *Ulex europeaus* lectin, UEA-I, or WGA were taken up by and subsequently internalized through the gastrointestinal tract of female Balb/c mice in higher amounts than the lectin-free liposomes (43).

VIII.   CONCLUSION

Lectins can increase the penetration and adherence of microparticles to the intestinal epithelium and on occasion can also increase the absorption of the microparticles. There is evidence that the passage time of some microparticles linked to suitable lectins through the gastrointestinal tract is significantly increased, leading to an increase in their systemic absorption. Furthermore, it may be possible to use lectins as selective targeting agents for different gut compartments and even for different cells (e.g., complex specific lectins for parietal cells or fucose-specific lectins for mouse M cells).

However, there are still considerable obstacles to establishing predictable and efficient oral delivery systems for encapsulated drugs, even with the use of lectin targeting. Thus, the mucus layer in most instances presents a first major barrier that prevents close enough contact between the microparticle and the epithelial cell membrane. Indeed, the extensive interaction of nanoparticles conjugated to lectins with mucins may be one of the main reasons for the disappointingly poor retardation of tomato lectin–conjugated particles in the intestine in vivo (1). One of the first priorities must therefore be to focus our research on overcoming this major impediment to successful delivery of oral drugs encapsulated in microparticles. However, the mucus layer does not present a significant obstacle to close enough penetration to epithelial cell membranes if the lectins do not react with mucins. These are not major impediments to the direct binding of these lectins, even if they are chemically conjugated with drugs, antigens, small nanoparticles, or liposomes. Therefore the targeted and possibly selective delivery of small quantities of drugs by lectin mediation to discrete locations in the gastrointestinal tract is a practical possibility even at present, with important implications for oral vaccination. Finally, as there are opportunities for using lectins (and type 2 RIP proteins) as drugs in their own right, the future of lectin mediation appears to be bright.

REFERENCES

1. CM Lehr, A Pusztai. The potential of bioadhesive lectins for the delivery of peptide and protein drugs to the gastrointestinal tract. In: A Pusztai, S Bardocz, eds. Lectins—Biomedical Perspectives. London: Taylor & Francis, 1995, pp 117–140.
2. A Pusztai. Transport of proteins through the membranes of the adult gastrointestinal tract—a potential for drug delivery? Adv Drug Delivery Rev 3: 225–238, 1989.
3. A Frey, KT Giannasca, R Weltzin, PJ Giannasca, H Reggio, WI Lencer, MR Neutra. Role of the glycocalix in regulating access of microparticles to apical plasma membranes of intestinal epithelial cells: Implications for microbial attachment and oral vaccine targeting. J Exp Med 184:1045–1059, 1996.
4. DT O'Hagan. Microparticles as oral vaccines. In: Novel Delivery Systems for Oral Vaccines. Boca Raton, FL: CRC Press, 1994, pp 175–205.
5. DE Chickering, JS Jacob, E Mathiowitz. Poly (fumaric-co-secacic) microspheres as oral drug delivery systems. Biotechnol Bioeng 52:96–101, 1996.
6. E Mathiowitz, JS Jacob, YS Jong, GP Carino, DE Chickering, P Chatuversi, CA Santos, K Vijayaraghaven, S Montgomery, M Bassett, C Morrell. Biologically erodible microspheres as potential oral drug delivery systems. Nature 386:410–414, 1997.

7. A Pusztai. Plant Lectins. Cambridge: Cambridge University Press, 1991.
8. A Pusztai. Lectin-targeting of microparticles to different parts of the gastrointestinal tract. Proc Int Symp Controlled Release Bioact Mater 22:161–162.
9. PJ Giannasca, JA Boden, TP Monath. Targeted delivery of antigen to hamster nasal lymphoid tissue with M-cell–directed lectins. Infect Immun 65:4288–4298.
10. A Pusztai, SWB Ewen, G Grant, DS Brown, JC Stewart, WJ Peumans, EJM Van Damme, S Bardocz. Antinutritional effects of wheat germ agglutinin and other *N*-acetylglucosamine specific lectins. Br J Nutr 70:313–321, 1993.
11. A Pusztai, F Greer, G Grant. Specific uptake of dietary lectins into the systemic circulation of rats. Biochem Soc Trans 17:81–82, 1989.
12. A Pusztai, SWB Ewen, G Grant, WJ Peumans, EJM van Damme, ME Coates, S Bardocz. Lectins and also bacteria modify the glycosylation of gut receptors in the rat. Glycoconj J 12:22–35, 1995.
13. S Bardocz, G Grant, SWB Ewen, TJ Duguid, DS Brown, K Englyst, A Pusztai. Reversible effect of phytohaemagglutinin on the growth and metabolism of rat gastrointestinal tract. Gut 37:353–360, 1995.
14. A Pusztai, G Grant, RJ Spencer, TJ Duguid, DS Brown, SWB Ewen, WJ Peumans, EJM Van Damme, S Bardocz. Kidney bean lectin–induced *Escherichia coli* overgrowth in the small intestine is blocked by GNA, a mannose-specific lectin. J Appl Bacteriol 75:360–368.
15. A Pusztai, G Grant, E Gelencsér, SWB Ewen, U Pfüller, R Eifler, S Bardocz. Effects of an orally administered mistletoe (type-2 RIP) lectin on growth, body composition, small intestinal structure and insulin levels in young rats. J Nutr Bioch, in press.
16. MA Jepson, MA Clark, N Foster, CM Mason, MK Bennett, NL Simmons, BH Hirst. Targeting to intestinal M cells. J Anat 189:507–516, 1996.
17. R Sharma, EJM van Damme, WJ Peumans, P Sarsfield, U Schumacher. Lectin binding reveals divergent carbohydrate expression in human and mose Peyer's patches. Histochem Cell Biol 105:459–465, 1996.
18. A Pusztai and S Bardocz. Biological effects of plant lectins on the gastrointestinal tract: Metabolic consequences and applications. Trends Glycosci Glycotechnol 8:149–165, 1996.
19. KH Herzig, S Bardocz, G Grant, R Nustede, UR Fölsch, A Pusztai. Red kidney bean lectin is a potent cholecystokinin releasing stimulus in the rat inducing pancreatic growth. Gut 41:333–338, 1997.
20. M Jordinson, PH Deprez, RJ Playford, TC Freeman, M Alison, J Calam. Soybean lectin stimulates pancreatic exocrine secretion via CCK-receptors in rats. Am J Physiol 270:G653–G659, 1996.
21. A Pusztai, WB Watt. Isolectins of *Phaseolus vulgaris*. A comprehensive study of fractionation. Biochim Biophys Acta 365:57–71, 1974.
22. S Bardocz, G Grant, A Pusztai, MF Franklin, A de FFU Carvalho. The effect of phytohaemagglutinin at different dietary concentrations on the growth, body composition and plasma insulin of the rat. Br J Nutr 76:613–626, 1996.

23. A Pusztai, G Grant, WC Buchan, S Bardocz, A Carvalho, SWB Ewen. Lipid accumulation in obese Zucker rat is reduced by inclusion of raw kidney bean in the diet. Br J Nutr, in press.

24. A Pusztai, S Bardocz, G Koteles, R Palmer, N Fish. Lectin compositions and uses thereof. International Patent Application PCT/GB97/01668, 1997.

25. G Varga, S Bardocz, K Baintner, A Pusztai. Effect of phytohaemagglutinin on gastric acid secretion in conscious rats. Gut 37(suppl 2):A221, 1995.

26. TC Tung. Orally administrable anti-metastatic lectin compositions and methods. US Patent 5,053,386, 1991.

27. T Hajto, K Hostanska, HJ Gabius. Modulatory potency of the beta-galactoside–specific lectin from mistletoe extract (iscador) on the host defense system in vivo in rabbits and patients. Cancer Res 49:4803–4808, 1989.

28. BM Vimer. Therapeutic immunostimulating effects of plant mitogens exemplified by the L4 isolectin of PHA. Cancer Biother Radiopharm 12:195–212, 1997.

29. IF Pryme, A Pusztai, G Grant, S Bardocz. Dietary phytohaemagglutinin slows down the proliferation of a mouse plasmacytoma (MPC-11) tumour in Balb/c mice. Cancer Lett 103:151–155, 1996.

30. SS Wong. Chemistry of Protein Conjugation and Cross-Linking. Boca Raton, FL: CRC Press, 1991.

31. T Kitao, K Hattori. Concanavalin A as a carrier of daunomycin. Nature 265: 81–82, 1977.

32. T Tsuruo, T Yamori, S Tsukagoshi, Y Sakurai. Enhanced cytocidal action of methotrexate by conjugation to concanavalin A. Int J Cancer 26:655–659, 1980.

33. FA Drobniewski. Immunotoxins up to the present day. Biosci Rep 9:139–156, 1989.

34. GR Thrush, LR Lark, BC Clinchy, ES Vitetta. Immunotoxins—an update. Annu Rev Immunol 14:49–71, 1996.

35. CM Lehr, JA Bouwstra, W Kok, ABJ Noach, AG De Boer, HE Junginger. Bioadhesion by means of specific binding of tomato lectin. Pharm Res 9: 547–553, 1992.

36. DC Kilpatrick, A Pusztai, G Grant, C Graham, S Ewen. Tomato lectin resists digestion in the mammalian alimentary canal and binds to intestinal villi without deleterious effects. FEBS Lett 185:5–10, 1985.

37. B Naisbett, J Woodley. The potential use of tomato lectin for oral drug delivery. 1. Lectin binding to rat small intestine in vitro. Int J Pharm 107:223–230, 1994.

38. JM Irache, C Durrer, D Duchene, G Ponchel. In vitro study of lectin-latex conjugates for specific bioadhesion. J Controlled Release 31:181–188, 1994.

39. AT Florence, AM Hillery, N Hussain, PU Jani. Nanoparticles as carriers for oral peptide absorption: Studies on particle uptake and fate. J Controlled Release 36:39–46, 1995.

40. M Hussain, PU Jani, AT Florence. Enhanced oral uptake of tomato lectin–conjugated nanoparticles in the rat. Pharm Res 14:613–618, 1997.

41. AJ Caston, SS Davis, P Williams. The potential of fimbrial proteins for delaying intestinal transit of oral drug delivery systems. Proc Int Symp Controlled Release Bioact Mater 17:313–314, 1990.

42. GJ Russel-Jones. Oral vaccination with lectins and lectin-like molecules. In: Novel Delivery Systems for Oral Vaccines. Boca Raton, FL: CRC Press, 1994, pp 219–236.

43. GJ Russel-Jones. The potential of receptor-mediated endocytosis for oral drug delivery. Adv Drug Delivery Rev 20:83–97, 1996.

44. H Chen, V Torchillin, R Langer. Lectin-bearing liposomes as potential oral vaccine carriers. Pharm Res 13:1378–1382, 1996.

15

Bacterial Invasion Factors and Lectins as Second-Generation Bioadhesives

James H. Easson, Eleonore Haltner, and Claus-Michael Lehr
Saarland University, Saarbrücken, Germany

Dieter Jahn
Universität Freiburg, Freiburg, Germany

I. CONCEPTS OF BIOADHESION IN DRUG DELIVERY

The delivery of macromolecular biopharmaceuticals is limited by their large size and hydrophilicity, as well as by their poor metabolic stability. The delivery through the gastrointestinal tract is especially impeded by several important biological barriers. The main task of the GI tract is to digest and absorb foodstuffs, and therefore various kinds of proteases can be found (Woodley, 1994). This enzymatic barrier is one of the most important of the multiple barriers limiting the absorption of macromolecular drugs because of the abundance and variety of proteases in the GI tract. All protein and peptide drugs in the GI tract are subject to massive enzymatic destruction and are simply considered to be unsuitable for oral delivery.

Today, the degradation of macromolecules can be avoided by encapsulation of these sensitive molecules in protective carrier systems such as stabilized liposomes and nanoparticles. Damgè et al. (1988) reported a reduction of glycemia in diabetic rats by 50–60% after intragastric administration of insulin in polyalkylcyanoacrylate nanocapsules. They assumed that this remarkable effect was achieved by paracellular transport of the 220-nm particles from the gut lumen into the systemic circulation. Although these carrier systems can hinder enzymes from attacking the incorporated drug, the problem of overcoming the barrier of a tight epithelium remains.

The uptake of macromolecules by using the space between the cells of the epithelium is regulated by the tightness of the cell junctions. The most important cell junction is the tight junction, which divides the epithelial cell into an apical (luminal) and a basolateral compartment and contains a meshwork of interconnecting strands. Within these strands are "pores," which have effective radii in human small intestinal epithelium of 3–4 Å for ileum and 7–9 Å for jejunum (Nellans, 1991). Under physiological conditions it seems to be impossible for 220-nm particles to use this paracellular pathway; however, the tightness of the cell-cell junctions is hormonally regulated by the body's water balance (Lowe et al., 1988), dependent on the intracellular calcium concentration (Llopis et al., 1991), and may vary considerably.

A. Mucoadhesion

Since the early 1980s, the concept of bioadhesion has been studied to improve both local and systemic drug delivery (Park and Robinson, 1984; Junginger, 1991). Bioadhesive delivery systems are able to adhere to epithelia representing absorption sites for drug molecules. As many epithelial tissues are covered by mucus, a viscoelastic hydrogel of 1–5% water-insoluble glycoproteins (Fogg et al., 1994), systems capable of adhering to such surfaces are also referred to as mucoadhesive. The first reasons for using bioadhesive drug delivery systems were (a) to prolong the residence time at the site of drug action or absorption, (b) to better locate the drug at a given target site, e.g., upper versus lower GI tract, and (c) to intensify the contact with the mucosa to increase the drug concentration gradient (Junginger, 1990; Lehr, 1994).

Mucoadhesive polymers, of either natural or synthetic origin, have the remarkable ability to "stick" to wet mucosal surfaces by nonspecific, physicochemical mechanisms, such as van der Waals forces or hydrogen bonding. Mucoadhesion in a more strict sense refers to just one approach to realizing bioadhesion, namely through adhesion to the mucus gel layer. This is partly due to the physicochemical properties of the polymers used (Juninger, 1991). The improved absorption of drugs linked to or encapsulated in bioadhesive polymers has lately been attributed to the calcium chelating potency of some polymers. This property of the polymers may (a) inhibit certain metalloproteases, resulting in decreased proteolytic activity of certain metalloproteases (Luessen et al., 1995), and (b) deplete the available extracellular calcium, which opens the intercellular spaces and thus facilitates the paracellular uptake (Kriwet and Kissel, 1996). Two major problems, even for an intrinsically excellent mucoadhesive polymer, will remain: (a) rapid inactivation by soluble mucins, as well as food and other contents of the GI lumen, and (b) the fast mucus turnover (Lehr et al., 1992). To overcome these problems, some new ideas have been brought up about achieving bioadhesion by approaches not based on mucoadhesive polymers.

B. Cytoadhesion

The new concepts of bioadhesion are based on highly specific interactions between adhesive agents and cell surfaces comparable to a receptor-ligand or antibody-antigen interaction. The mechanisms by which such cytoadhesives may help to overcome epithelial barriers are the following: (a) long-term fixation directly onto the surface of cells, independent of mucus turnover, (b) induction of and participation in specific vesicular transport processes (endo- or transcytosis), and (c) modulation of epithelial permeability by receptor-mediated opening of tight junctions. This novel concept of bioadhesion should be referred to as cytoadhesion in order to distinguish it from mucoadhesion.

Various plant lectins, such as tomato lectin (TL) or wheat germ agglutinin (WGA), have been demonstrated to interact specifically with certain epithelia of the GI tract (Neutra et al., 1987; Naisbett and Woodley, 1994). Also, model drug delivery systems comprising WGA-modified stabilized liposomes (Chen et al., 1996) or tomato lectin coupled to latex nanoparticles (Hussain et al., 1997) have been investigated in vivo. Both groups found that the lectin-modified systems not only attached to specific epithelial sites but also were detectable within epithelial cells and even in the systemic circulation. Not only plant lectins but also several bacterial lectins are capable of mediating such adherence. In particular, the intestinal *Escherichia coli* K99 fimbriae are known to adhere to the GI epithelium and are used to enhance the absorption of K99-methyprednisolone conjugates (Bernkop-Schnürch et al., 1995).

Lectins are not necessarily toxic and can resist proteolytic degradation; however, a major drawback may be their immunogeneity. Moreover, the use of cell cultures to assess the bioadhesion can be problematic. Because specific lectin-cell interactions are dependent on fitting receptors on the cell surface, the available intestinal cell lines may vary in their receptor repertoire. Gabor et al. (1997) found that the K99 fimbriae specifically adhere to enterocytes of the duodenum and the jejunum, but the well-established enterocyte-like cell line Caco-2 does not seem to have any affinity for the K99 fimbriae. Despite these methodological difficulties, lectin-based cytoadhesion is still a challenge for drug delivery of macromolecules. The more we learn about the lectin-cell interaction, the more we are enabled to limit the binding sites of the lectins to essential structural features and thus escape from immunological complications.

C. Bioinvasion

The use of ligands that are widely known to be taken up by endo- or transcytosis offers a new possibility for creating drug delivery systems with physiological mechanisms for entering or passing epithelial cells. Vitamins, toxins, serum components, viruses, and bacteria can induce such vesicular transport processes and therefore may be established to enhance uptake and transport of drug-containing particles or conjugates (Russell-Jones, 1996). Table 1 gives an overview of some of these systems currently under investigation.

Such drug delivery systems, which are based on ligands capable of inducing vesicular transport processes, may also be referred to as bioinvasive drug delivery systems. The term bioinvasive in a biopharmaceutical context refers to a subgroup of bioadhesive systems that utilize physiological mechanisms to enter cells. Systems that "only" adhere to the mucus or to cells and are not necessarily taken up should be referred to as mucoadhesives and cytoadhesives, respectively. By using this terminology we can differentiate between bacterial lectins that only adhere to the cell surface (Beachey, 1981) and lectins or proteins that induce specific uptake mechanisms.

The use of bacterial invasion factors has gained interest especially in the past couple of years. As a result of the rapid progress in cell and molecular biology, the main virulence factors of some enteropathogenic bacteria

Table 1 New Delivery Systems Based on Ligands Capable of Inducing Vesicular Transport Processes

Ligand	Delivery system	Drug	Reference
IgG	Liposomes	Insulin	Patel and Wild, 1988
Vitamin B_{12}	Conjugate	G-CSF	Habberfield et al., 1995
	Nanoparticle	No	Russell-Jones et al., 1996
Transferrin	Conjugate	Insulin	Wang et al., 1997
Folic acid	Liposome	Oligonucleotides	Wang et al., 1995
Riboflavin	Conjugate	BSA (model)	Wangensteen et al., 1996
Diphteria toxin	Poly-L-lysine	DNA	Fisner and Wilson, 1997
Invasin	Nanoparticle	No	Hussain and Florence, 1998
GAL4/invasin	Poly-L-lysine	DNA	Paul et al., 1997

Abbreviations: BSA, bovine serum albumin; G-CSF, granulocyte colony-stimulating factor; IgG, immunoglobulin G.

have become known. These enteropathogenic bacteria not only adhere to the gastrointestinal mucosa but also are able to invade the host organism. This group of bacteria comprises organisms such as enteroinvasive *E. coli* (EIEC), *Yersinia pseudotuberculosis, Yersinia enterocolitica, Shigella flexneri, Salmonella typhimurium*, and *Listeria monocytogenes* (Isberg, 1991). These species have developed specific mechanisms to induce their uptake by normally nonphagocytic epithelial cells. By examining the molecular mechanisms of this uptake, especially by identifying the invasive agents on the bacterium's surface and their corresponding receptors on the epithelial cell, we can exploit these natural mechanisms for the design of bioinvasive drug delivery systems (Smith et al., 1993). Until today, the best known uptake mechanisms has been that of the gram-negative *Yersinia* species.

Yersinia pseudotuberculosis infection is mostly initiated by ingestion of contaminated foodstuffs (Fukai and Maruyama, 1979) and causes a variety of illnesses, ranging from mild gastroenteritis to mesenteric lymphadenitis (Brubaker, 1991). The most extensively studied animals for this infection are the rabbit and mouse. The disease proceeds by translocation across the epithelium of the ileum or the colon to the submucosal regions (Carter, 1975). The white laboratory rat, however, is not susceptible to oral *Yersinia* infections. Using a ligated intestinal loop of mice, 75 minutes after initial contact of *Yersinia* with the GI mucosa, single bacteria located within the M cells of the Peyer's patch are detectable (Grützkau et al., 1990). *Yersinia pseudotuberculosis* is initially taken up by the M cells overlying the Peyer's patch epithelium (Fujimura et al., 1992). However, some bacteria can also be found in the neighboring resorbing enterocytes, so a subsequent lateral translocation of the bacterium can be assumed (Marra and Isberg, 1997). Other groups consider the normal enterocyte as the primary site of uptake (McCormick et al., 1997). After invasion, *Y. pseudotuberculosis* may replicate extracellulary in the liver and spleen (Simonet et al., 1990).

Yersinia pseudotuberculosis is capable of invading cultured cells (Bovallius and Nilsson, 1975). The invasion factors were identified by transformation of noninvasive *E. coli* with chromosomal DNA from *Y. pseudotuberculosis* chromosomal DNA. All *E. coli* clones, which were able to invade cultured cells, expressed invasin on their surface. Invasin was derived from the *inv* gene of *Y. pseudotuberculosis* (Isberg and Falkow, 1985). Subsequent studies showed that invasin consists of 986 amino acids (Isberg et al., 1987) and that successful invasion further depends on the presence of calcium ions and an intact COOH terminus of invasin comprising the last 192 amino acids (Leong et al., 1990). The NH_2 terminus signals the correct localization and anchoring in the bacterial outer membrane. Multiple β_1-integrins located on the eukaryotic cell surface were identified as invasin receptors, as they are capable of specifically binding to immobilized invasin (Isberg and

Leong, 1990). To understand these mechanisms fully it is necessary to have good knowledge of the structures and functions of the integrins.

D. Integrins as Receptors of Bioinvasive Ligands

The host cell receptors for invasin are multiple β_1 chain integrins. These integrins are members of the integrin superfamily of cell adhesion molecules (CAMs). Integrins are large, $\alpha\beta$ heterodimeric surface molecules that are able to bind extracellular ligands as well as cytoskeletal components (Hynes, 1987). Integrins are involved in a wide variety of adhesive interactions. Each β chain is capable of associating with a number of different α chains, and a classification scheme has been proposed that subdivides integrins into subfamilies based on the particular β chain (called β_1, β_2, β_3, etc.). The six known VLA (very large after activation) proteins are members of one such subfamily of integrins that have identical β_1 chains and distinct α chains. These chains determine the substrate specificity of the heterodimer. An overview of these heterodimers with their synonyms and ligand specificities is given in Table 2.

The $\alpha_2\beta_1$ chain integrin (VLA-2) is a receptor for collagen, and the $\alpha_5\beta_1$ chain integrin (VLA-5) is a receptor for fibronectin. Ligands that are able to recognize these chain integrins often carry several specific amino acid motifs. Fibronectin, vitronectin, collagen, osteopontin, and other extracellular matrix (ECM) proteins carry the RGD motif (arginine-glycine-aspartic acid) (Ruoslahti and Pierschbacher, 1987). However, this RGD sequence can be recognized by only some of the integrins, especially by CD49e and CD51. Regarding the induction of endocytotic processes, the association of the β_1 chain with cytoskeletal components (i.e., talin, actinin) is pivotal, as during bacterial invasion a reorganization of the host cell's cytoskeleton is observable (Finlay, 1988).

Table 2 Synonyms for Integrins and Their Different Ligand Specificities

Heterodimer	VLA synonym	CD	Ligands
$\alpha_1\beta_1$	VLA-1	CD49a	Collagen, laminin
$\alpha_2\beta_1$	VLA-2	CD49b	Collagen
$\alpha_3\beta_1$	VLA-3	CD49c	Laminin, collagen, fibronectin
$\alpha_4\beta_1$	VLA-4	CD49d	Fibronectin, VCAM-1
$\alpha_5\beta_1$	VLA-5	CD49e	Fibronectin
$\alpha_6\beta_1$	VLA-6	CD49f	Laminin
$\alpha_v\beta_1$	—	CD51	Vitronectin

Abbreviations: VCAM, vascular cell adhesion molecule; VLA, very late after activation.

Each of the integrin receptors bound by invasin has the β_1 chain, but invasin does not bind directly to this chain. The isolated β_1 chain cannot bind invasin, nor can the purified collagen receptor (Isberg and Leong, 1990). Invasin has a high affinity for the fibronectin receptor ($\alpha_5\beta_1$), and fibronectin provided in excess can competitively inhibit invasin binding. However, invasin does not carry an RGD sequence, and it is assumed that invasin and fibronectin bind to mutually exclusive sites on the CD49e (Tran Van Nhieu and Isberg, 1991). Moreover, invasin binds with 1000-fold higher affinity than fibronectin (Tran Van Nhieu and Isberg, 1993), but fibronectin-coated bacteria are not able to enter cells (Yang et al., 1992).

Today there are several complementary explanations for the requirement for high-affinity binding and subsequent uptake of invasin-expressing bacteria (Isberg, 1996). One explanation is that high-affinity binding increases the number of integrin receptors available. A second explanation is the "zipper" model (Griffin et al., 1975), which postulates that internalization takes place by circumferential binding of mammalian cell receptors around the surface of the bacterium. This model requires a larger surface area of contact for uptake than for adhesion and also implies that the rate of dissociation of the receptor from the bacteria-encoded ligand must be rather slow; otherwise, the zipper would fall apart before the internalization process was completed. A final explanation for the need for high-affinity binding assumes that the invasin binding can cause multimerization or clustering of the integrins, which is often required to send a proper intracellular signal for movement of the cytoskeleton (Hermanowski-Vosatka et al., 1988).

Further analysis of the structural requirements of the invasin molecule to induce uptake showed that the cysteine residues Cys-907 and Cys-982 are important for function and form a disulfide bond (Leong et al., 1993). In addition, site-directed oligonucleotide mutagenesis revealed that in this region the aspartate-911 is the only absolutely critical residue and a change of the aspartate to an alanine at this site caused total loss of integrin binding (Leong et al., 1995). A similar effect could be observed on addition of amino acids to the COOH terminus (Isberg et al., 1993), indicating the importance of a defined COOH-terminal end.

II. EXPERIMENTAL EVALUATION OF INVASIN-MODIFIED PHARMACEUTICAL CARRIER SYSTEMS

In our work we functionalized fluorescent latex particles with invasin derivatives and tested these systems (Fig. 1) in terms of binding and adhesion by

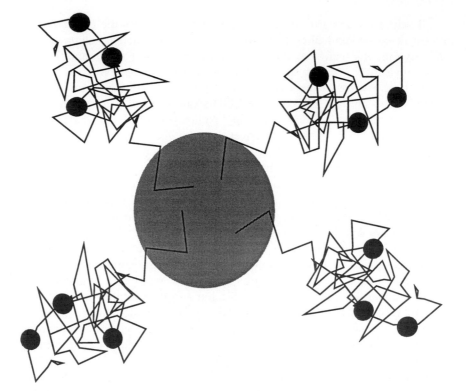

Figure 1   Bioinvasive test system resembling the bacterium: FITC-labeled, carboxylated latex particles with covalently attached Texas-Red–labeled invasin derivative.

using various kinds of epithelial cell cultures. In the first step, sufficient amounts of invasin had to be produced by biotechnology.

For the overproduction of recombinant proteins, a variety of expression vectors are commercially available. We used a pMAL-2 vector system (Guan et al., 1987) (New England Biolabs), which provides a method for readily synthesizing and purifying proteins as fusions to the maltose binding protein (MBP). The *inv192* gene, which encodes the COOH-terminal 192 amino acids of invasin, was inserted downstream from the *malE* gene, resulting in the expression of an MBP-invasin192 fusion protein (INV-MBP). This method uses the plasmid-encoded strong "tac" promotor and the *malE* translation initiation signals to give high-level expression of the cloned genes (Duplay et al., 1984). The expressed protein can easily be purified by an affinity chromatography step on an amylose resin using MBP's affinity for maltose (Kellermann and Ferenci, 1982).

A. Material and Methods

1. Production and Purification of Invasin

The vector carrying the *inv192* gene, pLHS11 (kind gift of Dr. Isberg, Boston), was transformed into *E. coli* DH5α by electroporation. *E. coli* carrying pLHS11 were grown to midexponential growth phase. Overexpression was started by adding the inducer IPTG to a final concentration of 1 mM. The bacteria were grown for another 2 hours, collected by centrifugation at 5000 *g* for 15 minutes at 4°C, and resuspended in 10 mL of HEPES buffer (25 mM, pH 7) containing 1 mM phenylmethylsulfonyl fluoride (PMSF). The bacterial suspension was lysed by French press and centrifuged for 15 minutes at 10,000 *g*. The cell-free extract was subsequently loaded onto an amylose resin (New England Biolabs) column. The elution buffer was 25 mM HEPES, 1 mM PMSF supplemented with 10 mM maltose. The purity of eluted proteins was checked by sodium dodecyl sulfate–polyacrylamide gel electrophoresis (SDS-PAGE). The proteins were kept at 4°C for storage. Protein content was determined using the BCA Pierce assay.

The pMAL-2 vector without insert encodes only MBP and served as a control.

2. Fluorescence Labeling of Invasin

For visualization of the proteins and for quantification of the amount of protein present on the latex particle, the proteins were dialyzed against bicarbonate buffer and adjusted to a protein concentration of 12 mg/mL by ultrafiltration (Centriplus, MWCO 30 kD, Amicon). Then 500 µL of protein solution was coupled to the fluorescent dye Texas-Red (Molecular Probes) at molar ratios (dye to protein) ranging from 0.5:1 to 10:1. The reaction mixture was incubated at 4°C for 1 hour. Unstable conjugates were removed by adding ethanolamine at a final concentration of 50 mM. Free labeling reagent was removed by gel filtration, and specific fluorescence of the conjugate (INV-MBP-RED) was determined at an excitation wavelength of 590 nm and an emission wavelength of 645 nm using a microplate fluorescence reader (CytoFluorII, Perspective Biosystems). The protein content was determined again.

3. Covalent Coupling of Invasin to Latex

INV-MBP-RED, INV-MBP, and MBP were covalently coupled to fluorescein isothiocyanate (FITC)-labeled carboxylated microspheres (Fluoresbrite,

Polysciences, Inc.) with nominal diameters of 0.5 and 2.0 μm. Carbodi-imide-activated latex particles were incubated with protein in varying concentrations. The amount of coupled protein was determined either by measuring the Texas-Red emission (590 nm, 645 nm) of protein attached to the particle's surface for the INV-MBP-TEX particles or by determining the amount of unreacted protein after the carbodiimide reaction for INV-MBP.

4. Cell Culture Binding and Uptake Assay

Caco-2 cells [passages 50–55, American Type Culture Collection (ATCC)] were cultured in flat-bottom 96-well or 24-well plates in Dulbecco's modified Eagle's medium (DMEM) until confluence. MDCK cells (canine epithelial kidney cells, passage 45–50, ATCC) were cultured in lactalbumin hydrolysate medium. Confluent monolayers of Caco-2 and MDCK cells were washed with Krebs-Ringer buffer (KRB). Differently modified particle suspensions were added to the monolayers and incubated for 1–3 hours at 37°C and 4°C. After incubation, the fluorescence emitted by the particles (excitation 485 nm, emission 530 nm) was measured and the monolayers were washed three times with KRB. The fluorescence intensity was measured again and calculated as remaining fraction bound to or taken up by the monolayers (Fig. 2). Binding of particles to integrins was blocked by adding 0.5 μg/mL flavoridin (snake venom disintegrin, Sigma) to the culture medium 24 hours prior to the experiment.

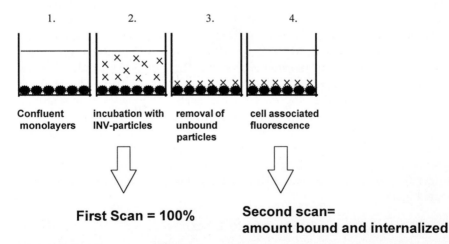

Figure 2 Binding and uptake assay with fluorescent particles and epithelial cell monolayers.

B. Results and Discussion

As shown in Fig. 3, the overexpressed proteins INV-MBP (lane A) and MBP (lane D) can be easily identified by their broad bands on the gel. The molecular masses of the recombinant proteins match their theoretical sizes, which are 64 kDa for INV-MBP and 51 kDa for MBP. After affinity chromatography on the amylose resin, the MBP fusion proteins were apparently pure (lane B). The yield for INV-MBP and MBP was about 15 mg per liter of culture, which is considerably higher than for other expression vectors (Leong et al., 1990).

To visualize the invasin we used Texas-Red as this fluorophore has a longer wavelength than either rhodamine or lissamine. Unlike the rhodamines, the Texas-Red fluorophore exhibits very little spectral overlap with FITC, so a direct determination of recombinant protein on an FITC-labeled particle is possible. However, the molar emission of Texas-Red–conjugated proteins is lower than that of other red dyes and an optimum in stability and specific emission depends strongly on the protein concentration and the molar ratios of dye to protein during the coupling procedure. The highest spe-

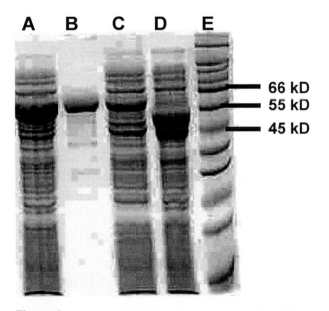

Figure 3 SDS-PAGE (12%). Lane A, whole bacterial extract after IPTG induction; lane B, eluted fraction after affinity chromatography; lane C, flow through after affinity chromatography; lane D, whole bacterial extract of only MBP-expressing control bacteria; lane E, molecular weight markers.

cific emission was obtained using 1 mol dye for 1 mol protein and the fluorescence was stable for 3 hours at 37°C (Fig. 4).

The conjugation resulted in 0.4 molecule of dye on 1 molecule of invasin. When larger ratios up to 10 mol dye per mol protein are employed in the conjugation procedure, the fluorescence signal decreases. This can be explained by quenching effects of the dye after conjugation to the free amino groups of the invasin.

By using Texas-Red–labeled invasin we are able to measure directly the amount of invasin coupled to the FITC-labeled carboxylated latex beads. As a possible high-affinity interaction between the integrins on epithelial cells and the invasin on the latex beads may depend on the ligand concentration, we used different amounts of invasin in the carbodiimide coupling procedure. In addition to directly determining the invasin concentration by measuring the Texas-Red emission, we determined the amount of unreacted protein after the coupling reaction.

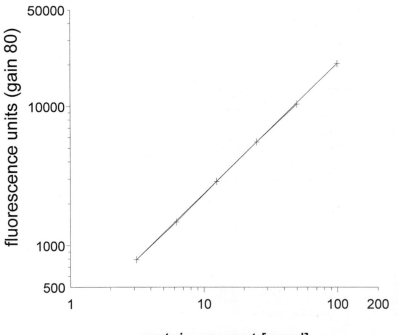

Figure 4   Emitted fluorescence of invasin conjugated to Texas-Red determined in the Cytofluor II microplate reader ($n = 3$) after incubation for 3 hours at 37°C.

By varying the protein concentration in the carbodiimide procedure, we obtained different amounts of protein present on the particle surface as shown in Table 3. The direct and indirect determinations of the protein bound to the particle yielded similar results for the invasin concentration. However, using highly fluorescent latex beads, the INV-MBP determination cannot be carried out by an enzyme-linked immunosorbent assay (ELISA) using MBP antibodies (Rankin et al., 1994) because the intensive FITC dye interferes with the ELISA color reaction. By using the carbodiimide reaction to covalently couple the invasin derivatives to the particles, one is able to specifically couple primary amino groups to the carboxyl residues on the particle. Amine-modified particles should not be used to immobilize invasin, because a free carboxyl terminal on the invasin molecule is necessary for high-affinity binding and is no longer available after coupling to amine-modified beads (Isberg et al., 1993).

The human colon adenocarcinoma cell line Caco-2 displays features similar to those of the absorptive intestinal cells such as microvilli, carrier-mediated transport systems, and tight junctions. Therefore, Caco-2 cells are an established model for studying drug absorption (Hilgers et al., 1990), binding characteristics (Lehr and Lee, 1993), and drug metabolism. More recent studies have also used Caco-2 cells as a model for the uptake of micro- and nanoparticles (Desai et al., 1997).

Figure 5 shows that the modification with invasin does not enhance the binding or uptake of 500-nm latex particles by Caco-2 cells, at either low or high particle concentrations. However, the same particle batches behave differently when incubated with MDCK cells (Fig. 6). MBP-modified

Table 3 Approximate Number of Invasin Molecules Covalently Coupled to Carboxylated Latex Particles as a Function of Protein Concentration[a]

Protein	Concentration (mg/mL)	Protein bound (molecules/particle)	
		500-nm particles	2000-nm particles
INV-MBP-RED	0.5	9,000	ND[b]
INV-MBP	2.0	35,000	ND
INV-MBP	1.0	11,000	52,000
INV-MBP	0.2	4,600	ND
MBP	1.0	13,000	73,000

[a]For INV-MBP-RED, the protein bound was determined by both fluorescence and protein determination and the unlabeled derivatives were determined indirectly.
[b]ND, not determined.
Abbreviations: INV, invasin; MBP, maltose binding protein; RED, Texas-Red.

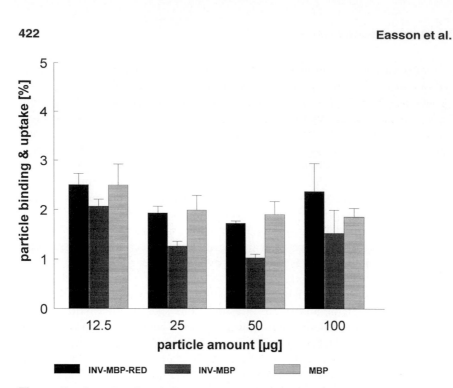

Figure 5 Caco-2 cell monolayers in 24-well plates were incubated with 500-μL latex particle (diameter 500 nm) suspensions in KRB at 37°C for 3 hours. Remaining green fluorescent particles with either red-labeled invasin (INV-MBP-RED) or un-labeled invasin (INV-MBP) were determined by fluorescence. MBP-modified parti-cles (MBP) were used as controls. Bars show number of particles bound and inter-nalized ($n = 3$).

particles, which were used as controls, show between 1.5 and 3% binding and uptake calculated from the initial particle mass before incubation with Caco-2 cells and with MDCK cells. Invasin-modified particles seem to in-teract only with MDCK cell monolayers.

Invasin-modified particles applied to MDCK cells seem to be specifi-cally bound and taken up. Between 3 and 5% of the initial particle mass can be detected after incubation and washing. The fluorescent label on the invasin molecule does not seem to hinder the interaction, as the results for the INV-MBP-RED and INV-MBP particles are similar. The reason for the different behavior of the Caco-2 cells may be the lack of expression of the corresponding β_1-integrins. Our results follow previous studies which have shown that upon confluence of the Caco-2 cells, the expression of the in-vasion receptor stops and *Y. pseudotuberculosis* is no longer able to invade the Caco-2 cells (Coconnier et al., 1994). Therefore, our further studies of

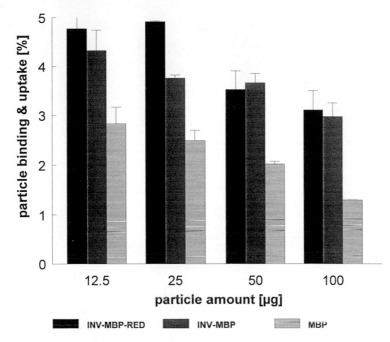

Figure 6 MDCK cell monolayers in 24-well plates were incubated with 500-μL latex particle suspensions (diameter 500 nm) in KRB at 37°C for 3 hours. Remaining green fluorescent particles with either red-labeled invasin (INV-MBP-RED) or un-labeled invasin (INV-MBP) were determined by fluorescence. MBP-modified parti-cles (MBP) were used as controls. Bars show number of particles bound and inter-nalized ($n = 3$).

particle interaction with cell cultures with respect to temperature dependence and invasin concentration were carried out with MDCK cells.

Figure 7 shows that the binding and uptake of invasin-engineered par-ticles by MDCK cells depends on the temperature during the incubation period. At 37°C we observed a three- to fourfold enhancement in binding and uptake, and when the temperature was lowered to 4°C we did not find any differences between the invasin and the control particles. This result can be explained by involvement of an energy-consuming process during bind-ing and uptake of invasin particles. Another clue to a specific invasin-de-pendent mechanism is the influence of the invasin concentration on the par-ticles. As shown in Fig. 8, there seems to be an optimum in binding and uptake of approximately 11,000 invasin molecules. Although lower invasin densities have an effect corresponding to the control, higher numbers do not enhance the interaction but rather seem to hinder an optimal invasin-integrin

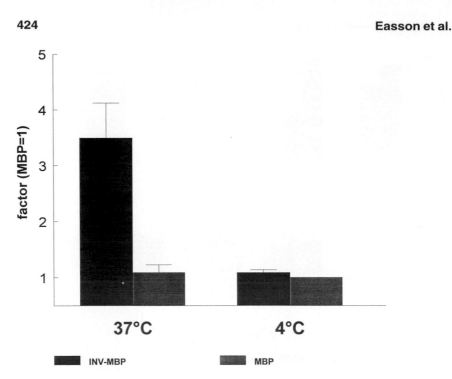

Figure 7 MDCK cell monolayers in 96-well plates were incubated for 2 hours with 100-μL latex particle (diameter 500 nm) suspensions (1 mg/mL) in KRB at 37°C or 4°C. The Y axis shows the enhancement of binding and uptake normalized to the binding and uptake of only MBP-modified control particles ($n = 3$).

interaction. Experiments involving invasin particles with a diameter of 2 μm did not show any enhanced binding or uptake to either Caco-2 or MDCK cells (data not shown). However, an alteration of invasin density on the 2-μm particles may have similar effects regarding binding and uptake in comparison with the 500-nm particles.

With the help of recently discovered disintegrins, it is possible to block certain integrins on cell surfaces. These peptides represent a new class of proteins that interfere with the interaction of adhesive ligands and their integrin receptors. The name of each disintegrin is derived from the name of the species or subspecies of the viper (Musial et al., 1990). We used flavoridin isolated from the viper *Trimeresurus flavoviridis* as an anticytoadhesive to block the invasin-integrin–mediated mechanisms. After preincubation with flavoridin, the binding and uptake of invasin particles were reduced to values similar to those of invasin particles incubated with MDCK cells at 4°C (Fig. 9). This also supports the hypothesis that not only the internalization but also the binding of invasin depends on the temperature, because

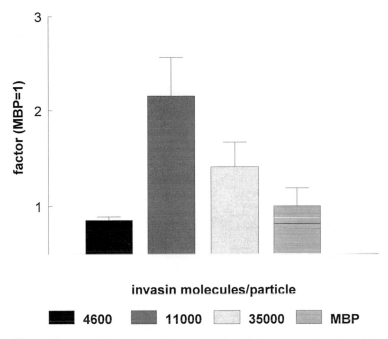

invasin molecules/particle

■ 4600   ■ 11000   □ 35000   ▤ MBP

Figure 8   MDCK cell monolayers in 96-well plates were incubated for 1 hour with 100-μL invasin-modified latex particle (diameter 500 nm) suspensions (500 μg/mL) in KRB at 37°C. The Y axis shows the enhancement of binding and uptake normalized to the binding and uptake of only MBP-modified control particles depending on the amount of invasin bound to the particles (see Table 3) ($n = 3$).

clustering of integrins and subsequent activation require a fluid cell membrane (Xiao et al., 1996). However, when the temperature is lowered, the rigidity of the cell membrane prevents lateral movement and clustering of the integrins.

III.   CONCLUSIONS

We have shown that a bacterial surface protein can confer bioinvasive properties on latex particles. The interaction of such invasin-modified particles depends on the temperature and the amount of invasin molecules bound to the surface of the particle. Our results confirm the in vivo data of Hussain and coworkers. They found 13% of an orally administered dose of 500-nm invasin-engineered particles in the systemic circulation of the rat (Hussain and Florence, 1998). Cell culture systems to test such systems have to be

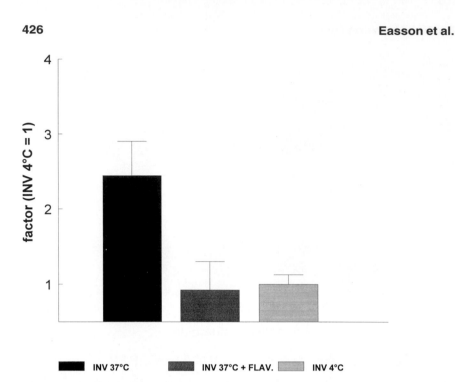

Figure 9 MDCK cell monolayers in 96-well plates were incubated for 1 hour with 100-μL invasin-modified latex particle (diameter 500 nm) suspensions (500 μg/mL) in KRB at 37°C in the presence or absence of the disintegrin flavoridin. The Y axis shows the enhancement of binding and uptake normalized to the binding and uptake of invasin particles at 4°C ($n = 3$).

chosen carefully with respect to their specific integrin receptor expression. On most epithelial cell cultures, the β_1-integrins are predominantly localized basolaterally. This also reflects the in vivo situation and contributes to the discussion of the initial uptake site of *Y. pseudotuberculosis* in the gastro-intestinal epithelium (enterocyte versus M cells). However, we assume that the enterocytes in the gastrointestinal tract are involved in the uptake of *Y. pseudotuberculosis* either as a portal for first entry or as a second target after uptake by M cells and subsequent lateral movement. As M cells are specialized to sample antigens and particulate material, one may speculate that *Y. pseudotuberculosis* would not need to use its sophisticated uptake mechanism in the case of exclusive uptake by M cells.

A major drawback, however, is the potential immunogenity of such invasin-engineered drug delivery systems. Although the bacterium itself has an impressive repertoire to escape from recognition by the immune system (Monack et al., 1997), a multiple application of invasin particles will probably cause an immune response. Therefore it is important to consider the scientific progress in minimizing the essential invasin structures to stimulate integrin-mediated uptake. Some research groups study use of RGD-containing peptides to induce uptake of nonviral gene delivery systems into cells and have already shown some promising effects with regard to enhanced transfection efficiency (Hart et al., 1997). Also, with the help of phage display techniques it seems possible to identify uptake-stimulating ligand structures (Koivunen et al., 1994), which in future may be used to achieve safe uptake of drug carrier systems or nonviral gene delivery systems.

REFERENCES

Beachey EH. 1981. Bacterial adherence: Adhesin-receptor interactions mediating the attachment of bacteria to mucosal surfaces. J Infect Dis 143:325–345.

Bernkop-Schnürch A, Gabor F, Szostak MP, Lubitz W. 1995. An adhesive drug delivery system based on K99-fimbriae. Eur J Pharm Sci 3:293–299.

Bovallius A, Nilsson G. 1975. Ingestion and survival of Yersinia pseudotuberculosis in HeLa cells. J Microbiol 7:1997–2007.

Brubaker RR. 1991. Factors promoting acute and chronic diseases caused by Yersinia. Clin Microbiol Rev 4:309–324.

Carter PB. 1975. Pathogenecity of Yersinia enterocolitica for mice. Infect Immun 11:167–170.

Chen H, Torchilin V, Langer R. 1996. Lectin-bearing polymerized liposomes as potential oral vaccine carriers. Pharm Res 13:1378–1383.

Coconnier MH, Bernet Camard MF, Servin AL. 1994. How intestinal epithelial cell differentiation inhibits the cell-entry of Yersinia pseudotuberculosis in colon carcinoma Caco-2 cell line in culture. Differentiation 58:87–94.

Damgè C, Michel C, Aprahamian M, Couvreur P. 1988. New approach for oral administration of insulin with polyalkylcyanoacrylate nanocapsules as drug carrier. Diabetes 37:246–251.

Desai MP, Labhasetwar V, Walter E, Levy RJ, Amidon GL. 1997. The mechanism of uptake of biodegradable microparticles in Caco-2 cells is size dependent. Pharm Res 14:1568–1573.

Duplay P, Bedouelle H, Fowler A, Zabin I, Saurin W, Hofnung M. 1984. Sequences of the malE gene and of its product, the maltose-binding protein of Escherichia coli K12. J Biol Chem 259:10606–10613.

Finlay BB, Falkow S. 1988. Comparison of the invasion strategies used by Salmonella cholerae-suis, Shigella flexneri, and Yersinia enterocolitica to enter cultured animal cells: Endosome acidification is not required for bacterial invasion or intracellular replication. Biochimie 70:1089–1099.

Fisner KJ, Wilson JM. 1997. The transmembrane domain of diphteria toxin improves molecular conjugate gene transfer. Biochem J 321:49–58.

Fogg FJ, Allen A, Harding SE, Pearson JP. 1994. The structure of secreted mucins isolated from the adherent mucus gel: Comparison with the gene products. Biochem Soc Trans 22:229S.

Fujimura Y, Kihara T, Mine H. 1992. Membranous cells as a portal for yp entry into rabbit ileum. J Clin Electron Microsc 25:35–45.

Fukai K, Maruyama T. 1979. Histopathological studies on experimental *Yersinia enterocolitica* infection in animals. Contrib Microbiol Immunol 5: 310–316.

Gabor F, Bernkop-Schnürch A, Hamilton G. 1997. Bioadhesion to the intestine by means of *E. coli* K99-fimbriae: Gastrointestinal stability and specificity of adherence. Eur J Pharm Sci 5:233–242.

Griffin FMJ, Griffin JA, Leider JE, Silverstein SC. 1975. Studies on the mechanism of phagocytosis. I. Requirements for circumferential attachment of particle-bound ligands to specific receptors on the macrophage plasma membrane. J Exp Med 142:1268–1282.

Grützkau A, Hanski C, Menge H, Riecken EO. 1990. Involvement of M cells in bacterial invasion of Peyer's patches: A common mechanism shared by *Yersinia enterocolitica* and other enteroinvasive bacteria. Gut 31:1011–1015.

Guan C, Li P, Riggs PD, Inouye H. 1987. Vectors that facilitate the expression and purification of foreign peptides in *Escherichia coli* by fusion to maltose-binding protein. Gene 67:21–30.

Habberfield AD, Jensen-Pippo K, Ralph L, Westwood SW, Russell-Jones GJ. 1995. Vitamin B_{12} mediated drug delivery systems for granulocyte stimulating factor and erythropoietin. Int J Pharm 145:1–8.

Hart SL, Collins L, Gustaffson K, Fabre JW. 1997. Integrin-mediated transfection with peptides containing arginin-glycine-aspartic acid domains. Gene Ther 4: 1225–1230.

Hermanowski-Vosatka A, Detmers PA, Götze O, Siöverstein SC, Wright SD. 1988. Clustering of ligand on the surface of a particle enhances adhesion to receptor-bearing cells. J Biol Chem 263:17822–17827.

Hilgers AR, Conradi RA, Burton PS. 1990. Caco-2 cell monolayers as a model for drug transport across the intestinal mucosa. Pharm Res 7:902–910.

Hussain N, Florence AT. 1998. Utilizing bacterial mechanism of cell entry: Invasin-induced oral uptake of latex nanoparticles. Pharm Res 15:153–156.

Hussain N, Jani PU, Florence AT. 1997. Enhanced oral uptake of tomato lectin–conjugated nanoparticles in the rat. Pharm Res 14:613–618.

Hynes OH. 1987. Integrins: A family of cell surface receptors. Cell 48:549–554.

Isberg RR. 1991. Discrimination between intracellular uptake and surface adhesion of bacterial proteins. Science 252:934–938.

Isberg RR. 1996. Uptake of enteropathogenic *Yersinia* by mammalian cells. Curr Top Microbiol Immunol 209:1–24.

Isberg RR, Falkow S. 1985. A single genetic locus encoded by *Yersinia pseudotuberculosis* permits invasion of cultured animal cells by *Escherichia coli* K-12. Nature 317:262–264.

Isberg RR, Leong JM. 1990. Multiple beta 1 chain integrins are receptors for invasin, a protein that promotes bacterial penetration into mammalian cells. Cell 60: 861–871.

Isberg RR, Voorhis DL, Falkow S. 1987. Identification of invasion: A protein that allows enteric bacteria to penetrate cultured mammalian cells. Cell 50:769–778.

Isberg RR, Yang Y, Voorhis DL. 1993. Residues added to the carboxyl terminus of the *Yersinia pseudotuberculosis* invasin protein interfere with recognition by integrin receptors. J Biol Chem 268:15840–15846.

Junginger HE. 1990. Bioadhesive polymer systems for peptide delivery. Acta Pharm Technol 36:110–126.

Junginger HE. 1991. Mucoadhesive hydrogels. Pharm Ind 53:1056–1065.

Kellermann OK, Ferenci T. 1982. Maltose binding protein from *E. coli*. Methods Enzymol 90:459–463.

Koivunen E, Wang G, Ruoslahti E. 1994. Selection of peptides binding to the alpha5-beta1 integrin from phage display library. J Cell Biol 124:373–380.

Kriwet B, Kissel T. 1996. Poly(acrylic acid) microparticles widen the intercellular spaces of Caco-2 cell monolayers: An examination by confocal laser scanning microscopy. Eur J Pharm Biopharm 42:233–240.

Lehr C-M. 1994. Bioadhesion technologies for the delivery of peptide and protein drugs to the gastrointestinal tract. Crit Rev Ther Drug Carrier Syst 11(2&3): 119–160.

Lehr C-M, Bouwstra JA, Schacht EH, H.E., J. 1992. In vitro evaluation of mucoadhesive properties of chitosan and some other natural polymers. Int J Pharm 78:43–48.

Lehr C-M, Lee VHL. 1993. Binding and transport of some bioadhesive plant lectins across caco-2 cell monolayers. Pharm Res 10:1796–1799.

Leong JM, Fournier RS, Isberg RR. 1990. Identification of the integrin binding domain of the *Yersinia pseudotuberculosis* invasin protein. EMBO J 9:1979–1989.

Leong JM, Morrissey PE, Isberg RR. 1993. A 76-amino acid disulfide loop in the *Yersinia pseudotuberculosis* invasin protein is required for integrin receptor recognition. J Biol Chem 268:20524–20532.

Leong JM, Morrissey PE, Marra A, Isberg RR. 1995. An aspartate residue of the *Yersinia pseudotuberculosis* invasin protein that is critical for integrin binding. EMBO J 14:422–431.

Llopis J, Kass GEN, Duddy SK, Moore GA, Orrenius S. 1991. Mobilization of hormone sensitive calcium pool increases hepatocyte tight junctional permeability in the perfused rat liver. FEBS Lett 280:84–86.

Lowe PJ, Mijai K, Steinbach J, Hardison WGM. 1988. Hormonal regulation of hepatocyte tight junction permeability. Am J Physiol 255:G454–G461.

Luessen H, Verhoef JC, Borchard G. 1995. Mucoadhesive polymers in peroral peptide drug delivery. II. Carbomer and polycarbophil are potent inhibitors of the intestinal proteolytic enzyme trypsin. Pharm Res 12:1293–1298.

Marra A, Isberg RR. 1997. Invasin-dependent and invasin-independent pathways for translocation of *Yersinia pseudotuberculosis* across the Peyer's patch intestinal epithelium. Infect Immun 65:3412–3421.

McCormick BA, Nusrat A, Parkos CA, D'Andrea L, Hofman PM, Carnes D, Liang TW, Madara JL. 1997. Unmasking of intestinal epithelial lateral membrane beta1 integrin consequent to transepithelial neutrophil migration in vitro facilitates inv-mediated invasion by *Yersinia pseudotuberculosis*. Infect Immun 65:1414–1421.

Monack DM, Mecsas J, Ghori N, Falkow S. 1997. *Yersinia* signals macrophages to undergo apoptosis and YopJ is necessary for this cell death. Proc Natl Acad Sci USA 94:10385–10390.

Musial J, Niewiarowski S, Rucinski B, Stewart GJ, Cook JJ, Williams JA, Edmunds LH Jr. 1990. Inhibition of platelet adhesion to surfaces of extracorporeal circuits by disintegrins. RGD-containing peptides from viper venoms. Circulation 82:261–273.

Naisbett B, Woodley J. 1994. The potential use of tomato lectin for oral drug delivery. 1. Lectin binding to rat small intestine in vitro. Int J Pharm 110:127–136.

Neutra MR, Phillips TL, Mayer EL, Fishkind DL. 1987. Transport of membrane-bound macromolules by M-cells in follicle associated epithelium of rabbit Peyer's patch. Cell Tissue Res 247:537–546.

Park K, Robinson JR. 1984. Bioadhesive polymers as platforms for oral controlled drug delivery: Method to study bioadhesion. Int J Pharm 19:107–127.

Patel HM, Wild AE. 1988. Fc receptor mediated transcytosis of IgG-coated liposomes across epithelial barriers. FEBS Lett 234:321–325.

Paul RW, Weisser KE, Loomis A, Sloane DL, LaFoe D, Atkinson EM, Overell RW. 1997. Gene transfer using a novel fusion protein, GAL4/invasin. Hum Gene Ther 8:1253–1262.

Rankin S, Tran Van Nhieu G, Isberg RR. 1994. Use of *Staphylococcus aureus* coated with invasin derivatives to assay invasin function. Methods Enzymol 236:566–577.

Ruoslahti E, Pierschbacher MD. 1987. New perspectives in cell adhesion: RGD and integrins. Science 238:491–497.

Russell-Jones GJ. 1996. The potential use of receptor-mediated endocytosis for oral drug delivery. Adv Drug Deliv Rev 20:83–97.

Russell-Jones GJ, Westwood SW, Habberfield AD. 1996. The use of vitamin B_{12} transport systems as a carrier for the oral delivery of peptides, proteins and nanoparticles. Proc Int Symp Controlled Release Bioact Mater 23:49–50.

Simonet M, Richard S, Berche P. 1990. Electron microscopic evidence for in vivo extracellular localisation of *Yersinia pseudotuberculosis* harboring the pYV plasmid. Infect Immun 58:841–845.

Smith PL, Wall DA, Wilson G. 1993. Drug carriers for the oral administration and transport of peptide drugs across the gastrointestinal epithelium. Rolland A. New York: Marcel Dekker, pp 109–134.

Tran Van Nhieu G, Isberg RR. 1991. The *Yersinia pseudotuberculosis* invasin protein and human fibronectin bind to mutually exclusive sites on the alpha5-beta1 receptor. J Biol Chem 266:24367–24375.

Tran Van Nhieu G, Isberg RR. 1993. Bacterial internalization mediated by beta 1 chain integrins is determined by ligand affinity and receptor density. EMBO J 12:1887–1895.

Wang J, Shah D, Shen WC. 1997. Oral delivery of an insulin transferrin conjugate in streptozolocin-treated cf1 mice. Pharm Res 14:S469.

Wang S, Lee RJ, Cauchon G, Gorenstein DG, Low PS. 1995. Delivery of antisense oligodeoxyribonucleotides against the human epidermal growth factor receptor into cultured KB cells with liposomes conjugated to folate via polyethylene glycol. Proc Natl Acad Sci USA 92:3318–3322.

Wangensteen OD, Bartlett MM, James JK, Yang ZF, Low PS. 1996. Riboflavin enhanced transport of serum albumin across the diastal pulmonary epithelium. Pharm Res 13:1861–1864.

Woodley JF. 1994. Enzymatic barriers for GI peptide and protein delivery. Crit Rev Ther Drug Carrier Syst 11(2&3):61–95.

Xiao J, Messinger Y, Jin J, Myers DE, Bolen JB, Uckun FM. 1996. Signal transduction through the beta-1 integrin family surface adhesion molecule VLA-1 and VLA-5 of human B-cell precursors activates CD19 receptor-associated protein tyrosine kinases. J Biol Chem 271:7659 7664.

Yang Y, Falkow S, Schoolnik GK. 1992. Cellular internalisation in the absence of invasin expression is promoted by the *Yersinia pseudotuberculosis yadA* product. Infect Immun 61:3907–3913.

16

Novel PEG-Containing Acrylate Copolymers with Improved Mucoadhesive Properties

Amir H. Shojaei* and Xiaoling Li
University of the Pacific, Stockton, California

I. INTRODUCTION

Adhesion is due to the interactions developed between two condensed phases. In general, an adhesion process is considered to occur in three major stages: "wetting," interpenetration or interdiffusion, and mechanical interlocking between a substrate and an adhesive. Therefore, adhesion can be described in terms of both thermodynamic and kinetic aspects (Bodde et al., 1990). Because the thermodynamic aspect deals with the driving force for adhesion, it provides information about the potential occurrence of adhesion. The major driving force for adhesion can be related to the surface energies of the adhesive and the substrate. The potential for an adhesive to wet or spread on a substrate can be measured by the spreading coefficient, which is a function of work of adhesion and work of cohesion. Hence, mucoadhesives can be designed in a rational way by analyzing the thermodynamic parameters of adhesion.

Optimization of the surface characteristics and composition of the mucoadhesive material is crucial for achieving intimate contact between the adhesive and the mucosal tissue. In order to optimize the properties of an

*Current affiliation: Texas Tech University Health Sciences Center, Amarillo, Texas.

adhesive for buccal mucosa, the adhesion must be assessed from a thermodynamic point of view at the molecular level. From the molecular interaction point of view, the attractive forces between two phases can range from strong bonds, such as covalent or ionic bonds, to weak secondary molecular interactions such as hydrogen bonds and van der Waals interactions. These molecular interactions are at a distance of a few angstroms. Also, the last stage of adhesion, physical and mechanical bonding, requires the adhesive to form entanglements with the extended mucus chains. On this basis, mucoadhesive polymers should have certain properties to achieve optimal mucoadhesive performance. The polymers should be hydrophilic and contain numerous hydrogen bond–forming groups (de Vries et al., 1988). The surface free energies of the polymer should be adequate so that wetting of the mucosal surface can be achieved (Peppas and Buri, 1985). Also, the polymer should have sufficient flexibility to penetrate the mucus network (Peppas and Buri, 1985; de Vries et al., 1988).

Diverse classes of polymers have been investigated for potential use as mucoadhesives. These include synthetic polymers such as poly(acrylic acid) (PAA) (Ch'ng et al., 1985), hydroxypropylmethylcellulose (Gandhi and Robinson, 1988), and poly(methyacrylate) derivatives (Leung and Robinson, 1990) as well as naturally occurring polymers such as hyaluronic acid (Sanzgiri et al., 1994) and chitosan (Lehr et al., 1992a). Among the various possible bioadhesive polymeric hydrogels, PAA has been considered as a good mucoadhesive (Ch'ng et al., 1985; Park and Robinson, 1987). Although the excellent mucoadhesive properties of PAA hydrogels are well known, they are obviously not the optimal mucoadhesive. The high glass transition temperature (T_g) of PAA makes the delivery system problematic in terms of flexibility in the dry state at room temperature, which could lead to suboptimal wetting and poor intimacy of contact. The high T_g also suggests that the lower mobility of the chain segment of polymers would result in decreased interpenetration and interdiffusion during the initial stages of adhesion in the glassy state. Considering the surface thermodynamics of mucoadhesion, one may find that the polarity of PAA imposes a rather high interfacial free energy, which also leads to poor wetting on the mucosal surface. Therefore, it is apparent that the mucoadhesive properties of PAA can be improved by designing a new mucoadhesive polymer that would eliminate these drawbacks.

In this chapter, the characteristics and mechanisms of mucoadhesion of the novel copolymers of AA and poly(ethylene glycol) monomethylether monomethacrylate [P(AA-co-PEG)] are discussed. The PEG moiety was incorporated in the copolymers to enhance the intrinsic mucoadhesive properties of PAA. The mucoadhesive performance of the copolymers was compared with that of cross-linked PAA (cr-PAA). The contributions of PEG to mucoadhesive performance were studied by surface energy analysis, mucoadhesive force measurements, and attenuated total reflection Fourier transform infrared (ATR-FTIR) spectroscopy.

II. SYNTHESIS OF PEG-PAA COPOLYMERS

Comb-shaped copolymers can be synthesized by grafting polymer side chains onto the polymer backbone or by a more direct method such as radical polymerization of PEG monomethacrylate. Bo et al. (1992) synthesized amphiphilic comb-shaped copolymers of methacrylic acid and poly(ethylene glycol) monomethacrylates and monomethylethers of different molecular weights using radical copolymerization. Klier et al. (1990) prepared polymer networks of poly(methacrylic acid-g-ethylene glycol) by copolymerizing methacrylic acid and poly(ethylene glycol) monomethylether monomethacrylates. Many kinds of comb-shaped copolymers have been synthesized by the use of macromonomers during the past several years (Ito et al., 1985; Tsukahara et al., 1989; Wesslen and Wesslen, 1989, Klier et al., 1990; Lee et al., 1990; Bo et al., 1992; Thermes et al., 1992; Yao and Sun, 1993; Chen and Hoffman, 1995; Choi et al., 1997). By this method, polymers containing various side chains with widely different properties may be synthesized, provided well-characterized macromonomers are available. The success of the method relies, to a large extent, on the reactivity of the macromonomers in copolymerization with other monomers.

The radical copolymerization method was used to synthesize copolymers of acrylic acid (AA) and poly(ethylene glycol) monomethylether monomethacrylate (PEGMM) (Shojaei and Li, 1995). Copolymers were synthesized by adding AA, PEGMM, and bromotrichloromethane (BTCM), a chain transfer agent, to a three-necked flask (equipped with a stirrer) in an oil bath. The free radical copolymerizations (Fig. 1) were carried out in tetrahydrofuran (THF) at 69°C under nitrogen upon dropwise addition of the initiator, 2,2′-azobisisobutyronitrile (AIBN) (dissolved in 1 mL of THF). After verification of the formation of copolymer by precipitating the reaction solution in petroleum ether, the crude copolymers were precipitated by adding the solution dropwise to a large amount (1000 mL) of cold petroleum ether under vigorous stirring. The precipitated P(AA-co-PEG) products obtained were further purified by reprecipitation twice using THF and petroleum ether as solvents. The copolymers were then dried under vacuum. Because a chain transfer agent was used in the copolymerization, non–cross-linked copolymers were obtained.

To synthesize the copolymer films, acrylic acid and PEGMM (PEG 200, 400, 1000) were dehibited for 24 hours using De-hibit 100 ion exchange resin. Appropriate amounts ([M]/[I] ratio of 1000) of AIBN were dissolved

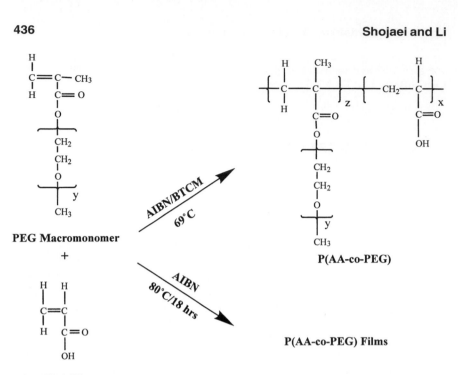

Figure 1 Synthesis of copolymers of acrylic acid and poly(ethylene glycol) mono-methylether monomethacrylate.

in 1 mL of THF. This solution was mixed with appropriate amounts of AA and PEGMM and purged with nitrogen for 2 minutes. The solutions were then degassed under vacuum and filled into molds constructed with two glass plates using silicone rod as a spacer. Copolymerization was carried out at 80°C for 18 hours (Fig. 1). The compositions of copolymers were varied by changing the feed ratio of AA to PEGMM. Because of the presence of ethylene glycol chains in the macromonomers, there is a high probability for chain transfer to polymer. This would yield a significant chain branching, and cross-linking would then occur through radical coupling reactions (Bo et al., 1992). Therefore, the copolymer films were slightly cross-linked due to chain transfer in the polymerization process. The resulting films were placed in THF (for 24 hours) followed by methanol (for 16 hours) and deionized water (for 24 hours). Solvents were constantly replaced by fresh solvent to remove unreacted monomer and initiator. Polymer films containing ethylene glycol dimethacrylate (EGDMA) as a cross-linker, including poly(acrylic acid) cross-linked with 0.3 wt % EGDMA (cr-PAA) and P(AA-

co-PEG) cross-linked with various amounts of EGDMA, were prepared using the same method.

III. CHARACTERIZATION OF PEG-PAA COPOLYMERS

A. Chemical Analysis

The compositions of all comb-shaped copolymers were determined by ^{1}H
nuclear magnetic resonance (NMR), ^{13}C NMR, and IR. The peak assignments of ^{13}C NMR spectra were verified using distortionless enhancement
by polarization transfer (DEPT). ^{1}H NMR (d-chloroform) spectra showed
peaks at δ 0.9 (t, 1H, for —CH of AA), 1.3 (s, 3H, for CH$_3$ of methacrylate),
3.4 (s, 3H for O—CH$_3$ of PEGMM), 3.65 (t, 4H for CH$_2$—CH$_2$—O of EG),
and 12.15 (s, 1H for RCOO—H of AA). The ^{13}C NMR (DMSO-d_6) peak
assignments were δ 24.9 and 69.5 (CH$_3$), 57.9 and 69.7 (CH$_2$), 66.9 (CH),
71.2 (C—O), and 175.9 (C—O). The number of PEG side chains attached
to the backbone was calculated from ^{1}H NMR spectra by comparing integrated signals from ethylene oxide groups in the PEG side chains with signals from the C—H groups in the AA units. This ratio was 1:1 for the
copolymer containing 16 mole % PEGMM; consequently, there was one
PEG side chain for every AA repeat unit. The IR spectra of linear P(AA-
co-PEG) showed typical absorption peaks attributed to the PEGMM moiety,
namely C—H stretching and C—O stretching. They were observed at about
2900 and 1100 cm^{-1}, respectively. The spectra also showed a broad absorption peak due to O—H and a sharp, strong absorption peak due to C═O
stretching mode attributed to the AA moiety at 3500 and 1700 cm^{-1}.

The molecular weights of the copolymers were determined through
viscosity measurements and gel permeation chromatography (GPC). The
molecular masses of the copolymers were measured as 21.9 to 73.0 kDa,
with the homopolymers (PAA and PEGMM) having the two largest molecular weights (Table 1).

B. Hydration Studies

Polymer films (0.5 cm in diameter) were placed in 25-mL scintillation vials
with 15 mL of swelling medium and the vials were immersed in a water
bath at 37°C. The weight of the film was measured periodically for 48 hours.
Three replications of each measurement were carried out. Equilibrium hy-
drations were determined as follows:

$$\%\text{Hydration} = \left(\frac{W_{\text{Swollen}} - W_{\text{Dry}}}{W_{\text{Swollen}}} \right) \times 100 \tag{1}$$

Deionized water (DIW) and pH 6.8 isotonic McIlvaine buffer (IMB)

Table 1 Intrinsic Viscosity and Molecular Weight of the Copolymers

PEGMM content (mole %)	Intrinsic viscosity (η)	Elution volume[a] (mL)	Estimated molecular weight (g/mol)
0.0	0.33	10.65	73,000
4.5	0.095	11.18	28,700
8.5	0.096	11.16	30,400
16.0	0.096	11.15	30,900
20.0	0.096	11.17	29,800
23.0	0.096	11.32	21,900
27.0	0.096	11.07	37,700
43.0	0.096	10.79	47,000
100.0	0.30	10.64	69,800

[a]Using gel permeation chromatography with a linear poly(styrene) standard kit (MW 4000 to 1,800,000).

(simulated gingival fluid without enzymes) were used as media for hydration studies. In addition, the pH-responsive swelling behavior of the copolymers was evaluated by determining the equilibrium degrees of hydration of the copolymers with a PEG molecular weight of 400 in buffer solutions of various pH values.

The P(AA-co-PEG) films contained two components known to exhibit interpolymer complexation (Bednar et al., 1984; Nishi and Kotaka, 1985). Because of this, under certain conditions, reversible complexes are formed, which enables the PAA-PEG copolymers to exhibit pH-responsive as well as thermoresponsive hydration behavior (Chen and Hoffman, 1995; Peppas, 1995). Equilibrium swelling for all film compositions was achieved in 1–2 hours. The hydration of polymer films with PEGMM (PEG molecular weight of 400) content ranging from 0 to 100 mole % is shown in Fig. 2. Equilibrium hydration decreased with increasing PEGMM content up to 16 mol %, after which the hydration increased proportionally with increasing PEGMM content up to 48 mole %. Beyond this point, further increases in PEGMM content did not change the equilibrium hydration of the copolymer films, yielding a steady plateau region. This behavior can be explained by considering the mole ratio of the PEG repeat unit, ethylene glycol (EG), to acrylic acid (AA). With increasing mole ratio of EG to AA intrapolymer complexation increases, leading to a decrease in hydration with a minimum seen at an EG:AA mole feed ratio of 0.6:0.4 (corresponding to 16 mole % PEGMM). When complexation occurs, the polymer gel collapses and the degree of equilibrium hydration decreases. The presence of hydrogen bond-

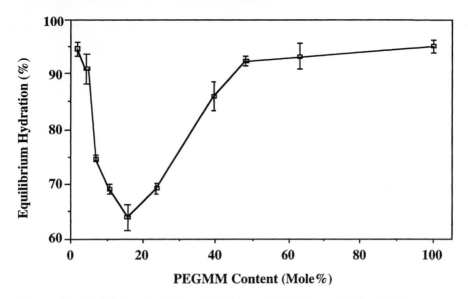

Figure 2 Equilibrium hydration of P(AA-co-PEG) films in DIW at 37°C.

ing in the complexes caused the network to become denser, because the H-bond formation led to a physically cross-linked network. Further studies were conducted with copolymer films containing 2.2 to 16 mole % PEGMM (PEG molecular weight of 400) based upon their hydration profile in DIW. The swelling behavior of these copolymers was also studied in IMB (Fig. 3). The degree of equilibrium hydration decreased with increasing molar ratio of PEGMM in the same pattern as observed in DIW.

Figure 4 shows the effects of pH on equilibrium hydration. At low pH, equilibrium hydration decreased because of proton association and complexation within the copolymer network. This behavior was seen up to about pH 4 (the pK_a of acrylic acid is 4.26). The equilibrium hydration of the films increased above pH 3, because the complexes dissociated as the carboxylic acid groups on the PAA became ionized. The increase in ionization resulted in electrostatic repulsion, which in turn led to the expansion of the network.

C. Glass Transition Temperatures (T_g) of P(AA-co-PEG)

The physical appearance of the P(AA-co-PEG) films changed with increasing PEGMM content at room temperature in the dry state. Pure PAA films were white and very rigid, whereas pure PEGMM polymer films were transparent and extremely elastic (with poor cohesion). Differential scanning calorimetry (DSC) studies revealed that the glass transition temperatures de-

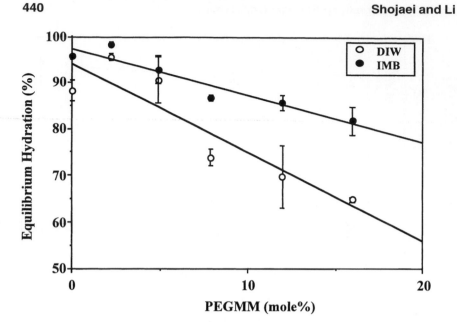

Figure 3 Effect of PEGMM content on equilibrium hydration of polymer membranes. Deionized water (DIW), isotonic McIlvaine buffer pH 6.8 (IMB).

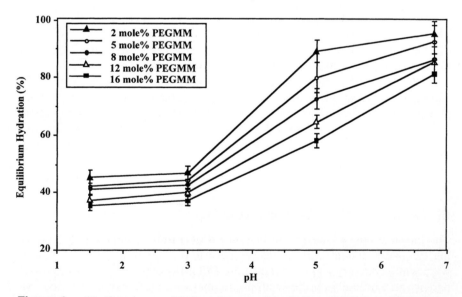

Figure 4 pH effects on equilibrium degrees of hydration of P(AA-co-PEG) films.

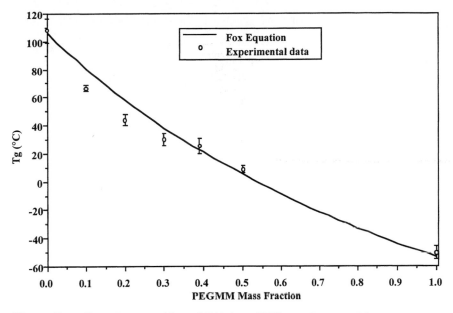

Figure 5 Effect of composition of P(AA-co-PEG) on glass transition temperature of copolymers.

creased with increasing PEGMM in copolymers. Fox's equation (Fox, 1956) provides a relationship for the dependence of the glass transition temperature on composition for a copolymer:

$$\frac{1}{T_g} = \frac{w_1}{T_{g(1)}} + \frac{w_2}{T_{g(2)}} \tag{2}$$

where $T_{g(1)}$ and $T_{g(2)}$ represent glass temperatures of the two corresponding homopolymers, and w_1 and w_2 refer to the weight fractions of the two co-monomers. The T_g of the copolymers was found to follow Fox's equation (Fig. 5). The agreement between experimental values of T_g and theoretical calculations suggested that the copolymers were random copolymers.

D. Adhesive Strength Measurements

The mucoadhesive strength of the P(AA-co-PEG) (containing 2.2 to 16 mol % PEGMM with PEG molecular weight of 400) and the cr-PAA films is shown in Table 2. For all films, the mucoadhesive forces measured in normal saline (NS) were higher than those measured in IMB. This can be attributed to the pH dependence of mucoadhesion with PAA containing hy-

Table 2   In Vitro Evaluation of Mucoadhesive Force of Polymer Films

AA (feed mole %)	PEGMM (feed mole %)	EG (feed mole %)	Mean force in NS (N/cm² ± SD)[a]	Mean force in IMB (N/cm² ± SD)[a]
84.0	16.0	63.0	4.35 ± 0.69[b]	3.93 ± 0.51[b]
88.0	12.0	55.0	4.29 ± 0.78[b]	3.78 ± 0.62[b]
92.0	8.0	44.0	3.56 ± 0.49	3.02 ± 0.34
95.1	4.9	32.0	3.09 ± 0.33	2.76 ± 0.58
97.8	2.2	17.0	2.56 ± 0.27[b]	2.21 ± 0.42[b]
100[c]	0.0	0.0	3.41 ± 0.44	2.82 ± 0.39
—[d]	—	—	0.61 ± 0.09[b]	0.74 ± 0.11[b]

[a]Normalized to the surface area of contact.
[b]$P < .05$ compared with cr-PAA film.
[c]Cross-linked with EGDMA, 0.3 wt %.
[d]Blank = Labsyn precleaned plain cut 3 × 1 inch slides.

drogels. It has been shown (Ch'ng et al., 1985; Park and Robinson, 1987) that the force of mucoadhesion increases with increasing degree of carboxyl group protonation in cross-linked PAA hydrogels. Higher forces of mucoadhesion were obtained in NS because the pH of this medium was lower than that of IMB. The mucoadhesive force increased with increasing PEGMM content of the films within the range investigated. Copolymer films containing 12 and 16 mole % PEGMM showed a significantly [$P < .05$, analysis of variance (ANOVA)] greater force of buccal mucoadhesion than the cr-PAA film. All films showed significantly greater ($P < .05$) forces of mucoadhesion than the blank control. These results were in good agreement with those reported by De Ascentiis et al. (1995) on improved mucoadhesion of poly(2-hydroxyethyl methacrylate) when linear PEO chains were introduced in the polymer network.

Two factors that influence the mucoadhesive force measurements obtained with our system were investigated: the raising rate of the platform and the initial contact time between the polymer film and the mucous membrane. The effects of these factors on mucoadhesive force measurements were studied using copolymer films with 16 mole % PEGMM (PEG molecular weight of 400). Only the raising rate of the platform was found to have a significant effect ($P < .05$) on the mucoadhesive force. The force of mucoadhesion increased with increasing rate of platform raising. This result suggested that the mucoadhesive bond was viscoelastic in nature, so that at higher rates of application of stress the bond has less time to deform (Fig. 6).

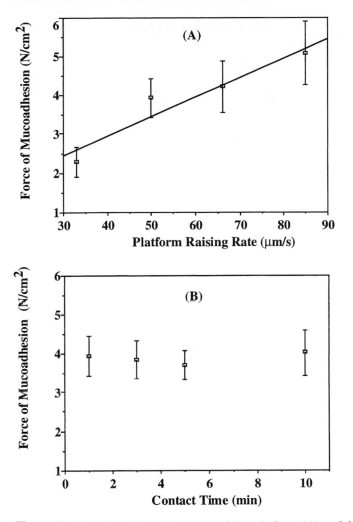

Figure 6 Effects of the raising rate of the platform (A) and initial contact time between polymer and mucous membrane (B) on the measured mucoadhesive force. The copolymers containing 16 mole % PEGMM were used for these experiments.

After the in vitro mucoadhesive force measurements, it was observed that, when films containing 12 and 16 mole % PEGMM were separated from the tissue, some tissue remains were found bound to the surface of films under an optical microscope. This suggested that the mucoadhesives had balanced adhesion and cohesion properties for a removable buccal mucoadhesion system.

IV. MECHANISMS OF BUCCAL ADHESION FOR PEG-PAA COPOLYMERS

Based on the feature of molecular interactions developed during adhesion, several mechanisms have been proposed, namely adsorption, electronic interaction, diffusion, "wetting," acid-base interaction, and fracture theories (Gu et al., 1988). When an adhesive is applied to a biological surface, the polymer should be able to wet the surface for adhesion to occur. Because wetting is the foremost event in the adhesion process, the interfacial interactions play an important role in the polymer adhesion to biological surfaces (Peppas and Buri, 1985). It is clear that a successful adhesive should be designed on the basis of the surface properties of both adhesives and biological tissues. Consequently, models (Van Oss et al., 1988; Lehr et al., 1992b, 1993; Rillosi and Buckton, 1995a, 1995b) based on interfacial interactions have been proposed.

A. Surface Energy Analysis

To probe the mechanism of buccal adhesion, dispersive and polar components of surface free energies for both buccal mucosa and polymer were calculated from contact angle data using the method described by Lehr et al. (1992b, 1993). For a captive air bubble at the adhesive-water interface, Young's equation can be written as

$$\gamma_{sv} - \gamma_{sw} = \gamma_{wv} \cos \theta_w \tag{3}$$

where s, v, and w refer to solid, vapor, and water, respectively. The equation for a drop of n-octane (o) at a solid-water interface would then become

$$\gamma_{sw} - \gamma_{so} = \gamma_{ow} \cos \theta_o \tag{4}$$

Using the geometric mean method, the following equations can be derived (Kaelble, 1974):

$$\gamma_{so} = \gamma_{sv} + \gamma_{ov} - 2(\gamma_{sv}^d \gamma_{ov}^d)^{0.5} - 2(\gamma_{sv}^p \gamma_{ov}^p)^{0.5} \tag{5}$$

$$\gamma_{sw} = \gamma_{sv} + \gamma_{wv} - 2(\gamma_{sv}^d \gamma_{wv}^d)^{0.5} - 2(\gamma_{sv}^p \gamma_{wv}^p)^{0.5} \tag{6}$$

The values of γ_{wv} and γ_{ov} as well as their respective dispersive and polar components (γ_{wv}^p, γ_{wv}^d, γ_{ov}^p, and γ_{ov}^d) are known. By using these values, Eq. (5) can be subtracted from Eq. (6) and substituted into Eq. (4) to solve for the polar component at the solid-vapor interface, γ_{sv}^p, using the octane contact angle (θ_o). The dispersive component of the solid-vapor interface, γ_{sv}^d, can be calculated by rearranging Eq. (6) and substituting it into Eq. (3) using the water contact angle (θ_w).

Because mucoadhesion occurs in the presence of a third interstitial phase, all possible interactions between the adherent and the substrate, the adherent and the surrounding medium, and the substrate and the surrounding medium should be considered. Inclusion of a third phase, a liquid phase, would be necessary to describe this system. Therefore, the spreading coefficients of all three phases are required. Assuming that the liquid phase would have a greater surface tension than the mucosal surface, the Griffith fracture energy will be at its maximum when the polymer's surface energy is at its lowest (Bodde and Lehr, 1991). However, such an assumption accounts for only S_L (the spreading coefficient of the liquid phase) and leaves both the mucosa spreading coefficient (S_M) and the polymer spreading coefficient (S_P) out of consideration. Under such condition, only a negative value for S_L, which in turn yields a positive Griffith fracture energy, is a requirement for spontaneous bonding. Moreover, either the polymer spreading coefficient, S_P, or the mucosa spreading coefficient, S_M, must be positive in order to have a spontaneous and stable bond formed. Because the mucosal phase will not spread over the hydrogel (Lehr et al., 1993), S_M will be negative and is ignored when S_P has a positive value. Therefore, the properties required for spontaneous bonding of a mucoadhesive hydrogel are a negative S_L and a positive S_P. These two spreading coefficients can be combined by their geometric mean to give a single term, the combined spreading coefficient (S_C) (Lehr et al., 1993) as follows:

$$S_C = \gamma_G^{0.5} \times \left(\frac{S_P}{2}\right)^{0.5} \tag{7}$$

Because part of the buccal mucoadhesion process can be considered as a surface phenomenon, interfacial properties play a significant role in achieving appreciable binding strength (i.e., mucoadhesive performance) (Gutowski, 1985a, 1985b, 1987; Lehr et al., 1993). The surface polarity of the adhesive polymer should be similar to that of the buccal mucosa to have a strong mucoadhesive according to the wetting theory of adhesion. In order to form an intimate contact between the mucus and the polymer, the mucoadhesive must wet the mucosal surface, which reflects on the importance of the hydrophilic properties of the polymer. A low water polymer contact angle would promote intimate contact in a three-phase system by enhancing the hydration of the polymer chains. However, in the case of an extremely hydrophilic polymer the water contact angle would be much lower than that of the mucosal surface, which would discourage intimate contact because of a high interfacial surface free energy. The interpenetration between polymer and mucus may occur only after intimate contact is established. As the foremost event in the development of the mucoadhesive bond is intimate contact between the interacting molecules, the wetting stage can be regarded as rate limiting for the mucoadhesion.

Buccal mucosa was shown to have appreciable hydrophobicity as compared with polymer films (Li and Shojaei, 1996). The surface free energy of porcine buccal mucosa was shown to be significantly different from that of the copolymers and the PAA polymer films ($P < .05$). Addition of PEG decreased the polarity compared with PAA; thus the interfacial free energy was lowered and the establishment of intimate contact was ensured. Therefore, introduction of PEG to PAA resulted in more favorable thermodynamic profiles than those of the cr-PAA film for buccal adhesion. In the case of the 16 mole % PEGMM copolymer (PEG molecular weight of 400), the most favorable wetting condition was achieved.

Surface energy parameters (α_p, square root of the dispersive component, and β_p, square root of the polar component) of the copolymers (PEG molecular weight of 400) and cr-PAA polymers, Griffith's fracture energy (γ_G), and combined spreading coefficients (S_C) were calculated to evaluate and predict their mucoadhesive behavior (Table 3). The copolymer containing 16 mole % PEGMM had the most favorable surface thermodynamic profile and the highest mucoadhesive force for buccal mucoadhesion. As shown in Fig. 7, the combined spreading coefficient correlated well with the force of mucoadhesion. These results indicated that mucoadhesion can be predicted using Eq. (7).

Other models based on surface energy aspects of polymer adhesion have been described. In the model of Kaelble (1974) the Griffith fracture energy is considered, and the model accounted for only the spreading coefficient of the interstitial liquid phase. As pointed out earlier, a negative value for S_L is only a requirement for spontaneous bonding and is not sufficient by itself. Hence, in a three-phase system this model did not account for all the possible interactions between the adherent, the substrate, and the surrounding liquid medium.

Another model based on interfacial forces has been presented by Van Oss et al. (1988) and by Rillosi and Buckton (1995a, 1995b). The model emphasized the significance of acid-base interactions in surface and interfacial tensions. These interactions were the result of the potential of molecules to act either as proton acceptor or donor (Brönsted theory) or as electron donor or acceptor (Lewis theory). In accordance with this model, all surfaces can be described by γ_S^{LW} for apolar (Lifshitz–van der Waals) molecules, γ_S^+ for electron acceptors, and γ_S^- representing electron donors. The values for these parameters were calculated from contact angle data measured in three different test liquids, one apolar and two polar (with known polar and dispersive nature). However, the use of this model for pH-dependent, ionizable mucoadhesive polymers (e.g., Carbopol) has been dis-

Table 3 Values of γ_{sv}^p, γ_{sv}^d, and γ_{sv}, Surface Energy Parameters (α_p, β_p), Griffith's Fracture Energy (γ_G), and Combined Spreading Coefficient S_c for Polymer Films in Normal Saline (NS)

Film composition AA:PEGMM (mol %)	γ_{sv}^p (mN/m)	γ_{sv}^d (mN/m)	γ_{sv} (mN/m)	γ_p	β_p	γ_G (mN/m ± SD)	S_c (mN/m ± SD)
84:16	43.2	18.1	61.3	4.25	6.57	1.39 ± 0.17	1.12 ± 0.16
88:12	45.5	18.0	63.5	4.24	6.74	1.07 ± 0.14	0.87 ± 0.09
92:8	46.5	17.7	64.2	4.21	6.82	0.93 ± 0.10	0.76 ± 0.09
95:5	47.5	18.8	66.3	4.33	6.89	0.71 ± 0.08	0.62 ± 0.07
98:2	48.7	20.0	68.7	4.47	6.97	0.46 ± 0.05	0.43 ± 0.05
100:0	48.3	19.2	67.5	4.38	6.95	0.56 ± 0.05	0.50 ± 0.04
Porcine buccal mucosa	28.4	14.2	42.6	—	—	—	—

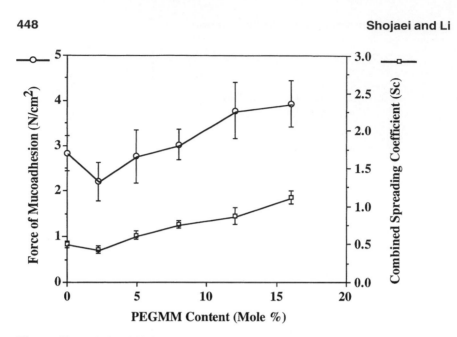

Figure 7   Relationship between mucoadhesive force, combined spreading coefficient, and polymer composition.

couraging, as a poor correlation was found between the total surface free energy of interaction and the force of detachment (Rillosi and Buckton, 1995b). Because differences in electrostatic interactions exist in the different test liquids (with different ionic strengths), this poor correlation can be expected.

In our investigation (Shojaei and Li, 1997), mucoadhesion was characterized by considering the interfacial free energy between the three phases (polymer, interstitial liquid, and mucosa) on the basis of the so-called geometric mean equation. Even though this model did not account for polymer interpenetration, acid-base interactions, ionic interactions, and hydrogen bonding, the mucoadhesive performance of the P(AA-co-PEG) films was correctly predicted. This suggests that the contribution of these forces to the mucoadhesive bond may be relatively small and that the assumption of interpenetration may not be needed a priori. Instead, the surface energy concept allows the explanation of a number of experimental findings in mucoadhesion, which would not be interpreted satisfactorily by the interpenetration theory alone.

B. Hydrogen Bonding

It has been noted that the capacity for formation of secondary bonds between the interacting molecules is another crucial factor in mucoadhesion (Leung and Robinson, 1988). In addition to the intimacy of contact and wetting, the strength of the hydrogen bond and its formation across interfaces may increase intrinsic adhesion (Gu et al., 1988). To investigate the effects of incorporation of PEG on hydrogen bonding, intra- and intermolecular hydrogen bondings were investigated using FTIR spectroscopy. The intramolecular hydrogen bonding (intrapolymer complexation) enhances the cohesive energy (W_c) of the polymer, whereas intermolecular hydrogen bonding between the polymer and mucus enhances the adhesive energy (W_a). Optimal bioadhesion is achieved when these two energies are in harmony with each other. If $W_c \gg W_a$, the polymer would have poor adhesive properties; and if $W_a \gg W_c$, the polymer would have little or no integrity.

The hydrogen-bonded O—H stretching absorption of carboxylic acids in hydrogen bond acceptor solvents such as ether or dioxane has been observed at 3100 cm^{-1} (Nishi and Kotaka, 1985). Figure 8 shows the FTIR spectra of P(AA-co-PEG) films with 5 mole % PEGMM (b), 9 mole % PEGMM (c), 16 mole % PEGMM (d), 20 mole % PEGMM (e), 23 mole % PEGMM (f), and 27 mole % PEGMM (g). The spectra for copolymers (b through f) exhibited a new absorption peak at about 3100 cm^{-1} as compared with those of corresponding cr-PEGMM (spectrum h) and cr-PAA (spectrum a). This absorption at 3100 cm^{-1} is characteristic of the stretching mode of carboxylic O—H groups hydrogen bonded with the ether oxygens in the PEG network, which confirms complexation within the copolymer network. As the mole ratio of EG:AA far exceeds 0.5:0.5 (as in the case of spectrum g), the intensity of the complex formation was diminished, this was also seen for copolymers containing higher molar ratios of PEGMM, up to 100 mole % pure PEGMM.

The frequency of the C=O stretching absorption of carboxylic acids is also known to be affected by hydrogen bonding. The monomeric state of the C=O stretching absorption is observed at $1730–1750$ cm^{-1}, attributed to the "free" C=O groups. The dimeric state, with hydrogen bonding, of the C=O stretching is observed at $1700–1715$ cm^{-1} (Nishi and Kotaka, 1985; Coleman et al., 1988). Therefore, the relative intensity of the absorption at $1730–1750$ cm^{-1} and $1700–1715$ cm^{-1} may be taken as the ratio of free C=O to H-bonded C=O. With increasing concentration of PEG in the copolymer, the relative fraction of free C=O increases, which was indicated by the shift of the C=O stretching absorption from 1710 cm^{-1} (in PAA) to 1731 cm^{-1} (in 43 mole % PEGMM containing P(AA-co-PEG) film) in the

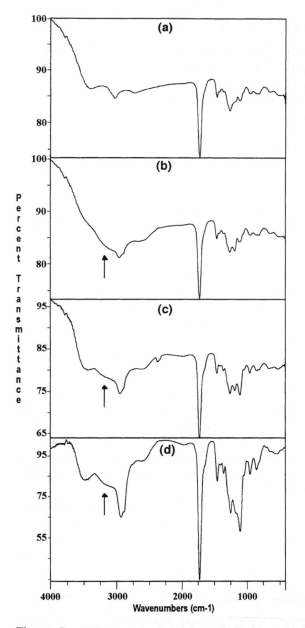

Figure 8 FTIR spectra of P(AA-co-PEG) films. (a) Pure cr-PAA; (b) 5 mole %
PEGMM; (c) 9 mole % PEGMM; (d) 16 mole % PEGMM.

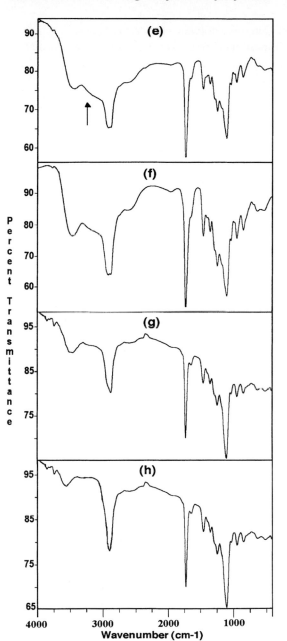

Figure 8 (Continued). (e) 20 mole % PEGMM; (f) 23 mole % PEGMM; (g) 27 mole % PEGMM; (h) pure cr-PEGMM.

presence of a model proton donating solvent such as acetic acid (Fig. 9). This was the result of the liberation of free C=O groups in the formation of hydrogen bonds between ether oxygens of ethylene glycol (EG) (repeat unit of PEG) and carboxylic acid. Figure 10 showed the relationship between PEGMM content and the relative increase in free C=O fraction, which was indicated by the shift in absorption frequency of C=O stretching. The relationship was linear with increasing PEGMM content up to 16 mole % PEGMM, which translated to a 0.6:0.4 feed ratio of EG to AA. With a further increase in PEGMM content, the linearity approached a plateau, which was due to the increase in the EG-to-AA ratio. As the EG-to-AA feed ratio far exceeded 0.5:0.5, the ether oxygens of PEG overwhelmed the system, which prevented the C=O moiety of AA from the formation of hydrogen bonds.

To determine the effect of the PEG moiety on increasing the potential for intermolecular hydrogen bonding, the intensity of the C—O stretching band was investigated. The region near 1100 cm^{-1} is unique to the C—O

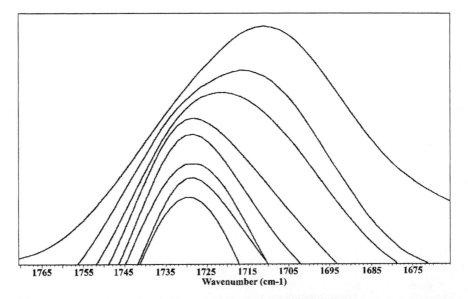

Figure 9   The C=O stretching region of P(AA-co-PEG) in the presence of acetic acid. From top to bottom: 0:100, 5:95, 9:91, 16:84, 20:80, 23:77, 27:73, and 43:57 mole % PEGMM to AA. Acetic acid absorption spectrum has been subtracted as background.

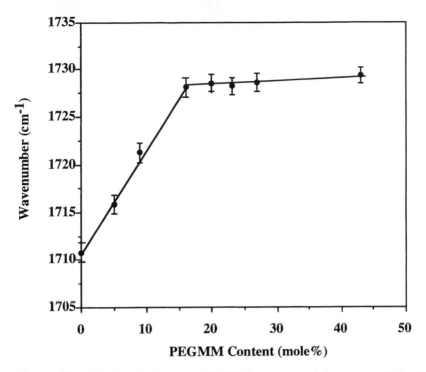

Figure 10 Relationship between PEGMM content and frequency of the C=O stretching absorption peak.

stretching of the EG unit of the copolymer. Formation of hydrogen bonds between PEG, as a hydrogen bond acceptor, and a proton donor acid was verified by using a proton-donating solvent such as acetic acid. Figure 11 shows a comparison of horizontal attenuated total reflection FTIR (HATR-FTIR) spectra of the P(AA-co-PEG) film with 16 mole % PEGMM (PEG molecular weight of 400) hydrated in mucin solution (spectrum a), the cr-PEGMM film in the presence of acetic acid (spectrum b), the cr-PEGMM film (spectrum c), and the cr-PAA film in the presence of acetic acid (spectrum d). There was an increase in intensity in this region. Therefore, these perturbations suggest the formation of hydrogen bonds between hydrogen donor groups of mucin and hydrogen acceptor groups (ether oxygens of the PEG moiety) in P(AA-co-PEG) films. The addition of the PEG moiety increased the potential of hydrogen bonding for P(AA-co-PEG) as compared with cr-PAA. FTIR studies revealed that the oxygen in the repeat unit of

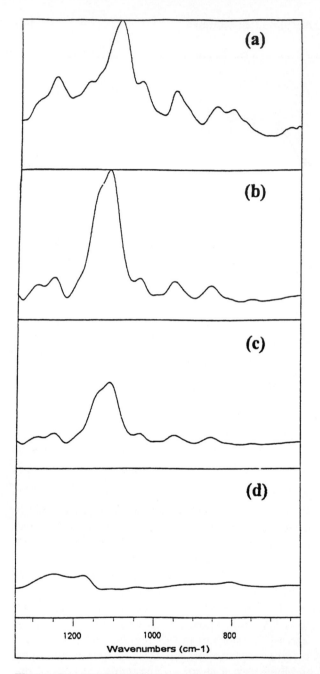

Figure 11 HATR-FTIR spectra for the ether oxygen (C—O) stretching region. (a) The 16 mole % PEGMM P(AA-co-PEG) copolymer in the presence of mucin solution (mucin solution absorption spectrum has been subtracted as background), (b) pure cr-PEGMM in the presence of acetic acid (acetic acid absorption spectrum has been subtracted as background), (c) pure cr-PEGMM, and (d) cross-linked PAA (cr-PAA).

PEG contributed to the formation of hydrogen bonding in the presence of the mucin solution.

V. EFFECTS OF PEG CHAIN LENGTH AND CROSS-LINKER ON BUCCAL ADHESION

To determine the effects of PEG chain length on mucoadhesion, copolymer films with PEG molecular weights of 200 and 1000 were further studied. Because the P(AA-co-PEG) copolymer containing 16 mole % PEGMM with PEG molecular weight 400 showed the best thermodynamic profile with the highest measured force of mucoadhesion, two compositions of copolymers were investigated for each of the PEG molecular weights. One copolymer composition was an EG:AA mole % of 60:40 and the other was 16 mole % PEGMM for both molecular weights of PEG. It was shown that chain length molecular weight does not contribute significantly ($P > .05$) to mucoadhesive force for the molecular weights of PEG investigated (Table 4). The films with EG:AA mole % of 60:40 showed the highest S_C values, which correlates well with their measured mucoadhesive forces. This further demonstrated that the mole ratio of the repeat units of PEG to AA in these copolymers is of the greatest importance for the mucoadhesive process.

The cross-linking density has been shown to influence the mucoadhesive performance of PAA by changing the effective number of PAA chains in a given volume (chain density) and the mobility of the PAA chains. Park and Robinson (1987) demonstrated that mucoadhesion of PAA hydrogels decreases with increased concentration of cross-linking agent. The introduc-

Table 4 Griffith's Fracture Energy (γ_G), Combined Spreading Coefficient (S_C), and Force of Mucoadhesion for Copolymer Films with Different PEG Side Chain Molecular Weight and Composition

Film composition (mol %)	PEG molecular weight	γ_G (mN/m)	S_C (mN/m)	Mucoadhesive force (N/cm^2)[a]
84:16 (AA:PEGMM)	200	1.01	0.65	2.10 ± 0.34
60:40 (EG:AA)	200	1.16	0.91	4.04 ± 0.52
84:16 (AA:PEGMM)	400	1.39	1.12	4.35 ± 0.69
84:16 (AA:PEGMM)	1000	0.56	0.51	2.34 ± 0.28
60:40 (EG:AA)	1000	1.33	1.01	4.11 ± 0.54

[a]Average of three experiments in NS ± standard deviation. Mucoadhesive force reported as normalized to the surface area of contact.

tion of a cross-linker resulted in a higher T_g, which decreased mucoadhesion through a decrease in chain mobility and interpenetration (de Vries et al., 1988). However, the mucoadhesive force of the EGDMA cross-linked P(AA-co-PEG) films showed no significant ($P > .05$) adhesion loss within the investigated range. This can be attributed to the fact that increased cross-linking density results in lower swelling, which in turn resulted in a higher number of carboxyl groups in a given surface area for mucoadhesion. Therefore, decreased chain flexibility was compensated for by increased carboxyl group density, which facilitated the formation of secondary hydrogen bonding.

VI. DOSAGE FORM DESIGN CONSIDERATIONS

Oral mucosal drug delivery systems must be easy to use and administer and should not cause any irritation when in place. The delivery system should be small enough so that it does not restrict eating, drinking, and speaking. Therefore, a rigid tablet formulation may be limited to 5–8 mm in diameter (Rathbone et al., 1994). On the other hand, a flexible patch system such as the 16 mole % P(AA-co-PEG) can be made a few times larger and still be well tolerated and comfortable. Also, because the PEG-PAA copolymers are hydrogels with a smooth surface, they have better compatibility in the mouth. Another factor to consider in terms of the dosage form design of the delivery system is the shape of the device. It has been reported that ellipsoid-shaped patches are most appropriate for the buccal area (Tucker, 1988). The thickness of the delivery system is another consideration and is limited to a few millimeters (Rathbone et al., 1994). Hence, factors such as tolerability, size, and thickness may restrict the extent of drug and/or inactive ingredients that can be loaded into a buccal mucoadhesive dosage form such as the PEG-PAA copolymers.

   The PEG-PAA copolymers can be employed for both local and systemic drug delivery. For local delivery the mucoadhesive system may be placed on or immediately adjacent to the intended site of drug action. For systemic drug delivery, the device may be placed on the upper labial sulcus of the buccal mucosa. At this site drug is released without much distribution to other parts of the oral cavity (Weatherell et al., 1994). The feasibility of systemic transbuccal delivery of acyclovir was explored in our laboratory using the P(AA-co-PEG) films. We were able to load 10 mg of acyclovir into a patch of 1.5 cm^2 in size and 2 mm thick. Drug release was controlled by a combination of diffusion of acyclovir from the matrix and macromolecular chain relaxation. In vitro permeation studies revealed the feasibility of buccal systemic delivery of acyclovir using the P(AA-co-PEG) copolymers in the presence of a penetration enhancer, sodium glycocholate (Shojaei and Li, 1996; Shojaei et al., 1998a; 1998b). Also, a mucoadhesive delivery system of metronidazole for local treatment of periodontitis is under investigation using the P(AA-co-PEG).

VII. CONCLUSIONS

A series of novel mucoadhesive polymers, copolymers of acrylic acid and poly(ethylene glycol) monomethylether monomethacrylate, was designed, synthesized, and characterized. The thermodynamic properties of copolymers were improved to afford a favorable surface characteristic for buccal adhesion by introducing PEG into the PAA backbone. The copolymer showed a significant increase in mucoadhesive force on porcine buccal mucosa compared with poly(acrylic acid). The mucoadhesive polymers can be used to fabricate buccal drug delivery systems for local or systemic delivery. These mucoadhesive polymers may represent a new generation of bioadhesives for transmucosal drug delivery.

REFERENCES

Bednar B, Morawetz H, Shafer JA. 1984. Macromolecules 17:1634–1636.
Bo G, Wesslen B, Wesslen K. 1992. J Polym Sci A Polym Chem 30:1799–1808.
Bodde HE, de Vries ME, Junginger HE. 1990. J Controlled Release 13:225–231.
Bodde HE, Lehr CM. 1991. Biofouling 4:163–169.
Ch'ng HS, Park H, Kelly P, Robinson JR. 1985. J Pharm Sci 74:399–405.
Chen G, Hoffman A. 1995. Nature 375:49–52.
Choi HK, Kim OJ, Jung CK, Cho YJ, Cho CS. 1997. Proceedings of International Symposium on Controlled Release of Bioactive Materials, 24:415–416.
Coleman MM, Skrovanek D, Hu J, Painter PC. 1988. Macromolecules 21:59–65.
De Ascentiis A, de Garczia J, Bowman C, Colombo P, Peppas N. 1995. J Controlled Release 33:197–201.
de Vries ME, Bodde HE, Busscher HJ, Junginger HE. 1988. J Biomed Mater Res. 22:1023–1032.
Fox TG. 1956. Bull Am Phys Soc 1:123.
Gandhi RE, Robinson JR. 1988. Ind J Pharm Sci 50:145–152.
Gu J, Robinson JR, Leung SS. 1988. Crit Rev Ther Drug Carrier Syst 8:21–67.
Gutowski WS. 1985a. J Adhes 19:29–49.
Gutowski WS. 1985b. J Adhes 19:51–70.
Gutowski WS. 1987. J Adhes 22:183–196.
Ito K, Tsuchida H, Hayashi A, Kitano T, Yamada Y. 1985. Polym J 17:827–839.
Kaelble DH. 1974. J Appl Polym Sci 18:1869.
Klier J, Scranton AB, Peppas NA. 1990. Macromolecules 23:4944–4949.

Lee JH, Kopeckova P, Kopecek J, Andrade JD. 1990. Biomaterials 11:455–464.

Lehr CM, Bodde HE, Bouwstra JA, Junginger HE. 1992b. Pharm Res 9:70–75.

Lehr CM, Bodde HE, Bouwstra JA, Junginger HE. 1993. Eur J Pharm Sci 1:19–30.

Lehr CM, Bouwstra JA, Schact EH, Juninger HE. 1992a. Int J Pharm 78:43–48.

Leung SS, Robinson JR. 1988. J Controlled Release 5:223–231.

Leung SS, Robinson JR. 1990. J Controlled Release 12:187–194.

Li X, Shojaei AH. 1996. Proceedings of International Symposium on Controlled Release of Bioactive Materials, 23:165–166.

Nishi S, Kotaka T. 1985. Macromolecules 18:1519–1524.

Park H, Robinson JR. 1987. Pharm Res 4:457–464.

Peppas NA. 1995. Seventh International Symposium on Recent Advances in Drug Delivery Systems, Salt Lake City, Utah, pp 47–50.

Peppas NA, Buri PA. 1985. J Controlled Release 2:257–275.

Rathbone M, Drummond B, Tucker I. 1994. Adv Drug Delivery Rev 13:1–22.

Rillosi M, Buckton G. 1995a. Int J Pharm 117:75–84.

Rillosi M, Buckton G. 1995b. Pharm Res 12:669–675.

Sanzgiri YD, Topp EM, Benedetti L, Stella VJ. 1994. Int J Pharm 107:91–97.

Shojaei AH, Li X. 1995. Pharm Res 12(S1):S210.

Shojaei AH, Li X. 1996. Proceedings of International Symposium on Controlled Release of Bioactive Materials, 23:507–508.

Shojaei AH, Li X. 1997. J Controlled Release 47:151–161.

Shojaei AH, Berner B, Li X. 1998a. Pharm Res 15:1182–1188.

Shojaei AH, Zhou S-L, Li X. 1998b. J Pharm Pharmaceutical Sci 1:66–73.

Thermes F, Grove J, Rozier A, Plazonnet B, Constancis A, Bunel C, Varion JP. 1992. Pharm Res 9:1563–1567.

Tsukahara Y, Ito K, Tsai H, Yamashita Y. 1989. J Polym Sci A Polym Chem 27:1099–1114.

Tucker IG. 1988. J Pharm Pharmacol 40:679–683.

Van Oss CJ, Chaudhury MK, Good RJ. 1988. Chem Rev 88:927–941.

Weatherell JA, Robinson C, Rathbone MJ. 1994. Adv Drug Delivery Rev 13:23–42.

Wesslen B, Wesslen K. 1989. J Polym Sci A Polym Chem 27:3915–3926.

Yao K, Sun S. 1993. Polym Int 32:19–22.

17

Bioadhesive, Bioerodible Polymers for Increased Intestinal Uptake

Gerardo P. Carino, Jules S. Jacob, C. James Chen, Camilla A. Santos, Benjamin A. Hertzog, and Edith Mathiowitz
Brown University, Providence, Rhode Island

I. INTRODUCTION

The oral route is clearly the most convenient way to administer therapeutics. However, many desirable drugs cannot be given orally because they are not transported across the gastrointestinal (GI) epithelium or, in the case of proteins, are digested by various enzymes located throughout the GI tract. It may be possible to incorporate these therapeutics agents in thermoplastic microspheres that will protect them from degradation and digestion, release them in a controlled fashion, and result in greater oral bioavailability. Bioadhesive drug delivery systems (BDDSs) made of thermoplastic microspheres have been shown to accomplish just this with the anticoagulant drug dicumarol (1). The bioadhesive properties of the microspheres allow greater interaction between the microsphere and intestinal wall, increase the residence time of the spheres in the gastrointestinal tract, and decrease the diffusion distance traveled by the drug. All these contribute to improved bioavailability when compared with unencapsulated drug. The same bioadhesive properties may also increase bioavailability of encapsulated drugs by allowing transfer of the BDDS, if small enough, across the gastrointestinal epithelium intact. This uptake allows the microspheres to travel systemically and release loaded drugs directly into the blood compartment or other tissues.

459

This chapter concentrates on describing the history of particulate uptake by the gastrointestinal epithelium as it evolved from a controversial
observation to a still not completely understood phenomenon. Parameters
that influence microsphere uptake will be discussed in greatest detail. Experiments that have related the observed intestinal uptake of thermoplastic
microspheres to the bioadhesive properties of the polymers will be described.

II. THE MUCOSAL BARRIER OF THE
GASTROINTESTINAL EPITHELIUM

The GI epithelium separates the internal tissues from the outside environment, as the intestinal contents are actually outside the body. Its purpose is
to protect from the external environment, act as the site for digestion, and
allow the balanced secretion or absorption of water and electrolytes. Like
other epithelial coverings, the GI tract is a barrier to unwanted substances
and must be crossed for successful drug delivery. This epithelium consists
of a single layer of simple, columnar epithelium lying above a collection of
cells termed the lamina propria (LP) and supported by a layer of smooth
muscle known as the muscularis mucosae (MM). The cells and LP form a
villous structure and the cells are lined with microvilli (or brush border), all
of which increase the area available for nutrient and electrolyte absorption.
The cells are held firmly together by tight junctions or zona occludens. As
a result, the epithelial layer has long been considered by many to be almost
impermeable to solid particles.

A special type of GI epithelium, the Peyer's patch, is also present and
not considered impermeable to particles. The Peyer's patch (PP) or gut-
associated lymphoid tissue (GALT) is an accumulation of lymphoid tissue
in the walls of the small intestine, most common in the terminal ileum, which
is important in sampling luminal antigens for developing both tolerance and
immunity. The PP is lined by a specialized epithelium, the follicle-associated
epithelium (FAE), containing membranous or microfold (M) cells, which
have the ability to phagocytize antigens, both soluble and particulate, in the
intestines. These antigens are then transported to antigen-presenting cells
(APCs) just below the M cells. The APCs will present the antigens to B
lymphocytes in the Peyer's patch, resulting in differentiation and proliferation of both B and T cells locally and in the regional lymph nodes, producing
a mucosal secretory immunoglobulin A (IgA) response against the antigen
(2). It is thought that polymer microspheres can also be phagocytized by
these M cells, retained in the PP, and deliver loaded antigens for vaccination
purposes (3).

In addition to the cells composing the gastrointestinal epithelium, mucus contributes to the barrier effect seen in the GI tract. It is constantly secreted by the epithelial lining, acts as a lubricant, and protects epithelial cells from abrasion, enzymatic destruction, and harsh pH levels. It is made up of mainly water (approximately 95%), proteins, lipids, and glycoproteins. Drugs, nutrients, and even particles must cross this additional layer before interacting directly with the epithelial lining.

III. HISTORY OF UPTAKE

With all these barriers to uptake present, observations of uptake have been controversial. However, the idea that orally administered microparticles could be taken up systemically is by no means a new one. More than 30 years ago, 2200 Å polystyrene (PS) microspheres fed to rats were shown to cross over the GI epithelium (actually through the cytoplasm of epithelial cells) and were identified in the liver by transmission electron microscopy (4). Although a very interesting result, this did not show much promise as a drug delivery system because PS is not biodegradable and would not be able to release any loaded drug. In fact, the authors related their results not to the development of drug delivery systems but to the potential negative health consequences associated with the buildup of small particles in the body. With the development and identification of biodegradable and biocompatible polymers, many investigators became much more interested in the fate of orally administered microspheres specifically for drug delivery purposes. However, because of its stability and availability, PS remained the model polymer that was evaluated most frequently.

Many years passed before the field of gastrointestinal uptake began to receive widespread attention. A large literature now exists in which investigators attempt to document the uptake of orally fed particles. Early work showed that commercially available 2-μm polyvinyltoluene particles chronically fed to mice would accumulate in the Peyer's patches (5). Light and fluorescence microscopy have been the techniques most extensively used to follow the translocation of fluorescent, most often polystyrene, microparticles through the intestinal wall of rabbits. The temporal movement of these 600- to 700-nm PS beads across the epithelial surface of the FAE was followed after intraluminal instillation. The translocation was shown to occur within 10 minutes, approximately the amount of time soluble tracers needed to cross the epithelium, implying an almost unimpeded pathway (6). In another study, chronically fed, radiolabeled polystyrene microspheres of different sizes (1 nm to 3μm) were also shown to appear in rat Peyer's patches, villi, liver, spleen, and mesenteric lymph nodes. The uptake was substantial, with 34% and 26% of the spheres of 50 nm and 100 nm, respectively, being detected in the tissues after feeding. No spheres were seen to accumulate in the heart, kidney, or lung (7). Other methods of microsphere identification and/or quantification have also been used, including confocal microscopy (8), electron microscopy (9–11), and extraction of polymer from macerated tissues (12). All of these studies came to similar conclusions; polymer microparticles indeed can enter the GALT and have promise as antigen carriers for controlled-release vaccine applications.

Although particle uptake was demonstrated using these hard, nonbiodegradable thermoplastic microspheres, it is clear that biodegradable polymers are more promising in the development of controlled-release delivery systems. Poly(D-L-lactic-co-glycolic acid) (PLGA) microspheres 1 to 10 μm in diameter have also been shown to be transported into rabbit Peyer's patches following intraluminal instillation. This system has advantages over the others mentioned in that PLGA is biodegradable and has already been studied extensively as the carrier in a number of controlled-release devices. Electron microscopy shows that microspheres of this copolymer are taken up by M cells and translocated toward the underlying lymphatic tissue within 1 hour (13). This rapid uptake of PLGA microspheres and biodegradation of PLGA have led to extensive work with this polymer in vaccine applications. Other work using confocal microscopy has emphasized that the majority of particle translocation across the GI epithelium occurs through the villous tissues adjacent to Peyer's patch tissue and implies that biodegradable microspheres that exhibit uptake can also be used for the systemic delivery of drugs (14,15). This was further supported with poly(fumaric-co-sebacic anhydride) [P(FA:SA)] microspheres, which have been shown to be taken up by both the normal absorptive epithelium and Peyer's patch and, when blended with PLGA, effective in the delivery of insulin (16).

With the existence of this work, it is now generally well accepted that particles can cross the epithelial barrier of the GI tract. However, the parameters that affect the extent and location of uptake are not completely understood. A review has summarized some parameters known to affect particle uptake by the intestine (most refer to uptake by the Peyer's patches) (17), among which are the following:

1. Size: particles smaller than 500 nm were taken up more efficiently than particles more than 1 μm in diameter (7).
2. Polymer composition of particle: polystyrene (PS), polymethylmethacrylate (PMMA), and poly(lactide-co-glycolide) (PLGA) particles all had "good" uptake, and polyhydroxybutyrate (PHB) uptake was described as being "very good" (18).
3. Surface charge: carboxylated PS particles had poorer uptake than charge-neutral PS particles (7).

4. Adsorbed hydrophilic poloxamer: reduced uptake by lymphoid tissue (17).
5. Covalently attached tomato lectin: increased uptake by nonlymphatic tissue (17).

The first parameter, size, is probably the most important and has been the focus of much of the uptake work. It appears that 10 μm is the absolute upper limit for any uptake to occur (13). However, no broad generalization has been made to identify polymers that have greater uptake potential. Besides size, the other parameters are related to either the polymer composition or surface properties of the microspheres. The properties that seemed to increase uptake (presence of carboxy groups, hydrophobicity, and attachment to lectins) are also properties that result in increased bioadhesion or mucoadhesion. Therefore, it is proposed that bioadhesive polymers will exhibit greater particle uptake when fabricated into small microspheres and administered orally.

IV. CORRELATION OF UPTAKE WITH BIOADHESIVE PROPERTIES

A number of methods have been proposed to quantify the bioadhesiveness of different thermoplastics. The simplest is based on contact angles. Polymer films that have a small contact angle with drops of water placed on them are said to be hydrophilic and are generally considered less bioadhesive. Similarly, small contact angles between polymer sheets and intestinal mucus indicate polymers that are mucophilic and therefore mucoadhesive (or bioadhesive). Another technique for measuring bioadhesive properties is based on a modified electrobalance technique (19) for estimating eight different mechanical forces associated with the contact of a single sphere with tissue under physiologic conditions. One of the most important quantities measurable by this technique is the fracture strength associated with removal of the single sphere from contact with the intestinal tissue. The final measure of bioadhesiveness considered here was obtained by a modified everted sac method that estimated the fraction of an initial load of microspheres that adhered to everted intestinal sacs after a short incubation period (20).

These three methods were used to evaluate the bioadhesive properties of four different polymers: polystyrene [PS, molecular weight (MW) = 50,000], polylactic acid (PLA, MW $\sim$ 2000), poly(lactide-co-glycolide) (PLGA, MW = 24,000), and poly(fumaric-co-sebacic) anhydride [P(FA:SA, MW $\sim$ 2000)].

A. Methods

1. Microsphere Fabrication

Three microsphere fabrication methods were used for these studies. Hot-melt microencapsulation (21,22) was used to form the P(FA:SA) spheres, and solvent evaporation was used to form the PS, PLA, and PLGA for bioadhesion measurements. A novel phase inversion nanoencapsulation (PIN) method (1) was used to form small microspheres from each of the polymers to be used in the everted sac and uptake studies.

P(FA:SA) microspheres for the bioadhesion force measurement experiments were fabricated using hot-melt encapsulation (21). Briefly, 2 g of polymer was heated above the polymer melting point and added to 250 mL of rapidly stirred silicone oil heated to approximately 10° above the melting point of the polymer. The emulsion was allowed to cool until particles solidified. The microspheres were collected, washed with petroleum ether, and separated into specific sizes using sieves. In contrast, solvent evaporation (23) was used to form the PLA, PLGA, and PS microspheres. With this method, a 20% solution of polymer in methylene chloride was added to 300 mL of 0.2% polyvinyl alcohol (PVA), forming an emulsion. This solution was vigorously stirred overnight, allowing evaporation of the solvent and formation of hardened microspheres. The spheres were collected, washed, lyophilized, and sieved to collect spheres between 600 and 720 μm in diameter.

For PIN microspheres each polymer was dissolved in methylene chloride at 3–5% w/v. This polymer solution was then quickly poured into an excess of bath-sonicated petroleum ether (a nonsolvent for the polymers). This results in the spontaneous formation of microspheres, which are then collected from the nonsolvent solution and lyophilized. The particles formed are typically in the size range 100 nm to 5μm. For light microscopy uptake studies, 1% rhodamine was added o the polymer solution to help in locating the spheres on histological slides.

All the microspheres were sputter coated with gold-palladium and observed by scanning electron microscopy. Microsphere size distributions were measured with a Coulter sizer.

2. Bioadhesive Force Measurements

Unfasted rats (250–400 g) were euthanized with an intraperitoneal (i.p.) injection of sodium pentobarbital (360 mg/kg). A 10- to 15-cm segment of the jejunum was excised, flushed with Dulbecco phosphate-buffered saline (DPBS), and maintained at 4°C. As mentioned earlier, bioadhesive force measurements were made using a modified Cahn Dynamic Contract Angle Analyzer (19). Briefly, microspheres sieved to between 710 and 850 μm were mounted on a rigid support wire by melt mounting and positioned on the tare loop of the microbalance. Tissue was kept in a specifically designed tissue chamber and kept under DPBS at 37°C. The tissue chamber was raised until the tissue made contact with the sphere. A compressive force between the two of 25 mN was maintained for 7 minutes, at which point the tissue was slowly moved away from the sphere and force measurements were recorded.

3. Contact Angle Measurements

Polymer films were prepared by casting 5% polymer solutions in methylene chloride onto a flat surface and allowing them to air dry. Drops (5 μL) of distilled water and rat mucin (PelFreeze) were carefully placed on the film surface. The angle the liquid made with the polymer film was immediately measured using a goniometer and taken as the contact angle.

4. Everted Sac Studies

Unfasted rats (250–400 g) were euthanized with an i.p. injection of sodium pentobarbital (360 mg/kg). Portions of small intestine were excised, flushed with 20 mL of ice-cold, phosphate-buffered saline, pH 7.2, containing 200 mg/dL glucose (PBSG), and everted using a stainless steel rod. Sacs were prepared from lengths of intestine about 6.0 cm in length by tying off the ends with 6–0 silk sutures and filling with approximately 3 mL of PBSG. Each sac was placed in 5.0 mL of PBSG in a 15-mL polypropylene, round-bottom tube into which approximately 100 mg of microspheres (w/v) was introduced. The everted sacs were incubated for 30 minutes at 37°C using end-over-end agitation provided by a Fisher Roto-Rack turning at 15 rpm. After the incubation, the sacs were gently washed in 5.0 mL of fresh PBSG to remove loosely bound microspheres. The 5.0-mL volume of incubation fluid, containing residual unbound spheres, and the 5.0-mL volume of wash fluid, containing loosely bound spheres, were combined and centrifuges for 20 minutes at 2000 rpm in an IEC tabletop centrifuge. The clear supernatant fluids were discarded, the microspheres were resuspended in 10 mL of distilled water, and the centrifugation step was repeated. Once again, the supernatant fluid was discarded and the beads were resuspended in 1 mL of distilled water, frozen in liquid nitrogen, and lyophilized for 24 hours. The purpose of the washing step was to remove residual solutes from the PBSG solution. The weight of the unbound beads was determined, allowing calculation of the percentage of bound beads.

5. Microsphere Uptake Studies

Fasted rats were anaesthetized with methoxyflurane and fed a 1-mL suspension of PIN microspheres (15 mg/kg) made of the four different polymer compositions. After 1 hour, the rats were sacrificed by an i.p. injection of pentobarbital (360 mg/kg). On the average, five samples of both jejunal and ileal Peyer's patches (with surrounding tissue), liver, and spleen sections were collected from each rat and fixed in 2% paraformaldehyde, 0.2% glutaraldehyde. tissue samples were prepared for examination by light microscopy by embedding in glycomethacrylate, sectioning at 5 μm thickness, and staining with 1% cresyl violet. In intestinal samples, the lumen, enterocytes, goblet cells, lamina propria, lacteals, and blood vessels within the intestinal wall were all examined carefully for evidence of microspheres. A minimum of three sections were observed for each tissue section in order to compare semiquantitatively the number of microspheres present in specific tissues.

B. Results and Discussion

1. Scanning Electron Microscopy

Figure 1 shows scanning electron micrographs of the spheres fabricated for the bioadhesive force measurements. The PLA spheres (Fig. 1A) show a very porous structure with many polymeric spherulites. The PLGA spheres (Fig. 1B) show a finer pore structure and a smoother surface. The PS spheres (Fig. 1C) are glassy, and the P(FA:SA) spheres (Fig. 1D) have an appearance similar to that of the PLA spheres. Only spheres that were sieved to between 710 and 850 nm were used for the bioadhesion studies.

Clearly, the microspheres produced by PIN for the uptake studies had different appearances (Fig. 2). The PLA spheres were the largest, from about 1 to 5 μm in size, and actually appeared to clump together (Fig. 2A). Both the PLGA and PS spheres were very small, apparently all less than 0.5 μm (Fig. 2B and C). The P(FA:SA) 20:80 spheres had a wider size distribution with spheres ranging from approximately 0.5 to 5 μm in size (Fig. 2D). The larger P(FA:SA) spheres also appeared to be porous. These visual observations were supported by the size distribution curves obtained with the Coulter sizer, which showed that PLA did indeed have the largest average size of approximately 2 μm while the other three polymers had average sizes of less than 1 μm (Fig. 3).

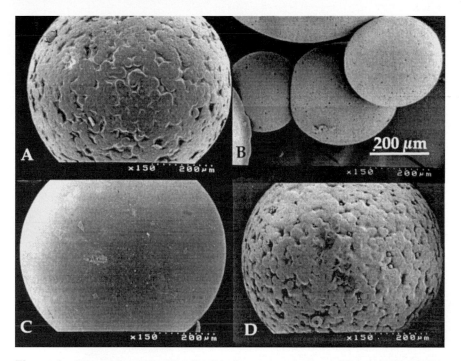

Figure 1 Microspheres for bioadhesive force measurements. PLA (A), PLGA (B), and PS (C) were made by solvent evaporation. P(FA:SA) (D) was made by hot-melt microencapsulation.

2. Bioadhesion Force Measurements

Figure 4 presents the fracture strength associated with removal of the different polymer spheres from the intestinal tissue. Each value has been normalized by dividing by a calculated surface area of sphere in contact with tissue. As shown, P(FA:SA) shows the greatest fracture strength (51.6 N/cm^2) with intestinal tissue, followed by PS (43.4 N/cm^2). PLA and PLGA had much lower fracture strengths (24.3 and 18.7 N/cm^2, respectively).

3. Contact Angle Measurements

Figure 5 shows the various contact angles of the four different polymers with both water and rat intestinal mucin. All of the polymers are relatively hydrophobic. As expected, because of its structure containing repeated aromatic rings, polystyrene is the most hydrophobic. The relative hydrophobicities of the other three polymers can be predicted by their chemical struc-

Figure 2 PIN microspheres of (A) PLA, (B) PLGA, (C) PS, and (D) P(FA:SA) 20:80. Bar is 6 μm.

tures. P(FA:SA) has long hydrocarbon chains and is therefore more hydrophobic than the two polyesters, which have much shorter repeat groups. Mucophilicity is a much harder parameter to predict because mucus is mostly water, yet there is a significant component made up of glycoproteins. Polystyrene is the most mucophilic of the polymers tested and PLA is the most mucophobic.

4.   Everted Sac Bioassay

Figure 6 shows the results of the everted sac bioassay, which measures the extent of spontaneous adhesion of polymer microspheres to viable intestine. The PS and P(FA:SA) spheres exhibited similar amounts of binding (45.4 and 43.4%, respectively), followed closely by the PLA microspheres (37.8%). The PLGA microspheres had the least binding (27.8%).

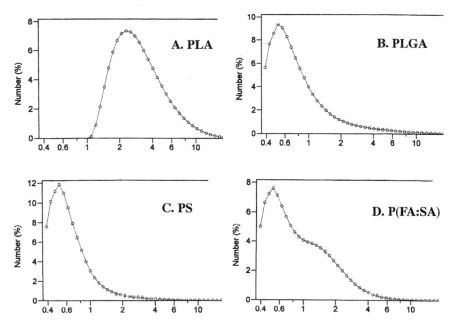

Figure 3 Coulter size distribution of microspheres used in uptake studies presented as number percent. A. PLA; B. PLGA; C. PS; D. P(FA:SA).

5. Uptake of Microspheres

The polymer microspheres withstood the low-temperature embedding in gly-comethacrylate well and were visualized in 5-μm sections of intestinal tissue under direct white light. Figure 7 shows the relative amount of microspheres observed in specific intestinal tissue in animals fed PS ($n = 2$), P(FA:SA) ($n = 4$), PLA ($n = 2$), and PLGA ($n = 2$). There were clear differences between the samples from animals fed the different polymer microspheres. The animals ($n = 2$) fed PLA microspheres showed very little or no evidence of microspheres in all the tissue slides (Fig. 8A). In most cases, the samples looked like control tissue. There was extensive uptake by the other three polymers, but in general PS showed the greatest uptake. It seemed to be taken up diffusely by both lymphoid and nonlymphoid tissue in the intestinal tract (Fig. 8B and C). Within 1 hour, PS microspheres appeared in great numbers in both the spleen and liver (data not shown). The uptake of P(FA: SA) microspheres was almost as extensive as that of PS microspheres. In all four rats fed P(FA:SA), there was extensive evidence of uptake in the villous epithelium (Fig. 8D), and in all but one many spheres were also

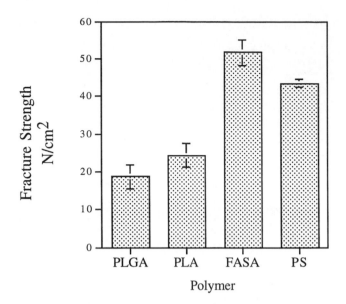

Figure 4 Fracture strengths of different polymer microspheres with intestinal tissue.

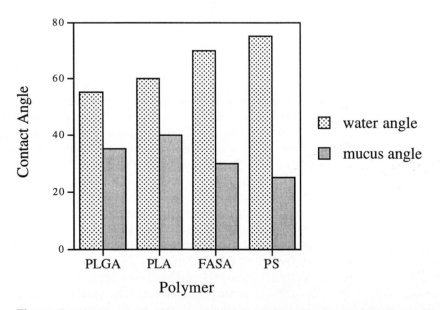

Figure 5 Contact angles of water and rat intestinal mucus on cast polymer films.

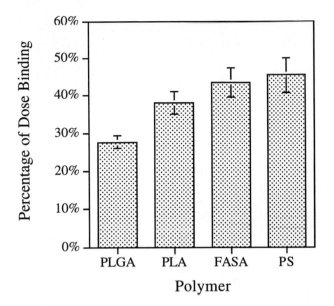

Figure 6 Percentage of PIN microscopes that spontaneously bind to everted intestinal sacs.

Intestinal Site	PS	FASA	PLA	PLGA
	n=2	n=4	n=2	n=1
Villous	****	***	*	****
Epithelium	****	***	-	***
		****		
		****		
Peyer's Patch	****	**	*	**
	****	***	-	*
		****		
		***		

****	very large number of microspheres present
***	large number of microspheres present
**	microspheres present
*	very rarely microspheres found
-	no microspheres found

Figure 7 Qualitative evaluation of observed uptake of microspheres by GI epithelium.

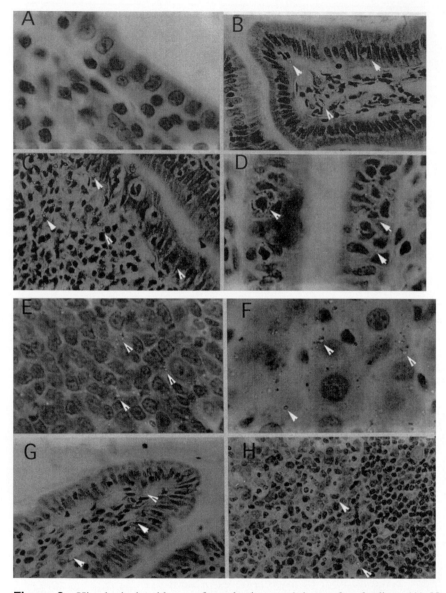

Figure 8 Histological evidence of uptake in rats 1 hour after feeding. (A) No evidence of uptake in rat fed PLA microspheres. Extensive uptake of PS microspheres by (B) an intestinal villus and (C) a Peyer's patch. (D) Villous epithelial uptake of P(FA:SA) microspheres. Many P(FA:SA) microspheres seen within (E) a Peyer's patch and (F) liver. (G) Extensive uptake of PLGA microspheres by an intestinal villus but much less (H) in a Peyer's patch. Arrows indicate microspheres.

present in the Peyer's patch 1 hour after feeding (Fig. 8E). Both liver (Fig. 8F) and spleen samples showed good evidence of microsphere uptake. Results for PLGA were interesting in that there was extensive uptake by the villous epithelial tissue (Fig. 8G) but only modest uptake by the lymphoid tissue (Fig. 8H). This is intriguing because it is the only polymer that seemed to show some sort of site-specific difference in uptake.

V. CONCLUSIONS

The uptake of particles by the gastrointestinal tract is a phenomenon that is still not completely understood. Size is clearly a limiting factor, but the other properties that influence uptake are unclear. A great number of investigators have attempted to elucidate the factors that lead to increased particle uptake, including such polymer properties as hydrophobicity, charge, or attached adhesion molecules. It is our purpose in this chapter to propose that the common denominator for all these properties is that they are all related to the bioadhesiveness of the polymer. More bioadhesive polymers, when appropriately sized (<10 μm), exhibit greater uptake. This is logical, as the polymer microsphere must first successfully interact with the epithelial surface before it can cross the barrier this surface forms. It is important, therefore, to recognize the utility of bioadhesive polymers in this application in order to identify and characterize polymers with these properties and use them in oral microsphere formulations when intending to achieve systemic effects. Also, it is important to recognize that uptake studies cannot be generalized from polymer to polymer as each will have its own extent of uptake based on both size and bioadhesiveness.

Further, the ability to identify bioadhesive polymers is an important step in the development of improved oral drug delivery systems. In this chapter we have used three different in vitro assays to characterize the bioadhesive properties of four different polymers. Contact angle measurements between cast sheets of polymer and either water or rat intestinal mucus elucidate the relative hydrophobicities and mucophobicities of the polymer surfaces. The advantages of the everted sac bioadhesion assay is that it actually uses the polymers in microsphere form to quantify the amount of spheres adhering to a segment of intestinal tissue. The modified Cahn microbalance is powerful because it is able to quantify the forces and works associated with the detachment of a single sphere from intestinal tissue. Taken together, these three techniques may prove helpful in identifying polymers that can be fabricated into bioadhesive drug delivery systems that can be taken up by the gastrointestinal epithelium.

Overall, the bioadhesive properties of polymers have been shown to correlate well with observed uptake. We have presented histological evidence that there are qualitatively different extents of uptake of different, similarly sized microspheres. PLA microspheres showed almost no evidence of uptake. PLGA showed some, most strikingly through the absorptive epithelium of the small intestine. Both PS and P(FA:SA) showed great uptake by both the absorptive and lymphoid cells. This observed uptake correlates very well to the polymers that demonstrate greater bioadhesiveness. These findings are particularly promising for P(FA:SA) as a polymer for use in oral controlled-release delivery systems, as it is also biodegradable and demonstrated this high degree of uptake even though it had a larger size distribution. The only polymers that did not seem to agree very well with the predictions of the different bioadhesive measurements were PLGA and PLA. According to the measured bioadhesive properties, PLA would be expected to have better uptake than PLGA. This was not found to be the case. This is most likely explained by the fact that the PIN method produced PLA spheres that were somewhat larger than PLGA spheres and therefore were less likely to be taken up by the gastrointestinal epithelium. Studies using smaller PLA microspheres are currently being conducted, and the results will be compared with those obtained thus far.

REFERENCES

1. D Chickering, J Jacob, E Mathiowitz. Biotechnol Bioeng 52:96–101, 1996.
2. WS Shalaby. Clin Immunol Immunopathol 74:127–134, 1995.
3. DT O'Hagan, K Palin, SS Davis, P Artursson, I Sjoholm. Vaccine 7:421–424, 1989.
4. E Sanders, CT Ashworth. Exp Cell Res 22:137–145, 1961.
5. ME LeFevre, JW Vanderhoff, JA Laissue, DD Joel. Experientia 34:120–122, 1978.
6. J Pappo, TH Ermak. Clin Exp Immunol 76:144–148, 1989.
7. P Jani, GW Halbert, J Langridge, AT Florence. J Pharm Pharmacol 41:809–812, 1989.
8. C Porta, PS James, AD Phillips, TC Savidge, MW Smith, D Cremaschi. Exp Physiol 77:929–932, 1992.
9. MA Jepson, NL Simmons, TC Savidge, PS James, BH Hirst. Cell Tissue Res 271:399–405, 1993.
10. T Landsverk. Immunol Cell Biol 66:261–268, 1988.
11. W Sass, HP Dreyer, J Seifert. Am J Gastroenterol 85:255–260, 1990.
12. JP Ebel. Pharm Res 7:848–851, 1990.
13. TH Ermak, EP Dougherty, HR Bhagat, Z Kabok, J Pappo. Cell Tissue Res 279:433–436, 1995.
14. GM Hodges, EA Carr, RA Hazzard, KE Carr. Dig Dis Sci 40:967–975, 1995.
15. GM Hodges, EA Carr, RA Hazzard, C O'Reilly, KE Carr. J Drug Target 3:57–60, 1995.

16. E Mathiowitz, JS Jacob, YS Jong, GP Carino, D Chickering, C Santos, P Chaturvedi, K Vijayaraghavan, M Bassett, S Montgomery, C Morrell. Nature 386:410–414, 1997.
17. AT Florence, AM Hillery, N Hussain, PU Jani. J Controlled Release 1995: 39–46, 1995.
18. JH Eldridge, CJ Hammond, JA Meulbroek, JK Staas, RM Gilley, TR Tice. J Controlled Release 1990:205–214, 1990.
19. DE Chickering, E Mathiowitz. J Controlled Release 1995:251–261, 1995.
20. J Jacob, C Santos, G Carino, D Chickering, E Mathiowitz. Proc Int Symp Controlled Release Bioactive Mater 22:312–313, 1995.
21. E Mathiowitz, R Langer. J Controlled Release 5:13–22, 1987.
22. E Mathiowitz, D Kline, R Langer. Scanning Microsc 4:329–340, 1990.
23. LR Beck, DR Cowsar, DH Lewis, JW Gibson, CE Flowers. Am J Obstet Gynecol 135:419–426, 1979.

18

Novel Formulation Approaches to Oral Mucoadhesive Drug Delivery Systems

Yohko Akiyama and Naoki Nagahara
Takeda Chemical Industries, Ltd., Yodogawa-ku, Osaka, Japan

I. INTRODUCTION

Pharmaceutical research has produced several oral mucoadhesive drug delivery systems (1–4). An oral mucoadhesive controlled-release delivery system can exert a positive influence on a drug's effectiveness by keeping the drug in the region proximal to its absorption window and allowing targeting and localization of the drug at a specific site in the gastrointestinal tract. All mucoadhesive drug delivery systems applied to the gastrointestinal tract require synthetic or natural bioadhesive polymers, which are usually macromolecular organic hydrocolloids with numerous hydrogen bond–forming groups such as carboxyl, hydroxyl, amide, and sulfate groups. Polyacrylic acid derivatives loosely cross-linked with allyl sucrose or divinyl glycol have been reported to have good mucosa-adhesive properties. The mucoadhesive properties are usually provided by mixing a drug with an adhesive polymer powder or granules or by coating it with an adhesive polymer (3). Longer et al. (1) confirmed that prolongation of the gastrointestinal transit and high bioavailability of chlorothiazide, which has bioavailability problems related to a specific window for absorption, was achieved in rats by administration as a mucoadhesive dosage form using a polyacrylic acid derivative. However, the gastrointestinal residence of mucoadhesive dosage forms was not prolonged in human volunteers (5,6). To improve the bioadhesive devices, various polyacrylic acid derivatives were synthesized (7). On the other hand, to keep drugs in the gastrointestinal tract for an extended period, we designed mucoadhesive microspheres (Ad-microspheres) referred to as an adhesive micromatrix system (AdMMS), which consists of a drug and an adhesive polymer such as a cross-linked polyacrylic acid derivative (carboxyvinyl polymer) dispersed in a spherical matrix of polyglycerol esters of fatty acids (PGEFs) with diameters of 177–500 μm (8).

In this chapter, first, functions of this AdMMS in vitro and in vivo are reported. In experiments using rats, prolongation of the gastrointestinal transit time and improvement of the bioavailability of furosemide with a narrow absorption window are shown. Second, the effectiveness of the AdMMS in human volunteers is shown in experiments using riboflavin (9). Third, the greater anti–*Helicobacter pylori* effect in Mongolian gerbils after oral administration of AdMMS containing amoxicillin is examined (10).

II. CONCEPTS, DESIGN, AND PREPARATION OF THE ADHESIVE MICROMATRIX SYSTEM (AdMMS) FOR THE GASTROINTESTINAL TRACT

An oral multiple-unit controlled-release drug delivery system in which a drug is dispersed in a spherical PGEF matrix was developed in our laboratory. In vitro drug release from the PGEF-microspheres, which were prepared by spraying and chilling the melted PGEF mixture using a rotating aluminum disk 15 cm in diameter (1000–3000 rpm) (11), was easily regulated by selecting PGEF with an appropriate hydrophile-lipophile balance (HLB). PGEF, a surfactant and a food additive listed in the Code of Federal Regulations (CFR), has various HLB values depending on the degrees of polymerization and esterification. The general chemical structure is shown in Fig. 1.

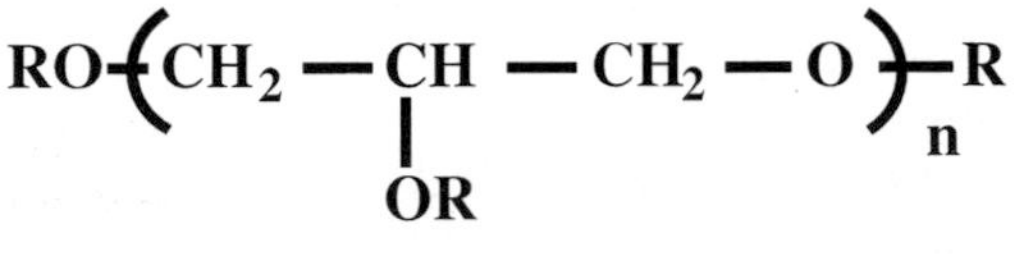

$$RO \left(CH_2 - \underset{\underset{OR}{|}}{CH} - CH_2 - O \right)_n R$$

n=2, 3, 4, ...10

R=H or fatty acid residue

Figure 1   Chemical structure of polyglycerol ester of fatty acid (PGEF).

Generally speaking, an oral sustained-release system does not always maximize the bioavailability of a drug or optimize clinical effectiveness if the absorption site is limited to a specific segment of the small intestine. To obtain higher bioavailability, it is necessary to keep such a drug in the vicinity of the absorption site for a longer period by modulating gastrointestinal transit. Mucoadhesive dosage forms as well as intragastric floating systems (12) and a controlled gastric emptying devices made of a shape memory plastic (13) can modulate the gastrointestinal transit of a drug. To prolong the gastrointestinal transit time, we added mucoadhesive properties to PGEF-microspheres. As the adhesive polymer, carboxyvinyl polymer (CP), (Carbopol 934P, BF Goodrich Company, Cleveland, OH), a swellable polymer of acrylic acid having loose cross-linking with allyl sucrose, was chosen because it has been reported to have good mucosa-adhesive properties (14,15) and is listed in the National Formulary XVII. A mixture of PGEFs [tetraglycerol pentastearate (TGPS) and tetraglycerol monostearate (TGMS)] was used as the microsphere matrix. Three types of microspheres, PGEF-microspheres, CP coated-microspheres, and CP dispersion-microspheres (Ad-microspheres referred to as AdMMS), were prepared. Formulations are shown in Table 1. Three types of microspheres were prepared as follows:

1. PGEF-microspheres: a mixture of TGPS (160 g) and TGMS (5 g) was melted at 85°C and dropped onto a rotating aluminum disk at 1500 rpm (spray-chilling method).
2. CP coated-microspheres: PGEF-microspheres (50 g) were placed in a centrifugal fluidizer (CF, Freund Industrial Co., Japan) and coated by spraying a CP suspension (5% w/v) in methanol from a spray nozzle 2 mm in diameter at a constant rate of 1 mL/min under the following conditions: rotation speed, 60 rpm; air spray rate, 120 L/min; air spray pressure, 2.0 kg/cm^2; slit air rate, 400 L/min. A slit air temperature of 46°C was used in order to achieve a microsphere bed temperature of 32–36°C.

Table 1 Formulations of PGEF-Microspheres, CP Coated-Microspheres, and Ad-Microspheres

Ingredient	PGEF-microspheres	CP coated-microspheres	Ad-microspheres
TGPS	160.0	160.0	160.0
TGMS	5.0	5.0	5.0
CP	—	50.0	18.0

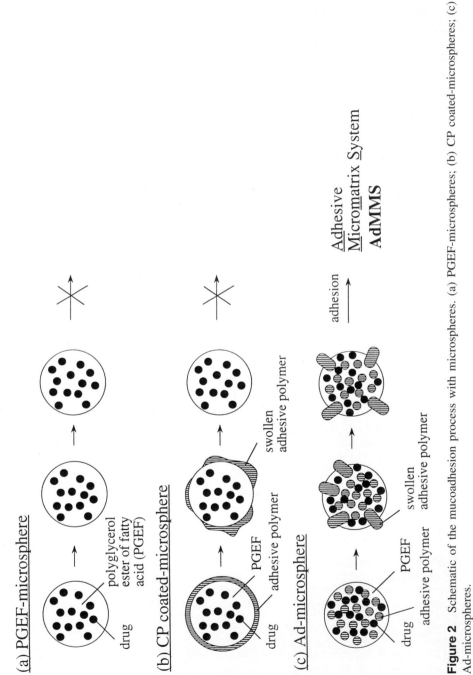

Figure 2 Schematic of the mucoadhesion process with microspheres. (a) PGEF-microspheres; (b) CP coated-microspheres; (c) Ad-microspheres.

3. Ad-microspheres: CP powder was dispersed in the melted TGPS and TGMS in a ratio of 160:5 and microspheres were prepared by the same spray-chilling method as used for PGEF-microspheres.

These three types of microspheres (177–500 μm in diameter obtained by sieving), which contain no drug, were used for the experiments to evaluate adhesiveness. A schematic of the three microspheres in which drug powder is contained is shown in Fig. 2.

III. IN VITRO EVALUATION OF MUCOADHESIVE MICROSPHERES

Swelling and gel-forming properties of the microspheres were examined microscopically after placing a drop of water on microspheres spread on a glass slide. When a drop of water was placed on the surface of the CP coated-microspheres, a gel layer of approximately 150 μm was formed around the microspheres in 5 minutes and the gel layer then swelled and expanded as shown in Fig. 3a and b. The PGEF-microspheres did not show any change in appearance (photograph not shown). In the case of Ad-microspheres, many small gelled CPs (10–20 μm in height) were found on the surface of the microsphere in 5 minutes as shown in Fig. 3c and d. The polymer particles dispersed inside each microsphere swelled, with part of them reaching the surface of each microsphere, extruding from the surface of the microsphere, and the remaining part staying within the microspheres. CP particles with a mean size of 10 μm gave uniform dispersion and good swelling properties.

Mucoadhesion, defined as an adhesion phenomenon occurring between a mucosal membrane covered with mucus and nonbiological materials, is a two-step process. The first step is spreading of a mucoadhesive hydrogel over the mucus layer. The second step involves diffusion or penetration of polymer chains into the mucus. This second step requires hydration of the polymers and is influenced by the molecular weight, molecular mobility, and viscosity of the adhesive polymer and swelling (and gel-forming) properties of both the adhesive polymer and the mucus (16). Swelling is a prerequisite for mucoadhesive drug delivery systems. Both the CP coated-microspheres and Ad-microspheres showed swelling properties. The in vitro mucoadhesiveness of these two types of microspheres, which are candidates for a mucoadhesive drug delivery system, was examined using the method designed by Ranga Rao and Buri (17). Briefly, 100 particles of PGEF-microspheres, CP coated-microspheres, or Ad-microspheres were placed uniformly on stomach tissue (2 × 1 cm) and jejunum tissue (4 cm in length) from

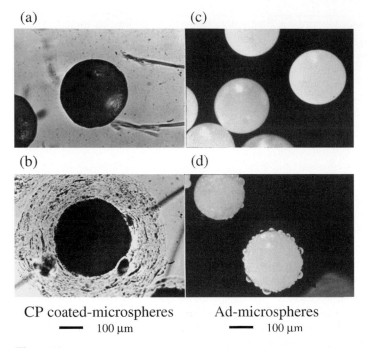

CP coated-microspheres   Ad-microspheres
━━ 100 μm   ━━ 100 μm

Figure 3 Micrographs of CP coated-microspheres before (a) and after (b) contact with water and Ad-microspheres before (c) and after (d) contact with water.

male Sprague-Dawley rats (300–400 g) that had been fasted overnight and dissected under ether anesthesia. The stomach and intestine tissues on a polyethylene support adjusted to an inclined position (45°) were rinsed with the JP XII 1st fluid (pH 1.2) and physiological saline for 5 minutes, respectively, at a rate of 22 mL/min. More than 90% of the Ad-microspheres remained on the gastric mucosa and small intestinal mucosa. The percentage of PGEF-microspheres or CP coated-microspheres remaining on the gastric mucosa was less than 10% after rising with the 1st fluid. Neither the PGEF-microspheres nor the CP coated-microspheres remained on the jejunum after rinsing with saline.

IV. IN VIVO EVALUATION OF MUCOADHESIVE MICROSPHERES

Approximately 50 mg of the three types of microspheres (PGEF-micro-spheres, CP coated-microspheres, and Ad-microspheres) placed in a polyethylene tube having one end covered with hydroxypropyl cellulose film were administered orally to male Sprague-Dawley rats (300–400 g), fasted for 24 hours, through the polyethylene tube attached to a gastric sonde with 0.2 mL of water (18). After 2.5 hours, the extent of the adhesion of these microspheres to the gastric mucosa was evaluated visually (Fig. 4). Most of the Ad-microspheres were found in the stomach, where very few of the PGEF-microspheres and the CP coated-microspheres were found (photograph with PGEF-microspheres not shown). Taking the results of the swelling test and adhesive properties into consideration, the process of in vivo adhesion is considered to occur as follows: As soon as the CP coated-microspheres came in contact with water, the CP particles formed a gel layer around the microspheres. The gel layer soon separated from the core of the microspheres because the hydrophilic gel lacked affinity for the core. On the other hand, the CP particles swelled from within the microsphere to the surface when the Ad-microspheres came in contact with water as shown schematically in Fig. 2. The swollen CP particles were strongly anchored to the Ad-microsphere and the extruded swollen CP particles in the Ad-microsphere adhered to the mucosa, leaving anchors within the Ad-microsphere. Therefore, the Ad-microspheres adhered tightly to the stomach wall.

Mucoadhesive drug delivery systems for the gastrointestinal tract require a stronger adhesive potential than a topical application system, such as those for the oral mucosal tissues (buccal, sublingual, or gingival), nasal cavity, rectum, or skin, because the system has to adhere to the mucosa under dynamic conditions. Therefore, the weight and volume of a mucoadhesive drug delivery system for the gastrointestinal tract and the area for

(a)   (b)

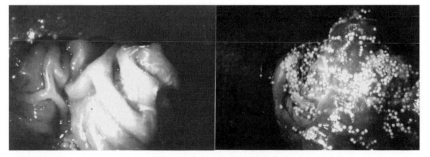

CP coated-microspheres   Ad-microspheres

Figure 4   Microspheres remaining in the stomach 2.5 hours after administration to a rat: (a) CP coated-microspheres; (b) Ad-microspheres.

adhesion are important factors. A heavy and bulky dosage form such as a tablet is apt to detach from the mucosa while traveling through the gastrointestinal tract even if the dosage form has attached to the mucosa. Because the appearance of the Ad-microspheres is that of a powder that is light and small, they are expected to fill the requirements for a mucoadhesive drug delivery system for the gastrointestinal tract.

V. QUANTITATIVE EVALUATION OF GASTROINTESTINAL TRANSIT OF MUCOADHESIVE MICROSPHERES

A. Percentage of Microspheres Remaining in the Rat's Gastrointestinal Tract

The gastrointestinal transit of the Ad-microspheres in rats was evaluated and compared with that of the PGEF-microspheres as a control. One hundred PGEF-microspheres or Ad-microspheres were administered to each fasted rat ($n = 3$). At specified time intervals, rats were sacrificed with ether, and the stomach and the small intestine were removed. The small intestine was cut into three equal segments and opened longitudinally. Both PGEF- and Ad-microspheres were found to distribute through the gastrointestinal tract. The swollen Ad-microspheres were found to adhere to the gastrointestinal mucosa all the time, although the Ad-microspheres adhered more strongly to the stomach than the small intestine.

The microspheres in the stomach and in each intestinal segment were counted. One hour after administration, the percentage of PGEF-microspheres remaining in the stomach was only 20.7%, and even 5 hours later few PGEF-microspheres were found (Fig. 5). On the other hand, the percentage of Ad-microspheres remaining in the stomach was 78.3% 1 hour after administration, and even after 5 hours, 5.3% was found in the stomach. The gastric emptying time of the Ad-microspheres was more prolonged than that of the PGEF-microspheres (Fig. 5). In addition, 89.6% of the Ad-microspheres still remained in the stomach or the small intestine 5 hours after administration, whereas only 43.0% of the PGEF-microspheres were found in the same region; that is, only 10.4% of the Ad-microspheres had reached the colon, while 57.0% of the PGEF-microspheres had reached the colon, assuming that the number of the microspheres that reached the colon (cecum) is the difference between the number of microspheres administered (100 microspheres) and the number of microspheres still in the stomach and the small intestine. The prolongation of the transit time of Ad-microspheres to the colon indicated that the Ad-microspheres hydrated rapidly in vivo,

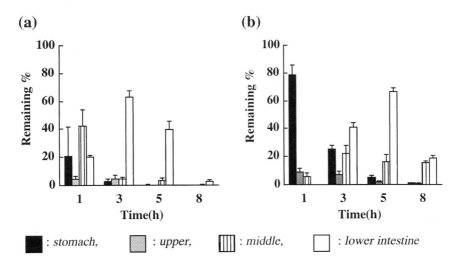

Figure 5 Distribution of (a) PGEF-microspheres and (b) Ad-microspheres in the stomach and upper, middle, and lower segments of the small intestine in rats. Data are shown as mean ± SEM.

adhered to the mucus-covered mucosa of the stomach or small intestine, and remained there for an extended period.

B. Model for Gastrointestinal Transit of the Microspheres

Transit of the microspheres through the gastrointestinal tract was evaluated according to a proposed model in which gastric emptying and intestinal transit patterns are expressed with relationships containing a lag time (Fig. 6a.

It was assumed that microspheres administered to a rat were emptied from the stomach monoexponentially after a lag time, T_s, and that transit of the microspheres through each segment of the small intestine was at zero order. The microspheres proceeded through the upper, middle, and lower segments of the small intestine in T_u, T_m, and T_l hours, respectively, and then reached the colon. R_s, R_{su}, R_{sm}, and R_{sl} represent the percentages of the microspheres remaining in the stomach, the stomach and the upper intestine (Region I), the stomach and upper and middle intestine (Region II), and the stomach and the entire intestine (Region III), respectively (Fig. 6b). The time at which 50% of the microspheres have been emptied from the stomach is expressed as T_{s50}. The time at which 50% of the microspheres have reached the colon is expressed as T_{50}. The percentage of the microspheres remaining in the stomach, R_s, at time t is written as

(a)   **(b)**

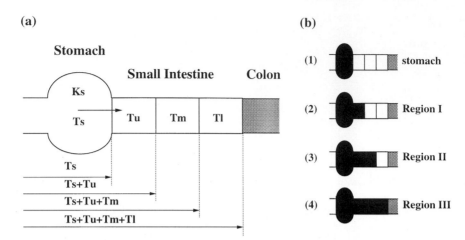

Figure 6 Gastrointestinal emptying model. (a) K_s, gastric emptying rate constant; T_s, lag time; T_u, T_m, and T_l, transit time through upper, middle, and lower small intestine. (b) (1) Stomach; (2) Region I; (3) Region II, (4) Region III.

$$R_s = 10^2 \qquad\qquad (t \leq T_s) \qquad\qquad\qquad\qquad\qquad (1)$$
$$R_s = 10^{2-K_s(t-T_s)} \qquad (t > T_s) \qquad\qquad\qquad\qquad\qquad (2)$$

where K_s is the gastric emptying rate constant.

   The pattern of gastrointestinal emptying is expressed as follows, assuming that the stomach and the upper segment of the small intestine belong to one compartment (Region I) and that a microsphere passes through the upper segment of the small intestine in time T_u:

$$R_{su} = 10^2 \qquad\qquad\qquad (t \leq T_s + T_u) \qquad\qquad\qquad\qquad (3)$$
$$R_{su} = 10^{2-K_s(t-T_s-T_u)} \qquad (t > T_s + T_u) \qquad\qquad\qquad\qquad (4)$$

where R_{su} is the percentage of the microspheres remaining in Region I, and the gastrointestinal pattern is expressed as Eqs. (3) and (4), which are obtained by shifting the gastric emptying pattern by time T_u. Similarly, assuming that the stomach and the upper and middle segments of the small intestine belong to one compartment (Region II) and that a microsphere passes through the middle part of the small intestine in time T_m, the following relationships are obtained:

$$R_{sm} = 10^2 \qquad (t \le T_s + T_u + T_m) \qquad (5)$$

$$R_{sm} = 10^{2 - K_s(t - T_s - T_u - T_m)} \qquad (t > T_s + T_u + T_m) \qquad (6)$$

where R_{sm} is the percentage of the microspheres remaining in Region II. Assuming that the stomach and the entire small intestine are regarded as one compartment (Region III) and that a microsphere passes through the lower part of the small intestine in time T_l, the following relations are obtained:

$$R_{sl} = 10^2 \qquad (t \le T_s + T_u + T_m + T_l) \qquad (7)$$

$$R_{sl} = 10^{2 - K_s(t - T_s - T_u - T_m - T_l)} \qquad (t > T_s + T_u + T_m + T_l) \qquad (8)$$

where R_{sl} is the percentage of the microspheres remaining in Region III.

The transit time of the small intestine, T_{si}, the time required for 50% of the microspheres to be emptied from the stomach (the mean gastric emptying time), T_{s50}, and the time required for 50% of the microspheres to reach the colon (cecum), T_{50}, are written as

$$T_{si} = T_u + T_m + T_l \qquad (9)$$

$$T_{s50} = T_s + (2 - \log 50)/K_s \qquad (10)$$

and

$$T_{50} = T_s + T_{si} + (2 - \log 50)/K_s \qquad (11)$$

respectively.

C. Quantitative Comparison of Gastrointestinal Transit of PGEF-Microspheres and Ad-Microspheres

To quantify the rate of gastric emptying and the small intestinal transit, the time courses of distribution of the microspheres (PGEF-microspheres and Ad-microspheres) in the stomach and different segments of the small intestine were calculated using the preceding model. Simultaneous fitting of the percentage of the microspheres remaining versus time curves to the equations using a nonlinear least squares program (MULTI) (19) resulted in the parameters listed in Table 2. The percentage of microspheres remaining in the stomach and each region was plotted on the computer-generated profiles in Fig. 7. Because the percentage generated by the computer agreed well with the percentage of the PGEF-microspheres and Ad-microspheres re-

Table 2   Rate and Time Constants for Gastrointestinal
Transit of PGEF-Microspheres and Ad-Microspheres

Parameter[a]	PGEF-microspheres	Ad-microspheres
K_s	0.6 1/h	0.2 1/h
T_s	0.0 h	0.5 h
T_u	0.1	0.3
T_m	0.7	1.2
T_l	3.7	4.1
$T_u + T_m + T_l$	4.5	5.6
T_{s50}	0.5	1.8
T_{50}	4.9	7.3

[a]K_s, gastric emptying rate constant; T_s, lag time; T_u, T_m, T_l, transit times
through upper, middle, and lower small intestine; T_{s50}, time required
for 50% of microspheres to leave stomach; T_{50}, time required for 50%
of microspheres to reach colon.

maining in the stomach and small intestine, the model seems to be appropriate for predicting the in vivo distribution of the microspheres.

The gastric emptying rate constant, K_s, after administration of the Ad-
microspheres to fasted rats was about one third of the value after administration of the PGEF-microspheres. When the Ad-microspheres were admin-

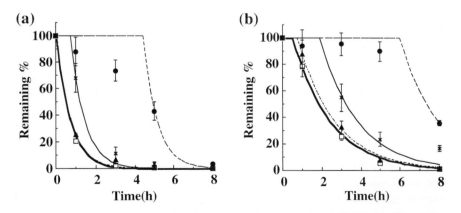

Figure 7   Percentage of (a) PGEF-microspheres and (b) Ad-microspheres remaining in the stomach, Region I, Region II, and Region III with computer-generated
gastrointestinal transit profiles. (□) Stomach; (▲) Region I; (×) Region II; (●) Region
III. Computer-generated profiles. (——), Stomach; (·–·–), Region I; (———), Region II; (- - - -), Region III.

istered to rats, the lag time, T_s, was 0.5 hour, whereas it was nearly zero when the PGEF-microspheres were administered. This means that the PGEF-microspheres were emptied from the stomach exponentially without a lag time as reported by Hunt and Spurrell (20), who examined the gastric emptying time of the human stomach and concluded that the gastric emptying of a standard pectin meal (750 mL) was exponential. On the other hand, the large positive T_s value of 0.5 when the Ad-microspheres were administered indicated that these microspheres adhered strongly to the gastric mucosa and remained in the stomach for a long period of time. The calculated transit times of the Ad-microspheres in the upper, middle, and lower segments of the small intestine, T_u, T_m, and T_l, were longer than those of the PGEF-microspheres; that is, the transit of the Ad-microspheres in the small intestine was prolonged by the incorporation of the CP particles. The mean gastric emptying time for the Ad-microspheres, T_{s50} was longer than that for the PGEF-microspheres by 1.3 hours. In conclusion, the gastrointestinal transit time was prolonged by 2.4 hours. This means that the Ad-microspheres provided a localized platform in the gastrointestinal tract for sustained drug release and prolonged the gastrointestinal transit time.

The reason for only a slight delay in the small intestinal transit compared with the delay in the gastric emptying rate may be explained by overhydration and overswelling of CP at higher (5–6) pH levels, which could have decreased mucoadhesion in the small intestine. Lehr et al. (21) reported that a cationic polymer, chitosan, showed the same degree of mucoadhesiveness as polycarbophil when they measured the force of detachment of swollen polymer film from pig mucosa in a saline medium. Irrespective of the intrinsic properties of the polymers, the maximal residence time is probably limited by the renewal of mucus gel layers. Turnover of mucin should also be investigated, because mucin is always being replaced and this is expected to affect the duration of mucoadhesion. Mucoadhesive drug delivery systems are not expected to adhere for more than 4–5 hours, considering the mucin turnover time (50–270 minutes) calculated by Lehr et al. (22).

Khosla and Davis (5) and Harris et al. (6) reported that polyacrylic acid did not retard the human gastric emptying of the pellets in their study using gamma scintigraphy, and Russel and Bass (23) reported that the large amount of polycarbophil may have elicited myoelectric migrating contractions (MMCs) of the fed stomach, which would result in a slower gastric emptying. The formulation used by Khosla and Davis (5) was prepared by mixing polycarbophil (of 0.5–1.0 mm in diameter) and pellets (Amberlite IRA410 anionic resin) and placing the mixture in a No. 0 hard gelatin capsule. The mucoadhesion of Ad-microspheres to human mucosa should be confirmed.

VI. APPLICATION OF THE AdMMS TO A DRUG HAVING A NARROW ABSORPTION WINDOW

A. Preparation and Characterization of PGEF-Microspheres and Ad-Microspheres

We selected furosemide and riboflavin (vitamin B_2) as model drugs whose absorption is limited to the upper part of the gastrointestinal tract (24–26). If a sustained-release dosage form containing furosemide or riboflavin having an absorption window in the upper small intestine is administered, it will quickly pass through the absorption window and reach the colon before releasing all of the furosemide or riboflavin content. However, if an adhesive dosage form could reside for longer periods in the stomach or the upper small intestine because of its adhesion to the mucosa, the bioavailability would be improved compared with that of a nonadhesive sustained-release dosage form. In other words, if the bioavailability of furosemide or riboflavin from an adhesive sustained-release dosage form is higher than that from a nonadhesive sustained-release dosage form that releases furosemide or riboflavin at the same rate, we could conclude that the adhesive sustained-release dosage form really does adhere to the gastrointestinal mucosa, resides in the gastrointestinal tract for an extended time, and allows the drug to be absorbed more effectively.

PGEF-microspheres, nonadhesive sustained-release microspheres, are prepared by spraying and chilling furosemide (10% w/w) and lactose (30% w/w) dispersed in melted PGEF (TGPS:TGMS ratio of 2:1) (60% w/w). Ad-microspheres, adhesive sustained-release microspheres, were prepared from Jurosemide (10% w/w) and CP (HIVISWAKO 104, Wako Pure Chem. Ind., Osaka, Japan) (15% w/w) dispersed in the melted PGEF [tetraglycerol hexabehenate (TGHBe):TGMS ratio of 65:10] (75% w/w). To prepare PGEF- and Ad-microspheres that release drugs at a similar rate, TGHBe, TGPS, and TGMS whose HLB values are 1.8, 2.6, and 8.4, respectively, were mixed in an appropriate ratio. The in vitro release of a drug from the Ad-microspheres was regulated by selecting an appropriate HLB of the PGEF, as is the case with the PGEF-microspheres. The size of Ad-microspheres and the content of a drug also affected the release rate.

Both PGEF- and Ad-microspheres in the size range 177 to 500 μm, obtained by sieving, were used for animal and in vitro release experiments. The remaining percentage of furosemide was determined as follows: each rat's stomach, removed 2 hours after microsphere administration at a dose of 10 mg/kg of furosemide, was put into a mixture of 100 mL of phosphate buffer solution (pH 7.2) and acetonitrile (65:35) and furosemide extracted at 80°C was determined by a high-performance liquid chromatographic (HPLC) method. The percentage of furosemide remaining in the rat stomach

(a) **(b)**

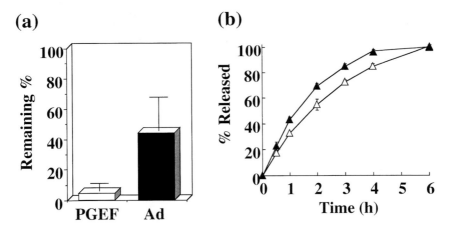

Figure 8 (a) Percentage of furosemide remaining in the stomach 2 hours after oral administration of PGEF-microspheres (□) and Ad-microspheres (■) to fasted rats (10 mg/kg, $n = 5$, mean $\pm$ SD). (b) Release profiles of furosemide from PGEF- (△) and Ad- (▲) microspheres ($n = 3$, mean $\pm$ SD).

2 hours after Ad-microsphere administration was higher than that after PGEF-microsphere administration (Fig. 8a). The in vitro release (measured using the USP XXII paddle apparatus, 100 rpm) profiles of furosemide from Ad-microspheres were similar to those from PGEF-microspheres in the JP XII 1st fluid (pH 1.2) (Fig. 8b). The difference in the remaining percentage is expected to result from the difference in adhesiveness of the two kinds of microspheres.

B. Enhanced Absorption of Furosemide After Ad-Microsphere Administration

The plasma profiles after administration of a 0.5% methylcellulose suspension, PGEF-microspheres, and Ad-microspheres containing 10 mg/kg of furosemide to fasted rats are shown in Fig. 9. In the case of the methylcellulose suspension, the plasma concentration of furosemide reached the C_{max} (2.4 $\pm$ 0.6 µg/mL) at 0.8 $\pm$ 0.4 hours and then rapidly decreased. On the other hand, PGEF- and Ad-microspheres produced a furosemide plasma profile characteristic of a sustained-release formulation. Mean residence time (MRT) values after PGEF- and Ad-microsphere administration were 6.1 $\pm$ 0.6 and 6.7 $\pm$ 07 hours, respectively, while the MRT value after methylcellulose suspension administration was 4.3 $\pm$ 04 hours. In the case of Ad-microsphere administration, the AUC_{0-24h} (11.57 $\pm$ 1.84 µg h/mL) was 1.8

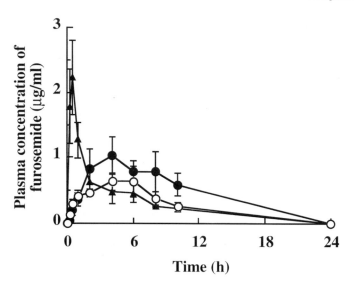

Figure 9   Plasma concentration of furosemide after oral administration of a methylcellulose suspension (▲), PGEF-microspheres (○), and Ad-microspheres (●) to fasted rats (10 mg/kg, $n = 5$, mean ± SEM).

times higher than that (6.56 ± 0.93 μg h/mL) after PGEF-microsphere administration. It seemed that furosemide was slowly released from the Ad-microspheres, which had adhered to a more proximal area of the gastrointestinal tract rather than the absorption window, and was thereby effectively absorbed from the absorption window. Lehr et al. (27) reported that in the in vivo absorption of 9-desglycineamide, 8-arginine vasopressin (DGAVP) from the rat small intestine was not improved after administration in the form of controlled-release microspheres coated with an adhesive polyacrylic acid derivative compared with the absorption after administration in the form of noncoated controlled-release microspheres, although DGAVP absorption was improved when administered in a liquid dispersion of the polyacrylic acid derivative. Therefore, there is little possibility that the polyacrylic acid derivative powder dispersed in the Ad-microspheres worked as an absorption enhancer. These results suggest that the improved absorption of furosemide after Ad-microsphere administration resulted from the prolonged gastrointestinal residence of the microspheres. The reason why the AUC value of furosemide was smaller after PGEF-microsphere administration might be that the microspheres passed through the absorption window before all of the furosemide was released. This means the preceding assumption is correct. These results indicate that the mucoadhesiveness of Ad-microspheres can be evaluated by a comparison of drug availability, as determined by AUC or urinary recovery, after the administration of PGEF- and Ad-microspheres.

VII. EVALUATION OF THE AdMMS IN HUMAN VOLUNTEERS

Considering the results obtained for the enhanced furosemide absorption in rats after Ad-microsphere administration, we proceeded to an experiment to confirm the ability of a mucoadhesive dosage form to adhere to the gastro-intestinal mucosa and prolong the drug's residence in the gastrointestinal tract in humans; i.e., we planned to compare the pharmacokinetics of ribo-flavin after administration in the form of nonadhesive PGEF- and adhesive Ad-microspheres. Formulations of capsules containing PGEF- or Ad-micro-spheres are shown in Table 3. Two kinds of human studies (step I and step II) were performed using a No. 3 capsule filled with PGEF- and Ad-mi-crospheres containing riboflavin. The remaining percentage of riboflavin as well as furosemide in rat stomach 2 hours after Ad-microsphere administra-tion (10 mg/kg) was higher than that after PGEF-microsphere administration (Fig. 10a) and the in vitro release of riboflavin from Ad-microspheres was similar to that from PGEF-microspheres (Fig. 10b).

Step I was carried out under fasted conditions for evaluation of the adhesiveness of Ad-microspheres. Step II was carried out for evaluation of the effect of food on the adhesiveness of Ad-microspheres. The protocols of both studies were approved by the Institutional Review Board (IRB) of Osaka Pharmacology Research Clinic. Freely given informed consent was obtained from every subject prior to participation. Ten healthy male volun-teers between 20 and 26 years old participated in both the step I and step II studies. In each step, absorption studies were performed in a 2×2 cross-over design. Analysis of variance on Recovery$_{0-24h}$, maximal urinary excre-

Table 3 Formulations of Capsules Filled with PGEF- or Ad-Microspheres Containing Riboflavin

Formulation	Contents in one capsule (total 100 mg)			
	Riboflavin (mg)	PGEF (mg)	Carboxyvinyl polymer (mg)	Lactose (mg)
PGEF/riboflavin-C	10	60	—	30
Ad/riboflavin-C	10	70	20	—

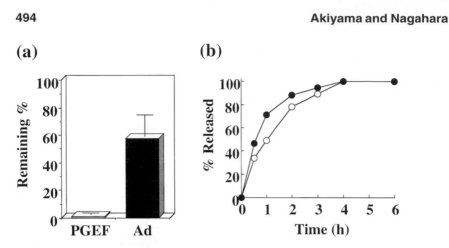

Figure 10   (a) Percentage of riboflavin remaining in the stomach 2 hours after oral administration of PGEF-microspheres (□) and Ad-microspheres (■) to fasted rats (10 mg/kg, $n = 5$, mean ± SD). (b) Release profiles of riboflavin from PGEF- (○) and Ad- (●) microspheres ($n = 3$, mean ± SD).

tion rate (R_{max}), the time required to reach R_{max} (T_{max}), and MRT based on the urinary excretion rate–time curve was performed using the SAS GLM procedure (SAS/STAT User's Guide, Version 6, 4th edition, SAS Institute Inc., 1990).

In step I, the pharmacokinetics of riboflavin after administration of a capsule filled with Ad-microspheres containing riboflavin (Ad/riboflavin-C) to each volunteer was compared with that after administration of a capsule filled with PGEF-microspheres containing riboflavin (PGEF/riboflavin-C). Urinary excretion rate versus time profiles are shown in Fig. 11. Recovery$_{0-24h}$ with Ad-microspheres was 2.27 ± 0.38 mg, whereas that with PGEF-microspheres was 0.96 ± 0.15 mg. Statistical comparison of Recovery$_{0-24h}$ parameters indicated a significant difference ($P < .01$) between PGEF- and Ad-microspheres. The higher bioavailability of riboflavin after Ad-microsphere administration than after PGEF-microsphere administration suggests that the Ad-microspheres adhered to the stomach and upper small intestine mucosa and released riboflavin slowly, and consequently the released riboflavin gradually passed through the absorption site and was absorbed more efficiently. Although urinary recovery of riboflavin administered in the form of the Ad-microspheres was more than twice as high as that seen with the PGEF-microspheres, there was no significant difference in T_{max} or MRT values. Levy and Jusko (28) reported that T_{max} was in the range 0.5 to 1.5 hours when riboflavin was administered to fasted human volunteers in the form of a solution. In comparison with the shorter T_{max} after solution

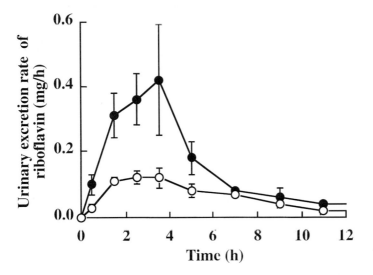

Figure 11 Urinary excretion of riboflavin after oral administration of PGEF- (○) and Ad- (●) microspheres to fasted human volunteers (10 mg/man, $n = 10$, mean ± SEM).

administration, riboflavin was absorbed more slowly after Ad-microsphere administration (T_{max} 2.1 hours). If riboflavin was absorbed only from the upper small intestine over prolonged periods, MRT after Ad-microsphere administration should be longer than that after PGEF-microsphere administration. However, the MRT value of riboflavin after PGEF-microsphere administration, 5.6 ± 0.4 hours was slightly longer than that after Ad-microsphere administration, 4.8 ± 0.3 hours. Middleton (29) reported that uptake of riboflavin occurred throughout the rat small intestine in vitro. Therefore, a possible reason why the MRT after Ad-microsphere administration was not longer than that after PGEF-microsphere administration is that absorption of riboflavin occurred in the lower part of the small intestine in addition to the upper part of the small intestine.

Generally, adhesion of microspheres to the gastric mucosa is expected to be reduced by food intake. In the step II study, urinary recovery of riboflavin after Ad/riboflavin-C administration under fasted conditions was compared with that under fed conditions. As shown in Fig. 12, higher urinary excretion rates of riboflavin were maintained from 3.5 to 11.0 hours, and both T_{max} and MRT were prolonged under fed conditions compared with fasted conditions. Recovery$_{0-24h}$ of riboflavin in fed volunteers was more than twice that after administration in fasted volunteers (Table 4). Statistical comparison of these three values, Recovery$_{0-24h}$, T_{max}, and MRT, indicated

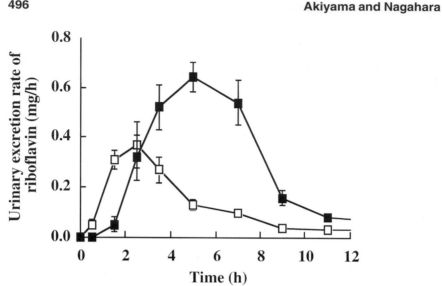

Figure 12   Urinary excretion of riboflavin after oral administration of Ad-microspheres to human volunteers under fasted (□) and fed (■) conditions (10 mg/man, *n*
= 10, mean ± SEM).

that there were significant differences in absorption profiles between fasted
and fed conditions in addition to greater absorption with the Ad-microspheres. Levy and Jusko (28) reported that food intake increased urinary
recovery of riboflavin after solution administration to humans. In the report,
the urinary recoveries of riboflavin under fasted and fed conditions were 30
and 63%, respectively. In the case of a solution, MRT under fed conditions,
2.2 hours, was not so significantly prolonged in comparison with that under

Table 4   Pharmacokinetic Parameters of Riboflavin in Ad-Microspheres Given Orally
at a Dose of 10 mg as Riboflavin to Human Volunteers (*n* = 10) Under Fasted and
Fed Conditions

Capsule containing microspheres	Recovery$_{0-24h}$ (mg)	R_{max}[a] (mg/h)	T_{max} (h)	MRT (h)
Ad/riboflavin-C-fasted	2.05 ± 0.26[b]	0.49 ± 0.06	2.3 ± 0.2	4.9 ± 0.2
Ad/riboflavin-C-fed	4.22 ± 0.21[c]	0.82 ± 0.03[c]	4.9 ± 0.5[c]	6.3 ± 0.3[c]

[a]Maximal urinary excretion rate.
[b]Each value is the mean ± SEM.
[c]$P < .001$.

fastcd conditions, 1.9 hours. On the other hand, the prolonged MRT after Ad-microsphere administration, 6.3 ± 0.3 hours, might have resulted from longer residence in the gastrointestinal tract. Consequently, these results indicate that Ad-microsphere administration under fed conditions caused neither reduced riboflavin absorption due to interaction with food nor rapid riboflavin absorption due to interaction with food nor rapid riboflavin absorption due to the system's destruction, although Hosny et al. (30) suggested that the drastic reduction of indomethacin absorption due to food intake might have resulted from the complexation of bioadhesive material with food contents and consequent lack of bioadhesion in their experiments using bioadhesive tablets containing indomethacin and polycarbophil (calcium salt of cross-linked carboxyvinyl polymer).

When riboflavin, with a narrow absorption window, was administered to human volunteers in the form of Ad-microspheres, enhanced absorption due to the prolongation of the gastrointestinal residence was obtained without destruction of the microspheres under fed conditions. It is concluded that the Ad-microspheres could adhere to the human gastrointestinal mucosa, resulting in their prolonged gastrointestinal residence.

VIII. POSSIBILITY OF USING THE AdMMS AS AN ORAL DELIVERY SYSTEM TARGETING THE GASTROINTESTINAL TRACT

The main functions of mucoadhesive drug delivery systems are to delay the gastrointestinal transit of the delivery system and increase the local drug concentration at the site of adhesion. Ad-microspheres were found to have prolonged residence in the stomach and to provide a high drug concentration in the stomach. Since the discovery of *H. pylori* in 1983 by Marshall and Warren (31), great attention has focused on its association with gastric or duodenal ulcers (32,33). It has become increasingly accepted that *H. pylori* is the major cause of peptic ulcers (34). However, clinical trials using single antimicrobial agents have not shown complete eradication of *H. pylori*, although the organism is susceptible to many antimicrobial agents (35,36) such as amoxicillin and clarithromycin. One of the reasons for incomplete eradication is that the residence time of antimicrobial agents in the stomach is so short that an effective antimicrobial concentration cannot be achieved in the gastric mucus layer or epithelial cell surfaces where *H. pylori* exists (36). Therefore, it is expected that if local delivery of antimicrobial agents from the gastric lumen into the mucus layer can be achieved, the eradication rate of *H. pylori* will be increased. In fact, a 1-hour therapy developed by Kimura et al. (37) provided more complete eradication of *H. pylori* than conventional therapy because of the extended gastric residence of the anti-microbial agents. Amoxicillin is known to take several hours to kill *H. pylori* inoculated in agar, compared with shorter times when using clarithromycin or metronidazole (38), but amoxicillin has a low minimal inhibitory concentration (MIC) for *H. pylori* and has therefore been used in many clinical trials aimed at eradication of *H. pylori* from the gastric mucosa (39–41). However, no in vivo eradication trials utilizing dosage forms that have a prolonged gastric residence time have been reported.

A. Preparation and Characterization of Ad-Microspheres Containing Amoxicillin

The AdMMS, which was confirmed to adhere to the stomach wall and to prolong gastric residence, was applied to amoxicillin. Ad-microspheres containing amoxicillin (amoxicilin-Ad-microspheres) were prepared by spraying and chilling amoxicillin (0.15 g), curdlan (β-1,3 glucan type polysaccharide, 1.35 g) and CP (HIVISWAKO® 104, 1.0 g) dispersed in melted hydroxygenated castor oil (HCO, 7.5 g) instead of PGEF (10). Amoxicillin-Ad-microspheres 250–335 μm in diameter obtained by sieving were used for further experiments.

The remaining percentage of amoxicillin as an index of mucoadhesiveness and prolongation of gastric residence was evaluated after amoxicillin-Ad-microsphere administration at a dose of 10 mg/kg amoxicilin to fed 7-week-old male specific pathogen–free (SPF) Mongolian gerbils (50–60 g) and compared with that after administration in the form of a 0.5% methylcellulose suspension (amoxicillin-suspension). Amoxicillin-Ad-microspheres were administered to Mongolian gerbils using the same method as used for rats except that a polyethylene tube with a smaller diameter was used. As shown in Fig. 13, the remaining percentages of amoxicillin 2 and 4 hours after amoxicillin-Ad-microsphere administration, 47.3 ± 14.4 and $20.4 \pm 11.5\%$, respectively, were about three times higher than those after amoxicillin-suspension administration, 17.3 ± 3.2 and $6.2 \pm 1.1\%$, respectively.

B. In Vivo Clearance of *H. pylori*

As an animal model for evaluating *H. pylori* clearance in vivo, an SPF Mongolian gerbil that was infected with the *H. pylori* strain, TN2GF4 (4th passage derivative of TN2 strain) isolated from human patients with gastritis or gastric ulcers was adopted. The *H. pylori* was inoculated into the stomach of each 4-week-old male Mongolian gerbil via an orogastric tube under fasted conditions. Fourteen days after infection, amoxicillin-Ad-micro-

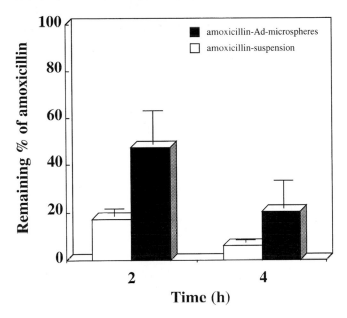

Figure 13 Percentage of amoxicillin remaining in the stomach of Mongolian gerbils 2 and 4 hours after oral administration of amoxicillin-Ad-microspheres (■) and amoxicillin suspension (□) (10 mg/kg, $n = 5$, mean ± SEM).

spheres and an amoxicillin suspension in addition to a methylcellulose suspension (10 mL/kg) as a vehicle control were orally administered to each *H. pylori*–infected Mongolian gerbil under fed conditions twice a day for 3 consecutive days at doses of 1, 3, 10, and 30 mg/kg. Viable cell counts per gastric wall 1 day after the final dose were calculated by counting colonies grown on agar plates after incubation for 4 days.

The mean bacterial count after oral administration of the amoxicillin suspension and amoxicilin-Ad-microspheres decreased as the dose of amoxicillin increased (Fig. 14). The *H. pylori* clearance rates at doses of 10 and 30 mg/kg were 20 and 60%, respectively, when an amoxicillin suspension was administered (Table 5). On the other hand, perfect clearance of *H. pylori* (clearance rate 100%) was obtained after amoxicillin-Ad-microsphere administration at doses of 10 and 30 mg/kg. Even at doses of 3 and 1 mg/kg, 40 and 20% clearance was obtained; i.e., the amoxicillin-Ad-microspheres at a dose of 1.0 mg/kg as amoxicillin provided the same clearance rate (20%) as the amoxicillin suspension at a dose of 10 mg/kg. This means that the amoxicillin-Ad-microspheres provided 10 times higher anti–*H. pylori* activity than that of the amoxicillin suspension. This higher activity would be a

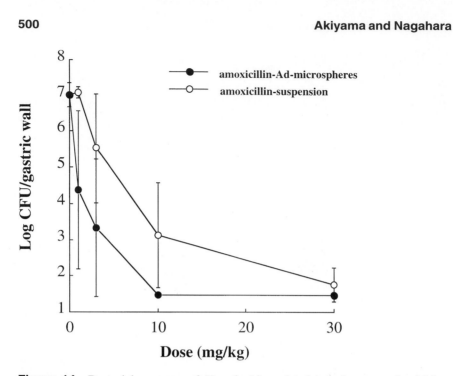

Figure 14 Bacterial recovery of *H. pylori* inoculated into the stomach of Mongolian gerbils after repetitive oral administration of amoxicillin-Ad-microspheres (●) and amoxicillin suspension (○) (*n* = 5, mean ± SEM).

result of the difference in gastric residence provided by the two dosage forms. These results indicate that the topical action of amoxicillin at the gastric mucus played an important role in the clearance of *H. pylori*. The results suggest that the amoxicillin-Ad-microspheres adhere to the *H. pylori*–infected gastric mucosa, which may not be covered with mucin, even under fed conditions.

Photomicrographs of the *H. pylori*–eradicated gastric mucosa in Mongolian gerbils 1 month after amoxicillin-Ad-microsphere administration as well as amoxicillin suspension administration revealed no histological changes in the pyloric region (photomicrographs not shown). On the other hand, dense cell infiltration, mainly of neutrophils and lymphocytes, and a remarkable increase in thickness of the mucosa were observed in the pyloric region of *H. pylori*–infected Mongolian gerbils without amoxicillin treatment. These results indicate that the prolonged adhesion of the amoxicillin-Ad-microspheres to the mucosa of the stomach did not cause any damage and that this dosage form consisting of mucoadhesive microspheres is useful in the treatment and healing of gastritis and peptic ulcers.

Table 5 Effect of Repetitive Oral Administration of Amoxicillin-Ad-Microspheres Containing Amoxicillin Against Gastric Infection Caused by *H. pylori* in Mongolian Gerbils

Preparation	Dose[a] (mg/kg)	Clearance rate		Bacterial recovery Log CFU/gastric wall, mean ± SD[b]
		Cleared/total	(%)	
Vehicle control[c]	0	0/5	(0)	7.02 ± 0.33
Amoxicillin suspension	1	0/5	(0)	7.08 ± 0.15
	3	0/5	(0)	5.53 ± 1.50
	10	1/5	(20)	3.12 ± 1.45
	30	3/5	(60)	1.76 ± 0.47
Amoxicillin-Ad-microspheres	1	1/5	(20)	4.37 ± 2.19
	3	2/5	(40)	3.32 ± 1.90
	10	5/5	(100)	ND
	30	5/5	(100)	ND

[a]Twice a day for 3 days.
[b]Bacterial counts less than $10^{1.48}$ CFU were taken as $10^{1.48}$ CFU to calculate the mean ± SD.
[c]0.5% methylcellulose suspension.

The present work is the first evidence demonstrating the in vivo usefulness of a mucoadhesive dosage form for clearing *H. pylori*. The 10 times stronger anti–*H. pylori* activity provided by the amoxicillin-Ad-microspheres in Mongolian gerbils suggests that drug delivery using Ad-microspheres, which strengthen the topical action of amoxicillin because of the prolonged residence in the stomach, could provide a high *H. pylori* eradication rate in humans.

IX. CONCLUSION

Ad-microspheres, which consist of a drug and an adhesive polymer (CP) powder dispersed in a waxy base such as PGEF and/or HCO, could strongly adhere to the stomach mucosa in rats and Mongolian gerbils and prolonged the drug's gastrointestinal residence after oral administration. The reason for the strong adherence of Ad-microspheres to the stomach might be that when the Ad-microspheres came in contact with water, each CP particle in each microsphere was hydrated and swelled, with part of it remaining within the Ad-microsphere and part extending to the surface and serving to anchor the Ad-microsphere to the mucus layer. When furosemide with a narrow absorption window was administered to rats and riboflavin with a narrow absorption window was administered to human volunteers in the form of Ad-microspheres, enhanced absorption due to prolongation of the gastrointestinal residence was obtained. In addition, it is unlikely that Ad-microspheres were destroyed even under fed conditions.

As a nonspecific targeting drug delivery system, the AdMMS was applied to amoxicillin, which shows anti–*H. pylori* activity on direct contact with the stomach. Amoxicillin-Ad-microspheres cleared *H. pylori* inoculated into the Mongolian gerbil stomach more effectively than amoxicillin suspension because of the increase in the extent and duration of contact of amoxicillin at the site of action. There is a possibility that amoxicillin-Ad-microspheres can provide more complete eradication (without recurrence) than conventional dosage forms in humans. All these results suggest that the AdMMS would work as a targeting drug delivery system when applied to a drug that exerts its pharmacological effect by direct contact with the gastrointestinal tract mucosa as well as an oral sustained-release delivery system for a drug with a narrow absorption window.

ACKNOWLEDGMENTS

The authors wish to thank Dr. H. Toguchi for his encouragement and critical comments through this work and Mr. J.A. Hogan for linguistic advice during preparation of the manuscript. The authors also thank Dr. Y. Ogawa, Dr. S. Hirai, Mr. T. Kashihara, Dr. S. Iwasa, Dr. E. Nara, Ms. M. Kitano, Dr. M. Nakao, and Mr. M. Tada for their valuable contributions to this work.

REFERENCES

1. MA Longer, HS Ch'ng, JR Robinson. Bioadhesive polymers as platforms for oral controlled drug delivery. III: Oral delivery of chlorothiazide using a bioadhesive polymer. J Pharm Sci 74:406–411, 1985.
2. V Lenaerts, P Couvreur, L Grislain, P Maincent. Nanoparticles as a gastroadhesive drug delivery system. In: V Lenaerts, R Gurny, eds. Bioadhesive Drug Delivery Systems. Boca Raton, FL: CRC Press, 1990, pp 95–104.
3. C-M Lehr, JA Boustra, JJ Tukker, HE Junginger. Intestinal transit of bioadhesive microspheres in an in situ loop in the rat—a comparable study with copolymers and blends based on poly(acrylic acid). J Controlled Release 13: 51–62, 1990.
4. E Mathiowitz, JS Jacob, YS Jong, GP Carino, DE Chickering, P Chaturvedi, CA Santos, K Vijayaraghavan, S Montogomery, M Bassett, C Morrell. Biolog-

ically erodable microspheres as potential oral drug delivery systems. Nature 386:410–414, 1997.

5. L Khosla, SS Davis. The effect of polycarbophil on the gastric emptying of pellets. J Pharm Pharmacol 39:47–49, 1987.

6. D Harris, JT Fell, HL Sharma, DC Taylor. GI transit of potential bioadhesive formulations in man: A scintigraphic study. J Controlled Release 12:45–53, 1990.

7. NA Peppas, AM Lowman, MD Little, A De Ascentiis. Poly(ethylene glycol)-tethered controlled release systems with improved mucoadhesive behavior: Preparation and studies with ATR-FTIR spectroscopy. Drug Delivery System. Vol 12. 13th Annual meeting of the Japan Society of Drug Delivery System, 1997, p 240.

8. Y Akiyama, N Nagahara, T Kashihara, S Hirai, H Toguchi. In vitro and in vivo evaluation of mucoadhesive microspheres prepared for the gastrointestinal tract using polyglycerol esters of fatty acids and a poly(acrylic acid) derivative. Pharm Res 12:397–405, 1995.

9. Y Akiyama, N Nagahara, E Nara, M Kitano, S Iwasa, I Yamamoto, J Azuma, Y Ogawa. Evaluation of oral mucoadhesive microspheres in man on the basis of pharmacokinetics of furosemide and riboflavin, Compounds with limited gastrointestinal absorption sites. J Pharm Pharmacol 50:159–166, 1998.

10. N Nagahara, Y Akiyama, M Nako, M Tada, M Kitano, Y Ogawa. Muco-adhesive microspheres containing amoxicillin for clearance of Helicobacter pylori. Antimicrob Agent Chemother 42:2492–2494, 1998.

11. Y Akiyama, M Yoshioka, H Horibe, S Hirai, N Kitamori, H Toguchi. Novel oral controlled release microspheres using polyglycerol esters of fatty acids. J Controlled Release 26:1–10, 1993.

12. PR Steth, J Tossonian. The hydrodynamically balanced system (HBS™): A novel drug delivery system for oral use. Drug Dev Ind Pharm 10:313–339, 1984.

13. R Chagrill, LJ Caldwell, K Engle, JA Fix, P Porter, CR Gardner. Controlled gastric emptying. I: Effect of physical properties on gastric residence times on nondisintegrating geometric shapes in beagle dogs. Pharm Res 5:533–536, 1988.

14. JD Smart, IW Kellaway, EC Worthington. An in-vitro investigation of mucosa-adhesive materials for use in controlled drug delivery. J Pharm Pharmacol 36: 295–299, 1984.

15. K Satoh, K Takayama, Y Machida, Y Suzuki, M Nakagaki, T Nagai. Factors affecting the bioadhesive property of tablets consisting of hydroxypropyl cellulose and carboxyvinyl polymer. Chem Pharm Bull 37:1366–1368, 1989.

16. HE Junginger. Mucoadhesive hydrogels. Pharm Ind 53:1056–1065, 1991.

17. KV Ranga Rao, P Buri. A novel in situ method to test polymers and coated microparticles for bioadhesion. Int J Pharm 52:265–270, 1989.

18. Y Akiyama, M Yoshioka, H Horibe, S Hirai, N Kitamori, H Toguchi. Anti-hypertensive effect of oral controlled-release microspheres containing an ACE inhibitor (delapril hydrochloride). J Pharm Pharmacol 46:661–665, 1994.

19. K Yamaoka, Y Tanigawara, T Nakagawa, T Uno. A pharmacokinetic analysis program (MULTI) for microcomputer. J Pharmacobiodyn 4:879–885, 1981.

20. JN Hunt, WR Spurrell. The pattern of emptying of the human stomach. J Physiol (Lond) 113:167–168, 1951.

21. C-M Lehr, JA Bouwstra, EH Schacht, HE Junginger. In vitro evaluation of mucoadhesive properties of chitosan and some other natural polymers. Int J Pharm 78:43–48, 1992.

22. C-M Lehr, FGJ Poelma, HE Junginger, JJ Tukker. An estimate of turnover time of intestinal mucus gel layer in the rat in situ loop. Int J Pharm 70:235–240, 1991.

23. J Russel, P Bass. Canine gastric emptying of polycarbophil: An indigestible, particulate substance. Gastroenterology 89:307–312, 1985.

24. B Stripp. Intestinal absorption of riboflavin by man. Acta Pharmacol Toxicol 22:353–362, 1965.

25. EH Graul, D Loew, O Schuster. Voraussetaung für die Entwicklung einer sinn-vollen Retard- und Diuretikakombination Therapiewoche 35:4277–4291, 1985.

26. LL Bontes Ponto, RD Scoenwal. Furosemide (Furusemide): A pharmacokinetic/pharmacodynamic review (Part 1). Clin Pharmacokinet 18:381–408, 1990.

27. C-M Lehr, JA Boustra, K Wouter, AG De Boer, JJ Tukker, JC Verhoef, DD Breimer, HE Junginger. Effects of the mucoadhesive polymer polycarbophil on the intestinal absorption of a peptide drug in the rat. J Pharm Pharmacol 44:402–407, 1992.

28. G Levy, CO Jusco. Factors affecting the absorption of riboflavin in man. J Pharm Sci 55:285–289, 1966.

29. HM Middleton III. Uptake of riboflavin by rat intestinal mucosa in vitro. J Nutr 120:588–593, 1990.

30. EA Hosny, YM El-Sayed, MA Al-Meshal, AA Al-Angaryl. Effect of food on bioavailability of bioadhesive-containing indomethacin tablets in dogs. Int J Pharm 112:87–91, 1994.

31. BJ Marshall, JR Warren. Unidentified cured bacilli on gastric epithelium in active chronic gastritis. Lancet 1:1273–1275, 1983.

32. E Hentschel, G Brandstatter, B Dragosics, AM Hirschl, H Nemec, K Schvtze, M Tanfer, H Wurzer. Effect of ranitidine and amoxicillin plus metronidazole on the eradication of *Helicobacter pylori* and the recurrence of duodenal ulcer. N Engl J Med 328:308–312, 1993.

33. EAJ Rauws, GNJ Tytgat. Cure of duodenal ulcer associated with eradication of *Helicobacter pylori*. Lancet 335:1233–1235, 1990.

34. DY Graham, GMA Borsch. The who's and when's of therapy for *Helicobacter pylori*. Am J Gastroenterol 85:1552–1555, 1990.

35. N Chiba, BV Rao, JW Rademaker, H Hunt. Metaanalysis of the efficacy of antibiotic therapy in eradicating *Helicobacter pylori*. Am J Gastroenterol 87:1716–1727, 1992.

36. DY Graham. *Helicobacter pylori:* Its epidemology and its role in duodenal ulcer diseases. J Gastroenterol Hepatol 6:105–113, 1991.

37. K Kimura, K Ido, K Saifuku, Y Taniguchi, K Kihira, K Satoh, T Takimoto, Y Yoshida. A 1-h topical therapy for the treatment of *Helicobacter pylori* infection. Am J Gastroenterol 90:60–63, 1995.

38. RK Flamm, J Beyer, SK Tanaka, J Clement. Kill kinetics of antimicrobial agents against *Helicobacter pylori*. J Antimicrob Chemother 38:719–725, 1996.

39. CS Goodwin, P Blake, E Blincow. The minimum inhibitory and bactericidal concentrations of antibiotics and anti-ulcer agents against *Campylobacter pyloridis*. J Antimicrob Chemother 17:309–314, 1986.

40. CAM McNulty, J Dent, R Wise. Susceptibility of clinical isolates of *Campylobacter pyloridis* to 11 antimicrobial agents. Antimicrob Agents Chemother 28:837–838, 1985.

41. EAJ Rauws, W Langenberg, HJ Houthoff, HC Zanzen, GNJ Tytgat. *Campylobacter pyloridis* associated chronic active antral gastritis: A prospective study of its prevalence and the effects of antibacterial and antiulcer treatment. Gastroenterology 94:33–40, 1988.

19
Bioadhesive Formulations for Nasal Peptide Delivery

Lisbeth Illum
DanBioSyst UK Ltd, Nottingham, England

I. INTRODUCTION

The nasal route of delivery has for many years been used mainly in therapeutic situations for the treatment of local diseases such as nasal allergies, e.g., allergic rhinitis, perennial rhinitis, viral infections, or inflammations. Nasal medication for stuffy noses in the form of decongestants is also common. Lately, there has been more and more interest in using the nose as a route for administration of systemically active drugs, especially in crisis treatment, because it is possible via this route to obtain a rapid onset of action. Hence, drugs for treatment of severe pain such as migraine [e.g., sumatriptan (Imigran)] and postoperational breakthrough pain [e.g., butorphanol (Stadol)] are now marketed in nasal formulations. Such conventional relatively lipophilic low-molecular-weight drugs are absorbed quite efficiently across the nasal cavity and achieve plasma level profiles close to those obtained after intravenous injection.

For more polar conventional drugs and larger hydrophilic drugs such as peptides and proteins, the absorption across the nasal membrane is considerably less efficient, with bioavailabilities of about 10% and less than 1%, respectively. The main reasons for the low degree of absorption of peptides and proteins via the nasal route are their hydrophilic nature, large molecular size, and physiological factors such as enzymatic degradation and rapid movement away from the absorption site in the nasal cavity due to the mucociliary clearance mechanism. In order to overcome the barrier to nasal absorption of these molecules, two main approaches have been utilized:

modification of the permeability of the nasal membrane by employment of absorption enhancers, such as surfactants, bile salts, cyclodextrins, phospholipids, and fatty acids (Fisher et al., 1991; Aungst, 1990; Hermens et al., 1990; Marttin et al., 1997), and the use of mucoadhesive systems such as bioadhesive liquid formulations (e.g., chitosan), microspheres, powders, and liquid gelling formulations that decrease the mucociliary clearance of the drug formulation and thereby increase the contact time between the drug and the site of absorption (Nagai and Machida, 1985; Farraj et al., 1990; Illum et al., 1994; Critchley et al., 1994; Aspden et al., 1997).

The objective of this chapter is to provide an overview of the bioadhesive systems that have been employed in the nasal delivery of drugs, especially peptides and proteins, and to discuss their effectiveness in improving transport across the nasal membrane, their mechanism of action, and the effect on the mucous membrane. A description of the nasal morphology, physiology, function, and structure will also be given in order to provide a basic understanding of the environment in which the bioadhesive formulations are employed.

II. THE HUMAN NOSE

Mygind (1978) has described very comprehensively the structure and the ultrastructure of the human nose including its applied function. Hence, only details that are necessary for the understanding of the morphological and physiological factors affecting the nasal absorption of drugs are provided here.

A. Nasal Anatomy and Morphology

The nasal cavity is divided longitudinally by the nasal septum into two halves and each half of the cavity is again divided into three regions; the nasal vestibule, the olfactory region, and the respiratory region. In the front the nose is open to the environment through the nares, and in the back the nasal cavity is connected to the pharynx and via the larynx to the trachea. The respiratory region of the nasal cavity comprises three large conchae, the superior, the middle, and the inferior, that ensure that the inspired air has good contact with the surface as it travels through the narrow passages in the nasal cavity. The cavity is about 12 cm long and has a total surface area of about 150 cm^2 and a volume of about 15 mL.

The anterior part of the nasal cavity is lined with squamous epithelium that gradually changes further back in the cavity to pseudostratified columnar epithelium that constitutes the respiratory epithelium. The latter epithelium consists of a layer of elongated columnar cells on the surface and below this two to four layers of basal cells on top of the basal membrane. The basal cells or "replacement cells" are, as the name indicates, able to differentiate into other cell types. Fifteen to 20% of the respiratory epithelium (including the columnar cells) is covered with a layer of cilia 2–4 μm long. Cilia are fine, hairlike structures that can move in a coordinated way to help the mucus flow across the epithelial surface. Each cilium comprises two central protein microtubules surrounded by nine microtubule pairs and is anchored below the luminal cell surface to structures called basal bodies. In addition, the cells of the respiratory epithelium are covered with about 300 villi per cell. Dispersed between the columnar cells are goblet cells, which are unicellular mucus-secreting cells. The mucus is rich in complex carbohydrates, which form important constituents of the protective surface mucus layer. The mucus layer that covers the respiratory epithelium consists of a sol layer, a low-viscosity periciliary fluid, which surrounds the cilia, and a more viscous gel layer on top of the sol layer and covering the tips of the cilia.

In the respiratory epithelium the epithelial cells are closely packed on the apical surface and surrounded by intercellular junctions, the specialized sites and structural components of which are commonly known as the junctional complex (Fig. 1). Each complex is composed of three regions: the zonula occludens, the zonula adherens, and the macula adherens. The zonula occludens (ZO), the region of the junctional complex closest to the apical surface, forms a tight band around the upper part of the cell and is therefore often known as the tight junction. The ZO contains the integral protein ZO-1 and controls the diffusion of ions and neutral molecules through the intercellular spaces.

The zonula adherens, just below the zonula occludens, is a relatively uniform space between neighboring cell membranes, filled by a diffuse material through which adherence is maintained. The macula adherens is typically located basal to the zonula adherens but can also be found in other sites along the lateral cell surface and is then termed a desmosome. The intercellular space is slightly larger than for the zonula adherens and is filled with a denser adhesive material.

Careful electrophysiological analysis has shown that the ZO is permeable to certain molecules, whose maximum size varies from tissue to tissue. The permeability is further dependent on the extracellular environment (Stevenson et al., 1988). It has been shown in two different epithelial tissues that the ZO limits the permeation of molecules with hydrodynamic radii greater than 3.6 Å and is impermeable to those with radii greater than 15 Å (Madara and Dharmsathaphorn, 1985; Madara et al., 1986). The integrity of the tight junction is dependent on the concentration of extracellular

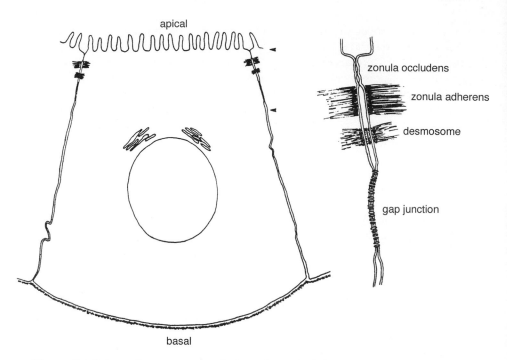

apical

zonula occludens

zonula adherens

desmosome

gap junction

basal

Figure 1 Schematic cross-sectional view of a typical simple epithelial cell. A portion of the lateral cell surface is shown to the right. (From Stevenson et al., 1988.)

calcium ions, although this effect may be related to changes in other junctional elements. Thus, it is known that compounds such as calcium chelators are able to open tight junctions by binding calcium ions. However, such calcium depletion induces global changes in the cells, including disruption of actin filaments, disruption of adherent junctions, and diminished cell adhesion (Citi, 1992). It has been observed in different systems that a spatial relationship exists between defined filamentous cytoplasmic elements and the ZO (Farquhar and Palade, 1963). Actin filaments have been observed in an end-on association with the points of the membrane contact at the ZO in intestinal cells, which is evidence of a relationship between a known cytoskeleton protein and the ZO (Madara, 1987). The zonula adherens is the site of a circumferential ring of actin and myosin that exhibits contractile activity (Rodewald et al., 1976). The exact interaction between ZO and the actomyosin ring at the zonula adherens has not been totally elucidated. However, exposure of intestinal epithelium to cytochalasin D, which interacts specifically with the cytoskeleton at the ZO, results in decreased transepithelial resistance due to diminished ZO resistance. Furthermore, this treatment results in condensation of microfilaments within segments of the perijunctional actomyosin ring, changes similar to those observed when the perijunctional actomyosin ring is stimulated to contract in isolated brush border preparations (Madara, 1988).

The respiratory epithelium in the nasal cavity is supplied with blood from the external and the internal carotid arteries through a dense network of capillaries in the lamina propria. The blood from the anterior part of the nasal cavity is drained via the facial vein, whereas the main part is drained via the sphenopalatine foramen into the pterygoid plexus or via the superior ophthalmic vein.

B. The Function and Physiology of the Nasal Cavity

The nose has four major distinct but interrelated functions: olfaction, mechanical defence, supply of air to the lungs, and conditioning of this air before it reaches the lungs in terms of humidity and temperature. Even when the inhaled air is at a temperature around 0°C, it can easily be warmed to about 31°C on its way through the respiratory tract. In humans the olfactory function is of less importance than in animals. This can easily be judged by the area of the olfactory region, which is 10–20 cm^2 in man but 170 cm^2 in dog. As part of the human defence system, the enzymes present in the nasal cavity can degrade inhaled bacteria or potentially toxic compounds. Bacteria and viruses can also be inactivated by immunological responses in the nasal-associated lymphoid tissue (NALT) in the nasal cavity. Furthermore, the mucociliary clearance mechanism plays an important part in removing unwanted material, caught by the mucus, from the nasal cavity.

It is important to realize that due to a change in blood volume in the mucosa of the conchae, the resistance to air flow in each half of the nasal cavity performs a cycle from low to high every 3–4 hours. This turbinate cycle is coordinated so that when the air resistance is low in one nostril it is high in the other, and hence the total nasal airway resistance is normally constant. This can be recognized most easily during colds, when a unilateral, partial obstruction changes to total and alternates from side to side (Stoksted, 1952).

As already mentioned, the mucociliary clearance mechanism in the respiratory tract provides the human organism with a very efficient defence system that protects the lungs against inhaled particles, bacteria, and irritants by transporting such agents sticking to the viscous mucus posteriorly in the nose and down the throat. The transport of the mucus is closely correlated with the movement of the cilia. The cilia beat with a frequency of about 1000 strokes a minute. Each stroke consists of a rapid forward movement of the fully stretched cilia, whose tip reaches into the viscous mucus and carries it forward, and a slow return movement in which the cilia are bent and move only in the pericellular fluid, not affecting the movement of the viscous mucus layer. In humans, the mucus flow rate has been found to be of the order of 5 mm per minute, with a range of 0–20 mm per minute. This means that the mucus layer in the nasal cavity theoretically is renewed about every 10–20 minutes, the new mucus being continuously supplied by the goblet cells and the mucosal glands (Proctor et al., 1973). This has been confirmed in saccharine clearance tests in humans, where a small fraction of a saccharine tablet placed in the front of the nasal cavity could be tasted after about 10 minutes (Aspden et al., 1997) and 20 minutes (Andersen et al., 1974).

A total of approximately 1500–2000 mL of mucus is produced daily in the respiratory tract. The mucus contains mainly water (95–97%), electrolytes (1–2%), and mucin (2.5–3%). Mucin consists mainly of glycoproteins (high-molecular-weight proteins possessing attached oligosaccharides such as sialic acid) and lipids. The mucus in the respiratory region in the nose moves with a main central stream in a posterior direction along the floor of the nasal cavity with streams of mucus joining from the maxillary sinus and ethmoid conchae, together reaching the nasopharynx, from which the mucus is drained into the esophagus. Movement of mucus from the anterior part of the nasal cavity, where no or few cilia are present, is by way of traction of the mucus from the large number of more posteriorly situated ciliated cells.

The mucociliary clearance rate can be affected by factors associated with (a) a quantitatively and qualitatively imbalanced mucus (i.e., the mucus can be too viscous or too watery, or there can be too little or too much mucus) and (b) impaired cilia movement due to topically applied agents such as cocaine, long-term exposure to toxic vapors, acute infections, and Kartagener's syndrome (Mygind, 1978). Also, studies have shown that active smoking has an effect on the baseline nasal mucociliary clearance. A comparison of nonsmokers and smokers showed a mean clearance time of 20.8 minutes in 28 smokers, which was significantly longer than the 11.8 minutes in 27 lifelong nonsmokers (Stanley et al., 1986).

C. Metabolism

It has been shown by various researchers that nasally applied materials such as conventional low-molecular-weight drugs and peptides and proteins can be metabolized to a great extent in the nasal cavity (Dahl, 1986; Lee, 1988; Sarkar, 1992). It was found that the content of cytochrome P-450, especially in the nasal olfactory epithelium in rat and man, was second to the content in the liver when expressed on the basis of per gram of tissue (Dahl, 1986; Sarkar, 1992).

A factor of importance for the low bioavailability often found for the nasal administration of peptides and proteins could be the proteolytic degradation of these drugs in the nasal cavity. The nasal tissue contains both exopeptidases that can cleave peptides at their NH_2 and COOH termini (e.g., mono- and diaminopeptidases) and endopeptidases, such as serine and cysteine, that can attack internal peptide bonds (Lee, 1988). The same author showed that the aminopeptidase activities in the nasal cavity are more or less equally divided between cytosol-bound and cell membrane–bound enzymes. Hence, peptides and proteins can be attacked by enzymes at several linkages in the molecules, not only while free in the nasal cavity but also during penetration of the nasal membrane whether this transport is transcellular or paracellular.

D. Pathology

The nasal cavity is susceptible to a range of pathological conditions such as allergic diseases, polyposis, acute infections, colds, and Kartagener's syndrome that could influence the performance of nasal drug delivery systems and thereby the absorption of drugs via this route. Hay fever, the most common of all allergic diseases and also known as seasonal allergic rhinitis or pollinosis, is a rhinoconjunctivitis caused by pollen or other allergens. The most common symptoms are tickling in the nose, sneezing, and watery rhinorrhea. Some patients also suffer predominantly from itching of the eyes and asthma. Perennial rhinitis is also a fairly common disease of the upper airways that can be classified as allergic, intrinsic, or autonomic, depending on the cause of the disease. The symptoms suffered by the patients are similar to those of hay fever: sneezing, rhinorrhea, and nasal blockage caused by the swelling of the mucous lining.

Polyposis is a disease in which round, smooth, soft, yellow to pale polyps attached to the mucosa by narrow pedicles are present in the nasal cavity of the sufferers. These polyps also commonly originate from the sinuses projecting into the nasal cavity. Polyps can grow so large that they block the entire nasal chamber. Polyposis is treated with surgery or medication, depending on the etiology of the disease.

Most people have suffered from a common cold several times in their life. This disease is caused by a viral infection of the respiratory tract and the symptoms are sneezing, coughing, and thin or thick excessive nasal secretion. The illness can, in many patients, progress to a stage of secondary bacterial infection, whose symptoms are purulent rhinorrhea, fever, and sore throat (Hilger, 1989).

It is important to consider the possible effect of nasal pathological conditions on the absorption efficiency of drugs given nasally. It has been shown by means of gamma scintigraphy that the clearance of a solution in the nasal cavity is affected by the disease state of patients with common colds. In patients with copious mucus flow (runny nose) rapid clearance of an albumin tracer solution was seen, whereas virtually no clearance took place in patients with nasal congestion (blocked nose) (Bond et al., 1984). In a similar gamma scintigraphy study, the presence of nasal polyps was found to increase the clearance time of a radiolabeled formulation, whereas the pattern of deposition was unaffected (Lee et al., 1984). In contrast to these results, Phillpotts et al. (1984) showed that the clearance times of solutions of interferon were similar both before and after challenge with virulent strains of human rhinovirus in the nasal cavity. Furthermore, no relationship was found between the amount of nasal secretion produced and the rate of clearance.

Although this is a question of importance for the therapeutic use of nasal delivery, few studies have been published on the influence of nasal pathological conditions on the absorption of drugs and especially peptides and proteins. Larsen et al. (1987) found that an experimentally induced rhinitis had little effect on the luteinizing hormone response in human volunteers given a nasal dose of buserelin. The authors concluded from this study that nasal administration of buserelin represented a reliable route of delivery and that modification of the administration route or a change in dosage schedule during naturally occurring nasal inflammations, such as the common cold and allergic rhinitis, was unnecessary in patients undergoing chronic treatment. Sandow and Petri (1985) showed similar effects in young boys suffering from common cold or intercurrent rhinitis. The work of Schafgen et al. (1983) on thyroid-stimulating hormone and Olanoff et al. (1987) on desmopressin similarly demonstrated no difference in absorption between healthy subjects and subjects suffering from common cold or rhinitis. Wood et al. (1995) evaluated the absorption of ipratropium bromide nasal spray in healthy volunteers and in patients with perennial rhinitis and the common cold. These authors found no significant pharmacokinetic differences in the absorption of the drug between various volunteer groups, the absorption being around 10%. Similarly, Argenti et al. (1994) explored the systemic absorption of triamcinolone acetonide after nasal administration in 12 patients with allergic rhinitis and in a similar number of healthy volunteers. There were no statistically significant differences in any of the derived pharmacokinetic parameters (C_{max}, T_{max}, $t_{1/2}$, AUC_{1-12} between the treatment groups. Finally, Humbert et al. (1996) reported on the use of dihydroergotamine for the treatment of migraine and found that the presence of acute viral rhinitis did not result in any change in nasal absorption of dihydroergotamine compared with absorption from the normal nasal mucosa. From these studies it is reasonable to conclude that rhinitis and common colds do not have a significant influence on the absorption of drugs from the nasal cavity. Whether patients receiving chronic therapy with a nasally administered formulation who experience acute nasal diseases, such as rhinitis or common cold, would prefer the use of another route of delivery in such situations remains to be seen.

III. NASAL DELIVERY OF PEPTIDES

As discussed earlier, low-molecular-weight lipophilic drugs such as propranolol and progesterone are readily absorbed via the nasal route with plasma profiles similar to those obtained after intravenous injection (Hussain et al., 1980a,b). For the large hydrophilic peptides and proteins, which furthermore can be degraded in the nasal cavity by peptidases, the absorption is considerably smaller. For peptides such as calcitonin and insulin, bioavailabilities of the order of less than 1% have been reported (Deurloo et al., 1989). The absorption of such molecules is dependent on a range of physicochemical and physiological factors, the most important of which are the site of deposition in the nasal cavity, the mucociliary clearance mechanism, passage through the mucus layer, transport across the epithelial membrane, and enzymatic degradation of the peptide or protein.

A. Site of Deposition and Mucociliary Clearance

The site of deposition and the mucociliary clearance of the nasal formulation after administration are distinct but interrelated factors of importance for the nasal absorption of peptides and proteins. The site of deposition of the formulation is determined by the nature of the delivery device in combination with the type of the formulation, i.e., liquid or powder. It is well known that large particles, including droplets (>10 μm), are deposited in the nasal cavity after inhalation, the larger the particle size, the more anterior the deposition (Mygind, 1978). For smaller particles, the site of deposition depends on the velocity at which the particles are inhaled and the turbulence of the air flow; however, particles of sizes smaller than 1 μm are not normally deposited in the nasal cavity but travel down the trachea to the lung region. The particle size distribution of the nasal powder or the formed nasal droplets therefore affects the deposition in the nasal cavity and the subsequent fate in the nasal cavity. The rate of clearance and subsequently the absorption from the nasal cavity have furthermore been shown to depend on the site of initial deposition of the formulation (Hardy et al., 1985; Harris et al., 1986).

Harris et al. (1986) administered desmopressin mixed with the radio-labeled human serum albumin nasally by means of a metered spray device, a rhinyl catheter, or a single-dose pipette. The deposition and clearance of the formulation were monitored by gamma scintigraphy and the level of systemically absorbed desmopressin by analysis of plasma samples. Sprays were found to be deposited more anteriorly, cleared more slowly, and absorbed to a higher degree than the formulation administered by rhinyl catheter or pipette, supporting results obtained by Hardy et al. (1985). The bioavailability of the desmopressin administered as a spray was two to three times as high as for the other types of administration. A mathematical model derived by Gonda and Gipps (1990) included clearance parameters and described the rate processes involved in the disposition of drugs in the nasal cavity, and literature data were consistent with this model. The model further suggested that the use of a bioadhesive formulation would improve bioavailability and reduce variation in absorption. Manipulation of formulation factors and choice of the delivery device may enable an increment in nasal absorption, although to obtain therapeutic blood levels other methods of absorption enhancement may have to be taken into consideration.

B. Passage Through Mucus Layer

Compared with the mucus layer in the gastrointestinal tract, the mucus layer in the nasal cavity is very thin, namely 4–5 μm. Apparently, no studies have been reported on the effect of the mucus layer as a barrier to peptide absorption, although it is likely that peptide and protein molecules could easily interact with the glycoproteins in this material and thereby be restrained in their passage to the cell membrane. It has been shown that a potent mucolytic agent, N-acetyl-L-cysteine, when used in a nasal formulation in combination with human growth hormone, increased the nasal absorption of this protein from 7 to 12% in a rat model (O'Hagan et al., 1990). However, it is not known whether the mucolytic agent also had a surfactant-like effect on the membrane, which could add to the apparent mucolytic effect.

C. Transport Across the Epithelial Membrane

It is generally accepted that two major pathways are available for transport of drugs across the epithelial membrane, namely the paracellular pathway and the transcellular pathway.

In using the paracellular pathway the drug passes between the cells through the intercellular tight junctions from the apical surface to the basolateral surface. Because the tight junction comprises an aqueous channel or pore, this pathway is available only for hydrophilic drugs such as peptides, which cross by passive diffusion. As discussed earlier, this channel, although dynamic in nature in terms of opening and closing with the influence of the environment, has a maximum size that normally allows only molecules of about 1000 daltons or smaller to pass (McMartin et al., 1987). The effect of molecular mass (1260–45,500 daltons) on the nasal absorption of a hydrophilic model drug (dextran) was investigated in rats (Fisher et al., 1992). A very high correlation was found between molecular mass and absorption, absorption being highest for the lowest molecular mass molecule.

Using the transcellular pathway, drugs can pass through the cell by a mechanism of concentration-dependent passive diffusion or endocytotic processes such as pinocytosis or receptor-mediated or adsorptive endocytosis. Normally, lipophilic drugs can diffuse through the cells, whereas large hydrophilic molecules such as proteins have been seen to pass through via an endocytotic process (Grass and Robinson, 1988; Inagaki et al., 1985).

D. Enzymatic Degradation

The metabolic capacity of the nasal mucosa may represent a significant barrier to the systemic absorption of peptides and proteins. Most of the work published on the effect of the nasal enzymes on the drugs has been performed using nasal tissue homogenates, where enzymes from all mucosal cells and subcellular fractions including lysosomes are liberated. It can be argued that such studies overestimate the amount and type of enzymes exposed in the nasal cavity and thereby the degree of degradation that can take place in the nasal cavity. It is shown by Hirai et al. (1981) that insulin was rapidly degraded with only 9% remaining after 60 minutes' contact with rat nasal tissue homogenate. Similarly, Kashi and Lee (1986) found that methionine enkephalin, leucine enkephalin, and (D-Ala^2) Met-enkephalinamide were all rapidly hydrolyzed in homogenates from the rabbit nasal tissue with half-lives of degradation in the order of 25 minutes. Supporting this point, Jørgensen and Bechgaard (1994) found that, although thyrotropin-releasing hormone was extensively degraded after incubation in rabbit nasal homogenate, no degradation was seen in human nasal wash. It was also found in this study that the degradation could be reduced by the addition of sodium glycocholate, which has an enzyme-inhibiting effect.

Hussain et al. (1985, 1989) initiated early in situ studies of the degradation of peptides in the nasal cavity of rats. The drugs, such as leucine enkephalin, were circulated for extensive times through the nasal cavity and found to undergo extensive hydrolysis. They also found that buffer, after recirculation for 30 minutes, was able to degrade the pentapeptide considerably. O'Hagan et al. (1990) reported that the aminopeptidase inhibitor amastatin, when given nasally to rats, was able to increase the peak height of human growth hormone in the plasma by 278% and the AUC by 291%, whereas no effect was seen with bestatin, another aminopeptidase inhibitor. For insulin, bestatin and not amastatin was found to increase the nasal absorption.

E. Strategies for Improving Absorption

Most researchers have exploited the use of absorption enhancers in order to improve the nasal absorption of peptides and proteins. A large number of absorption enhancers have been evaluated in animal models and in humans, including mucolytic agents such as N-acetyl-L-cysteine; transcellular and paracellular membrane modifiers such as surfactants, bile salts, bile salt an-alogues, EDTA, and cyclodextrins; and enzyme inhibitors such as aprotinin, bestatin, chymostatin, and amastatin. Generally, these enhancers are suc-cessful in improving the bioavailability of the drug, but although there are exceptions, quite often the degree of absorption enhancement is closely re-lated to the degree of damage encountered in the cell membrane.

As an alternative to the use of absorption enhancers, a strategy that has been employed by some workers (including our own group) is to use mucoadhesive delivery systems that enable prolonged contact between the drug and the absorptive sites in the nasal cavity by delaying the mucociliary clearance of the formulation. Systems that have been described in the lit-erature include liquid bioadhesive systems, self-gelling bioadhesive systems, bioadhesive powder systems, and bioadhesive microsphere systems (Nagai et al., 1984; Morimoto et al., 1985; Illum et al., 1994a; Edman and Bjørk, 1992).

IV. BIOADHESIVE NASAL DELIVERY SYSTEMS

A. Bioadhesion

The principles of bioadhesion are dealt with in detail in Sec. I of this book and hence will be only briefly reviewed here. A bioadhesive agent is defined as a compound that is capable of interacting with biological materials through interfacial forces and being retained on such material for prolonged periods of time. If the biological material is a mucous membrane, the bioad-hesive material is termed a mucoadhesive (Ahuja et al., 1997). For mu-coadhesion to occur, a succession of stages has to occur; (1) an intimate contact between the bioadhesive agent and the membrane has to be created, and (2) penetration of the bioadhesive agent into the crevices of the mem-brane or interpenetration of the chains of the bioadhesive agent with those of the mucus has to take place. On a molecular level, mucoadhesion can be

Table 1 Examples of Bioadhesive Polymers
Employed for Nasal Drug Delivery Systems

Carbopol (carboxy polymethylene)
Sodium carboxymethyl cellulose (SCMC)
Hydroxypropyl cellulose (HPC)
Hydroxypropylmethyl cellulose (HPMC)
Hydroxyethyl cellulose (HEC)
Methyl cellulose (MC)
Sodium hyaluronate
Guar gum
Sodium alginate
Polycarbophil
Starch
Dextran
Chitosan

---

explained on the basis of attractive molecular interactions involving forces
such as van der Waals, electrostatic interactions, hydrogen bonding, and
hydrophobic interactions.

The bioadhesive force of a polymer material is dependent on the nature
of the polymer (molecular weight, concentration, flexibility of the polymer
chain, spatial conformation), the surrounding medium (pH), swelling, and
physiological factors (mucin turnover, disease state). It has been shown that
in order to allow chain interpenetration, the polymer molecule must have
adequate length. However, the optimal size depends on the nature of the
material. Generally, the bioadhesive force increases with the molecular mass
up to 100 kDa, after which no additional effect is seen (Gurny et al., 1984).
Bremecker (1983) has suggested that for each polymer system an optimal
concentration exist for which best adhesion can occur. Also, the flexibility
of the polymer is important for interpenetration. Hence, the more cross-
linked the polymer, the less strong the mucoadhesive interaction. Furthermore, the spatial conformation of the polymer is important, as a contracted
conformation may shield any groups responsible for the adhesiveness of the
polymer. Table 1 lists the bioadhesive polymers mostly employed for nasal
drug delivery systems.

B. Liquid Bioadhesive Delivery Systems

A range of studies has been performed with liquid bioadhesive formulations
of variable viscosity. It has been shown that an increase in viscosity of a
solution by means of the bioadhesive material hydroxypropylmethyl cellu-

lose (HPMC) result in a prolonged clearance time from the nasal cavity (Pennington et al., 1988). Concentrations of 0.6, 0.9, and 1.25% HPMC resulted in clearance half-times of 0.47, 1.7, and 2.2 hours, respectively, in human volunteers.

Morimoto et al. (1985) improved the nasal bioavailability of eel calcitonin and insulin by means of formulations employing Carbopol 941 (carboxypolymethylene, a polymer of acrylic acid cross-linked with allylsucrose) and carboxymethyl cellulose (CMC). The concentration of polyacrylic acid was 0.1 and 1% and of CMC 1%, and the formulations had the consistency of an aqueous gel. The absorption of insulin was enhanced by the Carbopol gel formulation but not by the CMC formulation. The effect of the Carbopol gel on the absorption of calcitonin and insulin was not affected by the pH. Similarly, Ryden and Edman (1992a) found that insulin (5 IU/kg) administered in the nasal cavity of a rat in combination with 0.5% polyacrylic acid decreased the plasma glucose level with 18% as compared to 7% for the control. Critchley (1989) reported that the nasal administration to rats of desmopressin in a 2% Carbopol 934 gel solution increased the bioavailability from 15% to 77% as compared with a simple solution of desmopressin.

In a later study, Morimoto et al. (1987) evaluated the effect of Carbopol 941 gel with or without poly(ethylene glycol) (PEG) 400 on the nasal absorption of nifedipine in rats. Nifedipine is a poorly water soluble drug, but when it was administered nasally with PEG (viscosity 51 cP) in a soluble form the absorption was rapid with a high initial peak concentration. The addition of the PEG-Carbopol gel to the formulation (50% PEG 400) resulted in a lower peak concentration and more sustained absorption, which was similar to that seen for other small lipophilic drugs such as propranolol (Hussain et al., 1980c). The administration of nifedipine with Carbopol gels (viscosities 256 to 1054 cP) alone resulted in very low plasma concentrations. The effect of the PEG 400 was attributed to entrapment of the drug and the rapid dissolution in the mucosal fluid with PEG functioning as a cosolvent for nifedipine. These results were in line with results of Harris et al. (1988, 1989), who found that methyl cellulose did not increase the bioavailability of desmopressin when given nasally to human volunteers but gave a slower and more prolonged absorption.

Methyl cellulose (MC) and HPMC were shown by Critchley (1989) to increase the absorption of desmopressin in the rat to 27 and 50%, respectively, compared with 15% for a simple solution of desmopressin. Gel preparations of roxithromycin, prepared with the bioadhesive polymers sodium carboxymethyl cellulose (CMC), hydroxyethyl cellulose (HEC), HPMC, and Carbopol-PEG, administered into the nasal cavity of rabbits were shown to improve the absorption of the drug as compared with a simple aqueous formulation; the effectiveness was in the order CMC > HEC > HPMC >

Carbopol-PEG (Tuncel et al, 1994). The concentrations of the gels were 0.7, 0.5, 1, and 1%, respectively. In a study by Zhou and Donovan (1996) the effect of bioadhesive polymer gels on the nasal clearance of fluorescent labeled microspheres was studied in rats. The polymer gels investigated were MC (3%), HPMC (3%), CMC (3%), Carbopol 934P (0.2 and 0.4%), and PEG (5%). All gels effectively decreased the mucociliary clearance of the microspheres, and it was found that the clearance rate decreased mostly for MC, followed by HPMC (similar to PEG), CMC, and finally Carbopol. These results, in combination with results from the studies by Critchley (1989) and Tuncel et al. (1994), suggest that there is not direct relationship between the increase in clearance time and drug absorption. Also, the in vitro mucoadhesive forces, expressed as a percentage of a standard, for some of these polymers (Longer and Robinson, 1986) (CMC, 193%; Carbopol, 934–185%; PEG, 96%), suggest that there is also little direct relationship between degree of bioadhesiveness and effect on nasal absorption. Similarly, no study has conclusively shown that an increase in viscosity of a formulation alone has a significant effect on the absorption of drugs. This was supported by Olanoff and Gibson (1987), who measured the pharmocodynamic effect of desmopressin given nasally, with and without CMC, and found no improvement in systemic activity when using the bioadhesive agent. It should be noted that some of these discrepancies could be due to differences in the nature of the gel formulations as employed in the various studies. As indicated before, the viscosity and the mucoadhesiveness of the polymers are very dependent on the ionic strength, the concentration, and the pH of the formulation.

The effect of viscous solutions of hyaluronate sodium (HAS) of various average molecular weights on the nasal absorption of vasopressin and demopressin was studied in rats and compared with the effect of CMC (Morimoto et al., 1991). Hyaluronate is a natural polymer and a major component of intestinal tissue. Solutions of hyaluronate are viscous and mucoadhesive, with a detachment force of around 10,000 dyn/cm^2 (Hadler et al., 1982). The detachment force for CMC was measured as about 6000 dyn/cm^2. It was found that the HAS needed to reach a certain molecular mass ($>3 \times 10^5$ daltons) in order to improve absorption and the effect was concentration dependent. In all cases the HAS was found to give a better effect than CMC. The bioavailability increased about twofold for both molecules compared with simple solutions of the drugs as based on the pharmacodynamic effect of changing the urine volume. Similar effects were found by Ryden and Edman (1992a) for nasal administration of insulin in combination with HAS in rats. Because HAS may also have an effect on the membrane itself, a direct relationship between degree of mucoadhesiveness and effect on nasal absorption was not shown conclusively in these studies either.

It should be emphasized that the studies cited are generally lacking in information about the rate of release of drug from the formulation. Hence, it could be argued that although the formulations studied are based on the bioadhesive excipients, there is little proof that the drug stays in combination with the formulations and hence has an extended residence time in the nasal cavity. Only in the paper by Ryden and Edman (1992a) is it demonstrated that the insulin is released relatively slowly from the polyacrylic acid and the hyaluronate solutions, with 90% of the drug released in 60 minutes. However, it should be noted that 60–70% of the insulin was released within 10–15 minutes and hence only a limited controlled-release effect was found.

C.   Self-Gelling Bioadhesive Systems

A problem may be encountered in therapeutic use with application of the bioadhesive liquid gel systems in the nasal cavity, especially if a higher concentration of the polymer is used. The formulations are not likely to be readily delivered using a normal nasal spray device but rather will have to be applied by means of a tube. To overcome this problem, bioadhesive formulations that gel upon interaction with the nasal mucosa (due to either increase in temperature, increase in ionic strength, or the presence of calcium ions), so-called environmentally responsive polymers, have been exploited for the nasal delivery of drugs.

The thermogelling polymer Pluronic F127 is a polyoxyethylene-polyoxypropylene block copolymer that is liquid at a concentration of more than 25% in buffer at 4°C, whereas at room temperature and at higher temperatures it forms a clear viscous gel. Zhou and Donovan (1996) showed that this gel was able to decrease significantly the initial clearance rate of fluorescent microspheres from the nasal cavity of rats. The effect was very similar to that seen for methyl cellulose as discussed earlier. A Pluronic F127 formulation (27%) was used by Critchley (1989) for the nasal delivery of desmopressin and propranolol to rats. It was shown that this bioadhesive self-gelling system did not significantly alter the bioavailability of the desmopressin, whereas for the readily absorbable propranolol the plasma peak levels were decreased and the absorption profile sustained, resulting in an apparent decrease in bioavailability. Ryden and Edman (1992a) and Pereswetoff-Morath and Edman (1995a) studied the effect of ethylhydroxyethyl cellulose (EHEC), which showed thermal gelation without phase separation when low concentrations of ionic surfactants (e.g., sodium dodecyl sulfate (SDS)) were present. Binding of the surfactant to the polymer produced micelle-like aggregates along the polymer backbone. After administration of the EHEC in combination with insulin to rats, the plasma glucose levels were found to be decreased slightly (12%). It was later shown that when the EHEC was made hypotonic as compared with isotonic or hypertonic, the absorption of insulin was improved, with lowering of plasma glucose levels by 30% as compared with the control (Pereswetoff-Morath and Edman, 1995b). Toxicological studies showed that the EHEC caused irreversible ciliostasis of cilia situated in the tracheae as measured in in vitro studies after exposure for 15–30 minutes (Pereswetoff-Morath et al., 1996). It was suggested that the ciliotoxicity was due mainly to the hypotonicity of the formulation.

D. Bioadhesive Powder Systems

Nagai and coworkers investigated the use of bioadhesive powder dosage forms for the administration of peptides such as insulin to the nasal cavity (Nagai et al., 1984; Nagai and Machida, 1985). The bioadhesive agents studied, in combination with freeze-dried insulin, were crystalline cellulose, hydroxypropyl cellulose (HPC), and Carbopol 934. All formulations tested gave significant decreases in plasma glucose level when administered nasally to the dog and rabbit models. Freeze-dried insulin powders alone had relatively good bioavailabilities, which were not improved by the addition of the soluble and nonbioadhesive material lactose. A very effective formulation was crystalline cellulose blended with freeze-dried insulin, which resulted in a fast decline in the glucose level to 49% of the control value. Addition of HPC or Carbopol 934 to the freeze-dried insulin resulted in prolonged decreases of the plasma glucose levels but the extent of the decrease (the trough) was not affected. However, when Carbopol was freeze dried with the insulin and added to the crystalline cellulose powder, the effect of the formulation was considerably improved, leading to a hypoglycemia of the order of one third of the effect obtained after intravenous injection of the same dose of insulin. With the addition of increasing concentrations of Carbopol, the trough in the plasma glucose was increasingly delayed. The effect of the formulation was attributed to the formation of a gel in contact with the mucus, which prolonged the residence of the formulation in the nasal cavity. It was reported that the insulin-Carbopol-crystalline cellulose powder formulation was readily acceptable in terms of tolerability when administered to human volunteers; however, the plasma glucose levels obtained in the volunteers were quite variable (Nagai and Machida, 1985).

The potential of various powder formulations to enhance the nasal absorption of octreotide was studied in vivo in the anesthetized rat model (Oechslein et al., 1996). The powder formulations were also characterised in vitro in terms of calcium binding, water uptake, and drug release. The powder formulations were prepared by dry blending of octreotide with microcrystalline cellulose, semicrystalline cellulose, hydroxyethyl starch, cross-linked dextran, microcrystalline chitosan, pectin, and alginic acid. The bioavailabilities obtained for all of the powder formulations were low, with values ranging from 0.59% for the control to 5.56% for the cross-linked dextran powder formulation. The ranking of the formulations, in terms of absorption-enhancing effect, coincided with the ranking in terms of calcium binding properties, cross-linked dextran powder having the highest degree of calcium binding. No correlation was found between absorption-enhancing effect and water uptake. The release of drug from all powder formulations was complete within 10 minutes, which may explain the low bioavailabilities obtained for the various formulations. These results were surprising in light of the results obtained by Nagai and coworkers for insulin and also because it had been suggested by several authors that a reason for the improvement in absorption obtained for bioadhesive powder systems was their ability to take up water, form a gel, and increase the residence time in the nasal cavity. An additional factor that has been proposed is that the uptake of water by the powder can result in shrinkage of the cells and the opening of tight junctions (Edman et al., 1992; Illum et al., 1994b).

It was shown by Provasi et al. (1992) that the administration of salmon calcitonin in powder formulations containing lactose and colloidal silica (a swellable polymer) can improve the nasal absorption of the drug in rats as compared with a simple nasal solution of salmon calcitonin. It was also shown that a colyophilized formulation was more effective than a dry blend formulation. No bioavailability values were given in the paper, and hence the results are difficult to compare with the work of Oechslein et al. (1996). Sakr (1996) investigated the enhancing effect of dimethyl-β-cyclodextrin on the nasal absorption of glucagon when administered as a solution formulation or a freeze-dried powder formulation. The freeze-dried formulation was prepared from the solution formulation. No bioadhesive agent was added. No improvement was found when using the freeze-dried formulation as compared with the solution formulation, which is not surprising, because the formulation was readily soluble. Similarly, Vermehren et al. (1996) used freeze-dried powders consisting of methylcellulose and α-cyclodextrin, with or without the phospholipid didecanoylphosphatidylcholine (DDPC), to improve the nasal absorption of human growth hormone. There was no direct comparison with similar solution formulations; however, the paper mentions previously obtained bioavailabilities of about 20% in rabbits. This value is very similar to the result obtained for the powder formulations. Likewise, Marttin et al. (1997) found no difference in bioavailability for dihydroergotamine administered nasally to rabbits as a spray formulation or as a powder with methylated β-cyclodextrin. This is probably not surprising, considering that no bioadhesive material was added to the powder formulation.

Also of interest is the study by De Ascentiis et al. (1996) in which the effect of particle size of nasal powders of β-cyclodextrin containing progesterone on delivery behavior (including bioavailability) was studied). Progesterone–β-cyclodextrin powders were prepared by granulation with polyvinylpyrrolidone or mannitol and sieved into particle fractions of 0–45, 45–63, 63–88, 88–125, 125–180, 180–250, and 250–355 μm, and the various fractions administered nasally with a nasal powder insufflator (Miat S.p.A.). It was found that the rate of delivery from the insufflator decreased with decreasing particle size, whereas the time needed for full emission of the dose through the nose adapter increased with decreasing particle size in the range 50–150 μm. Furthermore, it was found that the particle size had an effect on the compactness of the cloud produced, which again would influence the impaction or sedimentation pattern of the particles after administration to the nasal cavity.

E. Bioadhesive Microsphere Systems

The use of bioadhesive microspheres for the nasal delivery of drugs that were poorly absorbed was first suggested in 1987 by our group (Illum, 1987; Illum et al., 1987). The rationale behind this suggestion was that the application of bioadhesive microspheres (in powder form) with good bioadhesive properties would permit such microspheres to swell in contact with the nasal mucosa to form a gel and control (decrease) the rate of clearance from the nasal cavity, thereby giving poorly absorbed drugs a longer time to be available at the absorptive surface for absorption. Microspheres made from DEAE-dextran (DEAE Sephadex), starch microspheres (Spherex), and albumin microspheres, about 25–50 μm in diameter, were radiolabeled, and it was shown in a gamma scintigraphy study in human volunteers that after nasal administration they were cleared significantly more slowly than solution and nonbioadhesive powder formulations. At 3 hours after administration, 50% of the initial amount of albumin and starch microspheres and 60% of the DEAE-dextran microspheres were still present in the nasal cavity, whereas the half-life of clearance of the control formulations was about 15 minutes (Illum et al., 1987). Similar results were later obtained by Ridley et al. (1995) for starch microspheres in human volunteers, with a half-life of clearance of about 20 minutes for a simple solution and nearly 2 hours for starch microspheres. No effect on the clearance pattern could be attributed to the posture (i.e., seated or supine) during administration.

The promoting effect of bioadhesive microspheres on the absorption of a poorly absorbable drug, when given nasally, was first demonstrated by Illum et al. (1988) using gentamicin. In this study gentamicin, as a solution formulation, with and without the enhancer lysophosphatidylcholine (LPC) and freeze dried with starch microspheres with and without the addition of LPC was administered nasally to sheep (Fig. 2). It was shown that the drug alone was absorbed to a negligible degree and that the addition of LPC only marginally improved the absorption from a solution. In contrast, the use of the freeze-dried bioadhesive microsphere formulation increased the bio-availability to 10% and the further addition of the LPC improved the bio-availability to 57%. This study showed that the bioadhesive microsphere concept worked in that the absorption was considerably improved. Surprisingly, the enhancer LPC improved the effect five- to sixfold, although when given as a solution the LPC had only a marginal effect on the absorption of gentamicin.

Further work was carried out on the effect of starch microspheres on the nasal absorption of peptides and proteins, both by our group and by Edman and others. Farraj et al. (1990) showed that the nasal absorption of insulin in sheep was improved from a bioavailability less than 1% to 11% and 32% when administered as a freeze-dried powder with starch micro-spheres without and with the addition of LPC, respectively. Similar effects were found for desmopressin (Critchley et al., 1994) and growth hormone (Illum et al., 1990). Bjørk and Edman (1988) obtained similar results for

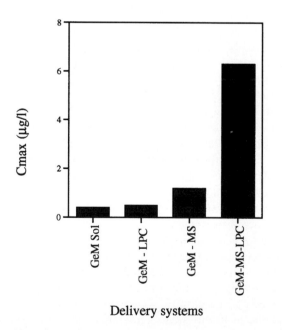

Figure 2   Nasal administration of gentamicin to sheep: effect of bioadhesive starch microspheres and lysophospholipid on the absorption across the membrane.

insulin in the rat model, with the bioavailability being as high as 30%. The reason for the better effect in the rat model is probably the impairment in mucociliary function observed in this anesthetized animal model. The same group later investigated the effect of degree of cross-linking (and thereby the degree of swelling of the starch microspheres) on the absorption-promoting ability (Bjørk and Edman, 1990). These authors found that although the release of insulin from the microspheres was affected by the degree of cross-linking, with the slowest release from highly cross-linked microspheres, there was no significant effect on the absorption-promoting ability. It was later suggested that the mechanism of action of the starch microspheres was a combination of the bioadhesive effect keeping the formulation for a longer time in the nasal cavity and the dry microspheres absorbing water in the nasal cavity and thereby opening up tight junctions (Bjørk and Edman, 1990; Edman et al., 1992; Bjørk et al., 1995). It was also shown by the same authors that the starch microspheres had little toxic effect on the nasal mucosa after up to 8 weeks of daily application in rabbits (Bjørk et al., 1991) and no adverse effect on the human mucociliary clearance mechanism (Bjørk et al., 1992; Holmberg et al., 1994).

A range of other bioadhesive microspheres has been used for the nasal administration of peptides and other drugs. Sephadex and DEAE-Sephadex (cross-linked dextran) microspheres, which are water insoluble and water absorbable, were shown to improve, to a lesser degree than starch microspheres, the nasal absorption of insulin (Edman et al., 1992; Ryden and Edman, 1992a). The same microspheres have also been suggested for the nasal delivery of nicotine (Cornaz et al., 1996) but no in vivo data have so far been presented. Pereswetoff-Morath and Edman (1995b) found an apparent difference in absorption-promoting effect between Sephadex G-25 and G-50 microspheres administered with insulin in rats; G-25 did not allow the absorption of insulin into the spheres whereas G-50 did. It was found that G-25 microspheres, where insulin was on the surface, had a greater absorption-enhancing effect, although no differences were seen in vitro between drug release rates. No apparent toxic effect was found from the Sephadex microspheres on cilia beat frequency (Pereswetoff-Morath et al., 1996), nor were they found to be immunogenic (Pereswetoff-Morath and Edman, 1996). Microspheres produced from polyacrylic acid and incorporating disodium cromoglycate were shown in a human clinical trial to decrease the clearance rate in the nasal cavity, with 50% remaining after 30 minutes compared with 27% for a control (Vidgren et al., 1991).

The use of hyaluronic acid ester microspheres for the nasal delivery of insulin in sheep was investigated by our group (Illum et al., 1994b). The increase in nasal absorption of insulin so achieved was found to be independent of the dose of microspheres in the range 0.5–2.0 mg/kg. The mean bioavailability of the system was found to be 10%, similar to what was previously achieved for starch microspheres in the sheep model. The mucoadhesive properties of these microspheres were not measured, but hyaluronate esters are known to have excellent adhesive properties. Because the microspheres expressed a very low degree of swelling in water, the full mechanism of action for the absorption improvement (including bioadhesion) was not known.

Thus, a wealth of literature shows the excellent absorption-enhancing properties of bioadhesive microsphere systems in combination with both low-molecular-weight drugs and peptides and proteins. However, very few studies have compared the effect of bioadhesive powders and microspheres made from the same material. Thus, it is difficult to conclude which would generally give the better effect. It was shown by Bjørk and Edman (1990) that microspheres made from starch and insoluble starch powder gave a similar decrease in glucose levels after nasal administration with insulin to rats. Soluble starch, however, showed no absorption enhancement. In comparison, our group has shown in sheep that starch microspheres gave a better absorption-enhancing effect than insoluble starch powder (unpublished data).

F.   Chitosan

Chitosan is a high-molecular-weight polysaccharide derived from naturally occurring chitin (in crab, shrimp, and lobster shells) by a process of deacetylation (Fig. 3). The primary unit in chitin is 2-deoxy-(acetyl-amino)glucose combined by glycosidic linkages into a linear polymer. Chitosan has one primary amino and two free hydroxyl groups for each C6 building unit. Chitosan can be dissolved in inorganic or organic acids to obtain protonization of the amino group resulting in positively charged chitosan salts. Chitosan is available in a broad range of molecular weights and salt forms such as glutamate, lactate, and chloride. The properties of chitosan, such as pK_a and solubility, can be be modified by changing the degree of deacetylation and the environment of the formulation, such as pH and ionic strength.

Chitosan has been used as a pharmaceutical excipient in oral drug formulations in order to improve the dissolution of poorly soluble drugs (Sawayanagi et al., 1982a,b; Imai et al., 1991) and for the sustained release of drugs (Takayama et al., 1990). Chitosan has also been successfully launched as a slimming agent, the mechanism of action probably being one of ionic interaction with the bile salts in the gastrointestinal tract, thereby decreasing fat absorption.

Chitosan has been shown to be mucoadhesive in nature, no doubt due to an interaction of the positive charges on the molecule with the negatively charged sialic acid groups of the mucin (Lehr et al., 1992; Fiebrig et al.,

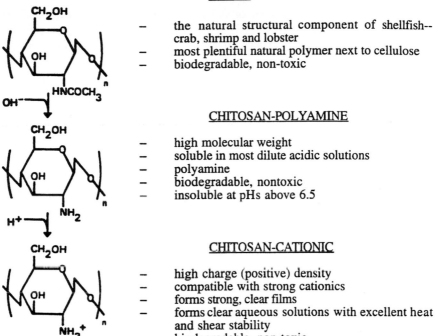

CHITIN

– the natural structural component of shellfish-- crab, shrimp and lobster
– most plentiful natural polymer next to cellulose
– biodegradable, non-toxic

CHITOSAN-POLYAMINE

– high molecular weight
– soluble in most dilute acidic solutions
– polyamine
– biodegradable, nontoxic
– insoluble at pHs above 6.5

CHITOSAN-CATIONIC

– high charge (positive) density
– compatible with strong cationics
– forms strong, clear films
– forms clear aqueous solutions with excellent heat and shear stability
– biodegradable, non-toxic

Figure 3 The structure of chitin and chitosan.

1994). The latter authors showed conclusively that the chitosan-mucin interaction was highly pH dependent with the highest interaction at a pH value at which both the chitosan and the sialic acid groups were well ionized. A study by He et al. (1998) has shown that microspheres made from chitosan were highly mucoadhesive; the binding to a rat intestinal loop in vitro was nearly 100% effective, whereas insignificant numbers of nonbioadhesive microspheres adhered to the intestinal wall.

Chitosan was first shown by us (Illum et al., 1994a) to enhance the nasal absorption of drugs such as polar, low-molecular-weight molecules as well as peptides and proteins. When insulin (2 IU/kg) was administered nasally to sheep in a simple chitosan solution, the plasma glucose fell to 43% of the control level within 90 minutes, whereas a solution of insulin without chitosan caused a fall to 83% of the control glucose level. The corresponding plasma insulin levels increased from a C_{max} of 34 to 191 mIU/ L and the AUC increased sevenfold. Similar effects have been found for a range of other peptides such as calcitonin, desmopressin, goserelin, and leuprolide. For sheep the concentration of chitosan salt necessary to obtain the optimal effect was of the order or 0.5%. A phase I clinical study has been performed in human volunteers with insulin and chitosan administered nasally. Bioavailabilities of 14.5% relative to subcutaneous administration were obtained. For the low-molecular-weight polar drug morphine, a nasal bioavailability of about 60% was obtained in human volunteers when the drug was administered with chitosan in an aqueous liquid formulation (Illum, 1998).

Chitosan can be formulated as a simple solution, as a spray-dried powder, or as a microsphere system. It has been shown by our group that a chitosan powder or microsphere system is more effective in providing enhancement of the nasal absorption of polar drugs than chitosan solutions. With the peptide drugs goserelin, leuprolide, and parathyroid hormone, bioavailabilities of the order of 20 to 40% after nasal administration, as compared with a parenteral injection control, were obtained in the sheep model. Studies in sheep have also shown that doses of chitosan powder from 2 to 0.3 mg/kg, in combination with insulin, are able to exert the same absorption-promoting effect. For more potent drugs, the dose of chitosan necessary may well be decreased. Studies in sheep with chitosan solution and chitosan powders and microspheres in combination with insulin have shown that the effect of the chitosan is reproducible on a daily basis when it is administered repeatedly for 5 days.

Chitosan has been subjected to a range of toxicity tests for effect on mucociliary clearance, effect on cilia beat frequency, immunogenicity, effect on nasal membranes, nasal histology, general toxicity, etc. In all cases the toxicity has been negligible (Table 2). A 10-day subacute nasal toxicity study

Table 2   Toxicity Evaluation of Chitosan

CaCo-2 cells—release of LDH
Frog palate model—mucociliary clearance
Mouse model—immunogenicity studies
Rat model—nasal histology
Rat model—transport of chitosan across membrane
Rat perfusion model—release of protein, 5'ND, LDH
Guinea pig model—cilia beat frequency, 28 days exposure
Rabbit model—10 days subacute toxicity
Human excised turbinates—mucociliary clearance
Human volunteers—mucociliary clearance, saccharine test
Human volunteers—nasal histology on biopsies
Human volunteers—tolerance studies

in rabbits showed neither microscopic nor macroscopic effects on tissues or organs. In a study using the frog palate model and human excised turbinates, the effect on mucociliary clearance of a model particle was investigated (Aspden et al., 1995, 1997b). It was found for both tissues that chitosan in solution had a transient, halting effect on the mucociliary clearance mechanism but that this returned to normal once the chitosan was removed. In the rat perfusion model the effect of chitosan on the release of cell membrane and cytosol-bound enzymes was evaluated (Aspden et al., 1996). Only negligible amounts of the two enzymes were released, as compared with a negative and a positive control, after perfusion for 90 minutes. The release of protein was slightly higher; this was considered to be due to the interaction of chitosan with the mucus that was washed away during the perfusion, after which new mucus was released from the goblet cells. The mucociliary clearance rate, as measured by a saccharine clearance test, was found to be unaffected in human volunteers who were given a nasal chitosan solution daily for 1 week (Aspden et al., 1997b). Biopsies from the same volunteers showed on histological difference in the appearance of the nasal tissue as compared with a control. Furthermore, repeated nasal administration of chitosan for 28 days in guinea pigs had no effect on the cilia beat frequency (Aspden et al., 1997a). The oral toxicity of chitosan has been reported as a medium lethal dose (LD_{50}) of 16 g/kg body weight in mice (Arai et al., 1968).

The mechanism of action of chitosan in improving drug absorption has been extensively studied by research groups interested in the nasal application of chitosan as well as its oral application. The clearance of chitosan formulations from the nasal cavity has been studied in eight healthy human volunteers by means of gamma scintigraphy. The disappearance of ^{99m}Tc-labeled chitosan was followed, and it was found that the times for 50% clearance were 25, 40, and 80 minutes for a control solution, a chitosan solution, and a chitosan powder, respectively (Soane et al., 1999). This work clearly demonstrated that chitosan is bioadhesive in the human nasal cavity and is able to increase the clearance time. It further showed that chitosan in powder form was more bioadhesive than in solution form, which may be reflected in the higher absorption-promoting effect for the powder formulations. An examination of the plasma profiles obtained for peptides when administered nasally with a chitosan solution has shown that bioadhesion is most likely not the only mechanism of action. Bioadhesion would be expected to give rise to a slow absorption profile, whereas the absorption from chitosan formulations is normally fast with a high peak in the plasma profile.

It has been shown by Granger et al. (1986) that the cationic peptide protamine enhances the lymph flow and the transcapillary lymph protein clearance after infusion into the intestinal lymph system of the rat. It has also been demonstrated by McEwan et al. (1993) in MDCK cell monolayers that polylysine, a positively charged polypeptide, is able to affect the tight junction opening. The effect was suggested to be due to neutralization of the fixed anionic sites on the capillary wall. It is likely that chitosan may have a similar effect on the mucosal membrane and, by neutralizing the anionic sites, causes the tight junction to open transiently. A pulse-chase study was performed in order to investigate the duration of the effect of chitosan on the membrane (Fig. 4). When insulin solution was administered to the nasal cavity of rats at time zero, 15 minutes and 30 minutes *after* the application of the chitosan the insulin was absorbed to the same degree. However, 45 minutes and 60 minutes *after* application of the chitosan the administered insulin was absorbed to a lesser degree. It was concluded that the effect on the membrane was transient, which is in line with an effect on the tight junctions (L. Illum et al., unpublished results).

The effect of chitosan on the permeability of monolayers of CaCo-2 cells to mannitol has been studied by several research groups (Artursson et al., 1994; Schipper et al., 1997). The penetration of mannitol was enhanced in a concentration and pH dependent manner. Moreover, immunofluorescent staining of tight junction proteins demonstrated that ZO-1 proteins were decreased or abolished and the cytoskeleton protein F-actin changed from a filamentous to a globular structure. These effects were all reversible and

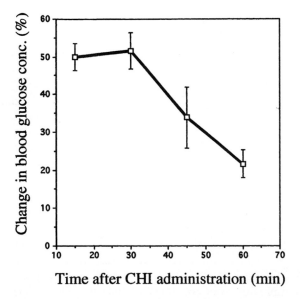

Figure 4   A pulse-chase study of chitosan and insulin in the rat model.

time dependent. The results clearly show that chitosan acts on the membrane by opening up the tight junctions.

V. CONCLUSION

The present chapter has given an overview of the function, anatomy, morphology, and physiology of the nasal passages as necessary for an understanding of the function of bioadhesive systems in nasal delivery of drugs. Furthermore, the barriers in the nasal cavity to successful transport of drugs across the membrane have been discussed. It has been shown that bioadhesive systems, in the liquid form, in the powder form, or as microspheres as well as in a self-gelling form, have great promise as nasal delivery systems. It has been demonstrated that powder formulations are better bioadhesives than liquids and often provide a better absorption-enhancing effect. It has also been shown that with nonsoluble bioadhesive systems in powder form including bioadhesive microspheres the likely mechanism of action is a combination of the bioadhesive effect with an effect on the tight junctions. The cationic polysaccharide chitosan has been shown to be an excellent and nontoxic nasal delivery system that is able to enhance the absorption of polar compounds including peptides and proteins across the nasal cavity to a high degree.

REFERENCES

Ahuja A, Khar RK, Ali J. 1997. Mucoadhesive drug delivery systems. Drug Device Ind Pharm 23:489 515.

Andersen I, Camner P, Jensen PL, Philipson K, Proctor DF. 1974. Nasal clearance in monozygotic twins. Am Rev Respir Dis 110:301–305.

Arai K, Kinumaki T, Fujita T. 1968. Toxicity of chitosan. Bull Tokai Reg Fish Lab 43:89–94.

Argenti D, Colligan I, Heald D, Ziemniak J. 1994. Nasal mucosal inflammation has no effect on the absorption of intranasal triamcinolone acetonide. J Clin Pharmacol 34:854–858.

Artusson P, Lindmark T, Davis SS, Illum L. 1994. Effect of chitosan on the permeability of monolayers of intestinal epithelial cells (CaCo-2). Pharm Res 11: 1358–1361.

Aspden TJ, Adler J, Davis SS, Skaugrud Ø, Illum L. 1995. Chitosan as a nasal delivery system: Evaluation of the effect of chitosan on mucociliary clearance rate in the frog palate model. Int J Pharm. 122:69–78.

Aspden T, Illum L, Skaugrud Ø. 1996. Chitosan as a nasal delivery: Evaluation of insulin absorption enhancement and effect on nasal membrane integrity using rat models. Eur J Pharm Sci 4:23–31.

Aspden T, Illum L, Skaugrud Ø. 1997a. The effect of chronic nasal application of chitosan solutions on cilia beat frequency in guinea pigs. Int J Pharm 153: 137–146.

Aspden TJ, Mason JDT, Jones N, Lowe J, Skaugrud Ø, Illum L. 1997b. Chitosan as a nasal delivery system: The effect of chitosan on in vitro and in vivo mucociliary transport rates. J Pharm Sci 86:509–513.

Aungst BJ. 1990. Transmucosal absorption promoters: Efficacy, mechanisms and safety concerns. Proc Int Symp Controlled Release Bioact Mater 17:10–11.

Bjørk E, Bjurström S, Edman P. 1991. Morphological examination of rabbit nasal mucosa after nasal administration of degradable starch microspheres. Int J Pharm 75:73–80.

Bjørk E, Edman P. 1988. Degradable starch microspheres as a nasal delivery system for insulin. Int J Pharm 47:233–238.

Bjørk E, Edman P. 1990. Characterisation of degradable starch microspheres as a nasal delivery system for drugs. Int J Pharm 62:187–192.

Bjørk E, Holmberg K, Bake B, Edman P. 1992. Effect of degradable starch microspheres on the human mucociliary clearance. Proc Int Symp Controlled Release Bioact Mater 19:417–418.

Bjørk E, Isakson U, Edman P, Artursson P. 1995. Starch microspheres induce pulsatile delivery of drugs and peptides across the epithelial barrier by reversible separation of the tight junctions. J Drug Target 2:501–507.

Bond SW, Hardy JG, Wilson CG. 1984. Deposition and clearance of nasal sprays. Proceedings of 2nd International Congress of Biopharmaceutics and Pharmacokinetics, Salamanca, pp 93–98.

Bremecker KD. 1983. Model to determine the adhesive tissue of mucosal adhesive ointments in vitro. Pharm Ind 45:417–419.

Citi S. 1992. Protein kinase inhibitors prevent junctional disassociation induced by low extracellular calcium in MDCK epithelial cells. J. Cell Biol. 117:169–178.

Cornaz A-L, De Ascentiis A, Colombo P, Buri P. 1996. In vitro characteristics of nicotine microspheres for transmucosal delivery. Int J Pharm 129:175–183.

Critchley H. 1989. Intranasal drug delivery. PhD thesis, University of Nottingham, Nottingham, UK.

Critchley H, Davis SS, Farraj NF, Illum L. 1994. Nasal absorption of desmopressin in rats and sheep. Effect of a bioadhesive microsphere delivery system. J Pharm Pharmacol 46:651–656.

Dahl AR. 1986. Possible consequences of cytochrome P-450 dependent monooxygenases in nasal tissues. In: Barrow GS, ed. Toxicology of the Nasal Passages. Washington, DC: Hemisphere, pp 263–273.

De Ascentiis A, Bettini R, Capoetti G, Catellani PL, Peracchia MT, Santi P, Colombo P. 1996. Delivery of nasal powders of β-cyclodextrin by insufflation. Pharm Res 13:734–738.

Deurloo MJM, Hermens AJJ, Romeyn SG, Verhoef JC, Merkus FWHM. 1989. Absorption enhancement of intranasally administered insulin by sodium taurodihydrofusidate (STDHF) in rabbits and rats. Pharm Res 6:853–856.

Edman P, Bjørk E. 1992. Routes of delivery: Case studies: (1) Nasal delivery of peptide drugs. Adv Drug Delivery Syst 8:165–177.

Edman P, Bjørk E, Ryden L. 1992. Microspheres as a nasal delivery system for peptide drugs. J Controlled Release 21:165–172.

Farquhar MG, Palade GE. 1963. Junctional complexes in various epithelia. J Cell Biol 17:375–412.

Farraj NF, Johansen BR, Davis SS, Illum L. 1990. Nasal administration of insulin using bioadhesive microspheres and lysophosphatidylcholine as a delivery system. J Controlled Release 13:253–261.

Fiebrig I, Harding SE, Davis SS. 1994. Sedimentation analysis of potential interactions between mucus and a putative bioadhesive polymer. Prog Coll Polym Sci 94:66–73.

Fisher AN, Farraj NF, O'Hagan DT, Jabbal-Gill I, Johansen BR, Davis SS, Illum L. 1991. Effect of L-α-lysophosphatidylcholine on the nasal absorption of human growth hormone in three animal species. Int J Pharm 74:147–156.

Fisher AN, Illum L, Davis SS, Schacht EH. 1992. Di-iodi-L-tyrosine-labelled dextrans as molecular size markers of nasal absorption in the rat. J Pharm Pharmacol 44:550–554.

Gonda I, Gipps E. 1990. Model of disposition of drugs administered into the human nasal cavity. Pharm Res 7:69–75.

Granger DN, Kvietys PR, Perry MA, Taylor AE. 1986. Charge selectivity of rat intestinal capillaries. Influence of polycations. Gastroenterology 91:1443–1446.

Grass GM, Robinson JR. 1988. Mechanisms of corneal drug penetration. II: Ultrastructure analysis of potential pathways for drug movement. J Pharm Sci 77:15–23.

Gurny R, Meyer JM, Peppas NA. 1984. Bioadhesive intraoral release systems: Design, testing and analysis. Biomaterials 5:336–340.

Hadler NM, Dourmashikin RR, Nermut MV, Williams LD. 1982. Ultrastructure of hyaluronic acid matrix. Biochemistry 79:307–309.

Hardy JG, Lee SW, Wilson CG. 1985. Intranasal drug delivery by spray and drops. J Pharm Pharmacol 37:294–297.

Harris AS, Nilsson IM, Wagener ZG, Alkner U. 1986. Intranasal administration of peptides: Nasal deposition, biological response and absorption of desmopressin. J Pharm Sci 75:1085–1088.

Harris AS, Ohlin M, Svensson E, Lethagen S, Nilsson IM. 1989. Effect of viscosity on the pharmacokinetics and biological response to intranasal desmopressin. J Pharm Sci 78:470–471.

Harris AS, Svensson E, Wagner ZG, Lethagen S, Nilsson IM. 1988. Effect of viscosity on particle size, deposition and clearance of nasal delivery systems containing desmopressin. J Pharm Sci 77:405–408.

He P, Davis SS, Illum L. 1998. In vitro evaluation of the mucoadhesive properties of chitosan microspheres. Int J Pharm 166:75–88.

Hermens WAJJ, Deurloo MJM, Romeyn SG, Verhoef JC, Merkus FWHM. 1990. Nasal absorption enhancement of 17β-oestradiol by dimethyl-β-cyclodextrin in rabbits and rats. Pharm Res 7:144–146.

Hilger PA. 1989. Diseases of the nose. In: Adams GL, Boies LR, Hilger PA, eds. Fundamentals of Otolaryngology. London: WB Saunders, pp 206–248.

Hirai S, Yashiki T, Mima H. 1981. Mechanisms for the enhancement of nasal absorption of insulin by surfactants. Int J Pharm 9:173–184.

Holmberg K, Bjørk E, Bake B, Edman P. 1994. Influence of degradable starch microspheres on the human nasal mucosa. Rhinology 32:74–77.

Humbert H, Cabiac M-D, Dubray C, Lavine D. 1996. Human pharamcokinetics of dihydroergotamine administered by nasal spray. Clin Pharmacol Ther 60: 265–275.

Hussain A, Faraj J, Aramaki Y, Truelove JE. 1985. Hydrolysis of leucine enkephalin in the nasal cavity of the rat—a possible factor in the low bioavailability of nasally administered peptides. Biochem Biophys Res Commun 133: 923–928.

Hussain A, Foster T, Hirai S, Kashihara T, Batenhorst R, Jones M. 1980a. Nasal absorption of propranolol in humans. J Pharm Sci 69:1240–1242.

Hussain A, Hirai S, Bawarshi R. 1980b. Nasal absorption of propranolol from different dosage forms by rats and dogs. J Pharm Sci 69:1411–1413.

Hussain MA, Shenvi AB, Rowe SM, Shefter E. 1989. The use of α-aminoboronic acid derivatives to stabilize peptide drugs during their intranasal absorption. Pharm Res 6:186–189.

Illum L. 1987. Microspheres as a potential controlled release nasal drug delivery system. In: Davis SS, Illum L, Tomlinson E, eds. Delivery Systems for Peptide Drugs, London: Plenum, pp 205–210.

Illum L. 1998. The nasal route for delivery of polypeptides. In: Frøkjaer S, Cristrup L, Krogsgaard-Larsen P, eds. Peptide and Protein Drug Delivery. Copenhagen: Munksgaard, pp 157–170.

Illum L, Farraj N, Critchley H, Davis SS. 1988. Nasal administration of gentamicin using a novel microsphere delivery system. Int J Pharm 46:261–265.

Illum L, Farraj NF, Davis SS. 1994a. Chitosan as a novel nasal delivery system for peptide drugs. Pharm Res 11:1186–1189.

Illum L, Farraj NF, Davis SS, Johansen BR, O'Hagan DT. 1990. Investigation of the nasal absorption of biosynthetic human growth hormone in sheep—use of a bioadhesive microsphere delivery system. Int J Pharm 63:207–211.

Illum L, Farraj NF, Fisher AN, Gill I, Miglietta M, Benedetti LM. 1994b. Hyaluronic acid ester microspheres as a nasal delivery system for insulin. J Controlled Release 29:133–141.

Illum L, Jorgensen H, Bisgaard H, Krogsgaard O, Rossing N. 1987. Bioadhesive microspheres as a potential nasal drug delivery system. Int J Pharm 39:189–199.

Imai T, Shiraishi S, Saito H, Otagiri M. 1991. Interaction of indomethacin with low molecular weight chitosan and improvements of some pharmaceutical properties of indomethacin by low molecular weight chitosan. Int J Pharm 67: 11–20.

Inagaki M, Sakakura Y, Itoh J, Ukai K, Miyoshi Y. 1985. Macromolecular permeability of the tight junction of the human nasal mucosa. Rhinology 23:213–221.

Jørgensen L, Bechgaard E. 1994. Intranasal permeation of thyrotropin-releasing hormone: In vitro study of permeation and enzymatic degradation. Int J Pharm 107:231–237.

Kashi SD, Lee VHL. 1986. Enkephalin hydrolysis in homogenates of various absorptive mucosae of the albino rabbit: Similarities in rates and involvement of aminopeptidases. Life Sci 38:2019–2028.

Larsen C, Niebuhr Jorgensen M, Tommerup B, Mygind N, Dagrosa EE, Grigokeit H-G, Malercyk V. 1987. Influence of experimental rhinitis on the gonadotropin response to intranasal administration of buserelin. Eur J Clin Pharmacol 33: 155–159.

Lee SW, Hardy JG, Wilson CG, Smelt GJC. 1984. Nasal sprays and polyps. Nucl Med Commun 5:697–703.

Lee VHL. 1988. Enzymatic barriers to peptide and protein absorption. CRC Crit Rev Ther Drug Carrier Syst 5:69–97.

Lehr CM, Bouwstra JA, Schacht EH, Junginger HE. 1992. In vitro evaluation of mucoadhesive properties of chitosan and some other natural polymers. Int J Pharm 78:43–48.

Longer MA, Robinson JR. 1986. Fundamental aspects of bioadhesion. Pharm Int 7: 114–117.

Madara JL. 1987. Intestinal absorptive cell tight junctions are linked to cytoskeleton. Am J Physiol 253:C171–C175.

Madara JL. 1988. Tight junction dynamics: Is paracellular transport regulated? Cell Vol 53:497–498.

Madara JL, Barnberg D, Carlson S. 1986. Effects of cytochalasin D on occluding junctions of intestinal absorptive cells; further evidence that the cytoskeleton may influence paracellular permeability and junctional charge selectivity. J Cell Biol 102:2125–2136.

Madara JL, Dharmsathaphorn K. 1985. Occluding junction structure function relationship in a cultured epithelial monolayer. J Cell Biol 101:2124–2133.

Marttin E, Romeijn SG, Verhoef JC, Merkus WHM. 1997. Nasal absorption of dihydroergotamine from liquid and powder formulations in rabbits. J Pharm Sci 86:802–807.

McEwan GT, Jepson MA, Hirst BH, Simmons NL. 1993. Polycation-induced enhancement of epithelial paracellular permeability is independent of tight junctional characteristics. Biochim Biophys Acta 1148:51–60.

McMartin C, Hutchinson LEF, Hyde R, Peters GE. 1987. Analysis of structural requirements for the absorption of drugs and macromolecules form the nasal cavity. J Pharm Sci 76:535–540.

Morimoto K, Morisaka K, Kamanda A. 1985. Enhancement of nasal absorption of insulin and calcitonin using polyacrylic acid gel. J Pharm Pharmacol 37: 134–136.

Morimoto K, Tabata H, Morisaka K. 1987. Nasal absorption of nifedipine from gel preparations in rats. Chem Pharm Bull 35:3041–3044.

Morimoto K, Yamaguchi H, Iwakura Y, Morisaka K, Ohashi Y, Nakai Y. 1991. Effects of viscous hyaluronate-sodium solutions on the nasal absorption of vasopressin and an analogue. Pharm Res 8:471–474.

Mygind N. 1978. Nasal Allergy. Oxford: Blackwell Scientific.

Nagai T, Machida Y. 1985. Mucosal adhesive dosage forms. Pharm Int 6:196–200.

Nagai T, Nishimoto Y, Nambu N, Suzuki Y, Sekine K. 1984. Powder dosage form of insulin for nasal administration. J Controlled Release 1:15–22.

Oechslein CR, Fricker G, Kissel T. 1996. Nasal delivery of octreotide: Absorption enhancement by particulate carrier systems. Int J Pharm 139:25–32.

O'Hagan DT, Critchley H, Farraj NF, Fisher AN, Johansen BR, Davis SS, Illum L. 1990. Nasal absorption enhancers for biosynthetic human growth hormone in rats. Pharm Res 7:772–776.

Olanoff LS, Gibson RE. 1987. Method to enhance intranasal peptide delivery. ACS Symp Ser 348:301–309.

Olanoff LS, Titus CR, Shea MS, Gibson RE, Brooks CD. 1987. Effect of intranasal histamine on nasal mucosal blood flow and the antidiuretic activity of desmopressin. J Clin Invest 80:890–895.

Pennington AK, Ratcliffe JH, Wilson CG, Hardy JG. 1988. The influence of solution viscosity on nasal spray deposition and clearance. Int J Pharm 43:221–224.

Pereswetoff-Morath L, Bjurstrom S, Khan R, Dahlin M, Edman P. 1996. Toxicological aspects of the use of dextran microspheres and thermogelling ethyl(hydroxyethyl) cellulose (EHEC) as nasal drug delivery systems. Int J Pharm 128:9–21.

Pereswetoff-Morath LP, Edman P. 1995a. Influence of osmolarity on nasal absorption of insulin from the thermogelling polymer ethyl(hydroxethyl) cellulose. Int J Pharm 125:205–213.

Pereswetoff-Morath LP, Edman P. 1995b. Dextran microspheres as a potential nasal drug delivery system for insulin—in vitro and in vivo properties. Int J Pharm 124:37–44.

Pereswetoff-Morath LP, Edman P. 1996. Immunological consequences of nasal drug delivery in dextran microspheres and ethyl(hydroxyethyl) cellulose in rats. Int J Pharm 128:23–28.

Phillpotts RJ, Davies HW, Willman J, Tyrell DAJ, Higgins PG. 1984. Pharamcokinetics of intranasally applied medication during a cold. Antiviral Res 4:71–74.

Proctor DF, Anderson I, Lundqvist G. 1973. Clearance of inhaled particles from the human nose. Arch Intern Med 131:132–139.

Provasi D, Minutello A, Catellani PL, Santi P, Massimo G, Colombo P. 1992. Nasal powders for calcitonin administration. Proc Int Symp Controlled Release Bioact Mater 19:421–422.

Ridley D, Perkins AC, Washington N, Wilson CG, Wastle ML, O'Flynn P, Blattman A, Ponchel G, Duchene O. 1995. The effect of posture on nasal clearance of bioadhesive starch microspheres. STP Pharm Sci 5:442–446.

Rodewald RS, Newman SB, Karnovski MJ. 1976. Contraction of isolated brush orders from the intestinal epithelium. J Cell Biol 70:541–554.

Ryden L, Edman P. 1992a. Effect of polymers and microspheres on the nasal absorption of insulin in rats. Int J Pharm 83:1–10.

Ryden L, Edman P. 1992b. A thermogelling polymer (EHEC) with different osmolarity as a delivery system for nasal administration of insulin. Proc Int Symp Controlled Release Bioact Mater 19:222–223.

Sakr FM. 1996. Nasal administration of glucagon combined with dimethyl-β-cyclodextrin: Comparison of pharmacokinetics and pharmacodynamics of spray and powder formulations. Int J Pharm 132:189–194.

Sandow J, Petri W. 1985. Intranasal administration of peptides: Biological activity and therapeutic efficacy. In: Chien YW, ed. Transnasal Systemic Medications. Amsterdam: Elsevier, pp 183–199.

Sarkar MA. 1992. Drug metabolism in the nasal mucosa. Pharm Res 9:1–9.

Sawayanagi Y, Nambu N, Nagai T. 1982a. Directly compressed tablets containing chitin or chitosan in addition to mannitol. Chem Pharm Bull 30:4216–4218.

Sawayanagi Y, Nambu N, Nagai T. 1982b. The use of chitosan for sustained release preparations of water soluble drugs. Chem Pharm Bull 30:4213–4215.

Schafgen W, Grebe SF, Schatz H. 1983. Pernasal versus intravenous administration of TRH: Effects on thyrotrophin, prolactin, triiodothyronine, thyroxine and thyroglobulin in healthy subjects. Horm Metab Res 15:52–53.

Schipper NGM, Olsson S, Hoogstraate JA, deBoer AG, Vårum KM, Artursson P. 1997. Chitosans as absorption enhancers for poorly absorbable drugs. 2: Mechanism of absorption enhancement. Pharm Res 14:923–929.

Soane RJ, Frier M, Perkins AC, Jones NS, Davis SS, Illum L. 1999. Evaluation of the clearance characteristics of bioadhesive systems in humans. Int J Pharm 178:55–65.

Stanley P, Wilson R, Greenstone M, MacWilliam L, Cole P. 1986. Effect of cigarette smoking on nasal mucociliary clearance and ciliary beat frequency. Thorax 41:519–523.

Stevenson BR, Anderson JM, Bullivant S. 1988. The epithelial tight junction: Structure, function and preliminary biochemical characterisation. Mol Cell Biochem 83:129–145.

Stoksted P. 1952. Rhinometric measurements for determination of the nasal cycle. Acta Otolaryngol Suppl (Stockh) 104:159–176.

Takayama K, Hirata M, Machida Y, Masada T, Sannan T, Nagai T. 1990. Effect of interpolymer complex formation on bioadhesive property and drug release phenomenon of compressed tablet consisting of chitosan and sodium hyaluronate. Chem Pharm Bull 38:1993–1997.

Tuncel T, Otuk G, Kuscu I, Ates S. 1994. Nasal absorption of roxithromycin from gel preparations in rabbits. Eur J Pharm Biopharm 40:24–26.

Vermehren C, Hansen HS, Thomsen MK. 1996. Time dependent effects of two absorption enhancers on the nasal absorption of growth hormone in rabbits. Int J Pharm 128:239–250.

Vidgren P, Vidgren M, Vainio P, Nuutinen J, Paronen P. 1991. Double labelling technique in the evaluation of nasal mucoadhesion of disodium cromoglycate microspheres. Int J Pharm 73:131–136.

Wood CC, Fireman P, Grossman J, Wecker M, MacGregor T. 1995. Product characteristics and pharmacokinetics of intranasal ipratropium bromide. J Allergy Clin Immunol 95:1111–1116.

Zhou M, Donovan MD. 1996. Intranasal mucociliary clearance of putative bioadhesive polymer gels. Int J Pharm 135:115–125.

20

Development of Bioadhesive Buccal Patches

Jian-Hwa Guo
Aqualon Division, Hercules Incorporated, Wilmington, Delaware

Karsten Cremer
LTS Lohmann Therapie-Systeme GmbH, Andernach, Germany

I. INTRODUCTION

Absorption of therapeutic agents from the oral mucosa overcomes premature drug degradation within the gastrointestinal tract, as well as active drug loss due to first-pass hepatic metabolism that may be associated with other routes of administration (1). The buccal mucosa was investigated as a potential site for drug delivery several decades ago, and interest in this area for transmucosal drug administration is still growing (2,3). A surface energy analysis of mucoadhesion has been made by Lehr et al. (4,5), and the contact angle and spreading coefficient have been used to predict the mucoadhesive performance of polymers in their reports. They found that the measured adhesive performance between polycarbophil and pig small intestinal mucosa was highest in nonbuffered saline medium, intermediate in gastric fluid, and minimal in intestinal fluid.

The oral cavity has a number of features that make it a desirable site for drug delivery, including a rich blood supply that drains directly into the jugular vein, bypassing the liver and thereby sparing the drug from first-pass metabolism (1). Successful buccal polymer patch delivery requires at least three things: (a) a bioadhesive to retain the drug in the oral cavity and maximize the intimacy of contact with the mucosa, (b) a vehicle that releases the drugs at an appropriate rate under the conditions prevailing in the mouth, and (c) strategies for overcoming the low permeability of the oral mucosa.

II. PATCH DESIGN

A.   General Therapeutic Aims and Requirements

1. Local Versus Systemic Action

Mucoadhesive patches for administration to the mucosa of the oral cavity may have a number of different designs depending on various considerations, such as the therapeutic aim and the physicochemical and pharmacokinetic properties of the active ingredient. Regarding the therapeutic aim, two different rationales for developing mucosal patches may be differentiated: patches can be intended to deliver a drug to the systemic circulation in a way that is superior to other routes of administration (6–8), or their purpose may be local therapy of the oral mucosa (6,9,10). As alternatives for both classes of patches, more conventional dosage forms are available. In the case of locally acting patches, the alternatives that are used most often today are oral gels, oral liquids, and lozenges. For systemic action, a number of dosage forms including sustained- or controlled-release oral technologies, transdermal patches, and injectable depot formulations exist. Compared with mucosal patches, the technologies related to the more conventional dosage forms including their production tend to be well established in most pharmaceutical companies and may be associated with reasonable development and production costs. Therefore, it is necessary for a successful mucosal patch to have clearly defined advantages over alternative products.

The advantages of patches for local therapy of the oral mucosa are easily perceived. Some inflammatory or infectious conditions of the mucosa, such as aphthae, herpes- or *Candida*-related stomatitis, or physical injuries, are most typically treated with local anesthetics, antimycotics, disinfectants, antiviral agents, or corticosteroids, which are administered as oral gels and liquids in some rather inefficient manner. To mention only one of many examples, oral *Candida albicans* infections, which frequently occur in very young infants, are treated with lozenges or mouth rinses with 400.000 IU of nystatin three to five times a day (11). This conventional therapy means that the affected tissues are exposed to but a minor fraction of the drug dose for very short periods of time. Most of the time, no effective drug level is present at the site of action.

In contrast, buccal patches that are applied directly to the affected mucosal region have the potential to supply the site of action with effective drug levels and sustain these levels over a long period of time (10). Their pharmacokinetic superiority over conventional dosage forms is accompanied by other advantages. For instance, the affected mucosa is covered and protected from contact with food and from other mechanical stress that may cause pain and further irritation. Furthermore, patches allow more exact dosing than the alternative gels or liquids.

The design of a mucosal patch will, therefore, take several considerations into account. One of the requirements will be rather gentle adhesion, as it would be undesirable to have to exert too much force to remove the patch after use; otherwise, the already affected mucosa could easily be injured. Alternatively, the patch can be designed in such a way that it dissolves or disintegrates completely during the application period. This is a good solution especially when short adhesion times, such as 20 to 40 minutes, are desired. Drug release may be an important aspect of the patch; sophisticated release rate control, on the other hand, will not play the major role in the design considerations as the tolerability of the drugs will not be critical in most cases. To ensure the correct and exact positioning of the patch on the affected site, it may also be desirable to include an application aid in the design.

A buccal patch for the systemic administration of a drug will, in general, be designed with much more emphasis on controlled-release rates and on achieving fairly even plasma levels over a predetermined period of time. Even more than with locally active drugs, drug release from systemic buccal patches should be unidirectional toward the mucosa, and release into the saliva should be avoided (6–8). In general, this type of buccal patch would require relatively long adhesion times, at least a few hours, to achieve the desired systemic effects.

In the scientific community with a focal interest in new dosage forms and drug delivery research, systemic buccal patches have clearly received more attention than locally active patches in the past (6,9,10). Because systemic delivery of drugs via the buccal route is very challenging and deserves detailed consideration of various parameters and requirements, most of the following text of this chapter is related primarily to systemic buccal patches, even though some aspects may also be relevant to local treatment of the oral mucosa.

2. Onset and Duration of Action

In most cases, the intended therapeutic function of a controlled-release product will result in some requirements concerning the onset and, far more important, the duration of action or, if there is a direct relationship between plasma levels and pharmacological effect, the duration of plasma levels over a specified minimum concentration. The time to the onset of action is of little importance if the patch is meant for multiple administrations or chronic treatment, in which a new patch is applied immediately after the old one has been removed. If a particularly long duration of action is needed, such as 12 or 24 hours, corresponding to one or two applications a day, an even release rate, which may be especially important in this case to avoid large variations of plasma levels, can be achieved by several methods. A common approach is to load the patch with much more drug than the dose it actually delivers (12). The price paid for this driving force for permeation into the mucosa may be literally high, depending on the cost of the active ingredient. Another method is to have a rate-controlling membrane at the side of the patch that adheres to the mucosa (13). A prerequisite of this design is a high mucosal flux potential of the drug relative to its dose, as the addition of a rate-controlling membrane may decrease the rate of the invasion process substantially compared with comparable designs without a membrane. Unfortunately, the rather low mucosal permeabilities of many drugs require special efforts, such as the use of absorption enhancers, to increase the invasion rate, so that a rate-controlling membrane would not be helpful (14).

If a quick onset of action is needed, a design must be chosen that allows direct contact between the drug reservoir or drug-containing layer and the mucosa. At the same time, the contact area should be as large as possible, a requirement that must be carefully balanced with compliance reservations related to large patches.

B. Pharmacokinetic Considerations

Pharmacokinetic rationales are the primary motivation for starting the development of a buccal patch for systemic action. If satisfactory plasma levels can be achieved with a conventional tablet or capsule, there is no need to develop a product that is more sophisticated in its design and manufacture, more expensive, more difficult to administer, and perhaps somewhat less acceptable to patients. If, on the other hand, conventional oral administration is not possible for pharmacokinetic reasons and the only alternative available is the parenteral route, with all its disadvantages in terms of compliance and medical staff involvement, a buccal patch may become a very attractive product idea (7,15–17).

The principal pharmacokinetic problems that may make the oral application of certain drugs impossible but that can potentially be solved with buccal patches are gastrointestinal degradation and presystemic elimination or, less often, very short elimination half-lives. Other restrictions to oral application, such as extremely low solubilities or low mucosal permeabilities often associated with high-molecular-weight compounds and low absorption rates, cannot normally be overcome with buccal dosage forms; if the gastrointestinal tract, having a few hundred square meters of mucosa especially adapted for absorption, is not able to absorb a specific dose of a substance, the oral cavity with only a few square centimeters of mucosa will be even less suitable for this task (18).

1. Gastrointestinal Degradation

Some drugs are not absorbed from the gastrointestinal tract to a significant extent because they are degraded before they have a chance to be taken up by mucosal cells. The mechanism of degradation may be chemical, catalyzed by acid, or, more frequently, it may be enzymatic. Enzyme activity is present to some degree throughout the gastrointestinal tract (17,19–22). Very high levels of activity are found in the lumen of the small intestine and in association with the apical membrane of its mucosa; this activity results from enzyme secretions of mucosal cells and is an important prerequisite for nutrient absorption by the body. Molecules with a peptide structure are very often degraded to small units before they can be absorbed. Another region with significant enzyme activity is the large intestine. Here the activity results primarily from the high bacterial content of the lumen (17,19–22). However, as an obstacle to the absorption of drugs it does not play a major role.

If a drug molecule capable of mucosal permeation is degraded in the upper gastrointestinal tract to such a degree that its oral bioavailability is too low to be acceptable, a buccal patch may be a good alternative. Typical examples are molecules with highly hydrolyzable peptide bonds that are at the same time small and lipophilic enough to cross biological membranes.

2. First-Pass Metabolism

Closely related to and sometimes summarized together with gastrointestinal degradation by the term presystemic elimination is the metabolic elimination of a compound once it has been absorbed but before it reaches the systemic circulation (23,24). This so-called first-pass effect refers to molecules that are taken up by the absorptive mucosa but are metabolized during their passage through the mucosa or, even more frequently, during their first passage through the liver via the portal vein to such a degree that the amount of drug appearing in the systemic circulation is significantly lower than the amount absorbed. This is a very common obstacle to oral administration and applies to compounds from a large number of therapeutic categories, including nitroglycerin (25), morphine (26), estradiol (27), lidocaine (28), ergotamine (29), scopolamine (30), and many others. Buccal administration has a high potential to solve the problem of presystemic hepatic elimination because the blood supply of the mucosa of the oral cavity is not drained into the portal vein but reaches the systemic circulation without liver passage (7). An extremely high hepatic first-pass effect of an essential drug for critically ill patients, such as nitroglycerin, whose oral bioavailability is less than 2% of the orally administered dose, may lead to a situation in which a peroral product is altogether ineffective and buccal administration is the principal route of application.

Apart from the overall increased bioavailability, an important advantage of a buccal patch for these drugs is also potentially better control of plasma levels. For instance, if the extent of the hepatic first-pass effect of a specific compound ranges from 97 to 99%, its bioavailability would range from 1 to 3%, which is an extremely wide and unacceptable range. Without a first-pass effect, there would still be variations in bioavailability because of biological factors such as different individual clearance values, but these variations are typically much lower. Another advantage of a buccal patch is the reduced costs of drug because of the application of much lower doses than those necessary for oral products.

Concerning the design of buccal patches with active compounds with high presystemic degradation or metabolism, it is important to pay much attention to the need to deliver the drug effectively to the oral mucosa and to avoid drug release into the saliva as much as possible (7,9,31). If a major fraction of the dose dissolves in the saliva and is subsequently swallowed, the disadvantages and limitations of gastrointestinal administration discussed before will still be present in spite of the buccal dosage form (see Sec. II.D.2).

3. Elimination Half-Life

If the duration of action of a drug with a short elimination half-life is shorter than desired, the method of choice for achieving a better product is to develop an oral sustained- or controlled-release dosage form. Thus, a short elimination half-life alone is probably not a sufficient reason to consider a buccal patch that may also provide controlled drug release and a long duration of action. However, in combination with one of the previously discussed pharmacokinetic properties—gastrointestinal degradation or a substantial first-pass effect—a short half-life may add a considerable rationale for the buccal patch alternative. For drugs with sufficiently slow elimination rates there is always the alternative of quick-release buccal or sublingual dosage forms, i.e., sprays or capsules, which are less sophisticated and costly than buccal patches; on the other hand, if the elimination rate constant is high, the buccal patch is a most attractive product.

C. Physicochemical Parameters

1. Drug Solubility

The solubility of a drug is of great importance to the pharmaceutical scientist for the development of a buccal patch formulation. However, when it comes to the basic design of the patch, e.g., before the formulation work has even started, it may be necessary to consider the solubility of a compound to evaluate the probable patch size and thickness resulting from the incorporation of the drug.

2. Drug Permeability

As discussed previously, insufficient gastrointestinal absorption related to low drug permeability is not a problem that can be solved with buccal patches (see Sec. II.B). In contrast, good mucosal permeability may be considered an important requirement for successful buccal administration because of the limited area available for absorption. However, it is not possible to express the mucosal flux requirement in terms of absolute values. Instead, it is important to consider the achievable mucosal flux rates in relationship to the dose to be delivered. For highly potent drugs that require plasma levels of the magnitude of pico- or nanomoles per liter, such as glucagon-like peptide I (GLP-I), even flux rates that would normally be considered very low may lead to systemically effective plasma concentrations (32). On the other hand, compounds with high mucosal permeability may still not be good candidates if a dose of several hundred milligrams is required.

Concerning the design of a buccal patch, there is an obvious need for a relatively large patch area to be in contact with the mucosa if the drug is poorly permeable. As the area of the buccal mucosa is limited and large buccal patches tend to be uncomfortable to wear, there is limited freedom for the formulation scientist to design patch sizes matching the needs resulting from physicochemical properties of the drug. If the flux rates are sufficiently high, the patch may be kept small, but not as small as possible: patches smaller than 0.5 to 1.0 mm^2 are difficult to handle and should be avoided.

3. Drug Stability

Buccal patches for the applications of drugs that are liable to degradation in the gastrointestinal tract may also present stability problems during the period of application. Highly hydrolyzable compounds, for instance, may be degraded by the saliva that is taken up by the patch. If such a compound is to be administered over a relatively long period of time, such as over 12 or 24 hours or even longer, it may need to be protected either by additives such as buffer salts, by antioxidants that create a stabilizing microenvironment within the patch, or by a design with a drug reservoir in which the major fraction of the dose is incorporated in the undissolved state.

D.   Patch Design and Geometry

1.   Acceptability Aspects

Patch size, geometry, and design depend not only on drug-related factors such as the dose, mucosal permeability, and physicochemical properties but also on considerations related to the acceptability of the product for patients. Again, acceptability cannot be translated into absolute definitions of maximum patch sizes or the like; it must be defined on a case-to-case basis in relation to the benefit a buccal patch has for patients and to existing alternative dosage forms. For instance, if the disease treated is severe and there is no comparably effective drug available in other than parenteral dosage forms, patients may be willing to accept much larger and more uncomfortable buccal patches than patients who take patches for minor diseases or conditions that could also be treated with peroral tablets or capsules.

This relativity also applies to other factors that influence the acceptability and patients' compliance. Among the most important potential factors are taste; mucosal irritation; impediment of lip and cheek movement while speaking, eating, or drinking; and an uncomfortable feeling in the mouth because of the continuous presence of a foreign body in the oral cavity (7,8). Frequently underestimated, the taste of many drugs is so bad that it can cause serious compliance problems if it is not either masked with appropriate additives or prevented from coming into contact with the taste receptors by the specific patch design, which is the more effective method.

A patch design to meet the needs of taste masking requires at least two layers. The mucoadhesive layer, which is applied to the mucosa, will in most cases also contain the active ingredient. The second layer functions as a backing layer covering the drug reservoir toward the lumen of the oral cavity; to be a good release barrier it has to be substantially less permeable to the drug and/or the saliva than the adhesive layer. For even more restricted access of saliva to the drug reservoir, the backing layer can be made larger than the drug-containing layer. Such a patch would be more difficult and expensive to make, but it may be necessary to choose this design to achieve a product that is acceptable to patients (31).

A patch with two or more layers, one of them being a backing layer, is also much more comfortable to wear than a single-layer patch (7,8). As the backing layer is not adhesive, it will not stick to the gums or teeth and thus will allow free lip and cheek movement around the jaws and teeth. Because of this, the adhesive layer can be formulated for much stronger mucoadhesion than with any one-layer design. Thus, nonadherence and dosage form failure can be largely avoided, and patches can be developed for much longer application times.

Mucosal irritation can hardly be avoided if the active ingredient itself is the irritating agent. In this case the outlook for an acceptable buccal patch is poor, unless an additive is found to reduce the degree of irritation. Often, however, irritation problems result from the formulation, especially the incorporation of permeation enhancers, which in most cases interact primarily with structures of the mucosa and alter its barrier function, substantially increasing the irritation potential of a patch (9,33). Therefore, efforts should be made to achieve acceptable bioavailabilities without the use of permeation enhancers and to consider permeation enhancement only if other options do not seem feasible. This certainly applies to products that are meant to be administered regularly in the long-term therapy of a chronic disease and to a lesser degree to buccal patches for acute or short-term treatment.

2. Avoidance of Gastrointestinal Administration

As mentioned before, a frequent rationale for the buccal administration of a drug is its liability to degradation or extensive first-pass metabolism leading to low and unacceptable plasma levels after conventional oral application (see Sec. II.B). If this is the case, the buccal patch must be designed in such a way that drug release into the saliva and subsequent swallowing (e.g., gastrointestinal administration) are prevented. This need and the ways to handle it are very similar to the taste-masking problem discussed in the preceding section, which is best solved by a two-layer design with an impermeable backing layer. Especially for drugs that are not taken up by the mucosa very rapidly, it is important to provide for unidirectional drug release; with compounds that have high buccal absorption rates, such as nitroglycerin or nicotine, it may not be necessary to prevent release into the saliva, as a major fraction of the drug dissolved in the saliva is absorbed before swallowing.

A schematic representation of some of the geometric patch design approaches discussed in this and the previous sections is given in Fig. 1.

3. Application Aids

Depending on the therapeutic aim of a buccal patch, it may be necessary to consider a design with an application aid. Buccal patches are innovative products, and most patients will be unfamiliar with the handling and application of a patch. If, for instance, a buccal patch is developed for the treatment of diseases that frequently occur in elderly people, the chances are high that the product will not be successful because of the patients' poor eyesight and limitation of the precise motor activity required for patch application. Another reason to consider a design with an application aid is that

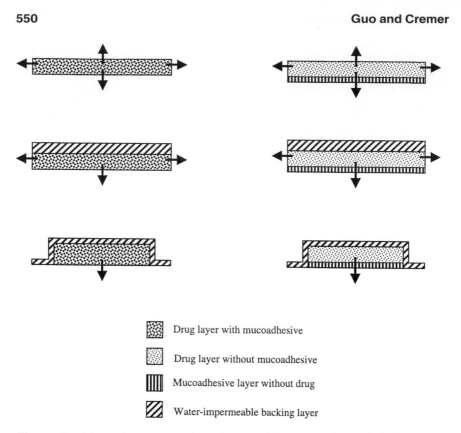

Drug layer with mucoadhesive

Drug layer without mucoadhesive

Mucoadhesive layer without drug

Water-impermeable backing layer

Figure 1   Schematic representation of some of the geometric patch designs.

the patch must be positioned exactly on a specified location of the mucosa, as in the case of local treatment of infected mucosal regions (see Sec. II.A.1). Especially small patches without application aids would present serious problems even for young and otherwise healthy people.

A good application aid should help a patient handle a thin and small patch in such a way that the patch itself does not have to be held with the fingers. As it may be difficult to put two fingers holding a patch deep into the mouth to reach an administration site in a distal region of the buccal mucosa, it is desirable for the application aid to have dimensions that allow the patient to hold it with one end outside the mouth while the other end is connected to the patch. An example of such an application aid is shown in Fig. 2.

Creativity is still needed to find appropriate designs for specific products. Again, it is important to design a specific new product on a case-to-case basis and take into consideration everything that will have a bearing

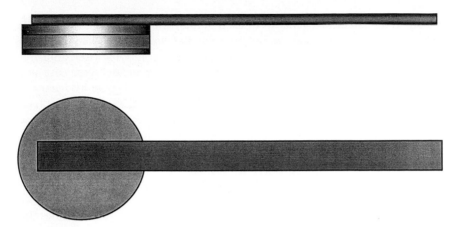

Figure 2 Schematic representation of buccal patch application aid.

on the requirements for the buccal patch, such as the physicochemical, or-
ganoleptic, and pharmacokinetic properties of the drug; the disease to be
treated; and the patients to be helped.

III. PHARMACEUTICAL PATCH DEVELOPMENT

A. Patch Formulation

1. Mucoadhesive Polymers

The first step in the development of a buccal patch is the selection and
characterization of an appropriate bioadhesive in the formulation. The polymers that are commonly used as bioadhesives in pharmaceutical applications
are acacia, chitosan, carboxy polymethylene, guar gum, hydroxypropylcellulose, hydroxyethylcellulose, hydroxypropylmethylcellulose, polycarbophil,
poly(vinylpyrrolidone), poly(vinylalcohol), sodium carboxymethylcellulose,
sodium alginate, etc.

   The in vitro characterization of a newly developed bioadhesive patch
for controlled drug delivery via the buccal mucosa was investigated by Guo
et al. (31,34–39). They developed patches composed of Carbopol 934P with
physical properties that are required for buccal controlled drug delivery. The
effects of different ratios of bioadhesive and supporting polymers on the
surface properties, adhesion, and swelling of buccal patches were investigated. The evaluation of laminated mucoadhesive patches for buccal drug
delivery was studied by Anders and Merkle (40), and the water-soluble hy-

drocolloid mucoadhesives that were used in their study are listed in the Table 1. Preparation of their adhesive polymer patches was as follows: given volumes of appropriately made aqueous polymer solutions (for drug-free patches) or drug-polymer solutions (for drug-loaded patches) were cast onto a backing layer sheet mounted on top of a stainless steel plate by means of a frame. They found hydroxyethylcellulose (Natrosol 250, Hercules Incorporated) to be the most effective bioadhesive in their study, because less Natrosol polymer per cm^2 was required to achieve similar durations of mucosal adhesion in the subject (Table 2). They also found that in vivo drug release could be significantly controlled by the choice of polymers, viscosity grades, and polymer loads per patch and that the performance of hydroxyethylcellulose was better than that of poly(vinylpyrrolidone) (Figs. 3 and 4).

The application of hydroxypropylcellulose (Klucel EF, Hercules Incorporated) for sublingual or buccal administration of therapeutic agents was investigated by Lu and Reiland (41). The compositions of liquid formulations and the plasma concentrations of leuprolide acetate are presented in Table 3 and Fig. 5, respectively. They found that hydroxypropylcellulose is a good bioadhesive for sublingual and buccal therapeutic formulations.

Table 1   Molecular Weights and Specific Viscosity of Water-Soluble Hydrocolloids

Polymer	Trade name	Molecular weight	Viscosity (mPa s)
Hydroxyethylcellulose	Natrosol 250 L	80,000	14 (2%)
(HEC)	Natrosol 250 G	300,000	300 (2%)
	Natrosol 250 K		2,000 (2%)
	Natrosol 250 M	650,000	600 (2%)
	Natrosol 250 H	900,000	30,000 (2%)
Hydroxypropylcellulose	Klucel EF (E)	60,000	500 (10%)
(HPC)	Klucel JF (J)		30 (2%)
	Klucel MF (M)		5,000 (2%)
	Klucel HF (H)	1,000,000	2,000 (1%)
Poly(vinylpyrrolidone)	Kollidon 17	9,500	2 (10%)
(PVP)	Kollidon 25	27,000	4 (10%)
	Kollidon 30	49,000	7 (10%)
	Kollidon 90	1,100,000	500 (10%)
Poly(vinyl alcohol)	Mowiol 4-88	23,300	4 (4%)
(PVA)	Mowiol 40-88	114,400	40 (4%)
	Mowiol 4-98	23,300	4 (4%)
	Mowiol 56-98	202,400	56 (4%)

From Ref. 40.

Table 2 Comparison of Adhesive Properties of Different Polymers in Vivo

Polymer/ trade name	Amount of polymer (mg/cm^2)	Duration of adhesion		
		Mean	SD	n
HEC/Natrosol 250 G	2.90	32.7	5.7	3
HPC/Klucel EF	5.82	32.3	7.6	3
PVP/Kollidon 90	8.81	36.3	8.1	3
PVA/Mowiol 44-88	8.75	30.3	6.7	3

From Ref. 40.

2. Supportive Materials

As with transdermal patch formulation, application of an impermeable backing layer on buccal patches has been considered to prevent drug loss and for the patient's application convenience. Guo and Cooklock (31) have studied the effects of backing layers on the swelling and adhesion of buccal patches. They found that ethylcellulose, a hydrophobic polymer, has very low water permeability and moderate flexibility; therefore, it is a good candidate for backing application. The effects of ethylcellulose on the hydration of polymer patches are significant, and the water uptake of polymer patches was delayed to about 24 hours by the application of ethylcellulose. The polymer patches that had a higher amount of ethylcellulose had a lower hydration rate. As alternative backing materials, polyvinylpyrrolidone and cellulose acetate mixture were also studied. The polyvinylpyrrolidone and cellulose acetate gel did swell with the polymer patch when the polymer patch was hydrated; however, this gel has very high water permeability and could allow the drug to pass through. Poly(ethylene-co-vinyl acetate) is another material that has been studied for backing layer application. Poly(ethylene-co-vinyl acetate) is a very hydrophobic and elastic polymer, and because most of the swelling force of the buccal patch was used to stretch the poly(ethylene-co-vinyl acetate) film, the swelling ratio of the buccal patch significantly decreased when the patch was coated with poly(ethylene-co-vinyl acetate).

B. In Vitro Characterization

1. Measurement of Mucoadhesion

Because dissolution of a bioadhesive occurs naturally during oral administration, it is important to establish the duration of adhesive force provided by the chosen polymer. A variety of in vitro methods have been employed

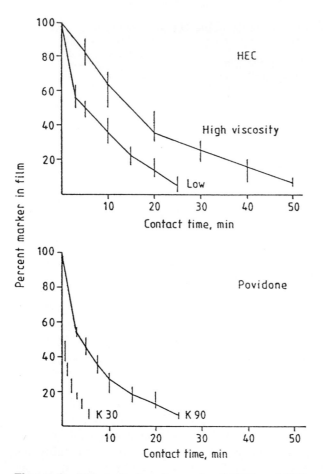

Figure 3 Effect of viscosity grade of HEC and PVP on buccal release of mucoadhesive patch; all data for the same subject; 2.0–2.1 mg polymer per patch; 10
mg sodium salicylate per patch as marker; bars indicate full range of data. HEC,
high-viscosity Natrosol 250 G, low viscosity 250 L. PVP, high-viscosity Kollidon
90, low viscosity Kollidon 30. Backing layer: Multiphor. (From Ref. 40.)

to measure these parameters (42,43). Scientifically, there are two simplifying
advantages of the peel test compared with other methods. It is the only
method in which failure proceeds at a controlled rate, and the peel force is
a direct measure of the work of detachment (44). The in vitro bioadhesion
between hydrated polyvinyl pyrrolidone/cellulose acetate hydrogel and

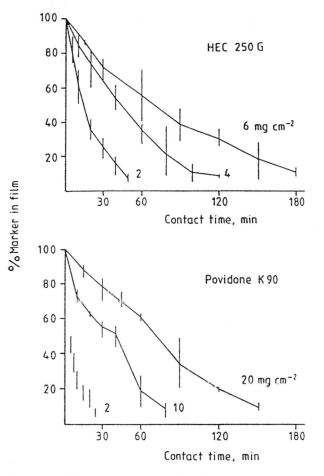

Figure 4   Effect of polymer load on buccal release of mucoadhesive patch; all data for the same subject; 10 mg sodium salicylate per patch as marker; bars indicate full range of data. Backing layer: Multiphor. (From Ref. 40.)

bioadhesive buccal patches could be assessed using an Instron (model 4201, Instron Co., Canton, MA). Adhesion between the patches and the test surface was expressed as the average peeling strength (kg/mm) or load (kg).

In vitro testing revealed that the average peeling strength of patches increased with increasing patch thickness and maximum bioadhesion was reached at a thickness of approximately 50 mil (36). Increases in thickness beyond this point did not alter bioadhesive strength. The double-logarithmic

Table 3 Formulations for Synthetic Polypeptide Leuprolide Acetate

Formulation number	Ingredients	Concentration
A	Leuprolide acetate	50 mg/mL
	Urea	10% (w/v)
	Klucel EF	2% (w/v)
B	Leuprolide acetate	50 mg/mL
	Benzoic acid	5% (w/v)
	Ethanol	50% (v/v)
	Klucel EF	2% (w/v)
C	Leuprolide acetate	50 mg/mL
	Klucel EF	2% (w/v)
D	Leuprolide acetate	50 mg/mL
	Hydroxypropyl cyclodextrin	20% (w/w)
	Klucel EF	2% (w/v)
E	Leuprolide acetate	50 mg/mL
	Ethanol	80% (v/v)
	Klucel EF	2% (w/v)
F	Leuprolide acetate	50 mg/mL
	Peppermint oil	10% (v/v)
	Ethanol	80% (v/v)
	Klucel EF	2% (w/v)
G	Leuprolide acetate	50 mg/mL
	Urea	10% (w/v)
	L-Arginine HCl	20 mg/mL
	Klucel EF	2% (w/v)

From Ref. 41.

plot between the average peeling strength and peeling velocity is presented in Fig. 6, and a sigmoid relation between these two parameters was found. When peeling energy was plotted logarithmically against peeling velocity, there was a regime of low-energy peeling (low velocity) and a transition to the normal high-energy peeling (high velocity).

For a viscous, uncross-linked elastomeric adhesive, the peeling energy could be expressed as

$$\Theta = \phi + (\kappa h)/2$$

where Θ is the peeling energy, ϕ is the surface energy, κ is the energy dissipated per unit volume of adhesive, and h is the adhesive layer thickness. The bioadhesive strength of buccal patches was found to increase with increasing thickness up to a maximum value, and this is in agreement with

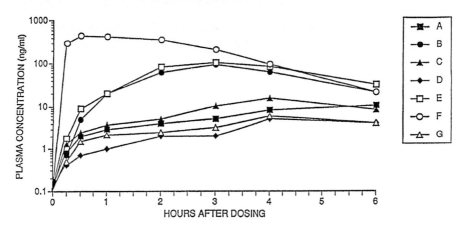

Figure 5 Plasma concentration of synthetic polypeptide leuprolide acetate. (From Ref. 41.)

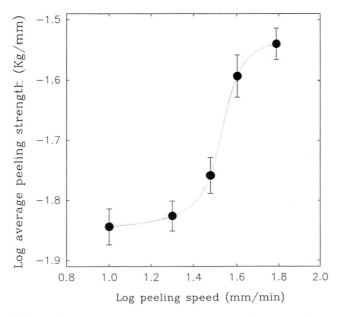

Figure 6 Double-logarithmic plot of peeling strength versus peeling velocity. (From Ref. 39.)

data provided in a review by Gent and Hamed (45). The phenomenon can be explained by an alteration of the dissipation energy of patch polymers of increasing thickness (i.e., increasing yield strength) under conditions of viscoelastic and plastic deformation. The peeling energy is a linearly increasing function of thickness when tearing apart highly hysteretic rubber samples (κ large). However, energy dissipated during peeling then becomes independent of the overall thickness of the adhesive because the dissipation process no longer involves the entire layer of adhesive ($\kappa \sim 0$).

The time-temperature superposition relationship between peeling energy, Θ, and peeling velocity, γ, can be expressed by the following equation:

$$\Theta = \Theta_0 \Phi(\gamma, T, \xi_0)$$

where Θ_0 is the interfacial bond energy and Φ is a loss function depending on velocity γ, temperature T, and strain ξ_0. A related observation is that time-temperature superposition of the plot of log Θ against log γ will be obtained only if the energy losses encompassed by the loss function Φ are thermorheologically simple (46). At very low peeling velocities $\Phi \approx 1$ and it is possible in principle to observe a region where Θ becomes independent of rate and temperature and assumes the value Θ_0. At low peeling rates failure is cohesive with peeling energy Θ increasing with the rate.

The surface properties of polymer patches were determined at ambient temperature by the pendant drop method (37) using a contact angle goniometer (model NRL-100, Rame-Hart, Mountain Lake, NJ) equipped with an environmental chamber. It was found that the bioadhesion of buccal patches could not be predicted from their surface properties only. Because the wetting angle and surface energy results could represent only the initial ability to adhere to substrate, however, the relation between bioadhesion and the morphologic structure of buccal patches should be considered as an important factor.

2. Modification of Drug Release

It has been suggested that drugs with biological half-lives in the range of 2–8 hours are good candidates for sustained-release formulations (47). The plasma half-life (3 hours), duration of action (4–5 hours), and other aspects of its pharmacokinetic profile (e.g., liver metabolism) make buprenorphine a suitable candidate for administration via a buccal patch that provides controlled drug delivery and bypasses first-pass hepatic metabolism. Buprenorphine is a partial μ agonist opioid analgesic that is well absorbed by both the intramuscular and sublingual routes and is 25–50 times more potent than morphine (48). The drug's onset of action is rapid, with maximum blood drug levels attained within 5 minutes of intramuscular injection. The duration of action of buprenorphine can be extended from the 4–5 hours obtained with a single intramuscular dose to 5–6 hours with sublingual administration.

The dissolution tests for buprenorphine buccal patches could be performed in phosphate buffer (pH 7). Buprenorphine patches were affixed to plexiglass sample blocks and placed in flasks containing 100 mL of buffer at 37°C. Aliquots were taken at various times up to 24 hours and assayed for buprenorphine by high-pressure liquid chromatography equipped with a variable-wavelength ultraviolet-visible detector. The gradient system used in this study consisted of the mobile phase CH_3CN/16.6 mM $CH_3(CH_2)5SO_3Na$ aqueous solution/CH_3COOH, 70%/30%/1% (v/v/v), at a flow rate of 1.5 mL/min. It was found that there was a high degree of correlation between drug dissolution and water uptake and that swelling is the major mechanism of buprenorphine release from these buccal patches (36).

The buprenorphine dissolution profile found in the present study is similar to that reported by Nagai and Konishi (3) for patches containing 30 mg of a freeze-dried 1:2 mixture of hydroxypropylcellulose and Carbopol 934P used for the buccal administration of lidocaine. The curve obtained with the current patches suggests sustained delivery of buprenorphine over a 24-hour period with the ultimate release of nearly 75% of the drug. The correlation between the drug dissolution and water uptake curves also suggests that patch swelling is the major mechanism of buprenorphine release.

IV. SUMMARY

In conclusion, bioadhesive patches for administration to the mucosa of the oral cavity may have a number of different designs depending on various considerations. Regarding the therapeutic aim, two different rationales for developing mucosal patches may be differentiated: patches can be intended to deliver a drug to the systemic circulation in a way that is superior to other routes of administration, or their purpose may be local therapy of the oral mucosa. As alternatives for both classes of patches, more conventional dosage forms are available. The first step in the development of a buccal patch is the selection and characterization of an appropriate bioadhesive in the formulation. As with transdermal patch formulation, an impermeable backing layer should be considered for buccal patches to prevent drug loss and for the patient's application convenience. Appropriate in vitro methods have to be employed to measure the duration of adhesive force provided by the chosen polymer. As shown by the information presented here, bioadhesive patch formulations are certainly promising alternatives for drug delivery systems.

REFERENCES

1. H Park, JR Robinson. Physico-chemical properties of water insoluble polymers important to mucin/epithelial adhesion. J Controlled Release 2:47–57, 1985.
2. R Gurny, HE Junginger. First International Joint Workshop of the Association for Pharmaceutical Technology and the Controlled Release Society, Leiden, 1989.
3. T Nagai, R Konishi. J Controlled Release 6:353, 1987.
4. C-M Lehr, JA Bouwstra, HE Boddé, HE Junginger. A surface energy analysis of mucoadhesion: Contact angle measurement on polycarbophil and pig intestinal mucosa in physiologically relevant fluids. Pharm Res 9:70, 1992.
5. C-M Lehr, HE Boddé, JA Bouwstra, HE Junginger. Eur. J Pharm Sci 1:19, 1993.
6. ME de Vries, HE Boddé, JC Verhoef, HE Junginger. Developments in buccal drug delivery. Crit Rev Ther Drug Carrier Syst 8:271, 1991.
7. MJ Rathbone, G Ponchel, FA Ghazali. Systemic oral mucosal drug delivery. In: MJ Rathbone, ed. Oral Mucosal Drug Delivery. New York: Marcel Dekker, 1996, p 241.
8. G DeGrande, L Benes, F Horrière, H Karsenty, C Lacoste, R McQuinn, J-H Guo, R Scherrer. Specialized oral mucosal drug delivery systems: Patches. In: MJ Rathbone, ed. Oral Mucosal Drug Delivery. New York: Marcel Dekker, 1996, p 285.
9. D Harris, JR Robinson. Drug delivery via the mucous membranes of the oral cavity. J Pharm Sci 81:1, 1992.
10. T Nagai, Y Machida, R Konishi. Bioadhesive dosage forms for buccal/gingival administration. In: V Lenaerts, R Gurny, eds. Bioadhesive Drug Delivery Systems. Boca Raton, FL: CRC Press, 1990, p 137.
11. B Millns, MV Martin. Nystatin pastilles and suspension in the treatment of oral candidosis. Br Dent J 181:209, 1996.
12. S Kim, W Snipes, GD Hodgen, F Anderson. Pharmacokinetics of a single dose of buccal testosterone. Contraception 52:313, 1995.
13. MM Veillard, MA Longer, TW Martens, JR Robinson. Preliminary studies of oral mucosal delivery of peptide drugs. J Controlled Release 21:123, 1987.
14. I Gonzalez-Younes, JG Wagner, DA Gaines, JJ Ferry, JM Hageman. Absorption of flurbiprofen through human buccal mucosa. J Pharm Sci 80:820, 1991.
15. RL McQuinn, DC Kvam, MJ Maser, AL Miller, S Oliver. Sustained oral mucosal delivery in human volunteers of buprenorphine from a thin non-eroding mucoadhesive polymeric disk. J Controlled Release 34:243, 1995.
16. HP Merkle, R Anders, A Wermerskirchen. Mucoadhesive buccal patches for peptide delivery. In: V Lenaerts, R Gurny, eds. Bioadhesive Drug Delivery Systems. Boca Raton, FL: CRC Press, 1990, p 105.

17. BJ Aungst. Novel formulation strategies for improving oral bioavailability of drugs with poor membrane permeation or presystemic metabolism. J Pharm Sci 82:979, 1993.

18. CA Squier, PW Wertz. Structure and function of the oral mucosa and implications for drug delivery. In: MJ Rathbone, ed. Oral Mucosal Drug Delivery. New York: Marcel Dekker, 1996, p 1.

19. JPF Bai, L-L Chang, J-H Guo. Targeting of peptide and protein drugs to specific sites in the oral route. Crit Rev Ther Drug Carrier Syst 12:339, 1995.

20. JF Woodley. Enzymatic barriers to GI delivery of peptides and proteins. Crit Rev Ther Drug Carrier Syst 11:61, 1994.

21. AP Sayani, YW Chien. Systemic delivery of peptides and proteins across absorptive mucosae. Crit Rev Ther Drug Carrier Syst 13:85, 1996.

22. SP Vyas, P Venugopalan, A Sood, N Mysore. Some approaches to improve bioavailability of peptides and proteins through oral and other mucosal routes. Pharmazie 52:339, 1997.

23. A Marzo. Metabolism of drugs: A reappraisal. Boll Chim Farm 13:139, 1992.

24. J Somberg, G Shroff, S Khosla, S Ehrenpreis. The clinical implications of first-pass metabolism: Treatment strategies for the 1990s. J Clin Pharmacol 33:670, 1993.

25. MG Bogaert. Clinical pharmacokinetics of nitrates. Cardiovasc Drugs Ther 8:693, 1994.

26. PA Glare, TD Walsh. Clinical pharmacokinetics of morphine. Ther Drug Monit 13:1, 1991.

27. RW Lievertz. Pharmacology and pharmacokinetics of estrogens. Am J Obstet Gynecol 156:1289, 1987.

28. JA Pieper, RL Slaughter, GD Anderson, MG Wyman, D Lalka. Lidocaine clinical pharmacokinetics. Drug Intell Clin Pharm 16:291, 1982.

29. V Ala-Hurula. Correlation between pharmacokinetics and clinical effects of ergotamine in patients suffering from migraine. Eur J Clin Pharmacol 21:397, 1982.

30. J Kanto, U Klotz. Pharmacokinetic implications for the clinical use of atropine, scopolamine and glycopyrrolate. Acta Anaesthesiol Scand 32:69, 1988.

31. J-H Guo, KM Cooklock. The effects of backing materials and multilayered systems on the characteristics of bioadhesive buccal patches. J Pharm Pharmacol 48:255, 1996.

32. MK Gutniak, H Larsson, SJ Heiber, OT Juneskans, JJ Holst, B Ahren. Potential therapeutic levels of glucagon-like peptide I achieved in humans by a buccal tablet. Diabetes Care 19:843, 1996.

33. MM Singer, RS Tjeerdema. Fate and effects of the surfactant sodium dodecyl sulfate. Rev Environ Contam Toxicol 133:95, 1995.

34. Jian-Hwa Guo. Investigating the bioadhesive properties of polymer patches for buccal drug delivery. J Controlled Release 28:272–273, 1994.

35. Jian-Hwa Guo. Preparation methods of biodegradable microspheres on bovine serum album loading efficiency and release profiles. Drug Dev Ind Pharm 20:2534–2545, 1994.

36. Jian-Hwa Guo. Bioadhesive polymer buccal patches for buprenorphine controlled delivery: Formulation, in-vitro adhesion and release properties. Drug Dev Ind Pharm 20:2809–2821, 1994.

37. Jian-Hwa Guo. Investigating the surface properties and bioadhesion of buccal patches. J Pharm Pharmacol 46:647–650, 1994.

38. Jian-Hwa Guo, KM Cooklock. Bioadhesive polymer buccal patches for buprenorphine controlled delivery: Solubility consideration. Drug Dev Ind Pharm 21:2013–2019, 1995.

39. Jian-Hwa Guo, KM Cooklock. Theoretical approaches and practical investigations in the Carbopol buccal patches for drug delivery. Drug Dev Ind Pharm 24:175–178, 1998.

40. R Anders, HP Merkle. Evaluation of laminated muco-adhesive patches for buccal drug delivery. Int J Pharm 49:231–240, 1989.

41. M-YF Lu, TL Reiland. Compositions and method for the sublingual or buccal administration therapeutic agents. US Patent 5,487,898, 1996.

42. R Gurny, J-M Meyer, N Peppas. Bioadhesive intraoral release systems: Design, testing and analysis. Biomaterials 5:336, 1984.

43. H Park, JR Robinson. Physico-chemical properties of water insoluble polymers important to mucin/epithelial adhesion. J Controlled Release 2:47, 1985.

44. RS Rivlin. The effective work of adhesion. Paint Technol. 9:215, 1944.

45. AN Gent, GR Hamed. Plast Rub Proc 3:17, 1978.

46. EH Andrews, TA Khan, HA Majid, J Mater Sci 20:3621–3630, 1985.

47. MA Longer, HS Ch'ng, JR Robinson. J Pharm Sci 74:406–411, 1985.

48. JH Jaffe, WR Martin. Opioid analgesics and antagonists. In: AG Gilman, TW Rall, AS Nies, P Taylor, eds. The Pharmacological Basis of Therapeutics. New York: Pergamon Press, 1990, pp 485–521.

21

Vaginal Delivery of Calcitonin by Hyaluronic Acid Formulations

Julie L. Richardson
Pfizer Ltd., Sandwich, Kent, England

Trevor I. Armstrong
Wyeth-Ayerst Research, Gosport, England

I. INTRODUCTION

The intravaginal route has frequently been used for the delivery of therapeutic and contraceptive agents to exert a local effect, such as antifungal agents, steroid hormones, and spermicides. The vagina also has considerable potential for the systemic delivery of drugs, particularly those used primarily to treat females (Table 1). Salmon calcitonin (sCT, Fig. 1) is a polypeptide hormone used in the treatment of postmenopausal osteoporosis and is normally administered by subcutaneous injection because of its low bioavailability after oral administration. To provide a more acceptable alternative to repeated injections, the delivery of calcitonin via nasal (Buclin et al., 1987), colonic (Beglinger et al., 1993), intrauterine (Golomb et al., 1993), and vaginal (Nakada et al., 1993) routes has been investigated and nasal sCT preparations are now commercially available.

In general, peptides and proteins show low bioavailability from mucosal sites because of their limited absorption and lability to mucosal enzymes. Consequently, there has been much research into methods of increasing peptide and protein uptake by the use of "penetration-enhancing" agents, enzyme inhibitors, and bioadhesive delivery systems (for reviews see Eppstein and Longenecker, 1988; Lee, 1988; Junginger, 1990; Wearley, 1991). The use of absorption-enhancing agents, such as surfactants and bile

563

Table 1 Conventional Drugs and Peptides and Proteins with Potential for Systemic Delivery by the Intravaginal Route

Therapeutic class or agent	Indication
Progestogens and estrogens	Hormone replacement therapy; contraception
Prostaglandins	Cervical dilatation; induction of labor
Bromocriptine, lisuride	Hyperprolactinemia
Oxytocin	Induction of labor; stimulation of lactation
Luteinizing hormone–releasing hormone (LH-RH) agonists	Hormone-dependent mammary tumors; induction of ovulation; contraception; idiopathic precocious puberty
Calcitonin	Postmenopausal osteoporosis

salts, while effective, may have adverse effects on mucosal integrity. Therefore, the development of biocompatible and bioadhesive delivery systems that improve retention of a drug at the site of application and thereby promote absorption offers a more attractive strategy for peptide and protein drug delivery.

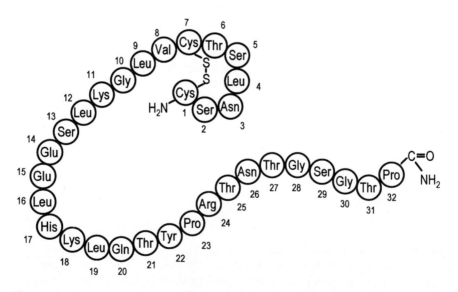

Figure 1 Salmon calcitonin, a 32-amino-acid, molecular weight 3400 polypeptide.

Fidia Advanced Biopolymers* has developed a novel class of biocom-patible and bioadhesive polymers (known as HYAFF polymers) by chemical modification of hyaluronic acid. These polymers may be processed into a wide variety of forms—microspheres, films, sponges, threads, and mem-branes—and therefore provide a versatile range of drug delivery systems. This chapter concerns the development of bioadhesive HYAFF microspheres as a novel vaginal delivery system for sCT. The potential of the vaginal route for systemic drug delivery is outlined, and some examples of thera-peutic agents and vaginal delivery systems that have been successfully used are highlighted. The characteristic features of HYAFF polymers and evi-dence for their bioadhesive properties are reviewed. The evaluation of the HYAFF/sCT vaginal delivery system is then described, from preclinical stud-ies in relevant animal models to phase I clinical studies in postmenopausal female volunteers.

II. POTENTIAL OF THE VAGINAL ROUTE FOR DRUG DELIVERY

In 1991, a survey of 185 female residents of Kingston, Rhode Island, found that nearly 50% of those questioned has used some form of intravaginal medication (Joglekar et al., 1991). Interestingly, of those who had never used an intravaginal formulation, 78% were willing to do so. In the females who used intravaginal medications, the frequency of use tended to be greater in those over the age of 45, which may reflect a greater incidence of vaginal infections in this population. With regard to the development of a vaginal sCT formulation for the treatment of postmenopausal osteoporosis, it is en-couraging to note that this route of administration was acceptable to the majority of females questioned and furthermore, that those from the older age group tended to use intravaginal medications more frequently. Although this survey sampled a relatively small number of females, the overall find-ings were not thought to be unrepresentative of attitudes of other women in the United States.

In addition to the widespread use of vaginal formulations to treat local conditions, over the past 20 years there has been significant interest in the vaginal route for the administration of drugs for systemic effects. It is now well known that the vagina is permeable to a wide range of compounds, and indeed, for drugs that are susceptible to hepatic or gut metabolism, vaginal administration may offer some advantages over oral delivery. For

---

*Fidia Advanced Biopolymers s.r.l., via Ponte della Fabbrica 3/A, 35031 Abano Terme, Italy.

peptide and protein delivery, a key advantage of the vaginal route is that prolonged contact of a drug delivery system may be more readily achieved than at other mucosal sites, such as the nose and gastrointestinal (GI) tract, where ciliary clearance and transit of food, respectively, can limit retention. The vaginal route does have some disadvantages, notably that changes in vaginal physiology may affect absorption. Some of the characteristics of the vaginal route and the implications for drug delivery are outlined in the following.

A. Physiology of the Vagina

This section focuses on the effects of changes in vaginal physiology on systemic drug absorption. Certainly, the vagina offers a substantial surface area for drug absorption. Typically, the length of the vagina is 6 to 10 cm and the surface area is increased by numerous folds in the epithelium (the rugae) and by microridges on the epithelial cell surface (Platzer et al., 1978). The rich vascular network surrounding the vagina would also be expected to facilitate drug absorption by delivering drug rapidly to the systemic circulation and thus maintaining a concentration gradient. The vaginal epithelium is covered by a film of moisture consisting mainly of cervical mucus and fluid exuded from the vascular vaginal wall. Because a drug must be in solution prior to absorption, the presence of some moisture in the vagina is advantageous. However, the volume, viscosity, and pH of vaginal fluids may all have an impact on drug absorption. For example, the secretion of thick cervical mucus may present a barrier to drug absorption. With regard to the effect of vaginal pH, many drugs are weak electrolytes and would be expected to show optimal absorption when nonionized. Usually, the pH of the vagina is between 4 and 5 and is maintained by bacterial conversion of glycogen from exfoliated epithelial cells to lactic acid. The presence of menstrual blood, semen, and cervical and uterine secretions may all serve to increase the vaginal pH.

Some of the factors affecting the absorption of drugs from the vagina are listed in Table 2. A number of interrelated changes in vaginal physiology occur throughout the lifetime that may be expected to affect the absorption of drugs. Before puberty, when the vaginal epithelium is relatively thin, the vaginal absorption of drugs may be easier to achieve. After puberty, as the thickness of the vaginal epithelium increases, the absorption of drugs may be lower and could also be affected by the ovarian cycle. Few studies have characterized the vaginal absorption of drugs at different stages of the menstrual cycle and the reports available tend to be somewhat conflicting. Some

Table 2 Factors Affecting Vaginal Absorption

Life Cycle
   Age (puberty, menarche, menopause)
   Menstrual cycle
   Pregnancy
Physiological factors
   Epithelial thickness
   Vaginal vasculature
   Vaginal secretions, cervical mucus (volume, viscosity, pH)
   Vaginal microflora
Pharmacological factors
   Hormone replacement therapy
Physicochemical properties of the drug
   Molecular weight
   Lipophilicity
   Ionization properties

early reports by Rock et al. (1947) attributed the reduced vaginal absorption of penicillin in women during the follicular phase of the menstrual cycle and the last months of pregnancy to the presence of a thicker vaginal epithelium. However, a later study by Schudmak and Hesseltine (1951) failed to show any correlation between age, the menstrual cycle, or pregnancy and penicillin absorption.

In postmenopausal females, the presence of a thin, atrophic vaginal epithelium may enhance the potential for vaginal absorption of drugs. Furthermore, the lack of a regular menstrual cycle may enable more consistent delivery. Furuhjelm et al. (1980) showed that vaginal administration of estrogens to pre- and postmenopausal women resulted in rapid increases in serum estrogen levels in the postmenopausal group but no significant change in the premenopausal females and postulated that this reflected the thinner vaginal epithelium in the former subjects. However, apart from the changes in epithelial thickness that occur after menopause, other physiological changes may adversely affect vaginal absorption. The reduced vascularity and glycogen content of the vaginal wall, which results in vaginal dryness, an increase in vaginal pH, and a greater propensity for local infections, may reduce absorption. Indeed, the vaginal absorption of progesterone in estrogen-deficient women was enhanced by concomitant estradiol therapy (Villanueva et al., 1981). Although the thickness of the vaginal epithelium may have been greater following estradiol treatment, the increased blood flow to the vagina could improve drug uptake. These observations highlight the complex interrelationship between vaginal physiology and absorption, which must be considered when developing a vaginal delivery system. From a safety perspective, it is crucial that the normal environment of the vagina, including the presence of local microflora, is not adversely affected by a vaginal drug delivery system.

B. Vaginal Absorption of Drugs

A number of reviews have described the vaginal absorption of drugs, including peptides and proteins (Benziger and Edelson, 1983; Okada, 1991; Deshpande et al., 1992; Richardson and Illum, 1992). Numerous studies have shown that steroid hormones used for hormone replacement therapy or for contraception are readily adsorbed from the vagina. Indeed, by avoiding first-class hepatic metabolism, intravaginal administration of contraceptive steroids can improve their bioavailability and reduce the risk of adverse effects on the liver (Cedars and Judd, 1987). The efficacy and acceptability of a contraceptive pill containing levonorgestrel and ethinyl estradiol were compared after oral and vaginal administration in a multicenter clinical trial in over 800 women (Coutinho et al., 1993). There were no statistically significant differences in involuntary pregnancies or treatment discontinuation rates in the two groups, indicating that vaginal administration was as acceptable and efficacious as the conventional oral route. The vaginal route has also been employed to avoid the GI side effects of bromocriptine and lisuride, both of which are usually given orally to women for the treatment of hyperprolactinemia (Kletzky and Vermesh, 1989; Tasdemir et al., 1995). Reductions in serum prolactin levels were similar after vaginal administration of lisuride (0.2 mg/kg) and oral administration (0.4 mg/kg) but no GI side effects were observed in those treated via the intravaginal route. In contrast, severe nausea and vomiting were experienced after oral dosing.

In early studies by Hwang et al. (1976, 1977) the permeability of the rabbit vaginal epithelium was assessed using model compounds exhibiting different physicochemical properties. The vaginal permeability of a series of alcohols increased with increasing chain length and lipophilicity (Hwang et al., 1976). Similarly, the permeability of relatively lipophilic steroids, such as progesterone and estrone, was greater than for the more polar molecules, hydrocortisone and testosterone (Flynn et al., 1976). The vaginal absorption of alkanoic acids in rabbits was also dependent on their lipophilicty and was affected by the pH of the dose solution (Hwang et al., 1977). From this work, a physical model was proposed for vaginal drug absorption, which described the vaginal epithelium as an aqueous diffusion layer in series with a membrane consisting of aqueous pores and lipoidal pathways. Lipophilic molecules, such as steroid hormones, were thought to be absorbed by transcellular diffusion, whereas more hydrophilic macromolecules, such as proteins and peptides, may gain entry into the systemic circulation via intercellular channels (Okada et al., 1982).

The absorption of peptides and proteins from the vagina, as from other mucosal sites, does tend to be low. Okada et al. (1982) compared the ovulation-inducing activity of the luteinizing hormone–releasing hormone analogue leuprolide after vaginal, rectal, nasal, and oral administration to rats. The absolute bioavailability of leuprolide by the vaginal route was low, approximately 3.8%, but was higher than that achieved with the other nonparenteral routes of administration. The median effective dose (ED_{50}) of leuprolide calculated was 58 ng/kg by the subcutaneous (s.c.), 1 µg/kg by the vaginal, 3.1 µg/kg by the rectal, 33.4 µg/kg by the nasal, and 112 µg/kg by the oral route. In a subsequent study in pregnant rats, serum leuprolide concentrations were compared after vaginal, rectal, nasal, and s.c. administration (Yamazaki, 1984). In this study, leuprolide was shown to be rapidly absorbed from the nasal cavity, but serum concentrations were not maintained and therefore the overall bioavailability was low. In comparison, after vaginal administration, low serum drug concentrations were sustained over a longer period. Thus, the enhanced bioavailability of leuprolide from the vaginal cavity is not necessarily indicative of greater permeability of the vaginal epithelium; instead, these results may reflect more prolonged retention of the peptide after vaginal instillation. Therefore, the use of an effective bioadhesive delivery system that maintains contact of a drug with the vaginal epithelium may further enhance absorption.

C. Intravaginal Delivery Systems

A range of pharmaceutical formulations and devices have been used to delivery drugs to the vagina. Drugs intended to act locally, such as antifungal agents, estrogens, and spermicides, are typically administered in formulations designed to give good coverage of the vaginal tract, such as creams, foams, gels, and pessaries. These formulations, however, are prone to leakage, and their efficacy may be limited by poor retention in the vagina. Furthermore, one of the findings of the survey referred to earlier (Joglekar et al., 1991) was that a principal reason for patients' discontinuation of a vaginal medication was that the product was perceived as "messy" and inconvenient to use. In general, a solid delivery system such as a vaginal tablet or pessary was considered to be easier to administer. Suffice it to say that a single-unit vaginal delivery system which disperses to cover the vaginal epithelium without undue loss via leakage, should improve the extent of drug absorption and enhance patient acceptability.

Over the past 20 years, numerous vaginal formulations have been developed for the systemic delivery of drugs. Vaginal rings made from silicone polymers, capable of releasing entrapped steroid hormones for several months, have been used for contraception (Rowe, 1990) and for hormone replacement therapy in postmenopausal women (Vartiainen et al., 1993). Although some problems have been reported, e.g., ring expulsion, interference with coitus, and localized erosion of the vaginal wall, overall the devices are well tolerated. Unfortunately, partly due to the conditions required for silicone polymerization and ring manufacture and partly due to low diffusion of hydrophilic drugs through silicone matrices, these delivery systems are unlikely to be suitable for peptide and protein drugs.

Various gel and pessary formulations have been evaluated for the vaginal delivery of prostaglandins. The efficacy and safety of two vaginal formulations of prostaglandin E_2, a low-dose gel and a controlled-release pessary, were assessed for cervical ripening and induction of labor (Smith et al., 1994). The controlled-release pessary was more effective than the lower dose gel and offered the possibility of removal in the event of uterine hyperstimulation. Another sustained-release pessary for progesterone (CRINONE) has been developed by Columbia Laboratories for a number of indications affecting pre- and postmenopausal females. The pessary, which comprises a proprietary bioadhesive delivery system, was reported to be 20-fold more bioavailable than oral progesterone and to show less variability than vaginal progesterone capsules (Columbia Laboratories, November 1996). CRINONE has been approved for marketing in the United Kingdom and the United States and is awaiting marketing approval for other European countries.

Bioadhesive gel formulations based on polyacrylic acid polymers have also been developed for local vaginal delivery of drugs. Polycarbophil, a cross-linked poly(acrylic acid) polymer, has been formulated as a gel (Replens) for the treatment of vaginal dryness in postmenopausal women. Robinson and Bologna (1994) have reported that the gel is capable of remaining on the vaginal tissue for 3 to 4 days and thus provides an excellent vehicle for the delivery of drugs such as progesterone and nonoxynol-9. A similar bioadhesive gel, prepared from a poly(acrylic acid) synthetic polymer (Carbopol® 934) has been evaluated in vitro for the controlled delivery of nonoxynol-9 and ethylenediaminetetraacetic acid (EDTA) (Lee and Chien, 1996).

The list of bioadhesive polymers that can potentially be used for the vaginal delivery of drugs is long and includes the polysaccharide hyaluronic acid (HA). The rest of this chapter will be used to describe the development of a bioadhesive vaginal delivery system based on HA.

III. HYALURONIC ACID POLYMERS AS NOVEL DRUG DELIVERY SYSTEMS

Hyaluronic acid is a naturally occurring, biocompatible and biodegradable, linear polysaccharide composed of repeating disaccharide units of glucuronic acid and N-acetylglucosamine linked by $\beta 1-3$ and $\beta 1-4$ glycosidic bonds (Fig. 2a). HA is present in all soft tissues of higher organisms and in particularly high concentrations in the synovial fluid of joints and the vitreous humor of the eye (Laurent and Reed, 1991). In addition to fulfilling structural roles by virtue of its lubricating and water-retaining properties, evidence is mounting that HA plays an important part in a number of biological processes such as cell motility, cell-cell interactions, control of the fluid content of the extracellular matrix, and scavenging of hydroxyl radicals during inflammation and angiogenesis (Laurent and Fraser, 1992; Knudson and Knudson, 1993).

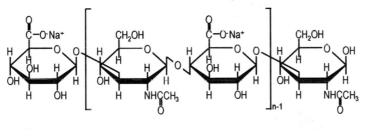

a

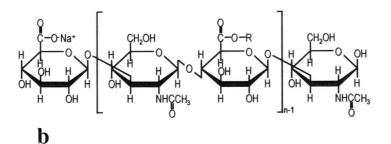

b

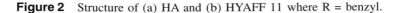

Figure 2 Structure of (a) HA and (b) HYAFF 11 where R = benzyl.

Since its discovery, much attention has focused on the possible biomedical applications of highly purified HA fractions. The biological properties of HA and its clinical use have been reviewed by Abatangelo et al. (1994). HA has been widely used in ophthalmic surgery and viscosupplementation and has been proposed for various other medical applications such as wound healing, adhesion prevention in abdominal and gynecological surgery, the development of artificial organs, and, more recently, drug delivery. To some extent, the widespread exploitation of HA has been limited by the fact that unmodified HA exists as an aqueous gel that cannot be processed into alternative physical forms and that has a short residence time because of biodegradation. In comparison, materials based on chemically modified HA, such as the HYAFF polymers, can retain the desirable biological properties of HA but present different physicochemical characteristics, which allow them to be processed into a variety of physical forms.

A. Physicochemical Properties

The HYAFF polymers (Fig. 2b) are obtained by a simple esterification reaction of the carboxyl group of the glucuronic moiety of HA with either linear or aromatic alcohols (Benedetti et al., 1990). In addition to varying the type of ester produced, the degree of esterification, i.e., the number of carboxylic groups that are esterified, can be controlled. For example, HYAFF 11 and HYAFF 11p75 denote benzyl esters of HA that have 100 and 75% of the carboxylic groups esterified, respectively. The physicochemical properties of the HYAFF polymers are thus governed by the hydrophobicity of the alcohol residue and the degree of esterification. As expected, the aqueous solubility of the polymers decreases with more hydrophobic alcohol residues and also with greater degrees of esterification, while, conversely, solubility in organic solvents increases.

HYAFF polymers are subject to enzymatic degradation and thus, like HA, they are totally biodegradable. In vitro studies have shown that degradation of HYAFF occurs via hydrolysis of the ester bond to release free alcohol and HA. Both species would then be cleared in vivo by normal metabolic pathways. The rate of degradation of the HYAFF polymer is again dependent on the hydrophobicity of the polymer. Studies in rats have shown that the choice of different HYAFF polymers for the fabrication of implants can provide residence times ranging from several weeks to several months (Benedetti et al., 1993).

One of the attractive features of the HYAFF polymer series is that they retain the good biocompatibility properties of HA itself. The biocompatibility of HYAFF polymers has been assessed in a number of in vitro assays and in vivo models. In vitro studies of the interactions of HYAFF materials with different cell types, such as fibroblasts, macrophages, and neutrophils, have demonstrated good cell adhesion and spreading on the polymers (Campoccia et al., 1993). Similarly, HYAFF polymers have shown good biocompatibility after intraperitoneal implantation in rats (Benedetti et al., 1993).

In contrast to native HA, HYAFF polymers may be processed into a versatile range of materials by exploiting their solubility in aprotic solvents (Benedetti, 1994). By extruding the polymers into organic solvents, fibers or sheets may be formed that are subsequently processed into woven or nonwoven materials and membranes. These novel materials have found a variety of biomedical applications. HYAFF membranes have been used as wound dressings (Davidson et al., 1993) and as supports for keratinocyte culture in the treatment of severe burns and leg ulcers (Andreassi et al., 1991; Hollander et al., 1996). Alternatively, spongelike materials may be obtained by lyophilization that have potential for ear, nose, and throat biomedical applications (Hellström et al., 1994). Furthermore, a combination of these technologies has been used to yield composite materials such as reinforced membranes and tubular structures for periodontal and urological applications. In addition to the biomedical applications of HYAFF polymers already described, these materials have been investigated as novel bioadhesive systems for drug delivery, mainly after preparation as microspheres.

B. Preparation and Characterization of HYAFF Microspheres

Microspheres of HYAFF may be prepared by a solvent extraction method (Benedetti et al., 1990), which is illustrated in Fig. 3. A number of peptide drugs have been successfully incorporated in HYAFF microspheres: nerve growth factor (Ghezzo et al., 1992), insulin (Illum et al., 1994) and sCT (Rochira et al., 1996). The preparation and characterization of HYAFF 11 microspheres containing sCT have been reported in detail by Rochira et al. 1996). In brief, the HYAFF 11/sCT microspheres obtained (Fig. 4) tend to be about 7 μm in diameter with smooth surfaces and a fairly narrow size distribution. Chemical characterization of HYAFF 11/sCT microspheres has shown that the polymer molecular weight and degree of esterification are unaffected by the microsphere preparation process and the concentrations of residual solvents are low. The amount of sCT recovered from the microspheres was found to be between 75 and 80% of the theoretical drug load-

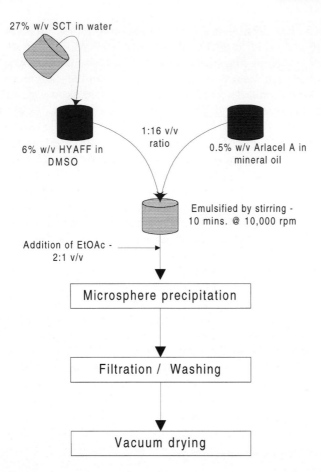

27% w/v SCT in water

6% w/v HYAFF in
DMSO

1:16 v/v
ratio

0.5% w/v Arlacel A in
mineral oil

Emulsified by stirring -
10 mins. @ 10,000 rpm

Addition of EtOAc -
2:1 v/v

Microsphere precipitation

Filtration / Washing

Vacuum drying

Figure 3   Schematic representation of preparation of HYAFF microspheres.

ing, confirming that the incorporation of the peptide was relatively high. Finally, the biological activity of the sCT after extraction from the microspheres has been assessed in rats (assay of the biological activity of sCT, British Pharmacopoeia 1980) to determine the effect of the manufacturing process on the integrity of the peptide. The biological activity of the sCT recovered from the microspheres correlated closely with the concentration assayed by high-performance liquid chromatography, confirming that there was no loss in potency of the peptide after exposure to the manufacturing conditions.

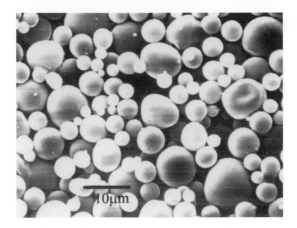

Figure 4 Photomicrograph of HYAFF 11 microspheres.

C. In Vitro Bioadhesive Properties

The in vitro bioadhesive properties of HA were demonstrated in 1984 by Park and Robinson who observed strong binding of HA to isolated human conjunctival cells. In further experiments by Saettone et al. (1989), the mucoadhesive potential of HA and an ethyl ester of HA was compared in vitro with that of a known bioadhesive, Carbopol. A tensile apparatus was used to measure the force required to separate two mucin-coated surfaces sandwiching the polymer samples. The detachment forces measured were high and were comparable for HA, the ethyl ester of HA, and Carbopol 941, the latter demonstrating the best adhesive performance of the poly(acrylic acid) series tested. The in vitro findings have been reflected in a number of subsequent studies in animal models, where HA has been found to increase the bioavailability of drugs delivered intranasally (Morimoto et al., 1991) and ophthalmically (Saettone et al., 1989, 1991).

In more recent studies, HA was again shown to exhibit bioadhesive properties similar to those of Carbopol (Pritchard et al., 1996) in several in vitro models. Detachment weight studies were carried out on various HA and HYAFF samples sandwiched mucus, gastric, small intestinal, colonic, rectal, and vaginal epithelia. In addition, mucociliary transport rate studies were performed using the excised upper palate of the frog. Overall, HA was found to be an excellent adhesive, showing effects on detachment weights and mucociliary transport rates comparable to those of Carbopol. The HYAFF polymers displayed more moderate bioadhesive properties, with significantly lower effects on detachment weight and mucociliary transport.

This decrease in the bioadhesive properties is thought to reflect the reduced potential of the HYAFF esters compared with HA for hydrogen bond formation. Nevertheless, the HYAFF polymers have clearly showed marked adhesive properties in several in vitro models.

IV. PRECLINICAL ASSESSMENT OF HYAFF/CALCITONIN MICROSPHERES

Following the successful entrapment of sCT in HYAFF microspheres without loss of biological activity, a series of experiments were conducted to assess their performance in relevant animal models. The relative merits of different animal models for the evaluation of intravaginal delivery systems have been discussed elsewhere (Richardson and Illum, 1992). The rat is a useful small animal model for preliminary studies of vaginal absorption. The main disadvantage of the rat and other rodents is that profound changes in the thickness and histology of the vaginal epithelium occur throughout the 4- to 5-day estrous cycle that can markedly affect the vaginal absorption of drugs (Okada et al., 1983, 1984). The sheep also provides an excellent large animal model for evaluation of intravaginal delivery systems because of similarities in vaginal anatomy and physiology to the human female (Richardson et al., 1992a). Furthermore, the docile nature of the sheep enables studies to be performed in the conscious animal without sedation or significant restriction of movement. In the experiments reported here, the vaginal absorption of sCT from different HYAFF 11 formulations was determined in the rat and sheep. In studies in rat, the animals were ovariectomized (OVX) and then treated with a fixed dose of estradiol to ensure consistent vaginal physiology. The resultant thin but multilayered vaginal epithelium was thought to provide a reasonable model for the postmenopausal female (Richardson et al., 1992b).

A. Absorption Studies in the Rat and Sheep

The absorption of sCT was compared after vaginal administration of an sCT solution and HYAFF 11/sCT microspheres to anesthetized rats. The HYAFF 11/sCT microspheres were administered as a dry powder and as a pessary formulation. Because of the difficulties in determining plasma sCT concentrations, particularly with limited volumes of blood samples, plasma calcium levels were measured to estimate sCT absorption. A reduction in plasma calcium levels is a specific marker of the pharmacological activity of calcitonin (Gennari et al., 1981) and therefore provides a useful and accurate pharmacodynamic measure of absorption (Di Perri et al., 1992). Vaginal administration of the HYAFF 11/sCT microspheres either as a dry powder or in a pessary resulted in more pronounced decreases in plasma calcium levels than the sCT solution (Table 3). The mean maximal change in plasma calcium concentrations (C_{max}) was approximately twofold greater after administration of the HYAFF 11/sCT microspheres than after an equivalent dose of sCT instilled as a solution.

The enhanced absorption of sCT from the HYAFF 11 formulations is thought to be due to the close contact achieved with the microspheres and the vaginal epithelium, resulting in high local drug concentrations and thus an increased gradient for absorption. Microscopic examination of the rat vagina 6 hours after administration of the HYAFF 11 formulations has shown that the microspheres are indeed closely attached to the epithelial surface (Fig. 5). Although some crenation of the surface epithelial cell layer was observed after vaginal administration in the rat, no adverse histological effects were observed. In addition to providing high local drug concentrations, it has been proposed that bioadhesive microspheres may enhance absorption (when applied nasally) by inducing transient widening of intercellular junctions (Illum and Davis, 1992). This effect has been demonstrated in Caco-2 cell monolayers by Ryden and Edman (1992), who found that tight junctions were wider after application of dry starch microspheres and the uptake of model compounds was increased. It was suggested that as starch microspheres imbibe water and swell, epithelial cells become dehydrated and drug absorption through intercellular junctions may be increased.

Table 3 Comparison of Hypocalcemia after Vaginal and Subcutaneous Administration of sCT in the Rat[a]

Administration route, formulation, and sCT dose	Mean maximal fall, C_{max} (% of basal ± SD)	Mean time of maximal fall, T_{max} (minutes ± SD)
Vaginal sCT solution, 100 IU/kg	12 ± 2.6	195 ± 30
Vaginal HYAFF 11/sCT dry powder, 100 IU/kg	24 ± 3.6	135 ± 30
Vaginal HYAFF 11/sCT pessary, 100 IU/kg	20 ± 3.6	252 ± 70
Subcutaneous sCT solution, 10 IU/kg	23 ± 3.4	264 ± 48

[a]Mean maximal fall in plasma calcium (expressed as percent change from basal levels) and time of maximal fall were calculated from individual rat data (groups of $n = 4$).

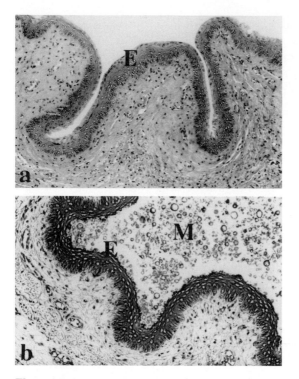

Figure 5 Photomicrographs of transverse sections through the vaginal tissues of the rat after administration of HYAFF 11/sCT microspheres. (a) Control group. The vaginal epithelium (E) comprises five to six cell layers organized in the form typical of a stratified squamous epithelium. (b) HYAFF 11/sCT microspheres administered in a Suppocire BS$_2$X pessary. Microspheres (M) are well dispersed through the vaginal lumen and form a layer on the epithelial surface (E). (Magnification ×200.) (From Richardson et al., 1995.)

The vaginal absorption of sCT was compared after administration of the HYAFF 11/sCT microspheres as a dry powder and in different pessary formulations. The vaginal pessaries were intended to provide a single-unit delivery system that would be easy to administer but would not interfere with the adhesion of HYAFF 11 microspheres to the vaginal epithelium. In addition, it was considered important that the delivery system would not aggravate vaginal dryness, a condition commonly experienced after menopause. A number of pessary vehicles were assessed for the delivery of HYAFF 11/sCT microspheres, but of those tested, Suppocire BS$_2$X (Gattefosse) was the only base that did not reduce sCT absorption (Richardson et al., 1995). Suppocire BS_2X consists of a mixture of semisynthetic polyethylene triglycerides that melt at $36-38°C$ and form a fine emulsion on contact with the aqueous environment of the vagina, thus encouraging the dispersion of the HYAFF 11 microspheres and the release and absorption of sCT. Incorporation of the HYAFF 11/sCT microspheres in a pessary tended to reduce aggregation of the microspheres and improve their dispersion throughout the vaginal tract. Figure 5 shows that the adhesion of the HYAFF 11 microspheres to the vaginal surface was not impaired by the Suppocire BS_2X vehicle. The absorption data collected also supported this observation because decreases in plasma calcium were similar after administration of HYAFF 11/sCT as a dry powder and in the Suppocire BS_2X pessary formulation (Table 3). The time of maximal hypocalcemic effects (T_{max}) was slightly longer for the pessary, which may reflect a short delay in melting and microsphere release. Interestingly, the profiles of plasma calcium concentrations with time were very similar after vaginal administration of the HYAFF 11/sCT pessary (100 IU/kg) and s.c. injection of sCT solution (10 IU/kg) (Fig. 6). These results suggest that absorption of sCT from the vagina

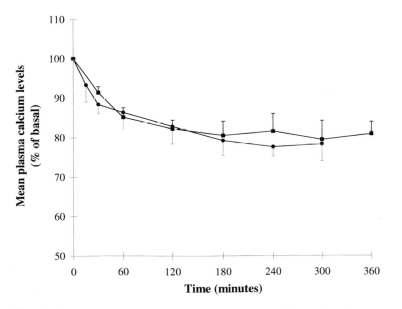

Figure 6 Mean changes in plasma calcium concentrations in rats ($n = 5$, $\pm SD$) after (■) intravaginal administration of HYAFF 11/sCT microspheres (100 IU/kg) in a Suppocire BS_2X pessary and after (●) s.c. injection of sCT solution (10 IU/kg). (From Richardson et al., 1995.)

is extremely rapid and that the vaginal HYAFF 11/sCT pessary can provide a pharmacodynamic profile similar to that of s.c. injection, albeit at a higher dose.

After the encouraging results obtained in the rat, further absorption studies were conducted in the sheep. The use of freely moving animals provided a more challenging and realistic model for the assessment of bioadhesive formulations than use of the anesthetised rat in the experiments described previously. In the first study, the vaginal absorption of sCT from a solution and from HYAFF 11/sCT microspheres was compared (Table 4, Fig. 7). As in the rat, the vaginal absorption of sCT in the sheep was significantly enhanced by the HYAFF 11/sCT microspheres. After administration of sCT solution, mean plasma calcium levels decreased by 13% of basal concentrations, and after administration of the HYAFF 11/sCT microspheres,

Table 4 Comparison of Hypocalcemia after Vaginal and Parenteral Administration of sCT in the Sheep[a]

Administration route, formulation, and sCT dose	Mean maximal fall, C_{max} (% of basal, $\pm$SD)	Mean time of maximal fall, T_{max} (minutes $\pm$ SD)
Study 1		
Vaginal "blank" HYAFF 11 powder, control	1.5 $\pm$ 1.3	106 $\pm$ 4
Vaginal sCT solution, 40 IU/kg	13 $\pm$ 3.2	165 $\pm$ 63
Vaginal HYAFF 11/sCT dry powder, 40 IU/kg	22 $\pm$ 3.4	390 $\pm$ 130
Intravenous sCT solution, 4 IU/kg	18 $\pm$ 8.2	360 $\pm$ 65
Study 2		
Intranasal HYAFF 11/sCT dry powder, 40 IU/kg	14 $\pm$ 3.1	375 $\pm$ 90
Vaginal HYAFF 11/sCT (100 mg) pessary, 40 IU/kg	20 $\pm$ 2.8[b]	480 $\pm$ 85
Vaginal HYAFF 11/sCT (200 mg) pessary, 40 IU/kg	24 $\pm$ 4.0[c]	495 $\pm$ 76
Vaginal HYAFF 11/sCT (300 mg) pessary, 40 IU/kg)	19 $\pm$ 5.6[b]	465 $\pm$ 158

[a]Mean maximal fall in plasma calcium (expressed as percent change from basal levels) and time of maximal fall were calculated from individual sheep data (groups of $n = 4$).
[b]Significantly more effective than the intranasal HYAFF 11/sCT formulation (unpaired Student's t-test, $P < .05$).
[c]Significantly more effective than the intranasal HYAFF 11/sCT formulation (unpaired Student's t-test, $P < .01$).

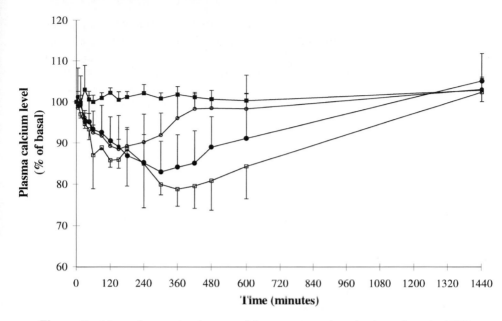

Figure 7 Mean changes in plasma calcium concentrations in sheep ($n = 4$, $\pm SD$) after (●) i.v. injection of sCT (4 IU/kg) solution and vaginal administration of (■) blank HYAFF 11, (○) sCT (40 IU/kg) solution, and (□) HYAFF 11/sCT (40 IU/kg) powder.

calcium concentrations fell by 22%. The profile of plasma calcium concentrations with time was similar after vaginal administration of 40 IU/kg HYAFF 11/sCT microspheres and intravenous (i.v.) injection of 4 IU/kg of sCT. Again, these results showed the vaginal absorption of sCT to be rapid and enhanced by incorporation in HYAFF 11 microspheres.

The vaginal absorption of sCT from the HYAFF 11/sCT formulations was investigated further in sheep after administration of vaginal pessaries (based on Suppocire BS_2X) containing different quantities of HYAFF 11 microspheres but the same dose of sCT. Mean maximal changes in plasma calcium after administration of HYAFF 11/sCT microspheres in a pessary were approximately 20% of initial concentrations, indicating that the pessary vehicle did not impair the performance of the bioadhesive microspheres (Table 4). Modification of the amount of HYAFF 11 microspheres incorporated in each pessary also produced some interesting results. Changes in plasma calcium levels were more pronounced after administration of pessaries containing 100 IU of sCT in 200 mg of microspheres than after administration of 100 IU of sCT in 100 mg of HYAFF 11 (Fig. 8). It is possible that the enhanced absorption with the former formulation was due to the increased surface area of the HYAFF 11/sCT microspheres. However, a third pessary formulation containing 300 mg of HYAFF 11 microspheres (100 IU of sCT) did not provide further enhancement of hypocalcemic effects, which may reflect the attainment of an optimal polymer-to-drug ratio.

In this second study, the absorption of sCT from the vaginal pessary formulations was also compared with that after nasal administration of the HYAFF 11/sCT dry powder. Intravaginal administration of the HYAFF 11/sCT formulations was found to be significantly more effective than intranasal administration in terms of the effects on plasma calcium concentrations (Table 4, Fig. 8). As a number of clinical studies have demonstrated the beneficial effects of intranasal calcitonin formulations in the treatment of postmenopausal osteoporosis (Overgaard et al., 1992; Perrone et al., 1992), our results showing enhanced absorption of sCT from the vagina relative to intranasal administration were highly encouraging.

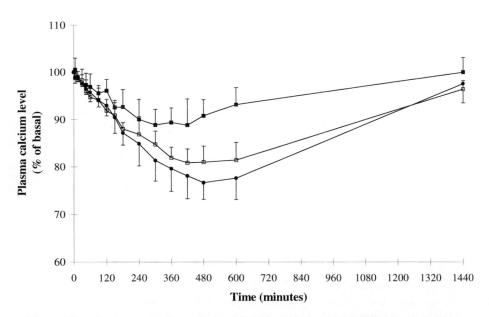

Figure 8   Mean changes in plasma calcium concentrations after intravaginal (ivg) and intranasal (i.n.) administration of different HYAFF 11/sCT (40 IU/kg) formulations in the sheep ($n = 4$, $\pm$SD); (■) i.n. HYAFF 11/sCT powder, (□) ivg HYAFF 11/sCT (100 mg) pessary, (●) ivg HYAFF 11/sCT (200 mg) pessary.

In order to investigate the potential efficacy of an intravaginal HYAFF 11/sCT delivery system further, the skeletal effects of sCT formulations were investigated in an experimental rat model for osteoporosis.

B. Effects on Bone Density in the Rat

The OVX rat has been extensively characterized as a model for bone loss in postmenopausal women and consequently is used to test the therapeutic efficacy of drugs against osteoporosis (Kalu, 1991; Wronski and Yen, 1991; Frost and Jee, 1992). Both OVX rats and postmenopausal females show rapid bone loss (osteopenia) and increased bone turnover in the early estrogen-deficient state. The treatment of OVX rats with sCT by s.c. injection over a 6-week period was shown to depress bone turnover and prevent the development of osteopenia, which was consistent with the protective effects of the peptide seen in early postmenopausal women (Wronski et al., 1991). Therefore, to investigate the potential therapeutic effects of intravaginal HYAFF 11/sCT formulations, the prevention of osteopenia development in OVX rats was assessed (Bonucci et al., 1995). In this study, the histomorphometry of the OVX rat tibia was characterized after daily treatment for 60 days with intravaginal or intramuscular (i.m.) sCT. The HYAFF 11/sCT microspheres were administered intravaginally as a dry powder and as a pessary at a dose of 50 IU/kg/day, and an sCT solution was injected i.m. at 10 IU/kg/day. In addition, two control groups were used that did not receive any sCT therapy: an OVX group and a sham-operated group.

At the end of the 60-day treatment, the proximal third of the right tibia of each animal was taken for histomorphometric analysis. A number of parameters were measured to assess bone loss: bone volume/tissue volume (BV/TV), i.e., the percentage of bone tissue occupied by trabeculae; the trabecular thickness; the trabecular number; and the trabecular separation. The numbers of polynucleated and mononucleated osteoclasts were counted to measure bone resorption, and bone formation was assessed by measurement of the daily mineral apposition rate.

The results of these studies were extremely encouraging. Daily treatment with the sCT formulations, regardless of the route of administration, largely prevented the bone loss associated with ovariectomy. Compared with the untreated OVX rats, the BV/TV ratio, the trabecular number, and the trabecular thickness were significantly increased in animals receiving sCT (Fig. 9). Both osteoclastic resorption and the increased mineral apposition rate seen after ovariectomy were significantly reduced after sCT treatment. There was no statistically significant difference in the effects of intravaginal and i.m. administration of sCT (at doses of 50 and 10 IU, respectively). It was concluded that vaginal administration of HYAFF 11/sCT formulations,

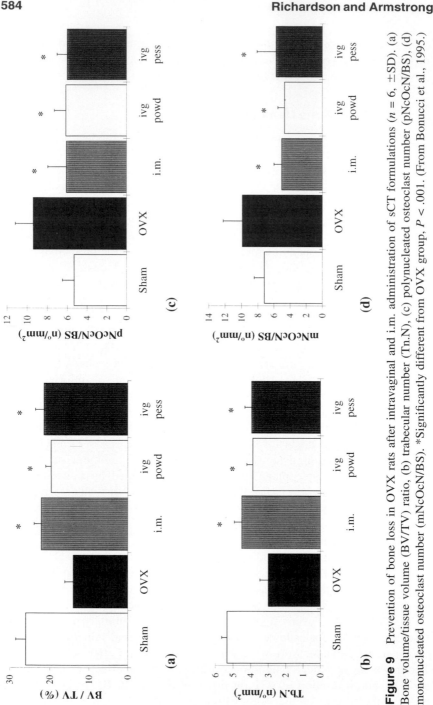

Figure 9 Prevention of bone loss in OVX rats after intravaginal and i.m. administration of sCT formulations ($n = 6$, ±SD). (a) Bone volume/tissue volume (BV/TV) ratio, (b) trabecular number (Tn.N), (c) polynucleated osteoclast number (pNcOcN/BS), (d) mononucleated osteoclast number (mNcOcN/BS). *Significantly different from OVX group, $P < .001$. (From Bonucci et al., 1995.)

as well as i.m. injection of sCT, prevented bone loss due to ovariectomy, principally by inhibiting osteoclastic resorption.

C.   Bioadhesive Studies in Sheep

The intravaginal absorption studies in the rat and sheep certainly provided some encouraging data on the bioadhesive properties of the HYAFF 11 microspheres. In further studies in sheep, the distribution and retention of HYAFF 11 microspheres after vaginal administration were assessed in more detail by gamma scintigraphy (Richardson et al., 1996). The principal objective of these experiments was to determine whether bioadhesive microspheres could be retained in the vagina for up to 12 hours after administration to a conscious, upright animal. In addition, it was of interest to investigate the distribution of the microspheres in the genital tract, particularly after administration as a dry powder. Previous studies in rats (Henderson et al., 1986) had demonstrated that talc particles placed in the vagina and uterine horns were capable of migrating to the ovary and were found embedded in ovarian tissue. Although similar experiments in the monkey failed to show any movement of talc from the vagina to the uterus or ovaries (Wehner and Weller, 1986), talc has been suspected to play an etiologic role in epithelial ovarian cancer in women (Whittmore et al., 1988). Therefore, although HYAFF polymers are biodegradable, yielding toxicologically acceptable breakdown products, it was of interest to characterize their distribution after intravaginal application.

In a preliminary experiment, gamma-scintigraphy methods were developed by DanBioSyst* to image and quantitatively measure the distribution of vaginal formulations in the sheep. Images of the entire vagina were taken after administration of a radiolabeled gel and the data were used as a reference for subsequent comparison with the radiolabeled HYAFF 11 formulations. Three regions of interest (ROIs) were defined for each scintigraphic measurement taken: ROI background; ROI equivalent to the area of the vagina; and ROI corresponding to the area of initial deposition of the microspheres. ^{99m}Tc-labeled HYAFF 11 microspheres were then administered, as a dry powder or in the Suppocire BS_2X pessary formulation, and images were taken at intervals over 12 hours. In each case, the formulations were administered to three animals.

Images of the distribution and spread of radioactivity showed that immediately after administration, the ^{99m}Tc-labeled HYAFF 11 microspheres

---

*Professor L. Illum, DanBioSyst UK Ltd., Albert Einstein Centre, Highfields Science Park, Nottingham, UK.

(dosed as both powder and pessary) were spread along the length of the vaginal tract. After 12 hours, the microspheres were still present in the vagina, although the dry HYAFF 11 formulation tended to occupy a slightly smaller area as the experiment progressed, which was thought to reflect some gelling and aggregation of the particles.

The relative intensity of radioactivity in each region of interest was calculated to assess quantitatively the distribution and retention of the HYAFF 11 formulations. To compare the distribution of the HYAFF 11 powder and pessary, the radioactivity in the initial deposition area with time was expressed as a percentage of the radioactivity in the total vaginal area (Fig. 10). The level of radioactivity with time was also expressed as a percentage of the reading taken immediately after dosing to determine the retention of the microspheres. After administration of the dry HYAFF 11 powder, approximately 93% of the initial radioactivity was evident 12 hours after dosing, demonstrating that the microspheres were essentially retained within the vagina for the duration of the experiment (Richardson et al., 1996). Figure 10 illustrates that 85% of this radioactivity was measured at the initial area of deposition, thus confirming minimal migration of the formulation from the application site. The retention of the ^{99m}Tc-HYAFF 11

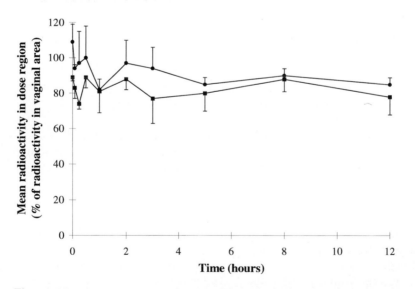

Figure 10   Retention of HYAFF 11 microspheres after vaginal administration as a powder (●) and pessary (■) to sheep. Mean radioactivity in dose region with time, expressed as a percentage of total vaginal radioactivity ($n = 3$). (From Richardson et al., 1996.)

microspheres administered in the pessary formulation was slightly lower than that measured for the dry powder. At 12 hours, approximately 64% of the initial radioactivity was measured, with 78% of this remaining within the initial dose area.

These experiments clearly demonstrated the bioadhesive properties of both HYAFF 11 microsphere formulations after intravaginal administration in a fairly challenging in vivo model. The retention of the microspheres was higher after administration as a dry powder compared with the pessary formulation (93% retained at 12 hours versus 64%), probably due to some leakage of the molten vehicle from the vagina. An increase in the ratio of microspheres to the pessary base and/or administration of the dosage form when supine could perhaps improve retention.

In terms of the distribution of the microspheres after dosing, there was no evidence of movement of the radiolabeled microspheres from the vagina to the upper areas of the genital tract. Indeed, a high percentage of the HYAFF 11 microspheres remained within the initial area of deposition. In this study, the DanBioSyst sheep model and the gamma-scintigraphy methodology provided a novel and highly effective tool for studying the spreading and bioadhesive properties of the HYAFF 11 microsphere system.

After the encouraging preclinical evaluation of HYAFF 11/sCT microsphere formulations, which demonstrated vaginal absorption of sCT in two animal models and prevention of bone loss in a rat model of osteoporosis, a series of clinical studies were conducted to assess the safety and toleration of this novel delivery system in healthy volunteers.

V. INTRAVAGINAL ADMINISTRATION OF HYAFF/CALCITONIN MICROSPHERES TO HEALTHY VOLUNTEERS

Osteoporosis is the most common metabolic bone disorder and is characterized by an increased tendency to fractures related to subnormal bone mass. Postmenopausal women are particularly prone to osteoporosis, and one treatment for established osteoporosis is the administration of calcitonin, which is thought to prevent further bone loss by inhibition of bone resorption. sCT is a synthetic peptide (Fig. 1) with a structure closely related to that of human calcitonin and is often used in preference because of its greater potency and longer duration of action (Singer, 1991). As discussed earlier, in common with other polypeptide drugs, sCT is normally administered parenterally. Unfortunately, unpleasant side effects have been reported in up to 65% of patients receiving this therapy (Wuster et al., 1991; Wimalawansa, 1993). The principal adverse effects of sCT treatment are vascular symptoms, such as flushing, and GI effects, such as nausea and vomiting. To some extent, this has limited the use of sCT and prompted investigations into novel routes of administration. A number of nasal spray formulations of sCT are commercially available but these products have not experienced wide success in some countries, mainly because they have been associated with rather low bioavailability (Lee et al., 1994). Indeed, in 1995, low-dose calcitonin nasal sprays were removed from the Italian market because of lack of evidence for efficacy (SCRIP, 1995).

In the following, studies to assess the intravaginal absorption of sCT in healthy volunteers by measurement of plasma sCT concentrations are described. As before, the HYAFF 11/sCT microspheres were formulated as a single-dose pessary based on Suppocire BS_2X.

At the time of writing, two clinical studies have been completed and a third is in progress. Initially, a dose finding study was performed to investigate the tolerability and bioavailability of the HYAFF 11/sCT pessary at two doses (200 and 400 IU), relative to a placebo pessary and s.c. injection (100 IU). The second study was designed to compare the tolerability and bioavailability of the HYAFF 11/sCT pessary (100 IU) with a commercially available nasal spray (100 IU). A third study is ongoing in a larger group of subjects (n = 18) to compare the HYAFF 11/sCT pessary (200 IU) with 100 IU of sCT administered by s.c. injection. The protocols and results of the first two clinical studies are outlined here.

A. Study Protocols

The clinical studies were carried out at Medeval Clinical Unit (Manchester, UK) in groups of 12 healthy postmenopausal female volunteers. All subjects were less than 70 years of age, and for the purposes of these studies, postmenopausal was defined as a period of two or more years since the volunteer's last menstrual period. Subjects were excluded from the investigation if they had a know intolerance or allergy to sCT or had any acute or chronic illness (including osteoporosis) or any genitourinary disease.

The first study followed a four-way, randomized crossover design, with each volunteer receiving a single dose of each formulation, with an interval of at least 1 week before dosing. The four treatments consisted of an intravaginal HYAFF 11/sCT at 400 IU and 200 IU, a placebo pessary containing blank HYAFF 11 microspheres, and an s.c. dose of 100 IU of sCT (Calsynar, Armour). In the second study, two treatments were evaluated: the HYAFF 11/sCT pessary (100 IU) and Calcitonina 100 IU nasal spray (Sandoz). The pessary formulations were inserted vaginally using an applicator to volunteers in a supine position; subjects then remained lying down for at least 1 hour after administration.

Blood samples were collected prior to dosing and at intervals up to 24 hours after administration. Plasma sCT determinations were carried out using a commercial radioimmunoassay (RIA) kit (Peninsula Laboratories). Although the kit was sensitive to sCT, there was some cross-reactivity for endogenous human calcitonin. In terms of pharmacokinetic analysis, for each subject, the mean of the two predose sCT determinations was used as the baseline value (C_0). The C_{max} and T_{max} were direct experimental observations. The area under the sCT concentration-time curve from 0 to 24 hours (AUC_{0-24}) was calculated using the linear trapezoidal approximation. Baseline-corrected C_{max} and AUC_{0-24} (BC C_{max} and BC AUC_{0-24}) were calculated by subtracting C_0 from C_{max} and by subtracting $C_0 \times 24$ from AUC_{0-24}, respectively. However, as will be discussed, there was a high degree of variability in baseline concentrations (C_0), which hindered an accurate calculation of relative bioavailabilities. In the first study, two methods were employed to estimate bioavailability: (a) the ratio of BC AUC_{0-24} for the vaginal versus s.c. treatments was calculated for each individual, or (b) the ratio of AUC_{0-24} for vaginal versus s.c. treatments was calculated after subtraction of the AUC_{0-24} obtained with the placebo. In both cases, the data were calculated for each individual subject and the ratios were normalized for the dose of sCT, which was 70, 35, and 25 μg for the vaginal 400 IU, vaginal 200 IU, and s.c. 100 IU treatments. In the second study, the relative doses were 17.5 and 20 μg for the vaginal 100 IU and nasal 100 IU treatments.

B. Dose-Finding Study

As reported in the literature, during this first clinical study, many subjects suffered gastric and vascular side effects following s.c. injection of sCT. Because of the severity of these side effects, only six subjects received all of the planned treatments; the remainder withdrew voluntarily or at the request of the clinical investigator. The incidence of adverse events is shown in Table 5. The most frequent adverse events reported were nausea and vomiting. The intravaginal sCT formulations tended to be better tolerated than s.c. dose and the incidence of adverse events was reduced. Somewhat surprisingly, one subject receiving the 400 IU vaginal pessary did experience severe vomiting. Examination of the plasma sCT concentrations in this volunteer did not show that sCT was unusually elevated; therefore the sensitivity of this volunteer to the sCT treatment was not explained. Encouragingly, the incidence of adverse events reported by volunteers receiving the 200 IU vaginal pessary was similar to that reported for the placebo pessary.

The plasma sCT concentration with time profiles following intravaginal and s.c. administration show notable differences that may help ex-

Table 5 Incidence of Adverse Events in Healthy, Postmenopausal Volunteers after Intravaginal and Subcutaneous Administration of sCT: Number of Events Experienced for Each Formulation

Adverse event	Subcutaneous 100 IU	Vaginal 400 IU	Vaginal 200 IU	Vaginal placebo	Total events
Nausea	4	2	1	0	7
Vomiting	6[a]	8[b]	1	0	15
Diarrhea	5	1	0	2	8
Abdominal pain	1	1	0	0	2
Headache	4	7[c]	3	4	18
Dizziness, light-headedness	5	1	1	0	7
Hot flush	2	1	1	1	5
Sweating	0	0	0	0	1
Vaginal discharge	0	1	0	0	1
Vaginal soreness, pain	0	1	0	0	1
Backache	1	0	0	0	1
Rash	1	0	0	2	3
Rigors	2	0	0	0	2
Dry skin	0	0	0	1	1
Feeling cold	0	1	0	0	1
Number of adverse events	31	24	7	10	72
Number of treatment exposures[d]	13	13	8	12	

[a]Event occurred in four subjects, with one of these subjects reporting vomiting three times.
[b]Event occurred in one subject only, who reported vomiting eight times.
[c]Event occurred in six subjects, with one of these subjects reporting headache two times.
[d]The total number of exposures to each treatment was increased when one study session was repeated after accidental thawing of the blood samples collected.

plain the lower incidence of adverse events with the former treatment (Fig. 11). After s.c. injection, peak plasma sCT concentrations were rapidly attained in all subjects within 15 to 30 minutes after dosing and then returned to near-basal concentrations after 5 hours. These sharp peak concentrations were avoided following intravaginal administration of the HYAFF 11/sCT pessaries. Peak plasma sCT concentrations were generally reached at approximately 2 hours after dosing, which probably reflected the time required for the pessary base to melt and for the peptide to diffuse from the HYAFF 11 microspheres across the vaginal epithelium. The sCT levels returned to near-basal concentrations after 10 hours. Thus, the vaginal pessaries provided a lower but more sustained plasma concentration of sCT.

The mean pharmacokinetic data for each treatment are shown in Table 6. Because of the number of volunteer dropouts and the rather variable

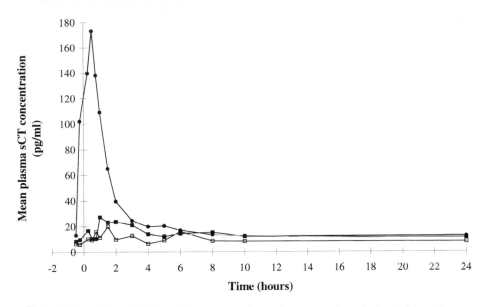

Figure 11 Mean plasma sCT concentrations after s.c. and vaginal administration of sCT to healthy, postmenopausal volunteers; (●) s.c. 100 IU, (■) ivg 400 IU, (□) ivg 200 IU, $n = 8-10$.

baseline plasma sCT concentrations measured, it was difficult to calculate the relative bioavailability of sCT from the intravaginal HYAFF 11/sCT delivery system with accuracy. Comparison of the BC AUC_{0-24} data for the intravaginal treatments and the s.c. dose gave an estimated bioavailability between 20 and 28%. However, the concentration of sCT in the predose samples (C_0) ranged from 2 to 43.8 pg/mL, which may be due to interference from exogenous calcitonin and background noise in the assay. In light of the variability in predose measurements, the bioavailability calculated from baseline corrected data should be viewed with some caution. Using an alternative means of approximation, the ratio of intravaginal and s.c. AUC_{0-24} values was calculated after subtraction of the AUC_{0-24} obtained for placebo. In this case, the bioavailability of the intravaginal sCT relative to s.c. administration was between 0 and 50%, with values of zero in the cases in which the placebo AUC was particularly high.

Despite these difficulties in assigning a value to the bioavailability of sCT from the intravaginal HYAFF 11/sCT delivery system, it was clear that plasma sCT concentrations were elevated after intravaginal administration relative to the placebo treatment. Moreover, both uncorrected and baseline corrected C_{max} and AUC were greater (and approximately dose proportional)

Table 6   Pharmacokinetic Parameters for Intravaginal and Subcutaneous Administration of sCT

Parameter	Vaginal 400 IU	Vaginal 200 IU	Vaginal placebo	Subcutaneous 100 IU
C_0 (pg mL^{-1})	8.5 (2.1)	6.2 (0.9)	20.9 (4.5)	11.6 (3.3)
C_{max} (pg mL^{-1})	46.9 (12.4)	31.9 (5.2)	27.3 (6.2)	180.0 (20.1)
T_{max} (h)[a]	1.75 (0.25–10)	1.5 (0.75–10)	1.0 (−0.5–10)	0.5 (0.25–0.5)
AUC_{0-24} (pg h mL^{-1})	319 (55)	219 (31)	306 (83)	507 (74)
BC C_{max} (pg mL^{-1})[b]	38.3 (12.6)	25.8 (5.0)	6.5 (2.0)	168.3 (20.5)
BC AUC_{0-24} (pg h mL^{-1})	115 (17)	71 (13)	−193 (67)	229 (61)
$F_{rel.sc}$[c] (a)	0.20 (0.04)	0.28 (0.13)	—	1.00
(b)	0.16 (0.18)	0.10 (0.18)	—	1.00
	($n = 9$)	($n = 6$)		

[a]All values shown represent the mean data with the standard deviation of the data shown in parentheses, with the exception of where the median of the data is reported and the range given in parentheses.

[b]Baseline corrected data.

[c]Relative bioavailability for vaginal versus s.c. route was estimated for each individual by (a) comparison of BC AUC_{0-24} data and (b) comparison of AUC_{0-24} data after subtraction of placebo data in the same subject. All data were normalized for dose.

for the 400 IU dose than for the 200 IU pessary. The mean BC AUC_{0-24} values compare well with literature reports of the absorption of sCT from intranasal formulations. For example, while accepting the high variability of baseline levels in the intravaginal study, the mean BC AUC_{0-24} values for the intravaginal 400 IU and 200 IU doses were 300 and 219 pg h mL^{-1}. Following nasal administration of 200 IU of sCT to volunteers, Buclin et al. (1987) obtained AUC values of approximately 83 pg h mL^{-1} and Thamsborg et al. (1990) reported an AUC of about 18.3 pg h mL^{-1}. In a subsequent study by Lee et al. (1994), AUC values reported after intranasal administration of a commercial sCT formulation (200 IU) were around 5 pg h mL^{-1} but were increased markedly by the use of an absorption-enhancing agent. In the same study, i.m. injection of 100 IU sCT resulted in an AUC of 190 pg h mL^{-1}, which was not dissimilar to our BC AUC value for 100 IU given subcutaneously (229 pg h mL^{-1}). Consequently, there is evidence that sCT is absorbed after vaginal administration of the HYAFF 11/sCT formulation in volunteers and, indeed, the flatter plasma profile after intravaginal delivery (mean T_{max} 180 minutes versus 15 minutes for the s.c. injection) may explain the improved tolerability of the intravaginal treatment

C. Comparison of Intravaginal and Intranasal Administration

In the second clinical study, the tolerability and pharmacokinetics of the vaginal HYAFF 11/sCT formulation were compared with those of a commercially available nasal spray, both administered at a dose of 100 IU. As expected from the first study and from previous data on intranasal sCT administration (Reginster and Franchimont, 1985), both treatments were well tolerated with few side effects reported (Table 7).

Table 7 Incidence of Adverse Events in Healthy, Postmenopausal Volunteers after Intravaginal and Nasal Administration of sCT: Number of Events Experienced for Each Formulation

Adverse event	Vaginal 100 IU	Nasal 100 IU	Total events
Nausea	1	0	1
Diarrhea	0	2	2
Headache	3	2	5
Vaginal bleeding	0	1	1
Number of adverse events	4	5	9
Number of treatment exposures	13	13	

The plasma sCT concentration with time profiles after vaginal and nasal administration are shown in Fig. 12 and the pharmacokinetic data are summarized in Table 8. Both vaginal and nasal administration resulted in a gradual absorption of sCT and mean T_{max} values were 1.5 and 5 hours, respectively. Again, administration of sCT by the vaginal and the nasal route avoided the initial high peak associated with s.c. injection and eliminated the unpleasant side effects experienced with the latter treatment.

As before, analysis of the pharmacokinetic data was hampered by variability in baseline sCT concentrations measured, which indeed tended to be higher and more variable than in the previous study. AUC values were highly variable and therefore were not used to calculate the relative bioavailability of vaginal versus nasal administration. The baseline corrected C_{max} data appeared to be more consistent for both treatments and could be used as a measure of drug absorption. As discussed by Bois et al. (1994), C_{max} can indicate the extent of drug absorption, although the absorption rate will also influence this parameter. Thus, on the basis of the BC C_{max} ratios (Table 8), absorption of sCT from the vaginal system would appear to be higher than from the nasal formulation.

A third clinical study is ongoing, utilizing a larger number of volunteers ($n = 18$), to compare the HYAFF 11/sCT pessary (200 IU) with s.c.

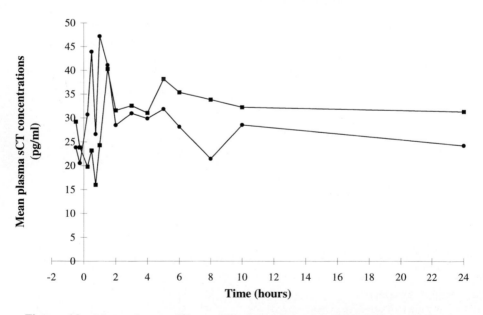

Figure 12   Mean plasma sCT concentrations after vaginal (●) and nasal (■) administration of 100 IU of sCT to healthy, postmenopausal volunteers, $n = 12$.

Table 8 Pharmacokinetic Parameters for Intravaginal and Nasal Administration of sCT

Parameter	Vaginal 100 IU	Nasal 100 IU
C_0 (pg mL^{-1})	21.1 (5.9)	29.5 (7.6)
C_{max} (pg mL^{-1})	59.4 (7.4)	48.6 (9.0)
T_{max} (h)[a]	1.5 (0.5–6)	5 (0.5–10)
AUC$_{0-24}$ (pg h mL^{-1})	662 (153)	777 (200)
BC C_{max} (pg mL^{-1})[b]	38.4 (5.5)	19.1 (4.9)
BC AUC$_{0-24}$ (pg h mL^{-1})[b]	157 (99)	69 (59)

[a]All values shown represent the mean data with the standard deviation of the data shown in parentheses, with the exception of where the median of the data is reported and the range given in parentheses.
[b]Baseline corrected data.

injection (100 IU). Plasma sCT levels will be determined using a novel antibody detection technique that offers better sensitivity than that seen previously with the RIA kit. It is anticipated that this will eliminate the problems associated with the variable baseline sCT concentrations. The data will be used to reinforce the encouraging yet variable pharmacokinetic data generated by the first two studies.

VI. CONCLUSIONS AND FUTURE PROSPECTS

In this chapter, the discussion has centered on the use of HYAFF 11 microspheres as a delivery vehicle for vaginal administration. The HYAFF 11 microspheres have demonstrated good bioadhesive properties both in vitro and in vivo. In an unconscious rat model, the microspheres maintained contact with the vaginal epithelium for at least 6 hours after administration. In a more challenging model, namely conscious and freely moving sheep, gamma-scintigraphy studies showed that HYAFF 11 microspheres were retained within the vagina for at least 12 hours. These bioadhesive properties have clear advantages for the promotion of drug delivery, supported by the vaginal absorption data with sCT. Hypocalcemic effects in the rat and the sheep confirmed that absorption of sCT was enhanced after administration of HYAFF 11/sCT microspheres compared with an aqueous sCT solution. Encouragingly, preliminary clinical data demonstrated vaginal absorption of sCT in postmenopausal volunteers with a reduced incidence of side effects. A further clinical study is ongoing to assess the performance of the HYAFF 11/sCT formulation in a larger number of volunteers.

The utility of the HYAFF polymers for drug delivery is certainly not restricted to the application described here. The number of clinically available peptide and protein drugs has increased dramatically over the past decade, but a patient-friendly and therapeutically effective system of delivery for these agents is an issue that remains open. It is well known that parenteral administration may fail to deliver an optimal pharmacokinetic profile for efficacy and certainly limits the patients' compliance. The use of bioadhesive delivery systems to enhance transmucosal absorption via nasal, buccal, pulmonary, rectal, and vaginal routes may provide an attractive alternative to parenteral therapy. In collaboration with other centers, HYAFF microspheres not only are being evaluated for the administration of other therapeutic peptides and proteins but also are under investigation as novel mucosal delivery systems for entrapped antigens, to enhance immunization by nasal, vaginal, and rectal routes.

The use of bioadhesive polymers to improve the delivery of locally acting drugs is also of great interest. The adhesive nature of HYAFF formulations is expected to increase the time of exposure of an incorporated drug and thus may improve efficacy while reducing the dose and frequency of application required. Some obvious choices for incorporation in topical formulations include steroids, analgesics, anti-inflammatory agents, and anti-infectives. With regard to the latter, there has been a great deal of interest in the development of safe and effective vaginal contraceptive–anti-infective formulations to control pregnancy and help prevent the spread of sexually transmitted disease. The favorable features of HYAFF polymers, such as their high biocompatibility, controllable degradation rate, and dosage form flexibility, make them highly attractive as bioadhesive delivery systems for localized drug delivery. Depending on the required application and site, the polymers can be formulated as microspheres, gels, sponges, films, or membranes as appropriate.

In conclusion, the HYAFF series are bioadhesive and biocompatible polymers that can be readily processed into a variety of physical forms and used to deliver systemically or locally acting drugs to different mucosal sites.

ACKNOWLEDGMENTS

The authors would like to acknowledge and thank the following: the researchers at Fidia Advanced Biopolymers, in particular, Drs. L. M. Benedetti, P. A. Ramires, E. Bigon, M. R. Rosaria, M. Rochira, and L. Callegaro; Dr. L. Bacelle at Ospedale Civile Padova; Drs. E. Bonucci and P. Ballanti at the University of Rome; Drs. L. Illum, N. F. Farraj, J. Whetstone, P. Watts, A. N. Fisher, M. Hinchcliffe, and H. Norbury at DanBioSyst; and Drs. J. D.

Davis and S. Toon at Medeval. Finally, our thanks to Mr. B. S. Williams for his help in preparing the manuscript.

REFERENCES

Abatangelo G, Brun P, Cortivo R. 1994. Hyaluronan (hyaluronic acid) and overview. Proceedings of Annual Meeting of European Society for Biomaterials, Pisa, Italy, pp 8–19.

Andreassi L, Casini L, Trabucchi E, Diamantini S, Rastrelli A, Donati L. 1991. Wounds 3:116.

Beglinger C, Born W, Muff R, Drewe J, Dreyfuss JL, Bock A, Mackay M, Fischer JA. 1993. Eur J Clin Pharmacol 43:527.

Benedetti L, Cortivo R, Berti T, Pea F, Mazzo M, Moras M, Abatangelo G. 1993. Biomaterials 14:1154.

Benedetti L. 1994. Med Device Tech 11:32.

Benedetti L, Topp EM, Stella VJ. 1990. J Controlled Release 13:33.

Benziger DP, Edelson J. 1983. Drug Metab Rev 14:137.

Bois FY, Tozer TN, Hauck WW, Chen M-L, Patnaik R, Williams RL. 1994. Pharm Res 11:715.

Bonucci E, Ballanti P, Ramires PA, Richardson JL, Benedetti L. 1995. Calcif Tissue Int 56:274.

Buclin T, Randin JP, Jacquet AF, Azria M, Attinger M, Gomez F, Burckhardt P. 1987. Calcif Tissue Int 41:252.

Campoccia D, Hunt JA, Doherty PJ, Zhong PJ, Callegaro L, Benedetti L, Williams DF. 1993. Biomaterials 14:1135.

Cedars MI, Judd HL. 1987. Obstet Gynecol Clin North Am 14:269.

Coutinho EM, Mascarenhas I, De Acosta OM, Flores JG, Gu ZP, Lapido OA, Adel-kunle AO, Otolorin EO, Shaaban MM, Oyoon MA, Kamal A, Plah A, Sikazwe NC, Segal SJ. 1993. Clin Pharmacol Ther 54:540.

Davidson JM, Beccaro M, Pressato D, Dona M. 1993. Wound Repair Regen 1:100.

Deshpande AA, Rhodes CT, Danish M. 1992. Drug Dev Ind Pharm 18:1225.

Di Perri T, Laghi Pasini F, Capecchi PL, Blardi P, Pasqui AL, Franchi M, Mazza S, Sodi N, Domini L, Ceccatelli L, Volpi L. 1992. Eur J Clin Pharmacol 43:229.

Eppstein DA, Longenecker JP. 1988. CRC Crit Rev Ther Drug Carrier Syst 5:99.

Flynn GL, Ho NFH, Hwang S, Owada E, Molokhia A, Behl CR, Higuchi I, Yot-suyanagi T, Shah Y, Park J. 1976. In: Paul DR, Harris FW, eds. Controlled Release Polymeric Formulations. Washington, DC: American Chemical Society, p 87.

Frost HM, Jee WSS. 1992. Bone Miner 18:227.

Furuhjelm M, Karlgren C, Carlstrom K. 1980. Int J Gynecol Obstet 17:335.

Gennari C, Chierichetti SM, Vibelli C, Francini G, Maioli E, Gonnelli S. 1981. Curr Ther Res 30:1024.

Ghezzo E, Benedetti L, Rochira M, Biviano F, Callegaro L. 1992. Int J Pharm 87: 21.

Golomb G, Ayramoff A, Hoffman A. 1993. Pharm Res 10:828.

Hellström S, Laurent C, Soderberg O, Spandow O. 1994. Endogenous and exogenous hyaluronan in otology. Proceedings of Annual Meeting of European Society for Biomaterials, Pisa, Italy, pp 38–43.

Henderson WJ, Hamilton TC, Baylis MS, Pierrepoint CG, Griffeths K. 1986. Environ Res 40:247.

Hollander D, Bernd A, Stein M, Muller J, Pannike A. 1996. Treatment of a circular
chronic ulcer of the lower leg after percutaneous transluminal angioplasty of
the popliteal artery in a compromised patient with cultured autologous keratinocytes on hyaluronic acid ester membranes: A case study. Proceedings of
6th European Conference on Advances in Wound Management, Amsterdam,
p 66.

Hwang S, Owada E, Suhardja L, Ho NFH, Flynn GL, Higuchi WI. 1977. J Pharm
Sci 66:781.

Hwang S, Owada E, Yotsuyanagi T, Suhardja I, Ho NFH, Flynn GL, Higuchi WI.
1976. J Pharm Sci 65:1574.

Illum L, Davis SS. 1992. Clin Pharmacokinet 23:30.

Illum L, Farraj NF, Fisher AN, Gill I, Miglietta M, Benedetti LM. 1994. J Controlled
Release 29:133.

Joglekar A, Rhodes CT, Danish M. 1991. Drug Dev Ind Pharm 17:2103.

Junginger HE. 1990. Acta Pharm Technol 36:110.

Kalu DN. 1991. Bone Miner 15:175.

Kletzky OA, Vermesh M. 1989. Fertil Steril 51:269.

Knudson CB, Knudson W. 1993. FASEB J 7:1233.

Laurent TC, Fraser JRE. 1992. FASEB J 6:2397.

Laurent UBG, Reed RK. 1991. Adv Drug Delivery Rev 7:237.

Lee C-H, Chien YW. 1996. J Controlled Release 39:93.

Lee VHL. 1988. CRC Crit Rev Ther Drug Carrier Syst 5:69.

Lee WA, Ennis RD, Longenecker JP, Bengtsson P. 1994. Pharm Res 11:747.

Morimoto K, Yamaguchi H, Iwakura Y, Morisaka K, Ohashi Y, Nakai Y. 1991.
Pharm Res 8:471.

Nakada Y, Miyake M, Awata N. 1993. Int J Pharm 89:169.

Okada H. 1991. In: Lee VHL, ed. Peptide and Protein Drug Delivery. New York:
Marcel Dekker, p 633.

Okada H, Yamazaki I, Ogawa Y, Hirai S, Yashiki T, Mima H. 1982. J Pharm Sci
71:1367.

Okada H, Yamazaki I, Yashiki T, Shimamoto T, Mima H. 1984. J Pharm Sci 73:
298.

Okada H, Yashiki T, Mima H. 1983. J Pharm Sci 72:173.

Overgaard K, Hansen MA, Jensen SB, Christiansen C. 1992. BMJ 305:556.

Park K, Robinson JR. 1984. Int J Pharm 19:107.

Perrone G, Galoppi P, Valente M, Capri O, D'Ubaldo C, Anelli G, Zichella L. 1992.
Gynecol Obstet Invest 33:168.

Platzer W, Poisel S, Hafez ESE. 1978. In: Hafez ESE, Evans TN, eds. The Human
Vagina. Amsterdam: Elsevier/North Holland Biomedical Press, p 39.

Pritchard K, Lansley AB, Martin GP, Helliwell M, Marriott C, Benedetti L. 1996. Int J Pharm 129:137.

Reginster JY, Franchimont P. 1985. Clin Exp. Rheumatol 3:155.

Richardson JL, Farraj NF, Illum L. 1992a. Int J Pharm 88:319.

Richardson JL, Illum L. 1992. Adv Drug Delivery Rev 8:341.

Richardson JL, Illum L, Thomas NW. 1992b. Pharm Res 9:878.

Richardson JL, Ramires PA, Miglietta MR, Rochira M, Bacelle L, Callegaro L, Benedetti L. 1995. Int J Pharm 115:9.

Richardson JL, Whetstone J, Fisher AN, Watts P, Farraj NF, Hinchcliffe M, Benedetti L, Illum I. 1996. J Controlled Release 42:133.

Robinson JR, Bologna WJ. 1994. J Controlled Release 28:87.

Rochira M, Miglietta MR, Richardson JL, Ferrari L, Beccaro M, Benedetti L. 1996. Int J Pharm 144:19.

Rock J, Barker RH, Bacon WB. 1947. Science 105:13.

Rowe PJ. (1990). Contraception 41:105.

Ryden L, Edman P. 1992. Int J Pharm 83:1.

Saettone MF, Chetoni P, Torracca MT, Burgalassi S, Giannaccini B. 1989. Int J Pharm 51:203.

Saettone MF, Giannaccini B, Chetoni P, Torracca MT, Monti D. 1991. Int J Pharm 72:131.

Schudmak M, Hesseltine HC. 1951. Am J Obstet Gynecol 62:669.

SCRIP. 1995. 2085:23.

Singer FR. 1991. Calcif Tissue Int 49:S7.

Smith CV, Rayburn WF, Miller AM. 1994. J Reprod Med 39:381.

Tasdemir M, Maral I, Tasdemir S, Tasdemir I. 1995. Lancet 346:1362.

Thamsborg G, Storm TL, Brinch E, Sykulski R, Fogh-Anderson N, Holmegaard SN, Sørensen OH. 1990. Calcif Tissue Int 46:5.

Vartiainen J, Wahlstroem T, Nilsson CG. 1993. Maturitas 17:129.

Villanueva B, Casper RF, Yen SSC. 1981. Fertil Steril 35:433.

Wearley LL. 1991. CRC Crit Rev Therap Drug Carrier Syst 8:331.

Wehner AP, Weller RE. 1986. Food Chem Toxicol 24:329.

Whittmore AS, Wu ML, Paffenbarger RS, Sarles DL, Kampert JB, Grosser S, Jung DL, Ballon S, Hendrickson M. 1988. Am J Epidemiol 128:1228.

Wimalawansa SJ. 1993. Calcif Tissue Int 52:90.

Wronski TJ, Yen C-F. 1991. Cells Mater Suppl 1:69.

Wronski TJ, Yen C-F, Burton KW, Mehta RC, Newman PS, Soltis EE, DeLuca PP. 1991. Endocrinology 129:2246.

Wuster C, Schurr W, Scharla S, Raue F, Minne HW, Ziegler R. 1991. Eur J Clin Pharmacol 41:211.

Yamazaki I. 1984. J Reprod Fertil 72:129.

22

Ocular Bioadhesive Drug Delivery Systems

Marco Fabrizio Saettone, Susi Burgalassi, and Patrizia Chetoni
University of Pisa, Pisa, Italy

I. INTRODUCTION

Since its early times (10–15 years ago), mucoadhesion has attracted the attention of ocular researchers, who sought to apply and to put to profit the concepts and techniques of this novel approach. Indeed, the ocular bioavailability of drugs administered by conventional eyedrops is low (2–10%) because of the small area available for absorption, the short time of contact of the medication with the eye, and other factors discussed in the next sections. Any modification of the vehicle resulting in increased time of contact with the absorbing surface can improve the drug bioavailability. Thus, the recognition of the mucoadhesive properties of some polymers and of the presence of a mucin-glycocalyx domain in the external portion of the eye was considered by many investigators as an interesting opportunity to prolong the residence time of medications in the preocular area, hence potentially increasing their bioavailability.

In this context, the first "intentional" study of ophthalmic mucoadhesion was that of Hui and Robinson, who in 1985 published a pioneering paper on the subject. However, conjunctival goblet cells and mucoadhesive polymers existed well before the "official" recognition of mucoadhesion, and in retrospect one can find in mucoadhesive phenomena an explanation for the unexpectedly good performance of some polymeric ocular formulations reported in the literature before 1985.

An exhaustive dissertation on all aspects of bioadhesion in ocular drug delivery would be a formidable task. Excellent reviews of this and related subjects, to which we refer the reader, have been published (Lee and Robinson, 1986; Middleton et al., 1990; Robinson, 1989, 1990; Greaves and Wilson, 1993; Krishnamoorthy and Mitra, 1993; Slovin and Robinson, 1993; Lehr, 1996; Sasaki et al., 1996). Instead, the present chapter aims at reappraising, both historically and scientifically, the work published to date. Only some necessary, brief introductory notes will precede the literature review. A final, necessary word of excuse is addressed to all authors unintentionally omitted in this presentation.

II. OCULAR ANATOMY AND PHYSIOLOGY RELEVANT TO BIOADHESION

The eye, also referred to as the ocular globe, is an isolated organ highly protected within the bony orbital cavity of the head. The external part of the eye is covered by the eyelids. These mobile folds protect the eye from mechanical or chemical injury (Robinson, 1993). The conjunctiva is a vascularized mucous membrane covering the anterior surface of the globe with the exception of the cornea (bulbar conjunctiva) and, continuing to the back and bending to the front (conjunctival fornix), it also covers the internal surface of the eyelids (palpebral conjunctiva) (Fig. 1).

The conjunctival epithelium is continuous with that of the cornea and with the epidermis of the lids: it is a multilayered nonkeratinized columnar

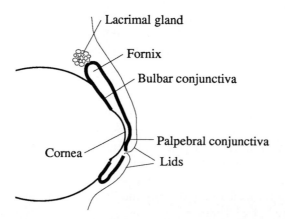

Figure 1   Schematic representation of the conjunctival membrane.

epithelium of five to seven layers, covering the highly vascularized substantia propria (Middleton et al., 1990).

The cornea is a five-layered structure made transparent by a special arrangement of cells, avascularity, and regularity and smoothness of the epithelium (Burstein, 1995). The corneal epithelium is a tight junction tissue that represents the most important barrier to invasion by foreign substances, including drugs (Slovin and Robinson, 1993). It has a cell turnover rate of approximately one layer per day. The anterior surface of the superficial epithelial cells displays microvilli and microplicae, which help to retain the tear film at the anterior surface of the eye.

The tear film covering the bulbar and palpebral conjunctiva and the cornea and defining the major optical surface of the eye is composed of three layers (Fig. 2). The outermost portion, a thin lipid monolayer, reduces evaporation and provides a continuous covering of the underlying portions. The middle portion, an aqueous layer, constitutes more than 95% of the total volume and contains electrolytes and various proteins. The inner, or basal, tear layer is composed mostly of mucus glycoproteins and coats the epithe-

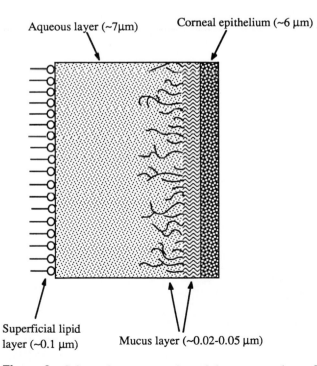

Aqueous layer (~7µm)   Corneal epithelium (~6 µm)

Superficial lipid layer (~0.1 µm)   Mucus layer (~0.02-0.05 µm)

Figure 2   Schematic representation of the precorneal tear film.

lial microvilli to help hydrate and maintain a continuous film over the pre-ocular surface (Holly, 1986; Holly and Lemp, 1977).

The cornea and conjunctiva are coated with a thin layer of mucin, secreted by approximately 1.5 million goblet cells located on the conjunctival surface and spread over the surface of the eye by the action of blinking (Adams, 1979). As shown in Fig. 3, the number of these cells increases in density in the lower fornices and inner canthus and decreases near the lid margins. The content of the goblet cells is only a precursor of mucus proper: essential differences have been demonstrated in staining properties between the goblet cell contents and excreted mucus (Norn, 1963).

Characteristically, mucus is composed of a number of elements: glycoproteins, proteins, lipids, electrolytes, inorganic salts, enzymes, mucopolysaccharides, etc. The mucus glycoproteins consist of hundreds of short polysaccharide chains, which usually constitute about 70% of the weight of the molecule, attached to a polypeptide backbone. The sequence of sugars in the side chains is precise; with few exceptions the principal sugars in mucus glycoproteins are galactose, fucose, acetylated amino sugars, and sialic acid (Berman, 1991). Each carbohydrate chain terminates in either a sialic acid ($pK_a = 2.6$) or an L-fucose group. As a result, at physiological pH the mucin molecules behave as anionic polyelectrolytes. Because of the rather large number of sugar groups, a mucin molecule is capable of picking up 40 to 80 times its weight in water (Holly and Lemp, 1977).

On the corneal mucosal surface the mucin molecules are found to be tightly packed, but as one moves outward from the epithelial surface, the mucin becomes less densely packed. This is accompanied by a correspond-

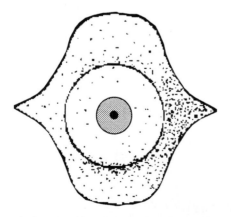

Figure 3   Distribution of goblet cells on the conjunctival surface. (From Calabria and Rolando, 1984.)

ing decrease in viscosity and ion content. It is important to note that while the mucus layers covering the cornea are thin, they are still thick enough for significant interpenetration of the bioadhesive polymer to occur. The residence time of mucin in the conjunctival fornices is very long and its production is very rapid: all mucus contained in the fornices can be secreted in 30 seconds. Continuous secretion of mucus is necessary to compensate for the loss due to digestion, bacterial degradation, and solubilization of mucin molecules (Robinson, 1989; Leung and Robinson, 1990).

The mucus layer acts as a wetting agent, reducing the interfacial tension between corneal epithelium and tears and stabilizing the extremely thin precorneal film between blinks (Holly and Lemp, 1971). Its principal functions are lubrication and protection of the underlying epithelial cells from dehydration and other challenges.

III. OCULAR BIOADHESIVE FORMULATIONS: INTRODUCTION

Topical administration of traditional ophthalmic formulations, such as aqueous solutions and ointments, is the common method for treatment of ocular diseases. As indicated in the introduction, the ocular bioavailability of drugs administered by conventional eyedrops is low (2–10%). This is due to the small area available for penetration; the presence of absorption barriers, constituted mainly by the lipophilic corneal epithelium; and a series of precorneal elimination factors that reduce the contact time of the medication with the corneal surface (Lee and Robinson, 1979, 1986; Maurice and Mishima, 1984). These elimination factors include

Drainage of instilled solutions
Lacrimation and tear turnover
Drug metabolism
Tear evaporation
Nonproductive absorption or adsorption
Possible binding by the lachrymal proteins.

The drainage of the administered dose via the nasolachrymal system into the nasopharynx and the gastrointestinal tract takes place when the volume of fluid in the eye exceeds the normal lachrymal volume of 7–10 μL. Thus, the portion of the instilled dose (one to two drops, corresponding to 50–100 μL) that is not eliminated by spillage from the palpebral fissure is quickly drained, and the contact time of the dose with the absorbing surfaces (cornea and sclera) is reduced to a maximum of 2 minutes. Repeated instillations, while inducing the desired therapeutic effect, are responsible for undesirable side effects resulting from systemic absorption through the nasolacrimal duct. Systemic absorption has been estimated as more than 50% of the instilled dose (Urtti and Salminen, 1993; Lee et al., 1993). The temporal pattern of drug concentration in tear fluid after topical administration of conventional eyedrops is characterized by "pulsed" profiles, with a short initial period of overdosing followed by a long period of underdosing. Better retained vehicles and/or constant-rate release devices can provide more favorable release profiles, as illustrated in Fig. 4.

During the past two to three decades, ointments, hydrogels, viscous liquid formulations, solid delivery devices (inserts, drug-soaked contact lenses), nano- and microparticulates, etc. have been investigated in the attempt to improve the efficacy of ocular medications. Although the idea of prolonging the contact time between drugs and the ocular surface is not new, only the past decade has witnessed serious efforts to apply the concepts and techniques of mucoadhesion to topical ophthalmic therapy. Consequently, different polymers have been evaluated for their ability to establish adhesive,

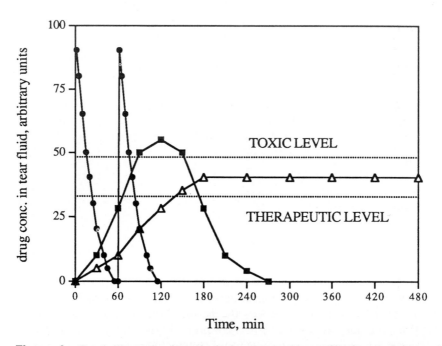

Figure 4 Graph illustrating hypothetical concentration profiles in tear fluid resulting from administration of drugs in different ophthalmic vehicles. (●) Concentration "pulses" produced by instillation of eyedrops at 60-minute intervals; (■) instillation of longer retained vehicles; (△) application of a constant-rate release insert.

noncovalent bonds with the mucin layer coating the corneal-conjunctival epithelium, thus giving ophthalmic drug delivery systems longer times of contact with the absorbing tissues. In general, mucoadhesive polymers are characterized by the following physicochemical properties: (a) strong hydrogen-bonding groups (—OH, —COOH), (b) strong anionic charges, (c) high molecular weight, (d) sufficient chain flexibility, (e) surface energy properties favoring spreading on mucous surfaces. Loss of a mucoadhesive polymer from the precorneal area will occur mainly as a consequence of mucin turnover, removal by blinking, and drainage subsequent to dissolution. The latter mechanism will prevail in the case of water-soluble polymers.

In the next sections, mucoadhesive ophthalmic formulations will be examined in the following sequence: semisolids (hydrogels), viscous liquids, inserts, and micro/nanoparticulates.

A. Semisolid (Hydrogel) Formulations

The year 1985, when Hui and Robinson published the first specific paper, "Ocular delivery of progesterone using a bioadhesive polymer," can be taken as the "official" date of birth of ocular bioadhesion. However, several investigations of semisolid ophthalmic vehicles prepared with poly(acrylic acid) (PAA, carbomer, Carbopol), subsequently recognized as an excellent mucoadhesive material, had been published prior to 1985. Thus, for historical reasons, semisolid mucoadhesive ophthalmic formulations will be examined first. An overview of the essential literature on these formulations is presented in Table 1.

Many investigators were attracted by the interesting characteristics of PAA, a polymer with a linear, branched, or cross-linked structure, available in a wide range of molecular weights. PAA was widely used as viscosity enhancer and gel former until its mucoadhesive properties were reported by Nagai et al. (1980) and subsequently by Smart et al. (1984).

Schoenwald et al. (1978), in a study of the miotic effect in rabbits of high-viscosity vehicles containing pilocarpine, found that PAA and ethylene maleic anhydride gels of comparable plastic viscosity significantly prolonged the drug's bioavailability, when compared with petrolatum and hydroxypropylcellulose vehicles. Analogous results were obtained by Schoenwald and Boltralik (1979) in a study in which prednisolone acetate and sodium phosphate were administered to rabbits in a PAA gel and in reference vehicles. The authors noticed the good ocular retention of the PAA hydrogel and, presumably unaware of mucoadhesion, speculated that the increased permanence "was a consequence of the gel's increased yield value, such that appreciable in vivo thinning of the gel does not take place with eyelid and/ or eyeball movements." The report of Schoenwald and Boltralik (1979)

Table 1 Synopsis of the Essential Literature on Mucoadhesive Semisolid Ophthalmic Vehicles

Polymers[a]	Drugs	Type of study	References
PAA	Pilocarpine, lidocaine, benzocaine, timolol, prednisolone, adrenaline	Tests in vivo (rabbits)	Schoenwald et al. (1978); Bottari et al. (1979); Schoenwald and Boltralik (1979); Kupferman et al. (1981); Habib and Attia (1984)
PAA, PAAm	Tropicamide Pilocarpine	Tests in vivo (humans) Mucoadhesion in vitro; tests in vivo (rabbits, humans)	Saettone et al. (1980) Saettone et al. (1986)
PAA, poloxamer	Flurbiprofen	Tests in rabbit inflammation model	Mengi and Deshpande (1992)
HA, PAA, and other polymers	Pilocarpine, tropicamide	Scintigraphy; tests in vivo (rabbits, humans)	Gurny et al. (1987); Saettone et al. (1989a)
Cross-linked PAA	Progesterone, fluorometholone, gentamicin	Mucoadhesion in vitro; tests in vivo (rabbits)	Hui and Robinson (1985); Middleton and Robinson (1991); Lehr et al. (1994)
XYL, XG, PEC, HPMC, PVA	Pilocarpine	Mucoadhesion in vitro; tests in vivo (rabbits)	Burgalassi et al. (1996)

[a]PAA, polyacrylic acid; PAAm, polyacrylamide; HA, hyaluronic acid; XYL, xyloglucan (tamarind) gum; XG, xanthan gum; PEC, pectin; HPMC, hydroxypropylmethylcellulose; PVA, polyvinyl alcohol.

prompted Kupferman et al. (1981) to test the prednisolone acetate PAA hydrogel in a rabbit model of ocular inflammation. The latter authors found that the gel formulations produced no greater anti-inflammatory effect than conventional suspensions; the duration of activity, however, was considerably longer.

Bottari et al. (1979) tested in rabbits the anesthetic activity of PAA hydrogels containing lidocaine and benzocaine in comparison with aqueous solutions and suspensions of the drugs in yellow soft paraffin. Obviously, the hydrogels produced significantly higher bioavailability and sustaining activity; particularly good results were obtained with a vehicle in which lidocaine, instead of diisopropanolamine, was used as a neutralizing agent for the polymer. The authors, also unaware of mucoadhesion, attributed the results to cooperation of different factors: vehicle viscosity, high diffusion coefficient of the drugs within the vehicle, and optimal miscibility of the gels with the tear film.

The capacity of a PAA hydrogel to increase the bioavailability of 0.2% tropicamide in humans with respect to other standard vehicles (a saline solution, a viscous vehicle, and a petrolatum ointment) was reported by Saettone et al. in 1980. An example of their results is given in Fig. 5.

Habib and Attia (1984), who studied the activity in rabbits of adrenaline bitartrate formulated in 2% PAA and in 25% poloxamer gels, also reported the superior activity of the gel formulations, in particular the PAA one.

On a chronological basis, the paper by Hui and Robinson (1985) must be cited at this point. These authors cross-linked PAA with divinyl glycol or 2,5-dimethyl-1,5-hexadiene and obtained water-insoluble materials with extensive capacity to attract and hold water and with good bioadhesion to freshly excised rabbit conjunctival membranes, as determined with a tensile apparatus. Progesterone was selected as the test drug. The bioadhesive hydrogels, tested in rabbits, showed an area under the curve (AUC) for progesterone in aqueous humor versus time 4.2 times greater than that obtained with a standard aqueous suspension. The authors concluded that their "relatively primitive dosage form is indicative of the powerful potential of ocular bioadhesives in drug delivery," thus paving the way to further, specifically oriented research in this field. A similar cross-linked water-insoluble PAA polymer of commercial origin (polycarbophil) was subsequently evaluated as vehicle for fluorometholone by Middleton and Robinson (1991). These authors reported that a hypotonic, slightly acidic polycarbophil vehicle could be administered as a drop but would gel in the precorneal pocket with subsequent attachment to the conjunctival fornix. The dosage form was well tolerated by human volunteers and produced much improved therapeutic

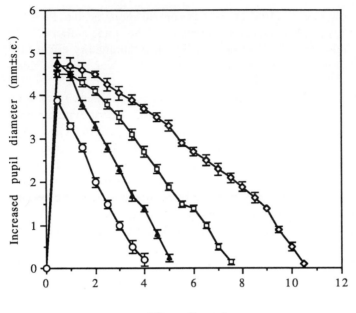

Figure 5 Mydriatic effect in humans of preparations containing 0.2% tropicamide. (○) Saline solution; (▲) viscous (0.7% HPC) vehicle; (◇) PAA (0.3%) hydrogel; (□) petrolatum ointment. (From Saettone et al., 1980.)

levels in the aqueous humor of rabbits compared with a standard aqueous suspension of the drug.

Polycarbophil was again tested by Lehr et al. (1994) as a vehicle for gentamicin. Two gentamicin formulations of this polymer (neutralized and nonneutralized) and an aqueous control solution were tested in rabbits. Both gel formulations increased the uptake of gentamicin by the bulbar conjunctiva, but only the acidic polymer increased significantly (approximately eight times) the AUC of the drug in the aqueous humor (Fig. 6).

The adhesive properties to pig intestinal mucosa of acidic polycarbophil gels (pH values lower than 4.5) had been reported previously by the same author (Lehr et al., 1992a). It is of interest to note that both formulations, neutralized and acidic, may be advantageous for pharmacological treatments of outer and inner ocular infections, respectively. The authors speculated, also on the basis of literature data relevant to nonocular epithelia, that the prolonged and intensified contact of the acidic mucoadhesive formulation would temporarily weaken the barrier properties of the corneal

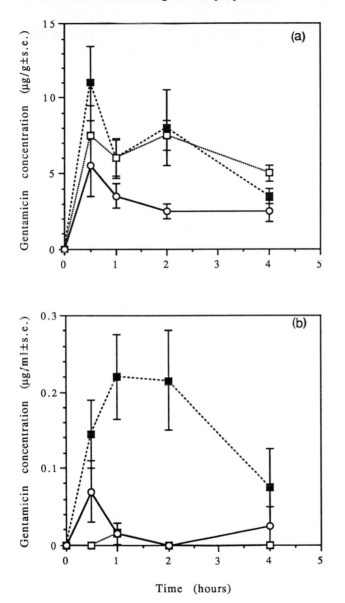

Figure 6 Gentamicin concentration versus time profiles in conjunctiva (a) and in aqueous humor (b) after administration to rabbits of acidic polycarbophil (■), neutralized polycarbophil (□), and saline solution (○). (From Lehr et al., 1994.)

epithelium, thus facilitating drug penetration. The indication of a possible complementary mechanism by which mucoadhesives may enhance the transport of poorly absorbed, hydrophilic drugs through epithelial diffusion barriers appears extremely interesting. A temporary opening of the tight intercellular junctions was mentioned by the authors as a likely mechanism (see also Lehr, 1996).

Going back to post-1985 investigations of hydrogels based on non–cross-linked PAA (carbomer) and of other gel-forming polymers, two papers by Saettone and coworkers should be mentioned. In a first study, the miotic activity in humans and in rabbits of four hydrogels containing pilocarpine was evaluated (Saettone et al., 1986). The vehicles were prepared with two different types of PAA (Carbopol 910 and 940), with poly(acrylamide) (PAAm), and with ethylene maleic anhydride (EMA). They showed a plastic type of flow (with the exception of PAAm, which was pseudoplastic) and apparent viscosities (measured at a rate of shear of 16 s^{-1}) ranging from 35.3 to 87.5 cps. In humans, the hydrogels showed more marked activity differences than in rabbits: one PAA vehicle (Carbopol 940) and PAAm produced an approximately threefold bioavailability increase over an aqueous solution, the other PAA (less viscous) vehicle only doubled the pilocarpine bioavailability, and EMA showed activity parameters not significantly different from those of the aqueous solution. Even if the mucoadhesion issue was not discussed (the awareness of ocular mucoadhesion was not widespread in those early times), the paper further evidenced the interesting properties of PAA and of another acrylic derivative, PAAm, and confirmed previous observations on the poor validity of rabbits for studies of ophthalmic vehicles. The latter theme will be discussed in ampler detail in the following section.

In a subsequent paper (Saettone et al., 1989a) the same group investigated a series of vehicles (solutions, hydrogels, solid matrices) prepared with different polymers: four types of PAA and three different hyaluronic acid (HA) derivatives. The vehicles, containing pilocarpine or tropicamide, were tested for mucoadhesion in vitro and for preocular retention and pharmacological activity in rabbits. The investigation, while confirming the good mucoadhesive properties of PAA gels, also evidenced outstanding mucoadhesive properties in some HA formulations, already noticed in a previous communication (Saettone et al., 1987). As evidenced by a fluorescent marker, the HA gels formed a stable precorneal film lasting over 2 hours and increased 2.9- and 3.7-fold, respectively, the miotic and mydriatic AUC values in rabbits when compared with aqueous solutions of pilocarpine and tropicamide. The inferior effect observed with pilocarpine was attributed to the higher solubility of pilocarpine than of tropicamide, which caused the drug to diffuse rapidly out of the vehicle.

In a previous study, Gurny et al. (1987) had investigated three formulations containing pilocarpine: two in situ gelling dispersions and one HA gel. These were evaluated in humans and in rabbits for miotic effect and for precorneal distribution by gamma scintigraphy. The study demonstrated the capacity of hyaluronic acid, even at low concentration, to prolong the precorneal residence time and the drug bioavailability.

A brief comment on hyaluronic acid might be of relevance at this point. This material, a natural polysaccharide belonging to the class of glycosaminoglycans (GAGs), is an important component of the extracellular matrix of connective tissues such as vitreous body, subcutaneous tissue, cartilages, umbilical cord, and synovial fluid and tissues. The mucoadhesive properties of HA, possibly deriving from its structural and functional similarities to mucopolysaccharides, have attracted the attention of many investigators. Data on the in vitro bioadhesive properties of HA, first presented by Park and Robinson (1984) and by Saettone et al. (1987), were subsequently substantiated by numerous in vitro and in vivo studies, of which a small sample follows: Camber et al. (1987), Gurny et al. (1987), Hazlett and Barrett (1987), Chang et al. (1988), Camber and Edman (1989), Madsen et al. (1989), Saettone et al. (1989a, 1989b), Snibson et al. (1990), and Huupponen et al. (1992). An excellent review by Bernatchez et al. (1993) on the use of HA in ocular therapy should also be mentioned.

The failure of the so-called mucoadhesive approach to provide increased ocular bioavailability in the case of highly soluble drugs such as pilocarpine and the relevant remarks of Davies et al. (1991), "delivery systems which would allow the drug to remain associated with the vehicle exhibiting precorneal retention may ... provide for an attractive ocular delivery system," stimulated Burgalassi et al. (1996) to test an alternative combination approach. This consisted of incorporating the drug in mucoadhesive hydrogels as a poorly soluble or diffusible complex. The gels were based on a series of polymers [xyloglucan gum, xanthan gum, pectin, hydroxypropylmethylcellulose, and polyvinyl alcohol (PVA)]; mucoadhesion tests in vitro evidenced the capacity of all gels to interact strongly with mucin. It is of interest to note that a 13% w/w PVA (molecular weight 9×10^4) gel showed a work of adhesion to mucin significantly higher than that of a 2% PAA (Carbopol 940) gel. The widespread assumption that PVA is not mucoadhesive should perhaps be reconsidered.

Pilocarpine was incorporated in the gels as a poorly soluble complex with tannic acid (Pi-TA); for comparison, gels containing the drug in solution as the nitrate ($PiNO_3$) were also tested. As indicated in Table 2, administration to rabbits of a xyloglucan-based (XYL) mucoadhesive hydrogel containing *suspended* Pi-TA resulted in more favorable pharmacokinetic parameters (apparent rate of elimination and half-life of pilocarpine in tear fluid,

Table 2   Pharmacokinetic Parameters of Mucoadhesive Vehicles Containing Pilocarpine Nitrate, $PiNO_3$, and Tannate, Pi-TA

Formulation	Ke_{tf}[a] $(min^{-1} \times 10^2)$	$t_{1/2tf}$[b] (min)	AUC^c (min mm)	Relative AUC
Reference solution	22.0	3.15	400.4	1
PVA gel $PiNO_3$	17.6	3.94	504.1	1.26
XYL gel $PiNO_3$	11.4	6.08	559.9	1.38
XYL gel Pi-TA	2.4	28.87	713.6	1.78

[a]Apparent rate constant for elimination of pilocarpine from tear fluid.
[b]Apparent half-time for elimination of pilocarpine from tear fluid.
[c]Area under the miotic activity versus time curve.
From data in Burgalassi et al. (1996).

AUC for miotic activity) with respect to other vehicles (i.e., the same hydrogel, a less mucoadhesive PVA hydrogel, and an aqueous solution) all containing dissolved $PiNO_3$. These findings confirmed the thesis that mucoadhesive hydrogels are insufficient per se to ensure good ocular bioavailability, unless the drug they contain is adequately retained within the vehicle.

PAA (Carbopol 940) and poloxamer hydrogels were also tested as vehicles for flurbiprofen (a nonsteroidal anti-inflammatory drug) by Mengi and Deshpande (1992). The anti-inflammatory effect of the hydrogels was investigated in a uveitis model in rabbits produced by intracameral administration of diphteria-tetanus-pertussis (DTP) vaccine. The study, involving measurement of intraocular pressure and total leukocyte count, indicated a sustained action of both gel formulations compared with eyedrops. However, the PAA vehicle showed superior results, ascribed by the authors to mucoadhesion.

B. Viscous Liquid Formulations

The addition of suitable polymers to liquid ophthalmic vehicles is a time-honored method for increasing the ocular contact time and hence the drug bioavailability. This effect, which has been widely investigated using different polymers, was ascribed in the past only to the increased viscosity induced by the macromolecules; no relevance to ocular bioavailability was attributed to the chemical nature of polymers and to their other physicochemical properties. Patton and Robinson (1975) actually stated that "It is intuitively clear that as long as two vehicles exhibit the same flow properties, solutions with the same viscosity should exhibit the same drainage behavior in the eye." Another concept, besides the "equal viscosity–equal activity" assumption, dominated the stage at the time. It was the contention that bioavailability data, generated from experiments carried out on rabbits, could also be assumed to be valid for humans. These implicit beliefs were disproved by Saettone et al. (1982, 1984a, 1989a), who tested the same series of isoviscous polymeric vehicles containing pilocarpine and tropicamide in rabbits and in humans. Some polymeric solutions (in particular, those containing PVA) were found to be more active both in humans and in rabbits than the other isoviscous solutions, and rabbits were found to be less sensitive than humans to polymer-mediated effects. The hypothesis was advanced at the time that the superior activity of PVA with respect to other isoviscous polymeric vehicles might be due to an influence on the spreading characteristics and on the thickness of the vehicle layer over the cornea. It was subsequently reported (Saettone et al., 1985) that benzalkonium chloride, even at a low concentration, significantly reduced the bioavailability enhancement produced in humans by PVA and polyvinylpyrrolidone (PVP) solutions containing tropicamide. This effect was ascribed to inhibition or reduction of the corneal spreading characteristics of the polymers by the cationic surfactant.

A different explanation for these findings might be offered today in light of the actual concepts of mucoadhesion. Depending on their molecular structure and configuration in solution, bioadhesive polymers may increase the vehicle viscosity, and this has been thought to be as a desirable property, as viscosity effects would contribute to ocular retention (Krishnamoorthy and Mitra, 1993; Slovin and Robinson, 1993; Duchêne et al., 1988). However, mucoadhesive polymers that induce modest viscosity changes might be preferable for ocular drug delivery because of greater ease of administration, dosing, and sterilization.

A synopsis of the essential literature on mucoadhesive liquid vehicles is presented in Table 3. Although these are examined separately from hydrogel vehicles, it must be remembered that many articles listed in Table 1 and 2 are concerned with both liquid and hydrogel preparations obtained with different concentrations of the same polymers.

Davies et al. (1991), using a dacryoscintigraphic method evaluated the precorneal clearance in albino rabbits of two isoviscous solutions (60 mPa s at unit shear rate) prepared with PAA and PVA. The precorneal retention of the PAA solution was significantly greater than that of the PVA solution (defined by the authors as nonmucoadhesive), and the retention of the latter was significantly greater than that of a saline solution. Accordingly, admin-

Table 3 Synopsis of the Essential Literature on Mucoadhesive Viscous Liquid Ophthalmic Vehicles

Polymers[a]	Drugs	Type of study	References
PAA, PVA (plus other polymers)	Pilocarpine	Precorneal clearance; activity (rabbits)	Davies et al. (1991)
	Timolol	Bioavailability (rabbits)	Thermes et al. (1992a)
	Naphazoline, atenolol	Tests ex vivo	Dittgen et al. (1992)
PULL	Timolol	Tests in vivo (rabbits)	Thermes et al. (1992b)
HA (plus other polymers)	Pilocarpine, timolol	Miotic effect; mucoadhesion in vitro; preocular retention (rabbits, humans)	Camber et al. (1987) Chang et al. (1988) Camber and Edman (1989) Saettone et al. (1989b, 1991)
PGA, HA, PAA, MG, CMCH, CMAM, HS, CS	Cyclopentolate, pilocarpine	Mucoadhesion in vitro; tests in vivo (rabbits)	Saettone et al. (1992, 1994)
PGA, HA	Cyclopentolate	Tests in vivo (rabbits) Tests in vivo (humans)	Huupponen et al. (1992) Lahdes et al. (1993)

[a]PAA, polyacrylic acid; PVA, polyvinyl alcohol; PULL, pullulan; HA, hyaluronic acid; PGA, polygalacturonic acid; MG, mesoglycan; CMCH, carboxymethylchitin; CMAM, carboxymethylamylose; HS, heparan sulfate; CS, chondroitin sulfate.

istration of 1% pilocarpine nitrate in the PAA solution produced a statistically significant increase in bioavailability (as determined by the miotic AUC) compared with the equiviscous PAA solution. Interestingly, Thermes et al. (1992a) evaluated two similar solutions (albeit of slightly lower viscosity: 45 mPa s at a shear rate range of $0-17$ s^{-1}) containing timolol for bioavailability in rabbits. An isoviscous solution containing timolol as the PAA salt was also tested. These workers found that the ocular bioavailability of timolol was increased by each of the viscous solutions compared with standard aqueous eyedrops (Timoptol). However, the largest AUC increases, assessed by measuring the $0-4$-hour drug concentration in cornea, aqueous humor, and iris plus ciliary body, were obtained with the "nonmucoadhesive" PVA solution, whereas the mucoadhesive PAA solutions provided higher concentrations in the iris plus ciliary body at later times after instillation (Fig. 7).

These data, according to the authors, would be consistent with slower release of timolol from PAA and longer retention of the vehicle in the conjunctival sac by mucoadhesion. The latter results raise the question of the effective mucoadhesive properties of PVA. The asserted "poor mucoadhesive properties" of PVA should also be reconsidered on the basis of the findings of Burgalassi et al. (1996).

A low-viscosity (12 mPa s) PAA solution, when tested by a tensiometric method by Dittgen et al. (1992), was also found to be more bioad-

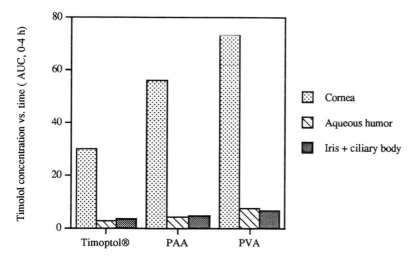

Figure 7 AUC ($0-4$ hours) of timolol concentration in ocular tissues of albino rabbits after instillation of various 0.5% formulations. (Drawn from data in Thermes et al., 1992a.)

hesive than other isoviscous polymeric solutions (sodium carboxymethyl-amylopectine, tragacant gum, hydroxyethylcellulose, PVA, and a polyacrylic ester). The solutions, containing naphazoline HCl or atenolol, were tested for elimination from the precorneal area of the isolated pig eye (wash-off technique) and for permeation through isolated pig cornea in vitro. The PAA solution also gave the best results in these cases, consisting of a lower rate of elimination and an increased amount of permeated drug.

Another paper by Thermes et al. (1992b) was concerned with pullulan, a naturally occurring mucoadhesive polysaccharide. Instillation in rabbits of timolol in a medium-viscosity (100 mPa s) pullulan solution increased the bioavailability of the drug over twofold in cornea, aqueous humor, and iris–ciliary body in comparison with Timoptol eyedrops. To distinguish effects due to viscosity from mucoadhesive effects, the authors again tested the formulations after removing precorneal-conjunctival mucus with N-acetylcysteine, a method proposed some years before by Wood et al. (1985) to verify the mucoadhesion of nanoparticles. On removal of the preocular mucin the bioavailability of the drug was lowered, but it was still higher than that of the aqueous eyedrops. The authors attributed this to viscous effects alone.

The mucoadhesive properties of hyaluronic acid have already been already mentioned in the previous section, dedicated to semisolid vehicles. This polymer has also been investigated in a variety of ways as component of liquid ocular formulations. Camber et al. (1987) and Camber and Edman (1989) reported an increased miotic response in rabbits after administration of pilocarpine in a high-molecular-weight (MW) (4.6×10^6) HA vehicle. Camber and Edman (1989) also observed a significant decrease in the area under the miosis-time curve with decreasing molecular weight of HA.

A series of papers by Saettone and coworkers was dedicated to the evaluation in vivo of liquid formulations containing salts (or polyanionic complexes) of the drug bases (pilocarpine, cyclopentolate) with different anionic polymers, including HA. The rationale for this approach was an attempt to favor the preocular retention and the bioavailability of the drug by ionically binding the drug itself to a polycarboxylic, mucoadhesive polymer. In a first study (Saettone et al. 1989b) several polymeric pilocarpine salts, tested for miotic activity in rabbits, showed significantly increased AUC values compared with an aqueous solution of pilocarpine nitrate. Interestingly, the activity of some low-viscosity vehicles (2 to 5.3 mPa s) prepared with mesoglycan, carboxymethylchitin, polygalacturonic acid, and low-MW (1.13×10^5) hyaluronic acid was of the same order as that of PAA vehicles of higher viscosity (54 to 97 mPa s). In particular, the vehicle containing the pilocarpine salt of mesoglycan (a mixture of mucopolysaccharides consisting mainly of dermatan sulfate and heparan sulfate) produced a 2-fold bioavailability increase over the aqueous solution and a 1.55-fold increase over a slightly more viscous (3.6 versus 2 mPa s) reference PVA vehicle. The authors, while attributing the performances of some of the tested polymers to their mucopolysaccharide or homopolysaccharide structure, favoring development of strong interactions with the glycoprotein structure of mucus, concluded that viscosity per se may not be relevant to ocular mucoadhesion.

In a subsequent study, the same authors (Saettone et al., 1991) investigated the influence of HA molecular weight and of ionic biding of pilocarpine to the polymer. A set of low- and high-MW hyaluronate vehicles with a wide range of viscosities (1.0 to 1054 mPa s), containing either pilocarpine nitrate or ionically bound pilocarpine, was evaluated by determining the ocular retention and the miotic effect in rabbits. The tested HA sodium salt fractions were Hyalastin, Hyalectin, and Healon, with MWs of 1.34×10^5, 6.2×10^5, and 4.6×10^6, respectively. The pilocarpine salt was prepared with HA acid of MW 1.13×10^5, obtained from Hyalastin by ion exchange. The results of this investigation indicated that one essential requirement for prolonged ocular permanence of HA, and hence for good performance of this material as an ophthalmic vehicle, is a high molecular weight. The longest ocular residence times, corresponding to the highest bioavailability of pilocarpine, were observed in the case of all high-MW HA-Na vehicles containing the drug nitrate. The performance of these vehicles, which was viscosity unrelated and particularly good at the lower tested concentrations, was attributed to mucoadhesive effects. Relatively good performances were also shown, however, by the vehicle containing pilocarpine bound to the low-MW hyaluronic acid, which provided higher AUC values than more viscous vehicles containing pilocarpine nitrate. According to the authors, the interest in this preparation was due to its low viscosity (2.3 mPa s), allowing sterile filtration and dispensing as eyedrops, two important factors in ophthalmic drug formulations. These results also stressed the validity of the drug-polymer salt approach.

The activity of low-viscosity vehicles containing drug bases ionically bound to mucoadhesive polymers was further examined in a series of investigations by the same group (Saettone et al., 1992, 1994; Huupponen et al., 1992; Lahdes et al. 1993). A first investigation (Saettone et al., 1992) was concerned with the evaluation of solutions containing cyclopentolate (CY), a mydriatic-cycloplegic drug, bound to polygalacturonic acid (PGA), mesoglycan (MG), carboxymethylchitin (CMCh) or carboxymethylamylose (CMAm). When tested for mydriatic activity in rabbits in comparison with appropriate reference vehicles containing cyclopentolate hydrochloride, the solutions containing the macromolecular CY salts were more active. In most cases, these results could be explained in terms of adhesive interactions of the polymers with preocular mucus. In a further study, two vehicles containing the cyclopentolate salts of PGA or of low-MW hyaluronic acid (see Saettone et al., 1991) were tested in rabbits for mydriatic effect and for ocular and systemic absorption. Both polymeric formulations increased (albeit not to a significant degree) the mydriatic effect in comparison with aqueous solutions containing the same amount of CY as the hydrochloride (CY-HCl). However, during the first half-hour after administration, the systemic absorption of CY was lower after CY-PGA than after CY-HCl. The ocular penetration of CY, based on the drug concentration in aqueous humor 30 minutes after instillation, was increased threefold after administration of the PGA complex with respect to the CY-HCl eyedrops. The increased ocular effects and decreased initial systemic absorption of CY from the CY-PGA salt were attributed to mucoadhesion. The authors concluded that the PGA salt, as like other polymeric salts, might offer the possibility of increasing the therapeutic index of cyclopentolate.

The latter results prompted further research on humans. Lahdes et al. (1993) evaluated the CY-PGA formulation in a group of eight volunteers for systemic absorption and for mydriatic-cycloplegic activity, in comparison with commercial CY-HCl eyedrops. The investigation, however, failed to confirm the superior effects of the CY-PGA complex observed in rabbits by Huupponen et al. (1992). As a possible explanation, the authors advanced the hypothesis that their results might be biased by the presence of borate ions, known to increase the ocular bioavailability of CY, in the reference CY-HCl eyedrops.

In a further study, Saettone et al. (1994) again investigated a series of low-viscosity vehicles containing polymeric salts of pilocarpine and cyclopentolate. The polymers, in addition to those tested in a previous study (Saettone et al., 1989b), were carboxymethylamylose, chondroitin sulfate, and heparan sulfate. The mucin-polymer adhesive bond strength was determined by the viscometric method proposed by Hassan and Gallo, which consisted of measuring viscosity changes induced in a mucin dispersion by the addition of the polymers. In vivo miosis and mydriasis tests were performed in rabbits, before and after removal of mucus from the corneal-conjunctival surfaces with N-acetylcysteine (see Thermes et al., 1992b). Small but significant bioavailability increases with respect to reference preparations, containing pilocarpine nitrate or cyclopentolate hydrochloride, were observed with most polymeric vehicles. Also, significant AUC decreases were found after removal of precorneal mucus. The correlations found between the bioavailability of the two drugs and the bioadhesive bond strengths of the corresponding polymers are illustrated in Fig. 8. A final conclusion was that the bioavailability enhancements observed with the low-viscosity polymer-salt vehicles were rather modest and corresponded at most

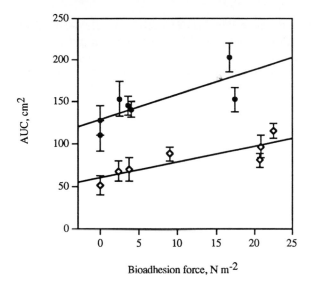

Figure 8 Correlation between bioavailability and bioadhesion force for a series of vehicles containing pilocarpine ($\Diamond$) and cyclopentolate ($\bullet$). (From Saettone et al., 1994.)

to a "prolonged pulse" type of release. Thus, in the words of the authors, "A proper combination, in the same polymer, of mucoadhesion ... and viscosity ... is probably desired for optimal activity."

At the end of this section, an investigation by Albasini and Ludwig (1995) should be mentioned. These workers evaluated the physicochemical characteristics and the rheological properties of different water-soluble polysaccharides (alginate, carrageenan, guar gum, locust bean gum, scleroglucan, xanthan gum) of possible use as adjuvants in ophthalmic dosage forms. The rheological synergism with gastric porcine mucin evidenced mucoadhesive properties in scleroglucan and xanthan gum. These polymers also had low human irritancy, low viscosity, viscoelastic behavior, and absence of surface-active properties and were considered by the authors as suitable adjuvants for ophthalmic formulations.

C. Solid Formulations (Inserts)

Solid ophthalmic delivery devices (inserts) are thin disks or small cylinders made with appropriate polymeric materials and fitting into the lower or upper conjunctival sac. Their long persistence in the preocular area can result in a greater drug bioavailability with respect to liquid and semisolid formulations. Some inserts, such as the now classical Ocusert, can release the drug at a slow, constant rate for at least 1 week. The advantages and drawbacks of these interesting, although not widely used, delivery systems have been discussed in several reviews (Richardson, 1975; Shell, 1980; Chiou and Watanabe, 1982; Chien, 1982; Mikkelson, 1984; Buri, 1985; Lee and Robinson, 1986; Salmien, 1987; Lee, 1990; Bawa, 1993; Saettone, 1993; Saettone and Salminen, 1995; Gurtler and Gurny, 1995). A synopsis of the essential literature on mucoadhesive inserts is presented in Table 4.

Mucoadhesive polymers can be profitably used as constituents of inserts to achieve prolonged contact with the conjunctival tissues and to alleviate the risk of expulsion from the cul-de-sac. These polymers can be the exclusive constituents of inserts, can be present as components of a polymeric mixture, or else can be used to coat a hydrophobic insert body.

Also in the case of inserts, early examples of unrecognized bioadhesive effects due to the presence of, e.g., PAA derivatives can be found in the literature.

Saettone et al. (1984b) evaluated on a physicochemical and biological basis a series of commercially available polymers as possible materials for the preparations of soluble, monolithic inserts. The inserts, prepared with different types of polyvinyl alcohol (PVA) and of hydroxypropylcellulose (HPC), contained pilocarpine as the nitrate ($PiNO_3$) or as a PAA salt. As shown in Fig. 9, a PVA insert containing the PAA salt was significantly more active in rabbits than an aqueous solution and an insert, both containing $PiNO_3$, or a hydrogel containing the pilocarpine PAA salt. This effect, attributed by the authors both to the physicochemical and biological characteristics of PVA and to an "optimal" delivery rate of the drug from the PVA matrix containing the PAA salt (see Bottari et al., 1979), can now be safely attributed to the mucoadhesive properties of PAA.

More recently, a PAA-based long-acting bioadhesive insert designed for animal health care and denominated Bioadhesive Ophthalmic Drug Insert (BODI) has been described by Gurtler and coworkers (Gurtler and Gurny, 1993; Gurtler et al., 1995a, 1995b). BODIs are prepared by extruding a dry mixture of hydroxypropylcellulose, ethylcellulose, cellulose acetate phthalate, PAA (as the mucoadhesive component), and gentamicin. These inserts are soluble and need not be removed at the end of activity. According to Gurtler and coworkers (1995a) the presence of PAA reduced the ocular expulsion in rabbits to less than 2%, as compared with 30% observed without the mucoadhesive polymer. In dogs and rabbits BODIs ensured effective gentamicin levels over 72 hours, versus 15 minutes observed with eyedrops. The drug release profiles were similar in both species: the gentamicin concentration in tears exhibited a plateau during the first 48 hours then declined in the following 24 hours (Gurtler et al., 1995b).

Table 4 Synopsis of the Essential Literature on Mucoadhesive Solid Formulations (Inserts)

Polymers[a]	Drugs	Type of Study	References
PAA (plus PVA, SIL, HPC, EC, CAP)	Pilocarpine, oxytetracycline, gentamicin	Release in vitro; tests in vivo (rabbits); mucoadhesion in vitro	Saettone et al. (1984b); Gurtler and Gurny (1993); Gurtler et al. (1995a; 1995b)
SIL (plus PAA, PMA)	Oxytetracycline	Release in vitro; tests in vivo (rabbits); mucoadhesion in vitro	Chetoni et al. (1996)
HA HA esters	Pilocarpine, methylprednisolone	Release in vitro; tests in vivo (rabbits)	Saettone et al. (1989a); Benedetti et al. (1991); Kyyrönen et al. (1992)
Collagen	Fluorescein	Release in vivo (humans)	Kaufman et al. (1994)

[a]PAA, polyacrylic acid; PVA, polyvinyl alcohol; SIL, silicone elastomer; HPC, hydroxypropylcellulose; EC, ethylcellulose; CAP, cellulose acetate phthalate; PMA, polymethacrylic acid; HA, hyaluronic acid.

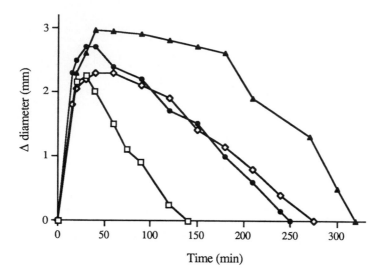

Figure 9   Mean change in pupillary diameter versus time for different vehicles containing pilocarpine (Pi). (□) PiNO₃ solution; (●) PAA-Pi salt, hydrogel; (◇) PVA/ PiNO₃ insert, (▲) PVA/PAA-Pi salt, insert. (From Saettone et al., 1984b.)

Silicone elastomer inserts, first described in patents by Darougar (1988, 1992), can be inadvertently expelled from the eye during long-term therapy because of the hydrophobic characteristic of silicone rubber. An approach to improving the ocular retention of silicone inserts by the use of acrylic polymers was reported by Chetoni et al. (1996). These workers coated rod-shaped silicone inserts containing oxytetracycline with a PAA or PMA (polymethacrylic acid) layer. In vitro mucoadhesion tests confirmed the increased mucoadhesive properties of the PAA- and PMA-coated inserts. Rabbit studies showed that the ocular retention time of the coated inserts was significantly longer than that of uncoated ones and that a constant, effective (20–30 μg/mL) oxytetracycline concentration in tear fluid could be maintained for over 24 hours. It should be pointed out in this connection that ophthalmic preparations producing a prolonged and constant drug level in tear fluid are particularly indicated in the treatment of ocular infections: common eyedrops ensure a constant level of antibiotic only with repeated administration, up to four times for hour (Baum et al., 1974).

The use of hyaluronic acid or of its derivatives for the preparation of long-retained inserts has been reported by different authors (Saettone et al., 1989a; Benedetti et al., 1991; Kyyrönen et al., 1992). In the previously mentioned study of HA gels, Saettone and coworkers (1989a) tested, together with gels, inserts prepared with two HA sodium salts (Hyalastin and

Hyalectin) and with a partial ethyl ester of HA. The inserts, prepared by casting or by compression, contained pilocarpine nitrate or tropicamide. When tested for pharmacological activity in rabbits, the pilocarpine inserts failed to produce significantly higher bioavailability than PAA or HA gels. The tropicamide inserts, however, produced better results, which the authors attributed to the lower solubility or diffusivity of the drug.

Benedetti et al. (1991) and Kyyrönen et al. (1992) evaluated in vitro and in vivo microspheres and films consisting of HA covalently bound to methylprednisolone (MP), and of HA partial ethyl and benzyl esters containing physically dispersed drug. In the MP ester, a nominal 50% of the carboxyl groups of HA were esterified to MP, and the remaining groups were present as the sodium salt. Significant increases in the tear film residence time of MP were observed after ocular administration of the films containing dispersed drug and of those containing MP chemically bound to the polymer. The authors concluded that drug delivery devices based on HA derivatives may prolong the absorption time and increase the ophthalmic availability of steroids.

The possibility of using biopolymers such as fibrin, collagen, or chitosan for the preparation of soluble or erodible inserts has been occasionally reported in the literature. Even if all of these materials have not been expressly defined as mucoadhesive, some of their applications deserve mention.

Fibrin, a biocompatible, biodegradable material prepared from human plasma, was used by Miyazaki et al. (1982) for the preparation of pilocarpine-loaded inserts. The inserts were evaluated in vitro for drug release and in vivo for miotic activity in rabbits. Although the in vitro data showed a faster release rate, with more than 90% of drug released in less than 1 hour, the in vivo activity profiles showed a prolonged effect (more than 8 hours versus 4–5 hours observed for an aqueous reference solution).

Collagen is a structural protein of bones, tendons, and ligaments that constitutes more than 25% of the total body protein in mammals. Collagen corneal bandages in the shape of a contact lens were first introduced by Fyodorov et al. (1985) as an alternative to soft contact lenses to protect the healing corneal epithelium after surgery. These devices, now called collagen shields, are manufactured by Bausch & Lomb (Bio-Cor), Chiron Ophthalmics (Medilens), and Alcon (ProShield) with collagen obtained from porcine scleral tissue or bovine corium tissue. Many studies have demonstrated the efficacy of collagen shields as tear substitutes (Bloomfield et al., 1977) and as drug delivery systems (Bloomfield et al., 1978; Mondino, 1991; Poland and Kaufman, 1988; O'Brien et al., 1988; Hwang et al., 1989; Pleyer et al., 1992; Chen et al., 1993; Hill et al., 1993). After ocular application, the shields absorb tear fluid from the ocular surface and begin to dissolve, "mimicking the surfactant-like properties of the mucin" (Marmer, 1988). The dissolution time of these devices can be prolonged by cross-linking the polymer to a variable extent, thus obtaining various types of shields with dissolution rates ranging between 12 and 72 hours (Aquavella et al., 1988). Kaufman et al. (1994) described a new application of collagen to ocular drug delivery, consisting of small collagen pieces, called Collasomes, suspended in a liquid viscous vehicle. Administration to human volunteers of Collasomes loaded with fluorescein as the model drug produced corneal and aqueous humor concentrations significantly higher than concentrations delivered by Collasome-free vehicles.

Chitosan is deacylated chitin, a cationic, cellulose-like biopolymer obtained industrially by alkaline hydrolysis of the aminoacetyl groups of chitin from crabs or shrimps (Greaves and Wilson, 1993). The mucoadhesive properties of chitosan were established by Lehr et al. (1992b) by measuring the force of detachment of polymer-coated cover glasses from pig intestinal mucosa. These authors suggested that a cationic polymer, such as chitosan, would probably be a superior adhesive especially in neutral or alkaline media, as desirable for adhesion to the eye. Indeed, positively charged polymeric hydrogels might develop additional attractive forces by electrostatic interactions with the negatively charged mucosal surfaces. Yomota et al. (1990) investigated the release of two chemicals, an acidic dye and pullulan, contained in chitosan films. These authors found that both compounds were released rapidly only after enzymatic degradation of the films by lysozyme present in tears. The release rates were controlled by the film degradation rates, which in turn depended on the number of acetyl groups on the polymeric chain and on pH: low pH values accelerated the degradation, whereas neutral pH decreased the degradation rate. Thus, chitosan, as well as the other biopolymers mentioned, appears to be an interesting prospective material for the preparation of mucoadhesive ocular dosage forms.

D.   Particulate Delivery Systems

We consider under this heading liposomes, nanoparticles, and microspheres, three different ocular delivery systems considered for the mucoadhesive approach. Liposomes are vesicles composed of a phospholipid membrane enclosing an aqueous volume. Depending on the method of production, these vesicles may consist of many or few layers, but they share the property of a liquid crystalline membrane, similar to the outer cell membrane. Liposomes are classified as multilamellar (MLVs) or unilamellar; the latter are further categorized as small unilamellar vesicles (SUVs, <100 nm) or large unilamellar vesicles (LUVs, >100 nm). Liposomes have been investigated as drug delivery systems for different routes, including ocular delivery (Davies et al., 1993; Mezei and Meisner, 1993).

Nano-microparticulates consist of polymeric dispersions containing entrapped, encapsulated, and/or adsorbed drugs. The difference between nanoparticles and microparticles (or microspheres) resides in their size: less or more than 1 μm, respectively. The upper size limit for solid microparticulates for ocular administration is about 5–10 μm, in analogy with traditional ophthalmic drug suspensions. Ocular applications of these systems have been described in several interesting reviews (Kreuter, 1983, 1990, 1993; Mezei and Meisner, 1993; Joshi, 1994; Zimmer and Kreuter, 1995). An overview of the essential literature on ophthalmic mucoadhesive particulate delivery systems is presented in Table 5.

1. Liposomes

The influence of PAA (whose mucoadhesive properties have been discussed in the previous sections) on the in vitro release and in vivo ocular bioavailability of liposome-entrapped drugs was investigated by Davies et al. (1992). Technetium-99m–labeled MLVs were coated with two PAA derivatives (Carbopol 934P and 1342) and evaluated for precorneal clearance in rabbits by dacrioscintigraphy, in comparison with uncoated vesicles. The coated vesicles demonstrated significantly enhanced precorneal retention with respect to uncoated ones at pH 5.0, and the precorneal drainage profiles at pH 7.4 were very similar. The same vesicles (coated with Carbopol 1342 and uncoated) containing tropicamide were subsequently tested for release in vitro and for mydriatic activity in rabbits. Even if prolonged activity was observed, the coated liposomes failed to increase significantly the mydriatic AUC of tropicamide with respect to uncoated ones and to the aqueous solution. The lack of a bioavailability-enhancing effect in vivo of the coated vesicles probably depended on the lachrymal pH, much closer to 7.4 than to 5.0, which is more favorable for retention.

The same group (Durrani et al., 1992) also evaluated the influence of PAA (Carbopol 1342) coating on in vitro release and in vivo bioavailability of pilocarpine-coated reverse-phase evaporation vesicles (REVs, a subtype of SUVs). The presence of coating reduced the drug release rate in vitro: the mucoadhesive film adsorbed onto the vesicles therefore provided a substantial barrier to drug release. When tested in rabbits, the coated vesicles showed a significantly prolonged duration of activity compared with uncoated ones and with an aqueous reference solution and a significantly increased miotic AUC with respect to the reference solution (Fig. 10). Both vesicular formulations also significantly reduced the maximal miosis intensity compared with the solution. The authors concluded that polymer-coated

Table 5 Synopsis of the Essential Literature on Mucoadhesive Particulate Formulations

Formulation	Drugs	Type of study	References
Liposomes	Tropicamide	Ocular clearance, mydriatic activity (rabbits)	Davies et al. (1992)
Microspheres	Pilocarpine	Miotic activity (rabbits)	Durrani et al. (1992)
	Methylprednisolone	In vitro and in vivo studies (rabbits)	Benedetti et al. (1991); Kyyrönen et al. (1992)
	—	Ocular clearance (rabbits)	Durrani et al. (1995)
	—	Ocular disposition (rabbits)	Wood et al. (1985)
Nanoparticles	—	Ocular distribution (rabbits, normal and inflamed eyes)	Diepold et al. (1989a)
	Amikacin sulfate	Ocular penetration (rabbits)	Losa et al. (1991)
	Progesterone	Ocular penetration (rabbits)	Li et al. (1986)
	Piloplex system	Clinical study (humans)	Klein et al. (1985)
	Pilocarpine	Pharmacological activity and/or ocular penetration (rabbits)	Harmia et al. (1986a, 1986b, 1986c); Diepold et al. (1989b); Zimmer et al. (1994, 1995)
	Timolol	Pharmacological activity and/or ocular penetration (rabbits)	Harmia-Pulkkinen et al. (1989)
	Betaxolol	Pharmacological activity and/or ocular penetration (rabbits	Marchal-Heussler et al. (1990, 1992)

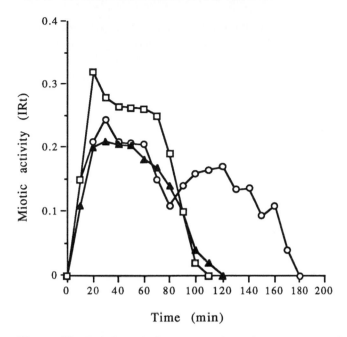

Figure 10 Relative miotic response intensity (IRt) versus time profiles after administration of three different formulations containing 0.5% pilocarpine nitrate at pH 5.0. (□) Phosphate buffer solution; (▲) uncoated REVs; (○) PAA coated REVs. (From Durrani et al., 1992.)

vesicles, or similar preparations, may provide a basis for improved ocular therapy.

2. Microspheres

Two papers by Benedetti et al. (1991) and by Kyyrönen et al. (1992) dealing with methylprednisolone esters of hyaluronic acid were mentioned in Sec. III.C (Inserts). These authors described, besides films, microspheres with diameters ranging from 1 to 10 μm and with an MP content of 22.2 ± 0.3%. Release tests in vitro and in vivo into rabbit tear fluid were carried out in comparison with MP suspensions and with microspheres containing MP incorporated in the polymer matrix. The microspheres (as well as the inserts) containing MP chemically bound to the polymer backbone through an ester linkage showed slower release of MP in vitro and produced sustained MP concentrations in the tear fluid of rabbits.

An investigation by Durrani et al. (1995) was concerned with the precorneal clearance of mucoadhesive microspheres. ^{111}In-labeled microspheres (size range 8–15 mm) composed of PAA (Carbopol 907) cross-linked with maltose were prepared by a water-oil emulsification process. In analogy with the previous liposome results, the prehydrated microspheres showed greater adhesion to a mucus gel at pH 5.0 than at pH 7.4. As indicated by the authors, at pH 5.0 the presence of protonated carboxyl groups permits enhanced mucoadhesion because of hydrogen bond formation with the hydroxyl groups of the glycoprotein molecules. For the precorneal clearance study, the radiolabeled microspheres were instilled in rabbits in suspension form at pH 5.0 and 7.4. The pH had no statistical effect on the preocular retention parameters (AUC, $t_{1/2}$) of the formulations. The clearance of microspheres administered in dry form was faster than in the hydrated form, probably due to incomplete hydration in the tear fluid. In any case, the results of this study were considered satisfactory: the retention on the ocular surface of approximately 25% of the instilled dose after 167 minutes suggested, according to the authors, that this technology may find future applications for controlled ophthalmic drug delivery.

3. Nanoparticles

The materials most commonly used for the preparation of ophthalmic nanoparticles are polyalkyl (e.g., polybutyl, polyhexyl) cyanoacrylates (Kreuter, 1993). These, because of their chemical structure of PAA derivatives, are mucoadhesive. Nanoparticles were first investigated as potential ophthalmic drug carriers by Wood et al. (1985). These authors prepared ^{14}C-labeled polyhexyl-2-cyanoacrylate nanoparticles and investigated their degradation in vitro in rabbit tears and the ocular disposition in tears, aqueous humor, cornea, and conjunctiva of rabbits. Degradation in tears was relatively rapid during the first hour (about 19%) and leveled off during the following 5 hours. The in vivo study showed that although the particles were rapidly removed from the precorneal area as the result of drainage, they were better retained than drug solutions, possibly due to adhesion to the mucin-epithelial surface of the cornea and conjunctiva. However, pretreatment of the eyes with the mucolytic agent N-acetylcysteine failed to reduce the preocular retention significantly, suggesting that the nanoparticles were also able to adhere directly to the corneal epithelium. Binding of polyalkylcyanoacrylate nanocapsules to the cornea and conjunctiva, particularly at early times after instillation, was confirmed by Kreuter (1990) and by Diepold et al. (1989a). The latter authors also reported that the concentration of nanoparticles in chronically inflamed eyes of rabbits is three to five times higher than in normal eyes, leading to interest in these systems for ocular delivery of anti-inflammatory steroids or NSAIDs.

Numerous studies (e.g., Harmia et al., 1986a, 1986b, 1986c; Diepold et al., 1989a, 1989b; Harmia-Pulkkinen, 1989; Marchal-Heussler et al., 1990;

Losa et al., 1991; Zimmer et al., 1994) have established the potential efficacy of polyalkylcyanoacrylate nanoparticles in ocular drug delivery. Pilocarpine appears to be the drug of choice for incorporation in these nanoparticles, but progesterone (Li et al., 1986), β-blocking agents (Harmia-Pulkkinen et al., 1989; Marchal-Heussler et al., 1990, 1992), and antibiotics (Losa et al., 1991) have also been tested.

Two particular nanoparticulate systems for delivery of pilocarpine deserve special attention. One is Piloplex, developed and patented in Israel (Ticho et al., 1978), consisting of pilocarpine ionically bound to poly(methyl)methacrylate-co-acrylic acid nanoparticles. Twice-daily instillations of Piloplex in glaucoma patients were claimed to be as effective as three to six instillations of traditional pilocarpine eyedrops (Klein et al., 1985).

The other system, described by Zimmer et al. (1995), consists of coad ministration of pilocarpine-loaded albumin nanoparticles with viscosity-enhancing polymers (e.g., PVA) or bioadhesive polymers such as mucin, PAA, hyaluronic acid, or sodium carboxymethylcellulose. These systems were tested in rabbits for miotic activity and for reduction of intraocular pressure (betamethasone model). The best results for miotic response and for intraocular pressure reduction were observed with bovine submaxillary mucin as the mucoadhesive polymer. Nanoparticles alone were equivalent in efficacy to solutions containing commercially available viscous or bioadhesive polymers; coadministration with mucin resulted in a 1.56- to 1.66-fold increase in hypotensive and miotic activity with respect to nanoparticles alone (Table 6).

Table 6 Summary of Some Miotic and Hypotensive Activity Parameters of Albumin Nanoparticles Coadministered with Mucin

Vehicle	Hypotensive AUC	Relative hypotensive AUC	Miotic AUC	Relative miotic AUC
Pilocarpine, 2% reference solution	858.32	1.00	254.37	1.00
Pilocarpine, 2% nanoparticles	1331.32*	1.55	383.37**	1.50
Mucin reference solution	1039.12	1.21	380.55	1.49
Nanoparticles plus mucin	2077.00**	2.42	631.80**	2.48

Statistically significant difference from the corresponding reference: $*P = .05$; $**P = .01$, Student's t-test.
From Zimmer et al. (1985).

The authors assumed that their results were due to improved adhesion of the pilocarpine-loaded albumin nanoparticles to the precorneal or conjunctival mucin layer and hence to prolongation of the residence time of the medication in the eye.

IV. CONCLUSIONS

The literature on ocular mucoadhesion definitely indicates that most if not all types of mucoadhesive vehicles meet the expectation of a prolonged or improved retention in the preocular area. The temporary weakening of the barrier properties of the corneal epithelium, indicated for some mucoadhesive polymers, also appears very interesting. However, a recurrent statement in papers dealing with ocular mucoadhesion is that a mucoadhesive vehicle is insufficient per se to secure improved drug bioavailability. If a very soluble or diffusible drug (e.g., pilocarpine) is not adequately retained within the vehicle, it will rapidly diffuse out of it, and the potential benefits provided by mucoadhesion will be lost. Enhanced retention can be obtained, in these cases, by incorporating the drug in the vehicle as a polymer salt, a poorly soluble complex, a micro- or nanoparticle preparation, etc.

It is unfortunate that most studies of ocular mucoadhesion have been performed with rabbits, animals repeatedly found to be poor models for vehicle effects on ocular bioavailability. The relatively few tests performed on humans have, however, confirmed the potential usefulness of the mucoadhesive approach.

Mucoadhesion as a science is still relatively young and in an experimental stage and does not appear to have yet reached, at least in the ophthalmic field, the stage of commercial applications. It should be remembered in this context that every innovation in ophthalmic drug delivery, before reaching the stage of clinical trials, must fight against the traditional reluctance of ophthalmologists and pharmaceutical companies to abandon traditional, well-established dosage forms such as eyedrops and ointments.

Even if manufacturers of ocular dosage forms appear to proceed at a very cautious pace, some drugs already in use have been revived in new, longer acting *liquid* presentations advertised for once-daily application. Although none of these delivery systems is explicitly indicated as mucoadhesive, some of them seem to have been developed in the light of the vast body of knowledge derived from studies on bio- or mucoadhesion. These are, e.g., an "in situ" gelling preparation of timolol (Timoptic XE, Merck), betaxolol adsorbed on an ion-exchange resin (Betoptic S ophthalmic suspension, Alcon Laboratories), levobunolol in a liquid viscous vehicle (Betagan in Liquifilm, Allergan Pharmaceuticals). *Semisolid*, gel-type preparations, such as the older Pilopine HS gel (Alcon Laboratories), in spite of the good mucoadhesive performances of polymeric hydrogels, do not seem to occupy an important position in ophthalmologists' preferences. The time appears ripe, however, for the appearance of innovative ocular delivery systems, expressly designed to incorporate the concepts and techniques of mucoadhesion.

REFERENCES

Adams AD. 1979. The morphology of human conjunctival mucus. Arch Ophthalmol 97:730–734.

Albasini M, Ludwig A. 1995. Evaluation of polysaccharides intended for ophthalmic use in ocular dosage forms. Farmaco 50:633–642.

Aquavella JV, Ruffini JJ, LoCascio JA. 1988. Effect of collagen shields as a surgical adjunct. J Cataract Refract Surg 14:492–498.

Baum L, Barza M, Shushan D, Weinstein L. 1974. Concentration of gentamicin in experimental corneal ulcers. Arch Ophthalmol 92:315–317.

Bawa G. 1993. Ocular inserts. In: Mitra AK, ed. Ophthalmic Drug Delivery Systems. New York: Marcel Dekker, pp 223–259.

Benedetti LM, Kyyrönen K, Hume L, Topp E, Stella V. 1991. Steroid ester of hyaluronic acid in ophthalmic drug delivery. Proceedings of International Symposium on Controlled Release of Bioactive Materials, Amsterdam, pp 497–498.

Berman ER. 1991. Tears. In: Blakemore C, ed. Biochemisry of the eye. New York: Plenum, pp 63–88.

Bernatchez SF, Camber O, Tabatabay C, Gurny R. 1993. Use of hyaluronic acid in ocular therapy. In: Edman P, ed. Biopharmaceutics of Ocular Drug Delivery. Boca Raton, FL: CRC Press, pp 105–120.

Bloomfield SE, Miyata T, Dunn MW, Bueser N, Stenzel KH, Rubin AL. 1977. Soluble artificial tear inserts. Arch Ophthalmol 95:247–250.

Bloomfield SE, Miyata T, Dunn MW, Bueser N, Stenzel KH, Rubin AL. 1978. Soluble gentamicin ophthalmic inserts. Arch Ophthalmol 96:885–887.

Bottari F, Giannaccini B, Peverini D, Saettone MF, Tellini N. 1979. Semisolid ophthalmic vehicles. II: Evaluation in albino rabbits of aqueous gel–type vehicles containing lidocaine and benzocaine. Can J Pharm Sci 14:39–43.

Burgalassi S, Chetoni P, Saettone MF. 1996. Hydrogels for ocular delivery of pilocarpine: Preliminary evaluation in rabbits of the influence of viscosity and of drug solubility. Eur J Pharm Biopharm 42:385–392.

Buri P. 1985. Voie oculaire. In: Buri P, Puisieux F, Doelker E, Benoit JP, eds. Formes Pharmaceutiques Nouvelles. Paris: Lavoisier, pp 411–438.

Burstein NL. 1995. Ophthalmic drug formulations. In: Bartlett JD, Jaanus SD, eds. Clinical Ocular Pharmacology, 3rd ed. Newton, MA: Butterworth-Heinemann, pp 21–45.

Calabria G, Rolando M. 1984. Struttura e funzioni del film lacrimale. Proceedings of the 64th Symposium of the Italian Ophthalmological Society (S.O.I.), Genoa, pp 9–35.

Camber O, Edman P. 1989. Sodium hyaluronate as an ophthalmic vehicle: Some factors governing its effect on the ocular bioavailability of pilocarpine. Curr Eye Res 8:563–567.

Camber O, Edman P, Gurny R. 1987. Influence of sodium hyaluronate on the miotic effect of pilocarpine. Curr Eye Res 6:779–784.

Chang SC, Chien DS, Bundgaard H, Lee VHL. 1988. Relative effectiveness of prodrug and viscous solution approach in maximizing the ratio of ocular to systemic absorption of topically applied timolol. Exp Eye Res 46:59–69.

Chen CC, Takruri H, Duzman E. 1993. Enhancement of the ocular bioavailability of topical tobramycin with use of a collagen shield. J Cataract Refract Surg 19:242–245.

Chetoni P, Di Colo G, Morelli M, Saettone MF. 1996. Mucoadhesive silicone inserts ensuring prolonged retention and sustained drug delivery: Preliminary in vitro/ in vivo evaluation. Proceedings of 3rd Jerusalem Conference on Pharmaceutical Sciences and Clinical Pharmacology, Jerusalem, p 50.

Chien YW. 1982. Ocular controlled-release drug administraiton. In: Swarbrick J, ed. Novel Drug Delivery Systems. New York: Marcel Dekker, pp 13–48.

Chiou GCY, Watanabe K. 1982. Drug delivery to the eye, Pharm Ther 17:269–278.

Darougar S. 1988. Ocular insert. European Patent Application 262-893-A2.

Darougar S. 1992. Ocular insert for the fornix. U.S. Patent 5,147,647.

Davies NM, Farr SJ, Hadgraft J, Kellaway IW. 1991. Evaluation of mucoadhesive polymers in ocular drug delivery. I. Viscous solutions. Pharm Res 8:1039–1043.

Davies NM, Farr SJ, Hadgraft J, Kellaway IW. 1992. Evaluation of mucoadhesive polymers in ocular drug delivery. II. Polymer-coated vesicles. Pharm Res 9:1137–1144.

Davies NM, Kellaway IW, Greaves JL, Wilson CG. 1993. Advanced corneal delivery systems: Liposomes. In: Mitra AK, ed. Ophthalmic Drug Delivery Systems. New York: Marcel Dekker, pp 289–306.

Diepold R, Kreuter J, Guggenbuhl P, Robinson JR. 1989a. Distribution of poly-hexyl-2-cyano[3-^{14}C]acrylate nanoparticles in healthy and chronically inflamed rabbit eyes. Int J Pharm 54:149–153.

Diepold R, Kreuter J, Himber J, Gurny R, Lee VHL, Robinson JR, Saettone MF, Schnaudigel OE. 1989b. Comparison of different models for the testing of pilocarpine eyedrops using conventional eyedrops and a novel depot formulation (nanoparticles). Graefes Arch Clin Exp Ophthalmol 227:188–193.

Dittgen M, Oestereich S, Eckhardt D. 1992. Influence of bioadhesion on the elimination of drugs from the eye and on their penetration ability across the pig cornea. STP Pharma 2:93–97.

Duchêne D, Touchard F, Peppas NA. 1988. Pharmaceutical and medical aspects of bioadhesive systems for drug administration. Drug Dev Ind Pharm 15:283–318.

Durrani AM, Davies NM, Thomas M, Kellaway IW. 1992. Pilocarpine bioavailability from mucoadhesive liposomal ophthalmic drug delivery system. Int J Pharm 88:409–415.

Durrani AM, Farr SJ, Kellaway IW. 1995. Precorneal clearance of mucoadhesive microspheres from the rabbit eye. J Pharm Pharmacol 47:581–584.

Fyodorov SN, Moroz ZI, Kramskaya ZI, Bagrov SN, Amstislavskaya TS, Zolotarevsky AV. 1985. Comprehensive conservative treatment of dystrophia endothelialis and epithelialis corneae, using a therapeutic collagen coating. Vestn Oftalmol 101:33–37.

Greaves JL, Wilson CG. 1993. Treatment of diseases of the eye with mucoadhesive delivery systems. Adv Drug Delivery Rev 11:349–383.

Gurny R, Ibrahim H, Aebi A, Buri P, Wilson CG, Washington N, Edman P, Camber O. 1987. Design and evaluation of controlled release systems for the eye. J Controlled Release 6:367–373.

Gurtler F, Gurny R. 1993. Insert ophthalmique bioadhésif. European Patent Application 561-695-A1.

Gurtler F, Gurny R. 1995. Patent literature review of ophthalmic inserts. Drug Dev Ind Pharm 21:1–18.

Gurtler F, Kaltsatos V, Boisramé B, Delaforge J, Gex-Fabry M, Balant LP, Gurny R. 1995b. Ocular availability of gentamicin in small animals after topical administration of a conventional eye drop solution and a novel long acting bioadhesive ophthalmic drug insert. Pharm Res 12:1791–1795.

Gurtler F, Kaltsatos V, Boisramé B, Gurny R. 1995a. Long-acting soluble bioadhesive ophthalmic drug insert (BODI) containing gentamicin for veterinary use: Optimization and clinical investigation. J Controlled Release 33; 231–236.

Habib FS, Attia MA. 1984. Comparative study of ocular activity in rabbits eyes of adrenaline bitartrate formulated in Carbopol and poloxamer gels. Arch Pharm Chem Ed 12:91–96.

Harmia T, Kreuter J, Speiser P, Boye T, Gurny R, Kubis A. 1986a. Enhancement of the miotic response of rabbits with pilocarpine loaded polybutylcyanoacrylate nanoparticles. Int J Pharm 33:187–193.

Harmia T, Speiser P, Kreuter J. 1986b. Optimization of pilocarpine loading onto nanoparticles by sorption procedures. Int J Pharm 33:45–54.

Harmia T, Speiser P, Kreuter J. 1986c. A solid colloidal drug delivery system for the eye: Encapsulation of pilocarpine in nanocapsules. J Microencapsul 3:3–12.

Harmia-Pulkkinen T, Tuomi A, Kristofferson E. 1989. Manufacture of polyalkylcyanoacrylate nanoparticles with pilocarpine and timolol by micelle polymerization: Factors influencing particle formation. J Microencapsul 6:87–93.

Hassan EE, Gallo JM. 1990. A simple rheological method for the in vitro assessment of mucin-polymer bioadhesive bond strength. Pharm Res 7:491–495.

Hazlett LD, Barrett R. 1987. Sodium hyaluronate eye drop. A scanning and transmission electron microscopy study of the corneal surface. Ophthalmic Res 19: 277–284.

Hill JM, O'Callaghan RJ, Hobden JA, Kaufman HE. 1993. Corneal collagen shields for ocular delivery. In: Mitra AK, ed. Ophthalmic Drug Delivery Systems. New York: Marcel Dekker, pp 261–273.

Holly FJ. 1986. Tear film formation and rupture. In: Holly FJ, ed. The Precorneal Tear Film in Health, Disease, and Contact Lens Wear. Lubbock, TX: The Dry Eye Institute, pp 634–645.

Holly FJ, Lemp MA. 1971. Wettability and wetting of corneal epithelium. Exp Eye Res 11:239–250.

Holly FJ, Lemp MA. 1977. Tear physiology and dry eye. Surv Ophthalmol 22:69–87.

Hui HW, Robinson JR. 1985. Ocular delivery of progesterone using a bioadhesive polymer. Int J Pharm 26:203–213.

Huupponen R, Kaila T, Saettone MF, Monti D, Iisalo E, Salminen L, Oksala O. 1992. The effect of some macromolecular ionic complexes on the pharmacokinetics and -dynamcis of ocular cyclopentolate in rabbits. J Ocul Pharmacol 8:59–67.

Hwang DG, Sern WH, Hwang PH, MacGowan-Smith LA. 1989. Collagen shield enhancement of topical dexamethasone penetration. Arch Ophthalmol 107:1375–1380.

Joshi A. 1994. Microparticulates for ophthalmic drug delivery. J Ocul Pharmacol 10:29–45.

Katz IM, Blackman WM. 1977. A soluble sustained-release ophthalmic delivery unit. Am J Ophthalmol 83:728–734.

Kaufman HE, Steinemann TL, Lehman E, Thompson HW, Varnell ED, Jacob-Labarre JT, Gebhardt BM. 1994. Collagen-based drug delivery and artificial tears. J Ocul Pharmacol 10:17–27.

Klein HZ, Lugo M, Shields B, Leon J, Duzman E. 1985. A dose-response study of Piloplex for duration of action. Am J Ophthalmol 99:23–26.

Kreuter J. 1983. Evaluation of nanoparticles as drug-delivery systems. I: Preparation methods. Pharm Acta Helv 58:196–207.

Kreuter J. 1990. Nanoparticles as bioadhesive ocular drug delivery systems. In: Lenaerts V, Gurny R, eds. Bioadhesive Drug Delivery Systems. Boca Raton, FL: CRC Press, pp 203–212.

Kreuter J. 1993. Particulates (nanoparticles and microparticles). In: Mitra AK, ed. Ophthalmic Drug Delivery Systems. New York: Marcel Dekker, pp 275–287.

Krishnamoorthy R, Mitra AK. 1993. Mucoadhesive polymers in ocular drug delivery. In: Mitra AK, ed. Ophthalmic Drug Delivery Systems. New York: Marcel Dekker, pp 199–221.

Kupferman A, Ryan WJ, Leibowitz HM. 1981. Prolongation of anti-inflammatory effect of prednisolone acetate. Influence of formulation in high-viscosity gel. Arch Ophthalmol 99:2028–2029.

Kyyrönen K, Hume L, Benedetti L, Urtti A, Topp E, Stella V. 1992. Methylprednisolone esters of hyaluronic acid in ophthalmic drug delivery: In vitro and in vivo release studies. Int J Pharm 80:161–169.

Lahdes K, Huupponen R, Kaila T, Monti D, Saettone MF, Salminen L. 1993. Plasma concentrations and ocular effects of cyclopentolate after ocular application of three formulations. Br J Clin Pharmacol 35:479–483.

Lee VHL. 1990. Review: New directions in optimization of ocular drug delivery. J Ocul Pharmacol 6:157–164.

Lee VHL, Robinson JR. 1979. Mechanistic and quantitative evaluation of precorneal pilocarpine disposition in albino rabbits. J Pharm Sci 6:673–684.

Lee VHL, Robinson JR. 1986. Review: Topical ocular drug delivery: Recent developments and future challenges. J Ocul Pharmacol 2:67–108.

Lee YH, Kompella UB, Lee VHL. 1993. Systemic absorption pathways of topically applied beta adrenergic antagonists in the pigmented rabbit. Exp Eye Res 57: 341–349.

Lehr C-M. 1996. From sticky stuff to sweet receptors—achievements, limits and novel approaches to bioadhesion. Eur J Drug Metab Pharmacokinet 21:139–148.

Lehr C-M, Bouwstra JA, Boddé HE, Junginger HE. 1992a. A surface energy analysis of mucoadhesion. In: Contact angle measurements on polycarbophil and pig intestinal mucosa in physiologically relevant fluids. Pharm Res 9:1051–1059.

Lehr C M, Bouwstra JA, Schacht EH, Junginger HE. 1992b. In vitro evaluation of mucoadhesive properties of chitosan and some other natural polymers. Int J Pharm 78:43–48.

Lehr C-M, Lee YH, Lee VHL. 1994. Improved ocular penetration of gentamicin by mucoadhesive polymer polycarbophil in the pigmented rabbit. Invest Ophthalmol Vis Sci 35:2809–2814.

Leung SHS, Robinson JR. 1990. Bioadhesives in drug delivery. Polym News 15: 333–344.

Li VHK, Wood RW, Kreuter J, Harmia T, Robinson JR. 1986. Ocular delivery of progesterone using nanoparticles. J Microencapsulation 3:213–218.

Losa C, Calvo P, Castro E, Vila-Jato L, Alonso MJ. 1991. Improvement of ocular penetration of amikacin sulphate by association to poly(butylcyanoacrylate) nanoparticles. J Pharm Pharmacol 43:548–552.

Madsen K, Schenholm M, Jahnke G, Tangblad A. 1989. Hyaluronate binding to intact corneas and cultured endothelial cells. Invest Ophthalmol Vis Sci 30: 2132–2137.

Marchal-Heussler L, Fessi H, Devissaguet JP, Hoffman M, Spittler J, Maincent P. 1992. Colloidal drug delivery systems for the eye: A comparison of the efficacy of three different polymers: Polyisobutylcyanoacylate, polylactic-co-glycolic acid, poly-epsilon-caprolacton. STP Pharma 2:98–104.

Marchal-Heussler L, Maincent P, Hoffman M, Spittler J, Couvreur P. 1990. Antiglaucomatous activity of betaxolol chlorhydrate sorbed onto different isobutylcyanoacrylate nanoparticles preparations. Int J Pharm 58:115–122.

Marmer RH. 1988. Therapeutic and protective properties of the corneal collagen shield. J Cataract Refract Surg 14:496–499.

Maurice DM, Mishima S. 1984. Ocular pharmacokinetics. In: Sears MC, ed. Pharmacology of the eye. Berlin: Springer-Verlag, pp 19–116.

Mengi S, Deshpande SG. 1992. Development and evaluation of flurbiprofen hydrogels on the breakdown of the blood/aqueous humor barrier. STP Pharma 2: 118–124.

Mezei M, Meisner D. 1993. Liposomes and nanoparticles as ocular drug delivery systems. In: Edman P, ed. Biopharmaceutics of Ocular Drug Delivery. Boca Raton, FL: CRC Press, pp 91–104.

Middleton DL, Leung SS, Robinson JR. 1990. Ocular bioadhesive delivery systems. In: Lenaerts V, Gurny R, eds. Bioadhesive Drug Delivery Systems. Boca Raton, FL: CRC Press, pp 179–202.

Middleton DL, Robinson JR. 1991. Design and evaluation of a ocular bioadhesive delivery system. STP Pharma 1:200–206.

Mikkelson TJ. 1984. Ophthalmic drug delivery. Pharm Technol 91:89–98.

Miyazaki S, Ishii K, Takada M. 1982. Use of fibrin film as a carrier for drug delivery: A long-acting delivery system for pilocarpine into the eye. Chem Pharm Bull 30:3405–3407.

Mondino BJ. 1991. Collagen shields. Am J Ophthalmol 112:587–590.

Nagai T, Machida Y, Suzuki Y, Ikura H. 1980. Method and preparation for administration to the mucosa of the oral or nasal cavity. U.S. Patent 4,226,848, October 7, 1980.

Norn MS. 1963. Mucus on conjunctiva and cornea. Acta Ophthalmol 41:13–24.

O'Brien TP, Sawusch MR, Dick JD, Hamburg TR, Gottsch JD. 1988. Use of collagen corneal shield versus soft contact lenses to enhance penetration of topical tobramycin. J Cataract Refract Surg 14:505–507.

Park K, Robinson JR. 1984. Bioadhesive polymers as platforms for oral controlled delivery: Method to study bioadhesion. Int J Pharm 19:107–127.

Patton TF, Robinson JR. 1975. Ocular evaluation of polyvinyl alcohol vehicle in rabbits. J Pharm Sci 65:264–276.

Pleyer U, Legmann A, Mondino BJ, Lee DA. 1992. Use of collagen shields containing amphotericin B in the treatment of experimental Candida albicans–induced keratomycosis in rabbits. Am J Ophthalmol 113:303–307.

Poland DE, Kaufman HE. 1988. Clinical uses of collagen shields. J Cataract Refract Surg 14:489–491.

Richardson KT. 1975. Ocular microtherapy: Membrane-controlled drug delivery. Arch Ophthalmol 93:74–86.

Robinson JC. 1989. Ocular drug delivery. Mechanism(s) of corneal drug transport and mucoadhesive delivery systems. STP Pharma 5:839–846.

Robinson JR. 1990. Mucoadhesive ocular drug delivery systems. In: Gurny R, Junginger HE, eds. Bioadhesion—Possibilities and Future Trends. Stuttgart: Wissenschaftliche Verlagsgesellschaft, pp 109–123.

Robinson JR. 1993. Ocular anatomy and physiology relevant to ocular drug delivery. In: Mitra AK, ed. Ophthalmic Drug Delivery Systems. New York: Marcel Dekker, pp 29–57.

Saettone MF. 1993. Solid Polymeric Inserts/Disks as Drug Delivery Devices. In: Edman P, ed. Biopharmaceutics of Ocular Drug Delivery. Boca Raton, FL: CRC Press, pp 61–79.

Saettone MF, Chetoni P, Torracca MT, Burgalassi S, Giannaccini B. 1989a. Evaluation of muco-adhesive properties and in vivo activity of ophthalmic vehicles based on hyaluronic acid. Int J Pharm 51:203–212.

Saettone MF, Giannaccini B, Chetoni P, Galli G, Chiellini E. 1984b. Vehicle effects on ophthalmic bioavailability: An evaluation of polymeric inserts containing pilocarpine. J Pharm Pharmacol 36:229–234.

Saettone MF, Giannaccini B, Chetoni P, Torracca MT, Monti D. 1991. Evaluation of high- and low-molecular-weight fractions of sodium hyaluronate and an ionic complex as adjuvants for topical ophthalmic vehicles containing pilocarpine. Int J Pharm 72:131–139.

Saettone MF, Giannaccini B, Guiducci A, La Marca G, Tota G. 1985. Polymer effects on ocular bioavailability. II. The influence of benzalkonium chloride on mydriatic response of tropicamide in different polymeric vehicles. Int J Pharm 25: 73–83.

Saettone MF, Giannaccini B, Guiducci A, Savigni P. 1986. Semisolid ophthalmic vehicles. III. An evaluation of four organic hydrogels containing pilocarpine. Int J Pharm 31:261–270.

Saettone MF, Giannaccini B, Ravecca S, La Marca F, Tota G. (1984a). Polymer effects on ocular bioavailability—the influence of different liquid vehicles on the mydriatic response of tropicamide in humans and in rabbits. Int J Pharm 20:187–202.

Saettone MF, Giannaccini B, Savigni P, Wirth A. 1980. The effect of different ophthalmic vehicles on the activity of tropicamide in man. J Pharm Sci 32:519–521.

Saettone MF, Giannaccini B, Teneggi A, Savigni P, Tellini N. 1982. Vehicle effects on ophthalmic bioavailability: The influence of different polymers on the activity of pilocarpine in rabbit and man. J Pharm Pharmacol 34:464–466.

Saettone MF, Giannaccini B, Torracca MT, Burgalassi S. 1987. An evaluation of the bioadhesive properties of hyaluronic acid. Proceedings of the 3rd European Congress of Biopharmaceutics and Pharmacokinetics, Clermont-Ferrand, France, Vol 1, pp 413–417.

Saettone MF, Monti D, Giannaccini B, Salminen L, Huupponen R. 1992. Macromolecular ionic complexes of cyclopentolate for topical ocular administration. Preparation and preliminary evaluation in albino rabbits. STP Pharma 2:68–75.

Saettone MF, Monti D, Torracca MT, Chetoni P. 1994. Mucoadhesive ophthalmic vehicles: Evaluation of polymeric low-viscosity formulations. J Ocul Pharmacol 10:83–92.

Saettone MF, Monti D, Torracca MT, Chetoni P, Giannaccini B. 1989b. Muco-adhesive liquid ophthalmic vehicles—evaluation of macromolecular ionic complexes of pilocarpine. Drug Dev Ind Pharm 15:2475–2489.

Saettone MF, Salminen L. 1995. Ocular inserts for topical delivery. Adv Drug Delivery Rev 16:95–106.

Salminen L. 1987. Pilocarpine inserts: Experimental and clinical experiences. In: Saettone MF, Bucci G, Speiser P, eds. Ophthalmic Drug Delivery; Biopharmaceutical, Technological and Clinical Aspects. Fidia Research Series. Padova, Italy: Liviana Press, pp 161–170.

Sasaki H, Yamamura K, Mishida K, Nakamura J, Ichikawa M. 1996. Delivery of drugs to the eye by topical application. Prog Retinal Eye Res 15:583–620.

Schoenwald RD, Boltralik JJ. 1979. A bioavailability comparison in rabbits of two steroids formulated as high-viscosity gels and reference aqueous preparations. Invest Ophthalmol Vis Sci 18:61–66.

Schoenwald RD, Ward RL, DeSantis LM, Roehrs RE. 1978. Influence of high-viscosity vehicles on miotic effect of pilocarpine. J Pharm Sci 67:1280–1283.

Shell JW. 1980. New ophthalmic drug delivery systems. In Robinson JR, ed. Ophthalmic Drug Delivery Systems. Washington, DC: American Pharmaceutical Association, pp 71–90.

Slovin EM, Robinson JR. 1993. Bioadhesives in ocular drug delivery. In: Edman P, ed. Biopharmaceutics of Ocular Drug Delivery. Boca Raton, FL: CRC Press, pp 145–157.

Smart JD, Kellaway IW, Worthington HEC. 1984. An in-vitro investigation of mucosa-adhesive materials for use in controlled drug delivery. J Pharm Pharmacol 36:295–299.

Snibson GR, Greaves JL, Soper NDW, Prydal JI, Wilson CG, Bron AJ. 1990. Precorneal residence times of sodium hyaluronate solutions studied by quantitative gamma scintigraphy. Eye 4:594–602.

Thermes F, Grove J, Chastaing G, Rozier A, Soper K, Plazonnet B. 1992b. Effect of the mucolytic agent, acetylcysteine, on bioadhesion and ocular penetration. STP Pharma 2:88–92.

Thermes F, Rozier A, Plazonnet B, Grove J. 1992a. Bioadhesion: The effect of polyacrylic acid on the ocular bioavailability of timolol. Int J Pharm 81:59–65.

Ticho U, Blumenthal M, Zonis S, Gal A, Blank I, Mazor Z. 1978. A new long-acting pilocarpine compound. Ann Ophthalmol 11:555–561.

Urtti A, Salminen L. 1993. Minimizing systemic absorption of topically administered ophthalmic drugs. Surv Ophthalmol 88:565–571.

Wood RW, Li VHK, Kreuter J, Robinson JR. 1985. Ocular disposition of poly-hexyl-2-cyano[3-^{14}C]acrylate nanoparticles in the albino rabbit. Int J Pharm 23:175–183.

Yomota C, Komuro T, Kimura T. 1990. Studies on the degradation of chitosan films by lysozyme and release of loaded chemicals. Yakugaku-Zasshi 110:442–448.

Zimmer A, Mutschler E, Lambrecht G, Mayer D, Kreuter J. 1994. Pharmacokinetic and pharmacodynamic aspects of an ophthalmic pilocarpine nanoparticle delivery system. Pharm Res 11:1435–1452.

Zimmer AK, Chetoni P, Saettone MF, Zerbe H, Kreuter J. 1995. Evaluation of pilocarpine-loaded albumin particles as controlled drug delivery systems for the eye. II. Co-administration with bioadhesive and viscous polymers. J Controlled Release 33:31–46.

Zimmer AK, Kreuter J. 1995. Microspheres and nanoparticles used in ocular delivery systems. Adv Drug Delivery Rev 16:61–73.

23
Bioadhesive Preparations as Topical Dosage Forms

Yoshiharu Machida and Tsuneji Nagai
Hoshi University, Ebara, Shinagawa-ku, Tokyo, Japan

I. MERITS OF BIOADHESIVE PREPARATIONS AS TOPICAL DOSAGE FORMS

Local treatment should be defined as "treatment of topical disease via direct administration of drug to specified affected parts of the body." The effects of local treatment, therefore, in addition to the therapeutic effect of the drug itself, are dependent on the abilities of excipients to protect the affected part and retain the drug within this region.

We produced a solid bioadhesive dosage form for a physician who was trying to treat uterine cervical cancer by topical chemotherapy (Machida et al., 1979). This was introduced on the market as the first oral mucosal adhesive dosage form Aftach, as will be mentioned later. Until the commencement of joint research, the physician used a stick-shaped suppository with Witepsol as a base. However, it melted quickly after insertion, and the anticancer agent was released and affected nonfocal parts such as the vaginal mucosa. After initial discussion with the physician and hospital pharmacists, we prepared a new dosage form for treatment of uterine cervical cancer by direct compression of powder mixtures consisting of polymer(s) and anticancer agent. Hydroxypropylcellulose (HPC) was chosen as the main polymer because of our experience with its use as an excipient with the characteristics of a binding agent (Machida and Nagai, 1974) and release-sustaining agent (Machida and Nagai, 1978).

The details of the dosage form for treatment of uterine cancer will be given later. This prototype bioadhesive preparation has the following merits as a topical dosage form.

1. Sufficient therapeutic effect with reduced side effects
2. Prolongation of drug effect by sustained drug release
3. Protection of diseased part from external irritation

The last point mentioned is not related to the dosage form for uterine cancer but to the first commercial bioadhesive preparation of aphthous stomatitis (Aftach, Teijin Ltd., Japan) in which the preparation adheres to and covers the painful diseased part. The Japan National Invention Prize was awarded for this product, as a new drug developed in pharmaceutical research on drug delivery systems (DDS).

All of the merits are due to the bioadhesive and sustained-release properties of polymers in the preparation. Usually, polymers combined in such preparations are called "excipients." However, the polymers combined in bioadhesive preparations function as more than excipients and are essential for the design of this new type of dosage form, which can be designated as a "topical drug delivery system." In this chapter, we will describe bioadhesive preparations designed for topical drug delivery to various parts of the body.

II. EXAMPLES OF BIOADHESIVE PREPARATIONS AS TOPICAL DOSAGE FORM

A. Conjunctival Sac

Because the eye is the most delicate sense organ, ophthalmic dosage forms have to be aseptic and nonirritant. Moreover, the size and materials used for the preparation should be limited so as not to disturb vision. Therefore, transparent gels or semisolid preparations and microparticulate preparations are possible dosage forms for bioadhesive ophthalmic preparations.

Sheardown et al. (1997) examined the healing effect of epidermal growth factor (EGF) on corneal epithelial wounds in New Zealand white rabbits using a semisolid dosage form. Gels containing 0.04–1% EGF were prepared using polyacrylic acid (Carbopol 940). Aliquots of 50 μL of the gel containing EGF or gel only were placed in the inferior fornix of the shallow circular anterior keratotomized eye. The gel remained in place until removal 8 hours after application. Table 1 summarizes the wound healing results. The highest healing rate was obtained with 0.40% EGF gel. The authors concluded that 0.40% is the optimal concentration of the gel, and slow release of EGF using this gel may be an effective delivery method.

Table 1 Summary of Healing Results of Gels Containing Epidermal Growth Factor (EGF) in Corneal Epithelial Wounds of Rabbits

Gel EGF concentration (%)	Healing rate (mm/h)	Enhancement factor
Placebo	0.0243 ± 0.0015	—
0.04	0.0275 ± 0.0025	1.13 ± 0.12
0.10	0.0340 ± 0.0028	1.40 ± 0.14^{a}
0.20	0.0315 ± 0.0021	1.30 ± 0.12^{a}
0.40	0.0438 ± 0.0044	1.81 ± 0.22^{a}
1.00	0.0266 ± 0.0022	1.10 ± 0.12

[a]Significantly different from placebos ($P = .05$).
From Sheardown et al. (1997).

Durrani et al. (1995) reported precorneal clearance of mucoadhesive microspheres consisting of cross-linked polyacrylic acid (Carbopol 907) in the rabbit eye. Microspheres ranging in size from 8 to 15 μm (90%) were prepared by the water-in-oil emulsion method and then labeled with indium-111. Clearance of the microspheres showed a biphasic pattern and was affected by the pH value of the solution used for their hydration. Clearance rates of the microspheres hydrated with a solution at pH 5.0 (0.28 ± 0.033 min^{-1}, 0.007 ± 0.001 min^{-1}) were slower than those at pH 7.4 (0.30 ± 0.036 min^{-1}, 0.034 ± 0.01 min^{-1}). This result agreed with the observed differences in detachment force of the microspheres from mucin gel.

B. Nasal Cavity

The nasal cavity has a complex structure because of the presence of turbinal and can often be an area of inflammation in nasal allergy, common cold, influenza, and other such conditions. Drugs such as antiallergic drugs, antihistamines, and steroids used to be administered as nasal drops or nasal sprays. However, mucociliary clearance disturbs the retention of the drugs and their prolonged effect on mucosal inflammation. Admixing of bioadhesive polymers with drugs can overcome the mucociliary clearance of the dosage form.

Zhou and Donovan (1996) examined the effects of bioadhesive polymers, i.e., methylcellulose (MC), sodium carboxymethylcellulose (CMC), hydroxypropylmethylcellulose (HPMC), chitosan glutamate, Carbopol 934P (CP), polyethylene oxide 600K (PEO), and Pluronic F127, on mucociliary clearance in rats. The clearance rate was monitored using fluorescently labeled latex microspheres (4 μm, FluoSpheres, Molecular Probes Inc., U.S.A.) and compared with that of a microsphere suspension (control). Table 2 shows the initial clearance rates of the gels in male Sprague Dawley rats administered 25 μL of each gel or microsphere suspension into the left nostril under slight sedation with 10% ketamine injection (subcutaneously, 25 μL every 30 minutes). All the gels showed decreases in mucociliary clearance. The most remarked decrease was observed with 3% MC gel and the lowest effect was observed with CP. Carbopols have been shown to display strong bioadhesive properties in powder dosage form, in the dry state, but not in preswollen gel form. In a study evaluating the bioadhesive properties of hyaluronan derivatives as microspheres, Pritchard et al. (1996) used Carbopol 974 as a positive control. They reported that hyaluronan and its auto-cross-linked esters displayed bioadhesion comparable to that of Carbopol in detachment force in rat mucosa and mucociliary transport rate using the excised upper palate of the frog (*Rana pipiens*).

Rhinocort (Teijin Ltd., Japan) is a commercial nasal spray, one capsule of which contains a powder mixture of beclomethasone dipropionate (50 μg) and HPC (30 mg), which was developed as an extension of our research following Aftach. Nasal allergy patients spray one capsule twice a day, 100 μg in total, into the nasal cavity using a sprayer (Publizer) equipped with a needle to pierce the capsule. The powder adheres to and swells on the nasal mucosa and remains in place until approximately 6 hours after administration (Kuroishi et al., 1984). In an ordinary nasal spray, 100 μg each of a similar drug must be administered four times a day (400 μg in total). The decrease in the dose of steroid is brought about by the combination with bioadhesive polymer, and this is preferable for avoidance of systemic side effects.

Efforts to find suitable new materials are important for the development of bioadhesive dosage forms. Nakamura et al. (1996) compared in vitro and in vivo adhesion of water-soluble polymers, xanthan gum (XG, Echogum T, Dainippon Pharmaceutical Co., Japan), tamarind gum (TG, Glyloid 3S, Dainippon Pharmaceutical Co., Japan), polyvinyl alcohol (PVA, PVA-224S, Kuraray Co., Japan), and hydroxypropylcellulose (HPC-H, Nippon Soda Co., Japan). When 3 or 5 mg of sample containing 5% w/v Brilliant Blue was applied to the nasal cavity of New Zealand white rabbits, XG showed the most remarked adhesion (Fig. 1). Mucoadhesion in vivo follows the order XG >> TG > HPC >> PVA, and this order was the same as that for in vitro adhesion to agar plates. Previously, Nakamura et al. (1993) reported that the solubility of halopredone acetate, a steroid drug, was improved by cogrinding with PVA (at room temperature) and HPC (at −5°C) because it underwent a change to the amorphous state. Similarly, the solu-

Table 2 Initial Clearance Rates of Gels from the Nasal Cavity of Rats

Polymer	Control	Rate constant $k \pm$ SE (min^{-1}) Gel formulation (% decrease from control)	24 h	48 h
0.2% CP	0.197 ± 0.017	0.113 ± 0.018[a] (43%)		
0.4% CP	0.164 ± 0.007	0.0692 ± 0.0060[a] (58%)	0.169 ± 0.008	
5% PEO	0.116 ± 0.007	0.0534 ± 0.005[a] (63%)	0.0647 ± 0.011[a]	0.160 ± 0.024[a]
3% Chitosan G	0.150 ± 0.011	0.0427 ± 0.014[a] (58%)	0.0689 ± 0.030[a]	0.103 ± 0.005
1.5% CP (PUIUF)	0.132 ± 0.013	0.121 ± 0.010 (8.3%)	0.060 ± 0.008[a]	
3% CMC	0.176 ± 0.013	0.0560 ± 0.0062[a] (68%)	0.213 ± 0.025[a]	
3% HPMC	0.161 ± 0.014	0.0448 ± 0.0074[a] (72%)	0.213 ± 0.025[a]	
3% MC	0.227 ± 0.016	0.0154 ± 0.0075[a] (93%)	0.102 ± 0.011[a]	
25% Pluronic	0.137 ± 0.019	0.0378 ± 0.0114[a] (72%)	0.101 ± 0.014[a]	

[a]Statistically different from control ($P \leq .05$).
From Zhou and Donovan (1996).

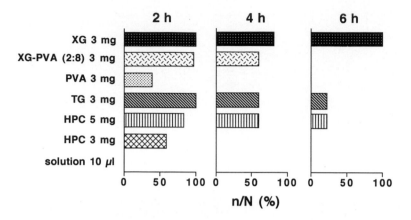

Figure 1 Mucosal adhesion of polymers in nasal cavity of rabbit: n, number of rabbits in which remaining dye was observed; N, total number of rabbits tested (*N* = 5–6). (From Nakamura et al., 1996.)

bility of halopredone acetate was improved by cogrinding with TG (Nakamura and Machida, 1997). These results suggest the future development of a halopredone acetate nasal spray using a new bioadhesive material such as TG or XG.

C. Oral Cavity

The oral cavity is an entrance of the digestive system and has important physiological functions such as mastication, phonation (speaking), and senses of smell (indirectly) and taste. Some of these functions are often disturbed by diseases such as aphthous stomatitis, microbial infection, and inflammation. Bioadhesive preparations are preferable for treatment of diseases in the oral cavity because they can adhere to the mucosa, protect the diseased part, and retain drug for the required period, resisting saliva and frictional stress between the buccal membrane and tongue.

A combination of HPC and Carbopol 934 (CP) found by us in a study of topical dosage forms for treatment of uterine cervical cancer (Machida et al., 1979) was utilized for production of bioadhesive tablets for aphthous stomatitis (Aftach, Teijin Ltd., Japan). This preparation is a double-layered tablet (7 mm in diameter, 1.1 mm thick) consisting of an adhesive layer (0.4 mm thick) containing HPC, CP, and 0.025 mg per tab of triamcinolone acetonide and a support layer (0.7 mm thick) containing a pigment, lactose, and HPC. The support layer disintegrates rapidly after application but the adhesive layer swells, gradually covering the diseased part, and releases the drug continuously (Teijin Ltd., 1982; Nagai and Machida, 1985). Later, a powder spray dosage form was developed and is available commercially (Salcoat, Teijin Ltd., Japan) for treatment of recurrent or multiple aphthae. Salcoat is a preparation similar to Rhinocort, and each capsule contains 50 μg of beclomethasone dipropionate and 200 mg of HPC. The larger amount of HPC in Salcoat than that in Rhinocort (30 mg) reflects the differences in area of diseased part and frictional stress. This preparation can cover the aphthae easily by spraying without pain on administration. Yamamoto et al. (1989) showed the superior retention of Salcoat using ^{3}H-labeled beclomethasone dipropionate and microautoradiography.

Mahdi et al. (1996) reported the healing effects of bioadhesive hydrogel patches against aphthous ulceration in recurrent aphthous stomatitis in 10 patients. The formula of the patches was not described in the paper, but a pharmaceutical grade cellulose derivative was used as the bioadhesive polymer. The patches were found to adhere for a long period to large ulcers in the early stage; the mean time of adhesion of the patches ranged from 20 to 54 minutes.

Burgalassi et al. (1996) prepared mucoadhesive patches (13 mm in diameter) containing benzydamine hydrochloride, an analgesic and anti-inflammatory drug, and lidocaine hydrochloride by direct compression of the components. Tamarind gum (TG, Glyloid 3S, Dainippon Pharmaceutical Co., Japan), polycarbophil (PCP, Noveon AA1, Goodrich Chemical Corp., U.S.A.), polyacrylic acid (PAA, Carbopol 940, Goodrich Chemical Corp., U.S.A.), and xanthan gum (XG, Keltrol TF, Kelco, U.S.A.) were examined as mucoadhesive polymers for buccal patches. TG was selected as a bioadhesive component, and it showed very low sensitivity to a cell line of human buccal epithelial origin in vitro. Each patch, consisting of benzydamine hydrochloride (10.0 mg), TG (200 mg), and lactose (100 mg), showed sustained release of the drug after application to the upper gums of healthy volunteers ($n = 6$), and the release profile in vivo fitted well the in vitro release profile obtained using the USP XXI rotating-basket method at 75 rev/min in pH 6.8, 66.7 mM isotonic phosphate buffer, $36 \pm 0.5°C$ (Fig. 2).

Backing in the multilayered bioadhesive dosage forms acts as protective layer and prevents adhesion and drug release to the opposite side of the preparation. 3M Pharmaceuticals developed bioadhesive buccal patches (Cydot) consisting of Carbopol 934P, polyisobutylene, and polyisoprene. Guo and Cooklock (1996) studied the effects of backing materials on the characteristics of this dosage form. Ethylcellulose (ethoxy content 48.5%, viscosity 45 cps, Sigma, U.S.A.), PVA (average molecular weight 40,000, Fisher Chemical, U.S.A.) and cellulose acetate ($MW_w = 177,000$, FMC Corporation, U.S.A.) mixture, or poly(ethylene-co-vinyl acetate) (Elvax 40W, E. I. Dupont, U.S.A.) was spray coated onto the patches as backing material.

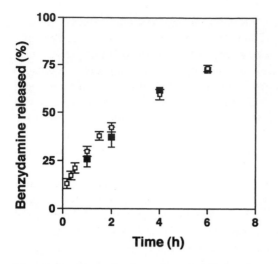

Figure 2 Benzydamine released (%) in vitro (□) and in vivo (■) from patches consisting of benzydamine hydrochloride (10.0 mg), TG (200 mg), and lactose (100 mg), mean ± SE, n = 6. (From Burgalassi et al., 1996.)

Ethylcellulose, a hydrophobic polymer, delayed the water uptake of the patches and prolonged the time to reach the maximum adhesive strength. The mixture of PVA and cellulose acetate could not prevent drug permeation. On the other hand, Elvax, a very hydrophobic and elastic polymer, significantly decreased the swelling ratio of buccal patches. The authors explained these differences in effect of backing materials using Fig. 3.

Film-type preparations were also studied as bioadhesive dosage forms. Saito et al. (1990) prepared double-layered films 250 μm thick and 7 or 10 mm in diameter. The adhesive layer (150 μm thick) consisted of HPC, HPMC, karaya gum, polyethylene glycol (PEG) 400, TiO$_2$, and 250 μg (7-mm films) or 500 μg (10-mm films) of benzydamine hydrochloride. The nonadhesive protective layer (100 μm thick) consisted of ethyl cellulose, HPC, CMC-Ca, PEG 4000, and Brilliant Blue FCF. The films were prepared by stepwise casting, first of the adhesive layer and then the protective layer. Table 3 shows the composition of the adhesive layer, adhesive strength, and adhesive time of three different films. The concentration of benzydamine hydrochloride in tongue tissue of rats was maintained at more than 5 μg/g for 2 hours after application of formula No. 3. On the other hand, the maximum plasma concentration was approximately 30 ng/mL, suggesting no incidence of a systemic effect of the drug.

Nagai and Konishi (1987) reported adhesive gingival plaster containing prostaglandin F$_{2\alpha}$ for orthodontic tooth movement. This film is composed of

a. Ethylcellulose

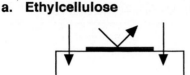

b. PVA/cellulose acetate mixture

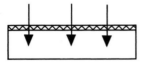

c. Elvax®

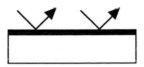

Figure 3   Schematic illustration of buccal patches with different backing materials. (a) Ethylcellulose did not swell with the patch, and water could diffuse into the polymer patch through the part of surface that was not covered by ethylcellulose. (b) Water penetrated the mixture and diffused into the patch. (c) Water could not penetrate the Elvax film. (From Guo and Cooklock, 1996.)

a backing layer (100 μm thick) and an adhesive layer (120 μm thick) containing 50 μg/cm² prostaglandin $F_{2\alpha}$ and accelerated tooth movement in at least 70% of the patients examined (Kawata and Yamashita, 1983). The main merit of film-type preparations is fittability to the mucosal membrane, and two kinds of double-layered films containing 0.025 mg of triamcinolone acetonide, Waplon-P (Kowa Co.) and Aphthaseal S (Taisho Pharmaceutical Co.), are commercially available in Japan.

Oral candidiasis is another significant disease of the oral cavity. Khanna et al. (1996) studied bioadhesive and bioerodible tablets containing 10 mg per tab of clotrimazole. HPC-M, HPMC-K4M, CMC, guar gum, and sodium alginate were examined as bioadhesive materials. Directly compressed tablets 13 mm in diameter were prepared and tested for in vitro drug release, adhesive strength, in vivo evaluation (placebo tablets), etc. The tablets consisting of 100 mg of guar gum and 90 mg of mannitol showed the longest mean adhesion time, 76.3 minutes in vitro and 66.3 minutes in vivo (12 healthy volunteers). A good correlation was observed between the mean

Table 3 Composition and Adhesive Properties of Double-Layered Films

	No. of formula		
Property	3	6	8
Composition of adhesive layer			
HPC-M	—	—	45
HPC-H	50	45	—
HPMC (K-4M)	25	20	20
HPMC (K-100LV)	25	20	20
Karaya gum	—	15	15
PEG 400	10	10	10
TiO_2	3	3	3
Adhesive strength	4^a	5^b	5
Adhesive time (min)			
Inside cheek	40	80	60
	D or P^c	D	D
Inside lower lip	70	80	70
	D or P	D	D

[a] Not peeled off by tongue but by finger.
[b] Not peeled off easily even by finger.
[c] D, dissolved; P, peeled off.
From Saito et al. (1990).

adhesion times in vitro and in vivo. Bouckaert et al. (1996) prepared bioadhesive slow-release tablets containing miconazole nitrate and studied the influence of the application site on the buccal level of the drug in cancer patients with candidiasis. Patients who receive irradiation in the neck region often suffer from oral candidiasis and decreased salivary flow. These directly compressed bioadhesive tablets contained 10.0 mg of miconazole nitrate, 82.8 mg of thermally modified maize starch, 5.0 mg of Carbopol 934, 2.0 mg of sodium benzoate, and 0.2 mg of silicon dioxide. The tablet was placed on the gingiva in the region of the right upper canine or on the right cheek buccal mucosa in the region of the last upper molar. The values of area under the curve (AUC), adhesion time, t_{max}, C_{max}, and the time period above the minimal inhibitory concentration (MIC) value ($T^{>MIC}$) are shown in Table 4. The t_{max}, adhesion time, and $T^{>MIC}$ with application to the gingiva were significantly higher than those with application to the cheek.

Another antifungal preparation, buccal bioadhesive slow-release tablets containing miconazole nitrate, was reported by Van Weissenbruch et al. (1997). The preparation was clinically examined by a double-blind randomized placebo-controlled study in 36 laryngectomees concerning the effect on

Table 4 Mean AUC, t_{max}, C_{max}, and $T^{>MIC}$ Values After Application of the Bioadhesive Tablets to Both Application Sites ($n = 6$)

Site	AUC (mg min^{-1} mL^{-1})	Adhesion time (h)	t_{max} (h)	C_{max} (μg mL^{-1})	$T^{>MIC}$ (h)
Gingiva	33.72 (17.4)[a]	10.2 (3.1)[a]	4.7 (2.1)[b]	86.4 (48.8)[b]	12 (0)[b]
Cheek	29.02 (16.8)	4 (0)	2 (0)	123.2 (69.7)	5.3 (1.6)

[a]Mean (SD).
[b]Significantly different ($P < .05$, two-tailed Wilcoxon test).
From Bouckaert et al. (1996).

the life and function of the Provox tracheoesophageal voice prosthesis. The bioadhesive tablets (100 mg, 7 mm in diameter, 2 mm thick), consisting of starch, 5% polyacrylic acid, and 10 mg of miconazole nitrate (Bouckaert et al., 1992), significantly extended the life of the prosthesis and the patients' compliance was good.

Other microbial infections in the oral cavity could be treated with bioadhesive dosage forms. Collins and Deasy (1990) prepared two- and three-layered bioadhesive lozenges containing cetylpyridinium chloride. The bioadhesive layer of these preparations contained HPC or a mixture of HPC and Carbopol. Jones et al. (1996) examined bioadhesive semisolid systems for treatment of periodontal diseases with regard to release properties, hardness, compressibility, adhesiveness, elasticity, cohesiveness, and syringeability. The semisolid systems containing 5.0% w/w tetracycline hydrochloride, hydroxyethylcellulose (5, 10, 20% w/w), PVP (5, 10, 20% w/w), and polycarbophil (1% w/w) were prepared using pH 6.8 phosphate-buffered saline as a solvent.

D. Digestive Tract

Orally administered preparations move down along the digestive tract, releasing the drug continuously or in the specified organ. This movement of the drug or the dosage form is not preferable from the viewpoint of topical therapeutic effectiveness. Hence, the bioadhesive dosage forms exhibit their usefulness in oral administration of drugs with the purpose of achieving a higher effect on a specified part of the digestive tract.

1. Esophagus

We developed a new dosage form for topical chemotherapy of esophageal cancer: magnetic granules containing bioadhesive polymers (Ito et al., 1990; Nagano et al., 1997). The magnetic granules were prepared from Brilliant Blue FCF (model drug), ultrafine ferrite (γ-Fe_2O_3), and adhesive polymers in an weight ratio of 1:10:9 by the wet granulation method using a 20-mesh sieve and ethanol as a kneading solvent. From the results of in vitro release tests and our previous experience (Machida et al., 1979), a 6:4 w/w mixture of hydroxypropylcellulose-H (HPC) and Carbopol 934 (CP) was selected as the bioadhesive material. The granules showed excellent targeting ability in vitro in a study using a model esophagus made from agar gel and a magnetic field of about 1700 G. When the granules were administered to nonanesthetized rabbits via a catheter with approximately 2 mL of 0.65% HPC solution, almost all granules were retained in the region of magnetic guidance ($\leq$1900 G) at 2 hours after administration, as shown in Fig. 4 (Ito et al., 1990).

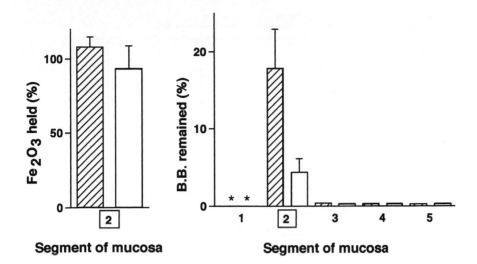

Figure 4 Effects of application time of the magnetic circuit on holding ratios of Fe_2O_3 and Brilliant Blue FCF (B.B.) in the bioadhesive magnetic granules at 2 hours after administration. Magnetic circuit was applied for (hatched bars) the initial 2 minutes and (empty bars) 2 hours. *Not detected. Each value is the mean ± SE. Segment 2 corresponds to the target. (From Ito et al., 1990.)

Magnetic granules containing bleomycin (BLM) (BLM:polymer:ferrite = 2:5:3) were prepared and their growth inhibitory effect against transplanted VX_2 cancer fragment in the esophagi of rabbits was examined with magnetic guidance using a magnetic circuit as shown in Fig. 5 (Nagano et al., 1997). The BLM granules were retained at the target site but the cancer growth was not markedly inhibited by administration of the granules once a day for 2 weeks. This result could be explained by the large variations in cancer growth, low permeation of BLM to the cancer tissue through the normal mucosa, and insufficient retention period of the drug on the target. Further investigations using a more powerful drug and bioadhesive materials and a more reproducible model of cancer are required for optimization of this dosage form.

2. Rectum

The rectum, as the end of the digestive tract, seemed to have less importance than stomach and intestine. However, it is a valuable route for administration of drugs that have severe gastrointestinal side effects and for administration to patients who cannot take medicines orally, such as unconscious patients

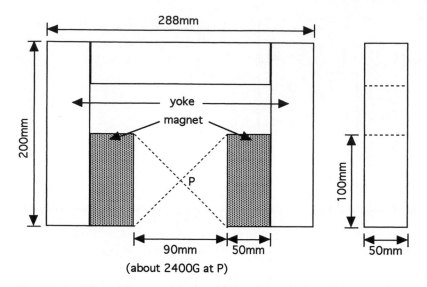

Figure 5 Illustration of permanent magnetic circuit used for the in vivo study of the growth-inhibiting effect of bioadhesive magnetic granules containing bleomycin against model esophageal cancer. (From Nagano et al., 1997.)

or infants. Because drugs absorbed from the lower part of the rectum can escape hepatic metabolism in the first pass (first-pass effect), mucoadhesive suppositories have been developed. Huang et al. (1987) prepared double-layered suppositories for prolonged retention in the lower rectum. These suppositories were composed of a bioadhesive front layer and a terminal layer containing a drug. The front layer consisted of a base (PEG 1000: PEG4000 = 96:4) containing 5 or 10% of the ammonium salt of carboxy-vinyl polymer (Hiviswako 204, Wako Pure Chemicals, Japan) to prevent upward spread of the drug in rabbits.

This type of suppository is applicable for the topical treatment of diseases such as hemorrhoids and rectal cancer. A bioadhesive suppository for the treatment of hemorrhoids is commercially available in Japan as an over-the-counter drug (Taisho Pharmaceutical Co.).

E.   Vagina and Uterus

As described in the beginning of this chapter, the development of bioadhesive dosage forms for the local chemotherapy of uterine cervical cancer was the starting point of studies in this area. Excision of the entire uterus was the most common method of treatment for uterine cervical cancer. However, this prevents future pregnancy. Bioadhesive tablets were developed due to the wish of a gynecologist, the late Prof. Dr. Hiroshi Masuda, to improve patients' quality of life.

After screening of polymers, mixtures of hydroxypropylcellulose-H (HPC) and Carbopol 934 (CP) in weight ratios of 1:1 to 1:2 were selected as the base of bioadhesive sustained-release tablets. Flat-faced tablets 300 mg in weight, 13 mm in diameter, and approximately 2 mm thick were prepared by direct compression combining 30 mg per tab of bleomycin (BLM). The tablets were administered with informed consent to patients with uterine cervical cancer at stage 0 to Ib. Administration of 90 to 195 mg of BLM via bioadhesive tablets brought about complete disappearance of cancerous foci in three of the nine patients. Despite this remarkable anticancer effect, the normal mucosa was not affected by BLM, reflecting retention of the drug on foci due to the bioadhesiveness of the preparation (Machida et al., 1979). As the next step in this study, stick-type preparations containing BLM, carboquone (CQ), or 5-fluorouracil (5-FU) were prepared and examined both in vitro and clinically (Machida et al., 1980; Masuda et al., 1981). The stick-type preparations were more advantageous than the tablets because they could be inserted into the cervical canal, where remnant foci were often found. When the stick-type preparation containing 30 mg of BLM was administered twice a week more than five times, cancer cells were not found in pathological samples taken from 90% of patients at stage zero. However, the cancer cells did not disappear completely with administration once a week (Masuda et al., 1981). So that patients would not have to visit the hospital twice a week, and to obtain a sufficient therapeutic effect by once-a-week administration, a double-layered stick-type preparation capable of delivering the drug continuously for 1 week was developed (Iwata et al., 1987). The double-layered sticks shown in Fig. 6 were prepared by two-step dry coating of the core stick. This preparation released approximately 30 and 70% within 24 and 72 hours, respectively, as determined by the Kerami filter method (Machida et al., 1980) using 400 mL of saline solution. When the core stick was coated with a 10% ethanol solution of 2-methyl-5-vinyl-pyridine methylacrylate−methacrylic acid copolymer (MPM, Tanabe Pharmaceutical Co., Japan) three times by the dipping and drying procedure, release of BLM within 24 and 72 hours was suppressed to 24 and 39%, respectively. The results suggest a possibility of once-a-week treatment of uterine cervical cancer using a double-layered stick-type preparation of BLM.

Woolfson et al. (1995) developed a novel bioadhesive cervical patch containing 5-FU. A bioadhesive layer 0.1 mm thick was formed by casting a gel consisting of 2% w/w Carbopol 981, 1% w/w glycerin (plasticizer), 5-FU (20 mg/patch), and an ethanol-water mixture (30:70) as a solvent on a

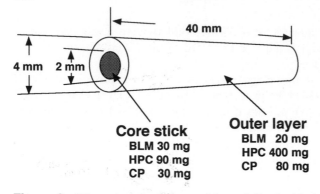

Core stick
BLM 30 mg
HPC 90 mg
CP   30 mg

Outer layer
BLM   20 mg
HPC 400 mg
CP   80 mg

Figure 6 Dimensions and composition of the double-layered bioadhesive stick containing bleomycin (BLM) designed for "once-a-week" treatment of uterine cervical cancer. (From Iwata et al., 1987.)

backing layer made of thermally treated PVA. The patch was cut to 26 mm in diameter and glued to a PVA tag and a 12-cm section of white linen thread, as shown in Fig. 7. The patches were examined for their plasticity, bioadhesive properties, release characteristics, etc. Sidhu et al. (1997) evaluated this patch in a randomized controlled trial in patients with cervical intraepithelial neoplasia (grades 1 and 2). The patches fulfilled the require-

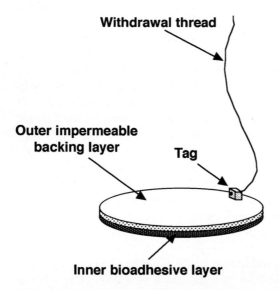

Withdrawal thread

Outer impermeable backing layer

Tag

Inner bioadhesive layer

Figure 7 Representation of the bioadhesive cytotoxic drug delivery system. (From Woolfson et al., 1995.)

ments for treatment of the disease but no therapeutic effect was provided by 5-FU, the chemotherapeutic agent chosen as an active ingredient for the study.

As a topical bioadhesive dosage form for the vagina, Lee and Chien (1996) prepared gels consisting of nonoxynol 9 (spermicidal agent) and Carbopol 934P. Good bioadhesion to lamb vaginal mucosa, prolonged release of the drug, and increased bioadhesiveness with the addition of ethylenediaminetetraacetic acid (EDTA) were reported.

III. CONCLUSION

Precise delivery of a required amount of drug to a required site of the body at or for a required time is the ultimate aim of drug delivery systems. Topical bioadhesive dosage forms potentially fulfill this aim. Further practical investigations, including searches for and production of novel bioadhesive materials, are expected in the future. We hope that the topical bioadhesive drug delivery system designed by our group will contribute to the improvement of the quality of life of patients worldwide.

REFERENCES

Bouckaert S, Schautteet H, Lefebvre RA, Remon JP, Van Clooster R. 1992. Eur J Clin Pharmacol 43:137.

Bouckaert S, Vakaet L, Remon JP. 1996. Int J Pharm. 130:257.

Burgalassi S, Panichi L, Saettone MF, Jacobsen J, Rassing MR. 1996. Int J Pharm 133:1.

Collins AE, Deasy PB. 1990. J Pharm Sci. 79:116.

Durrani AM, Farr SJ, Kellaway IW. 1995. J Pharm Pharmacol 47:581.

Guo J-H, Cooklock KM. 1996. J Pharm Pharmacol 48:255.

Huang CC, Tokumura T, Machida Y, Nagai T. 1987. J Pharm Sci Tech Jpn 47:42.

Ito R, Machida Y, Sannan T, Nagai T. 1990. Int J Pharm 61:109.

Iwata M, Machida Y, Nagai T, Masuda H. 1987. Drug Des Delivery 1:253.

Jones DS, Woolfson AD, Djokic J, Coulter WA. 1996. Pharm Res 13:1734.

Kawata T, Yamashita N. 1983. Nippon Dent Rev 484:10.

Khanna R, Agarwal SP, Ahuja A. 1996. Int J Pharm 138:67.

Kuroishi T, Asaka H, Okamoto M. 1984. Jpn Pharmacol Ther 12:4055.

Lee C-H, Chien YW. 1996. J Controlled Release 39:93.

Machida Y, Masuda H, Fujiyama N, Ito S, Iwata M, Nagai T. 1979. Chem Pharm Bull 27:93.

Machida Y, Masuda H, Fujiyama N, Iwata M, Nagai T. 1980. Chem Pharm Bull 28:1125.

Machida Y, Nagai T. 1974. Chem Pharm Bull 22:2346.

Machida Y, Nagai T. 1978. Chem Pharm Bull 26:1652.

Mahdi AB, Coulter WA, Woolfson AD, Lamey P-J. 1996. J Oral Pathol Med 25: 416.

Masuda H, Sumiyoshi Y, Shiojima Y, Suda T, Kikyo T, Iwata M, Fujiyama N, Machida Y, Nagai T. 1981. Cancer 48:1899.

Nagai T, Konishi R. 1987. J Controlled Release 6:353.

Nagai T, Machida Y. 1985. Pharm Int 6:196.

Nagano H, Machida Y, Iwata M, Imada T, Noguchi Y, Matsumoto A, Nagai T. 1997. Int J Pharm 147:119.

Nakamura F, Fujitani M, Machida Y, Nagai T. 1993. J Pharm Sci Techn Jpn 53:161.

Nakamura F, Machida Y. 1997. J Pharm Sci Techn Jpn 57:132.

Nakamura F, Ohta R, Machida Y, Nagai T. 1996. Int J Pharm 134:173.

Pritchard K, Lansley AB, Martin GP, Helliwell M, Marriott C, Benedetti LM. 1996. Int J Pharm 129:137.

Saito S, Sadamoto K, Ishikawa Y, Machida Y, Nagai T. 1990. J Pharm Sci Techn Jpn 50:347.

Sheardown H, Clark H, Wedge C, Apel R, Rootman D, Cheng YL. 1997. Curr Eye Res. 16:183.

Sidhu HK, Price JH, McCarron PA, McCafferty DF, Woolfson AD, Biggart D, Thompson W. 1997. Br J Obstet Gynaecol 104:145.

Teijin Ltd. 1982. Aftach®: Adhesive topical preparation for treatment of aphthous stomatitis. Teijin Ltd., Pharmaceutical Division, Uchisaiwaicho, Chiyoda-ku, Tokyo, Japan.

Van Weissenbruch R, Bouckaert S, Remon JP, Nelis HJ, Aerts R, Alberts FWJ. 1997. Ann Otol Rhinol Laryngol 106:329.

Woolfson AD, McCafferty DF, McCarron PA, Price JH. 1995. J Controlled Release 35:49.

Yamamoto M, Okabe K, Kubo J, Naruchi T, Ikura H, Suzuki Y, Nagai T. 1989. STP Pharma 5:878.

Zhou MP, Donovan MD. 1996. Int J Pharm 135:115.

Index